Fatigue Damage, Crack Growth and Life Prediction

Fatigue Damage, Crack Growth and Life Prediction

Fernand Ellyin

Professor of Mechanical Engineering
NOVA CORP and NSERC Senior Industrial Research Chair
University of Alberta
Edmonton, Alberta, Canada

CHAPMAN & HALL
London · Weinheim · New York · Tokyo · Melbourne · Madras

Published by Chapman & Hall, 2–6 Boundary Row, London SE1 8HN, UK

Chapman & Hall, 2–6 Boundary Row, London SE1 8HN, UK

Chapman & Hall, GmbH, Pappelallee 3, 69469 Weinheim, Germany

Chapman & Hall USA, 115 Fifth Avenue, New York, NY 10003, USA

Chapman & Hall Japan, ITP-Japan, Kyowa Building, 3F, 2-2-1 Hirakawacho, Chiyoda-ku, Tokyo 102, Japan

Chapman & Hall Australia, 102 Dodds Street, South Melbourne, Victoria 3205, Australia

Chapman & Hall India, R. Seshadri, 32 Second Main Road, CIT East, Madras 600 035, India

First edition 1997

Typeset in 10/12 Palatino by Thomson Press (India) Ltd, New Delhi

Printed in Great Britain by St Edmundsbury Press Limited, Bury St Edmunds, Suffolk

ISBN 0 412 59600 8

A catalogue record for this book is available from the British Library

Library of Congress Catalog Card Number: 96–86212

∞ Printed on permanent acid-free text paper, manufactured in accordance with ANSI/NISO Z39.48-1992 and ANSI/NISO Z39.48-1984 (Permanence of Paper).

To my wife, Suzanne, and our children, Lise and Christopher I dedicate this book

Contents

Preface

Fatigue failure is a multi-stage process. It begins with the initiation of cracks, and with continued cyclic loading the cracks propagate, finally leading to the rupture of a component or specimen. The demarcation between the above stages is not well-defined. Depending upon the scale of interest, the variation may span three orders of magnitude. For example, to a material scientist an initiated crack may be of the order of a micron, whereas for an engineer it can be of the order of a millimetre. It is not surprising therefore to see that investigation of the fatigue process has followed different paths depending upon the scale of phenomenon under investigation.

Interest in the study of fatigue failure increased with the advent of industrialization. Because of the urgent need to design against fatigue failure, early investigators focused on prototype testing and proposed failure criteria similar to design formulae. Thus, a methodology developed whereby the fatigue theories were proposed based on experimental observations, albeit at times with limited scope. This type of phenomenological approach progressed rapidly during the past four decades as closed-loop testing machines became available.

Developments in linear and nonlinear fracture mechanics provided another impetus for research on fatigue crack growth. In parallel, studies at micro-level shed light on the mechanisms of crack initiation and growth. The availability of high-powered microscopes (scanning electron and transmission electron) during the same period was instrumental in identifying the governing fatigue mechanisms for various conditions. The advent of high-powered computational capabilities are now enabling a merger between the various approaches into the fatigue failure process.

It is the objective of this book to present a unified approach to the problem of fatigue damage. The emphasis is on identification of the fatigue damage process and methods whereby the life of a component can be predicted based on limited experimental data. To this end various approaches are critically reviewed and the one deemed most appropriate is fully described. The aim has been to provide sound bases from which quantitative results for a given problem can be obtained. Multiaxial loading, which is of practical interest, is extensively treated and the transition from uniaxial to multiaxial loading is achieved in a consistent manner. This is a distinct feature of this unified approach.

The book is comprised of three parts. The first part, Chapters 1 and 2, deals with the changes which occur in a material as a result of cyclic loading. The mechanisms involved in the initiation and growth of cracks, and factors which affect the fatigue life of a component are discussed in Chapter 1. This introductory chapter presents a broad view of the fatigue damage process and the breadth of the subject area. Chapter 2 is concerned with the changes in material properties due to cyclic loading and attainment of a stable cyclic response. The stable cyclic behaviour is modelled through a convenient cyclic stress–strain curve, depending on the material classification, e.g. Masing or non-Masing behaviour. This distinction becomes important when one attempts to calculate the absorbed (hysteresis) energy.

The second part, Chapters 3 to 6, is mainly concerned with the identification of an appropriate damage parameter which describes the initiation of a macro-crack. It is shown that up to this stage, the bulk material response (stress–strain relation) is not significantly affected by the formation and propagation of micro-cracks. In Chapter 3, it is demonstrated that an energy quantity, the sum of the plastic strain energy and the positive elastic strain energy, correlates fairly well with a variety of experimental data. The damge approach presented in Chapter 3 for uniaxial loading and in Chapter 4 for multiaxial loading, is an alternative to the initiation and growth of short cracks. (The behaviour of crystallographic cracks which is microstructure-sensitive is rather complicated and is discussed in Chapter 9). In other words, the phenomenological approach of Chapters 3 and 4 is an alternative way of treating the short crack regime through an averaged process. Chapter 5 describes the various experimental methods available to obtain multiaxial experimental data. A test facility developed in the author's laboratory for the multiaxial loading of tubular specimens is presented along with various biaxial test data.

A comprehensive constitutive model is required to calculate the energy-based damage parameter for complex multiaxial loading paths (non-proportional paths). There is also a need to predict the rate-dependent material response (creep and relaxation) and ratcheting deformation due to mean-stress, etc. Such a constitutive model is presented in Chapter 6, and its predictions are compared to a variety of test data for different industrial materials. It is believed that Chapters 3 to 6 will provide researchers and practising engineers with appropriate tools to predict the time it takes for a crack to initiate and grow to a size of the order of a millimetre. The unifying feature of the strain energy parameter is seen through Chapters 3 and 4. The comprehensive treatment of the multiaxial deformation and fatigue behaviour in Chapters 4 to 6 is one of the distinct features of this book.

The third part, Chapters 7 to 10, deals with the mechanics of crack propagation. Chapter 7 contains a full treatment of crack growth and factors which affect it. To make the chapter sufficiently self-contained, a synopsis of the theoretical foundation of linear and nonlinear fracture mechanics is included. Chapter 8 is focused on notch analysis, initiation of cracks from notches, and on their growth. The behaviour of physically small cracks is described in Chapter 9. The mechanisms involved and the modelling of small crack growth are outlined in this chapter. In a sense, this chapter bridges the crack initiation phase described in Chapters 1 and 3 and the long crack growth behaviour explained in Chapter 7. Finally, the variability associated with the material microstructure and random distribution of defects and impurities in both micro- and macro-scales are considered through a probabilistic treatment in Chapter 10. This approach leads to the determination of reliability of a component during its service life.

It was mentioned earlier that the fatigue failure process is a complex one and is affected by various factors such as environment, temperature, prior history, manufacturing process, etc. There are several periodicals dedicated to publishing research efforts on this subject as well as annual conferences and symposia dealing with this technologically important area. Therefore, the literature on this subject is so extensive that one cannot expect to treat all the issues influencing fatigue failure within the confines of this monograph. The topics chosen here are those that in the author's view are of interest in engineering practice and have also interested him. The book could have been organized differently, for example, the small crack growth section (Chapter 9) could have followed Chapter 2 or preceded Chapter 7. However, the present order was chosen to introduce the subject within the context of its historical development, for ease of the mathematical treatment, and for application by practising engineers. As far as possible, the book is made to be self-contained, and the necessary background is introduced at the beginning of each chapter. The list of references for each chapter is by no means complete, and any omission of other contributions is purely unintentional. This book can be used by researchers as a reference guide and by practising engineers as a design or investigative tool. It could also serve as a reference text in a graduate course on fatigue. It has been tried in my graduate course on fatigue of engineering materials.

A book of this size cannot be written without help and advice from a number of people. I am indebted to my research associates, post-doctoral fellows and graduate students who worked diligently to make original contributions. Their names appear in the list of references as co-authors of our joint papers. However, I wish to mention my longtime co-investigators, Dr. Daniel Kujawski and Dr. Zihui Xia, who have made significant contributions in this area. They have helped me with the preparation of some figures and have made valuable comments throughout the text. Professor Timothy Topper of the University of Waterloo offered helpful comments on the third part of the book – the mechanics of crack propagation. I have received advice from Professor K.-T. Rie, Braunschweig University, Professor Y. Asada, The University of Tokyo, and Professor J.F. Williams, The University of Melbourne, on certain topics. I express my sincere thanks to the above. I must, however, take sole responsibility for any omission or misinterpretation which may still exist.

Various versions of these chapters were typed by competent and co-operative secretaries at the Department of Mechanical Engineering, and I thank them for their work. Finally, I wish to express my appreciation to the editorial staff of Chapman & Hall, in particular Dr P. Hastings, senior editor, and Ms E. Broomby, production editor, for their diligent efforts in overseeing the publication of this book.

FERNAND ELLYIN

Some general concepts concerning fatigue 1

1.1 INTRODUCTION

Interest in the prevention of fatigue failure in metallic components began with the advent of the industrial revolution where mechanical devices were subjected to repeated loading by the nature of their operation. As a result of numerous full-scale component and laboratory specimen tests, a design methodology emerged which limited the maximum operating stress to a level whereby the life of a load-bearing component was very long, i.e. essentially infinite. This approach required that the material response in highly stressed regions remain elastic. With the development of elasticity theory and notch analysis, the design of components and systems to withstand alternating loads became a matter of keeping the maximum stress below a safe stress level – a material's fatigue limit (Fig. 1.1).

This design philosophy continued until the development of metallic aircraft structures where a high premium is placed on low weight. A series of failures forced a re-examination of the above-mentioned philosophy, reaching the conclusion that economical structures for aircraft could not be designed unless the peak operating stresses were allowed to reach levels where inelastic deformation occurs in the highly stressed regions. It should be pointed out that the ensuing plastic deformation is constrained by the surrounding elastic material, thus the overall structural response remains elastic. However, the plastically deformed regions are potential sites for crack initiation resulting in a finite life of a component. The research and development work initiated by the aerospace industry spread to other fields of application such as ground transportation vehicles, power plant equipment, pressure vessels, etc. A field of study was thus initiated under the general heading of 'low-cycle' fatigue.

Research in the area of low-cycle fatigue accelerated with the commercial availability of closed-loop loading systems in the mid-fifties. Phenomenological life prediction and design rules were proposed in the late fifties and early sixties. High-temperature studies coupled with low-cycle fatigue

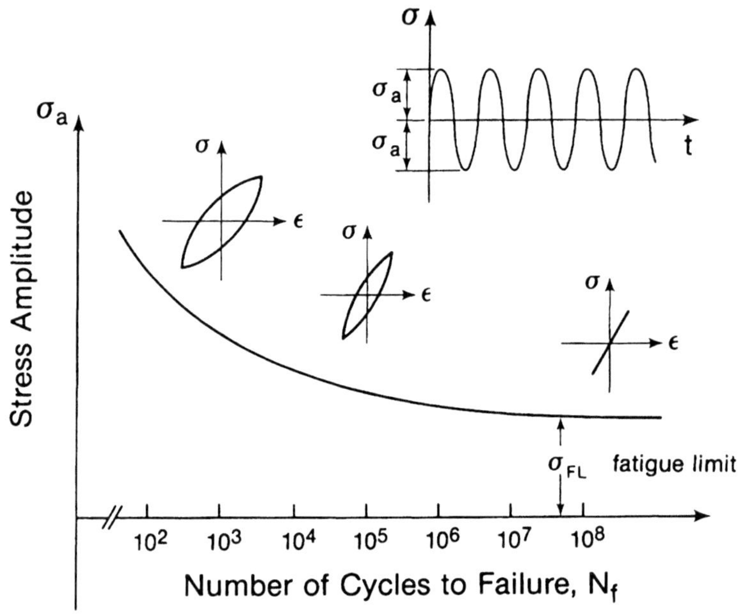

Fig. 1.1 Stress amplitude versus number of cycles to failure, *S–N* curve, and schematics of stable hysteresis loops at various fatigue lives.

were spurred on by the accelerated development of nuclear power plants in the sixties. The microstructural and environmental effects and multi-axial behaviour then followed. The focus of attention in the past decade has been on elastic–plastic modelling and damage monitoring. These activities may well continue into the next decade with the emphasis being on more advanced modelling and damage monitoring techniques. Another area of activity could be improved life prediction and life extension aided by microstructural studies of the underlying phenomena. In the following chapters, whenever appropriate, a brief historical development of the research efforts will be given.

It is the objective of this chapter to provide an overview of various factors which affect the fatigue life of a component. As one reads through the chapter, it will become evident that fatigue is a complex subject and it is difficult to cover all aspects in a monograph of reasonable size. In the following an attempt has been made to identify the major factors which influence fatigue. Some of these factors are discussed in later chapters and this has been indicated. However, appropriate references are provided for those topics which are not discussed beyond this chapter, e.g. types of loading in section 1.2. An interested reader can consult cited references for further details. Depending upon a reader's background, some of the topics in this chapter may appear rather advanced. This, however, should not be viewed as an impediment since the basic notions will be explained in the appropriate chapters to follow. The objective here is to provide an appreciation of the overall fatigue failure process, and the major factors which influence it.

1.2 TYPES OF LOADING

Designing against fatigue failure implies that the lifetime operating loads are either known or have been specified as a design condition, i.e. expected history. For certain machine components, e.g. rotating shafts of stationary equipment, the operating load history is almost uniform with a small variation from cycle to cycle. By contrast, in moving vehicles and aerospace components, the real load histories vary in both shape and magnitude with time. For example, load histories representative of ground vehicle components are given in [1]. In the case of a suspension component, the history is a mixture of random excitation with superimposed manoeuvre forces having a mean compressive load. In the case of a mounting bracket, the loading is a narrow-band random vibration type with an almost constant mean load.

In some cases, the loads experienced by a component are of a random nature. This implies that the load will vary in such a manner that its magnitude at any time, t, could not be predicted *a priori*. However, the ensemble of the load history may follow a statistically definable distribution. A description of a class of random variables that have finite mean values and variances and which are continuous can be found in most textbooks on statistics and probability, e.g. [2]. It suffices here to mention that within this class of distributions, the exponential and Rayleigh distributions are of a one-parameter family; the normal, gamma and log–normal are two parameters and Weibull is a three-parameter distribution. That is, in addition to the type of distribution, the appropriate parameters have to be specified for each distribution. The normal and exponential distributions tend to be a good approximation of dynamic loads.

If a sufficient record of the dynamic load is available, then the expected central tendency, e.g. mean or root mean square (RMS) values, could be calculated from the appropriate statistical formulae. A measure of scatter of data about the expected value is then found from calculating the variance or standard deviation of the data. Note that the standard deviation, mean value and RMS have the same units.

There is a special class of random processes which are called **stationary processes**. To this category belong all those processes where the sample distributions are independent of the time of sampling, i.e. the frequency distributions are time invariant. In several cases the spectrum of loading may be approximated as a stationary random process, and this leads to a relatively simple analysis. For example, any wavy sample taken from a real stationary random process with a zero mean value can be decomposed into its frequency components using a standard Fourier analysis [3]. In such a process the most useful quantity is the mean square value which is related to the spectral density, $S(\omega)$, through

$$E[x^2] = \int_{-\infty}^{+\infty} S(\omega)\, d\omega \qquad (1.1)$$

i.e. the mean square of the loading process is equal to the sum of all the frequencies of $S(\omega)$, the area under the spectral density.

A limiting case of a stationary process is a narrow-band process where the band of significant frequencies is small compared with the magnitude of the central frequency of the band. In a sample of this type of process only a narrow range of frequencies is observed. Figure 1.2 shows a narrow-band process which has the appearance of a sinusoidally varying wave form with a frequency of ω_0 and slowly varying random amplitudes. When the process is a normal (Gaussian) one, we can calculate the mean frequency, ω_0, of the cycles and the probability distribution of the envelope, provided

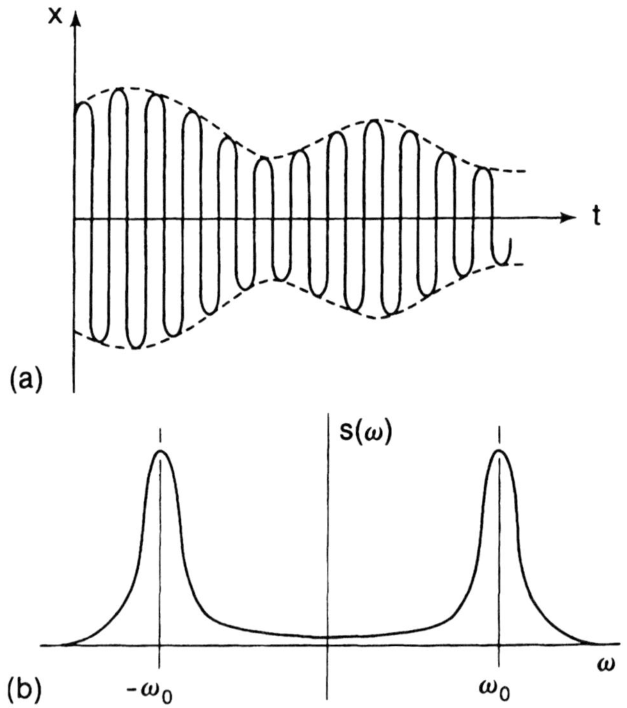

Fig. 1.2 A narrow-band stationary process. (a) time history of loading; (b) spectral density.

the spectrum density $S(\omega)$ is given, e.g. Crandall and Mark [3]. Other useful information such as the number of level x crossings with positive slope, the number of zero crossings, the peak frequency, etc. can also be derived.

Having determined the amplitude and frequency of the service loading, a component could be tested to find its fatigue life. However, it is worthwhile mentioning here that the fatigue life of a component can be strongly influenced by a small number of large events which may not have an appreciable effect on the central behaviour. (During these peak 'overloads' small cracks may initiate which could grow under smaller amplitude loads.) Thus, when statistical analysis is used, it would be more appropriate to use the root mean square as the expected value for the fatigue life prediction. The fatigue life of a component is also affected by the loading sequence, i.e. for the same statistical distribution, the fatigue life will be different depending upon whether the high loads follow the low ones or *vice versa*.

1.2.1 APPROXIMATE METHODS OF FATIGUE LOAD DESCRIPTION

It was mentioned above that when loading is described by statistical means, two features of significance for the fatigue damage process may be lost, namely the sequence of loading and infrequent but large peaks and valleys (overloads). The sequence of loading may have a major influence on the fatigue life, especially in the case of notched or cracked components, as will be seen in Chapters 3, 7 and 8. For example, infrequent tensile overloads tend to retard crack growth while compressive overloads, when large enough to cause local yielding, tend to accelerate crack propagation. Similarly a block of large amplitude cycles followed by a smaller amplitude one is more damaging than when the same loading sequence is reversed, i.e. the large amplitude follows a smaller one (see the discussion on cumulative damage in Chapter 3).

Various methods of cyclic counting have been proposed to convert the irregular load histories to blocks of constant amplitude cycles. The rainflow method is currently the most popular one and it will be described later. When there are no major deviations in a load history, a uniformly applied cyclic loading would yield a reasonable approximation. On the other hand, when there are appreciable variations between the cyclic ranges then we have to account for the sequence effect.

For example, consider a stationary variable loading with a zero mean stress as shown in Fig. 1.3(a). If a sufficient amount of sampling is available, we can construct a dimensionless cumulative frequency distribution. The abscissa in Fig. 1.3(b) is the cumulative distribution and the ordinate is the dimensionless stress ratio, a specified stress level divided by the maximum stress in the spectrum.

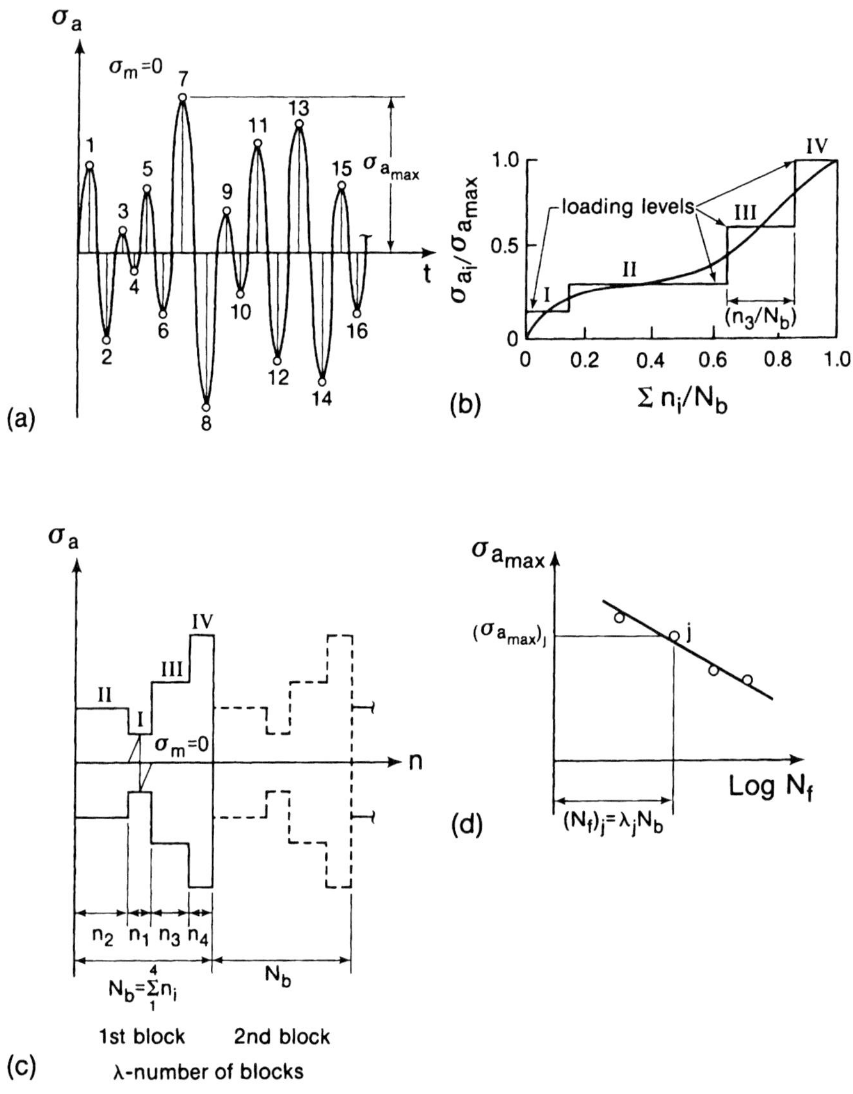

Fig. 1.3 (a) Spectrum of loading with a zero mean; (b) dimensionless cumulative distribution and approximation by four stress levels; (c) a block of constant amplitude stresses used to replace the load spectrum in (a); (d) construction of a *S–N* curve for the predetermined block form in (c).

Most approximate methods attempt to describe the loading shown in Fig. 1.3(a) by blocks of constant amplitude as depicted in Fig. 1.3(b). There are two reasons for doing this. Firstly, it considerably simplifies testing of components in the sense that instead of applying the loading spectrum in

Fig. 1.3(a), we can apply much simpler blocks of constant amplitude as shown in Fig. 1.3(b). Secondly, we can use available data from the constant amplitude tests in conjunction with a cumulative damage criterion (Chapter 3) to predict the life of a component subjected to variable amplitude loading.

It is worth noting that if the number of cycles in a block, N_b, is comparable to the number of cycles to failure, i.e. $N_b \approx N_f$, then the sequence in which these constant amplitude levels are applied has a significant effect on the cyclic life. On the other hand, if several blocks, as shown in Fig. 1.3(c), have to be applied to simulate the actual spectrum, then the sequence effect may not be as important.

Referring to Fig. 1.3(d), we can construct a *S–N* curve similar to that shown in Fig. 1.1, but in terms of the maximum stress amplitude of the block. Through changing σ_{max} and all other stress levels in the block in a proportional manner, we can obtain other points on the σ_{max} vs. N_f curve, Fig. 1.3(d). The reason for constructing a *S–N* curve in terms of the maximum stress amplitude in the block is to facilitate the design process in selecting a component cross-section, e.g. a shaft cross-section.

An example of the above-mentioned procedure is a European standard gust load sequence for flight simulation of transport aircraft wing structures, known as Mini-TWIST [4]. The load sequence in the block is from high to low stress amplitude with a positive mean stress (the most critical loading sequence). An interested reader should consult [4] for further details.

Referring to Fig. 1.4, the first step in any data reduction is to define peaks and valleys in the loading history. A band is subsequently defined to eliminate small peaks and valleys. (Large amounts of small amplitude

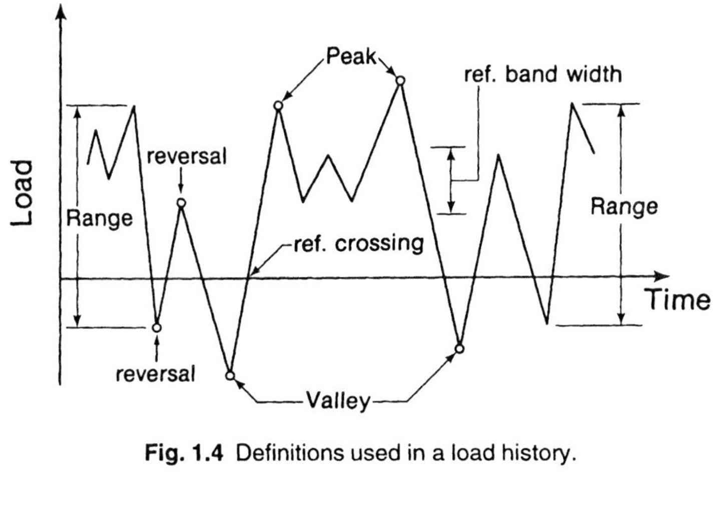

Fig. 1.4 Definitions used in a load history.

loads by themselves may not be damaging, however, when they follow a large overload they could influence the crack growth rate.) Several methods have been proposed to convert irregular load histories to blocks of uniformly repeated simple load cycles as shown in Fig. 1.3 (see for example [5, 6]). These counting methods can be subdivided into two groups: firstly a simple one-parameter description, e.g. level crossing count, peak count and range count (see Conle and Topper [7] for a description). Secondly, to properly specify a cycle, we would require at least two parameters, hence the two-parameter counting methods have been devised, e.g. range–mean count. Present practices for cycle counting are given in an ASTM standard [8].

1.2.2 THE RAINFLOW COUNTING METHOD

The 'rainflow' method was first proposed by Matsuishi and Endo [9]. The method is best illustrated with reference to an irregular loading pattern shown in Fig. 1.5 where the strain–time loading history and the corresponding stress–strain response are depicted [10]. In the rainflow method we start from the highest peak, point 1, and proceed along the descending branch until the point of reversal, 2, is reached. From point 2, we move horizontally to the next descending branch point 2′ and continue as before until the next reversal point, 4 (Fig. 1.5(a)). Since there is no other branch going further down than point 4 (deepest valley), we then reverse the direction and move along the ascending branch until we reach a peak point, 5. We repeat the process of climbing by moving horizontally and reaching another ascending branch at point 5′, and continue this procedure until we arrive at the highest peak, in this instance, point 1′. By this procedure we first identify the cycle with the largest range, illustrated by double lines in Fig. 1.6(a). Proceeding in a similar manner we identify the cycles with lower ranges. The result of the rainflow counting method is summarized in Fig. 1.6(b) with four cycles. Note that each cycle shown in Fig. 1.6(b) corresponds to the individual hysteresis loops shown in Fig. 1.5(b). Since damage is related to the hysteresis energy (Chapter 3), it then follows that smaller amplitude cycles such as 7–8–7′ in Figs 1.5(b) and 1.6(b) could be neglected in comparison to the larger ones.

Digitized data of load–time histories could be programmed for the purpose of cyclic counting. There are several versions of the rainflow cyclic counting method which are discussed in [11].

1.3 FATIGUE FAILURE MECHANISMS

The fatigue process begins with the accumulation of damage at a localized region or regions due to alternating loads, which eventually leads to formation of cracks and their subsequent propagation. When one of the cracks

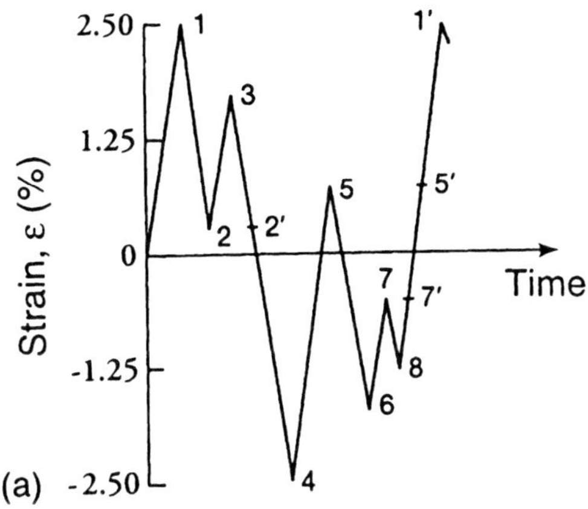

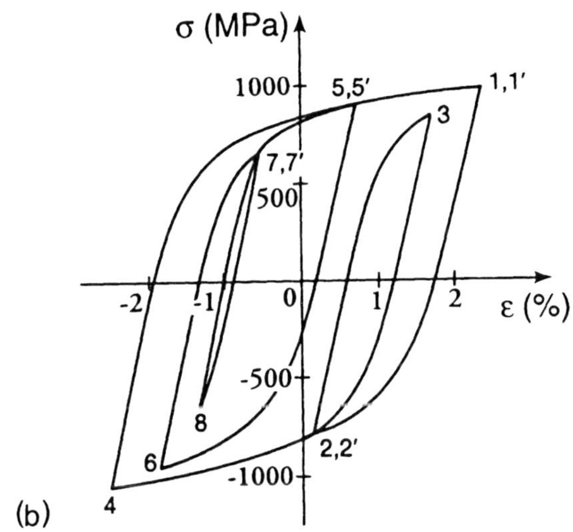

Fig. 1.5 A strain–time history and the corresponding stress–strain relation, from [10].

has grown to such an extent that the remaining net-section is insufficient to carry the applied load, a sudden fracture takes place. Therefore, when viewing a fatigue fractured surface we may distinguish three features:

- an initiation site (or sites);
- a crack growth surface area with distinct features;
- a final fractured surface.

In the absence of major internal flaws (large inclusions) the fatigue cracks usually initiate at the surface of a specimen and their initial growth

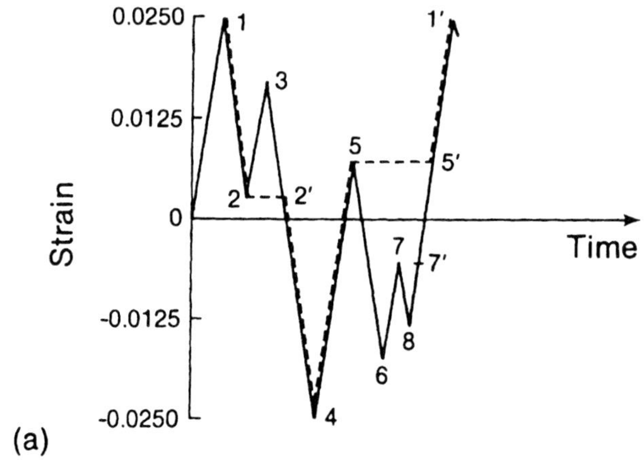

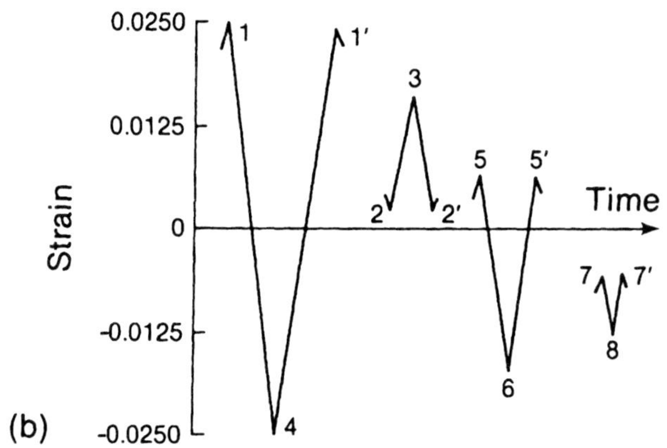

Fig. 1.6 Illustration of the rainflow method. (a) the given strain history; (b) the corresponding strain range cycles each corresponding to a closed loop shown in Fig. 1.5.

is generally in the direction of the maximum shear stress. The more likely sites for a crack initiation are either pre-existing highly stressed regions such as notches, pits and scratches, or surface notch-like valleys, large inclusions on the surface and discontinuities resulting from a slip along the crystallographic planes.

1.3.1 MICROSTRUCTURAL CHANGES DURING DEFORMATION PROCESSES

From a microscopic point of view, the fatigue process can be divided into two main regimes [12]. In an early stage microstructural changes occur in the bulk of the material due to cyclic deformation, eventually leading to some form of strain localization. The basic mechanism at this stage in metallic materials is caused by irreversible dislocation movement under cyclic loading. (A dislocation is a flaw in the lattice structure of the material which causes a slip to occur along crystallographic planes when a stress is applied to the lattice.) These dislocations agglomerate into bundles almost perpendicular to the active Burgers vector within a dislocation free matrix. (The Burgers vector, $\vec{b}$, represents both the magnitude and direction of slip of the dislocation.) Strain localization occurs when the dislocation pattern in a few veins (bundles) becomes locally unstable at a critical stress or strain and thin lamellae of persistent slip bands (PSBs) are formed (e.g. [12–14] where further references are listed). The subsequent deformation is mainly concentrated in these slip bands, and during a stable cyclic response of the material, the number of PSBs increases and may fill the entire volume of the crystal provided the applied plastic strain is high enough [12].

Most of the metallographic studies of fatigue mechanism have been performed on metallic single crystals to obtain some fundamental features. For example, in face-centred cubic single crystals (such as copper) it has been noted that during a stable cyclic response, the strain in the matrix is accommodated by the quasi-reversible to and fro bowing of screw dislocations in channels between the veins. However, in the softer persistent slip bands which experience larger strains, the edge dislocations bow out of the walls and traverse the channels [12]. In polycrystals the PSBs are generally found in the surface grains which have suitable orientation for the slips to occur.

The experimental observation of the PSB surface profile is quite complicated. Recent studies by the Basinskis [15], Mughrabi and co-workers [12], Hunsche and Neumann [14], and Ma and Laird [16] have illustrated that for the single crystal copper, the extrusions have an approximately triangular shape. The surface of these extrusions is covered with 'intrusion-like' features (roughness) which is highly temperature dependent. The mean extrusion height is about 3–4 μm. The roughness profile superimposed on the extrusion which forms valleys between them has a mean width (peak to valley) of about 0.3–0.4 μm [12]. Hunsche and Neumann [14] have reported a regular sawtooth-like PSB profile with a dominant wave length of 2 μm, superimposed on a triangular protrusion of the original crystal surface. For single crystals the extrusion profile may be dominant and the surface topography shows intrusions of 'V' type shape [14]. However, there are some questions regarding the depth of intrusions, i.e. whether they penetrate below the crystal surfaces, cf. [14, 15].

The PSB profile appears to be rather sensitive to the prior strain-hardening technique (stress- or strain-controlled mode) as well as sectioning technique. For example, in a study by Ma and Laird [16] PSB encroachments (negative protrusions) were observed in copper single crystals which have not been seen by other researchers, e.g. [12–15].

In summary, the deformation process in copper single crystals is comprised of two stages. The matrix substructure is formed during the initial rapid strain-hardening stage. At this stage the deformation is almost homogeneous, and surface markings are fine and uniformly distributed. The persistent slip bands are formed throughout the saturation period, and the cyclic strain is carried by the active PSBs. An irreversible dislocation mechanism gives rise to the rough profile of extrusions and intrusions. Basinski and Basinski [15] have provided a detailed study of the structure of the PSBs.

In the case of copper polycrystals, the net extrusion effect is much smaller than that of single crystals. Also, the PSB profile appears to be about evenly divided above and below the original surface. Dickson *et al.* [17] have reported profile surface topography for copper polycrystals which differs considerably from that reported for copper single crystals. They show the formation of thin ($\approx 0.1\ \mu$m) tongue-shaped extrusions at an early stage of cycling which often grow into relatively long ribbons. At higher strain amplitudes, hills and valleys topography (regularly-spaced wide extrusions and intrusions) was observed, resembling that of the single crystals.

Figure 1.7 shows a schematic illustration of slip during the monotonic and cyclic deformation process for a polycrystalline material, where the slips initially lead to extrusions of triangular shape. The subsequent profile as cyclic loading proceeds is strongly temperature dependent. For a review of the microstructural changes at elevated temperature, see for example Plumbridge [18]. Further discussion of the microstructural evolution during cyclic deformation will be given in Chapter 2.

It is worthwhile mentioning that the slip bands as shown in Fig. 1.7 create a rough surface topography, and intrusions within a surface are possible sites for crack initiation. There is ample experimental evidence dating back to the 1956 work of Thompson, Wadsworth and Louat [19] which indicates that removing intrusions and extrusions by electropolishing of the surface increases fatigue life.

1.3.2 INITIATION OF FATIGUE CRACKS

There is general agreement that the persistent slip bands are major nucleation sites for cracks in macroscopically isotropic materials (in which the material is highly inhomogeneous at the microscopic level, e.g. the plastic strain in the PSB lamellae is at least an order of magnitude higher than that in the matrix [18]). The cracks tend to form at the interface of the PSB and matrix. These cracks generally initiate after a localized saturation of the dislocation effects. Based on experimental observations,

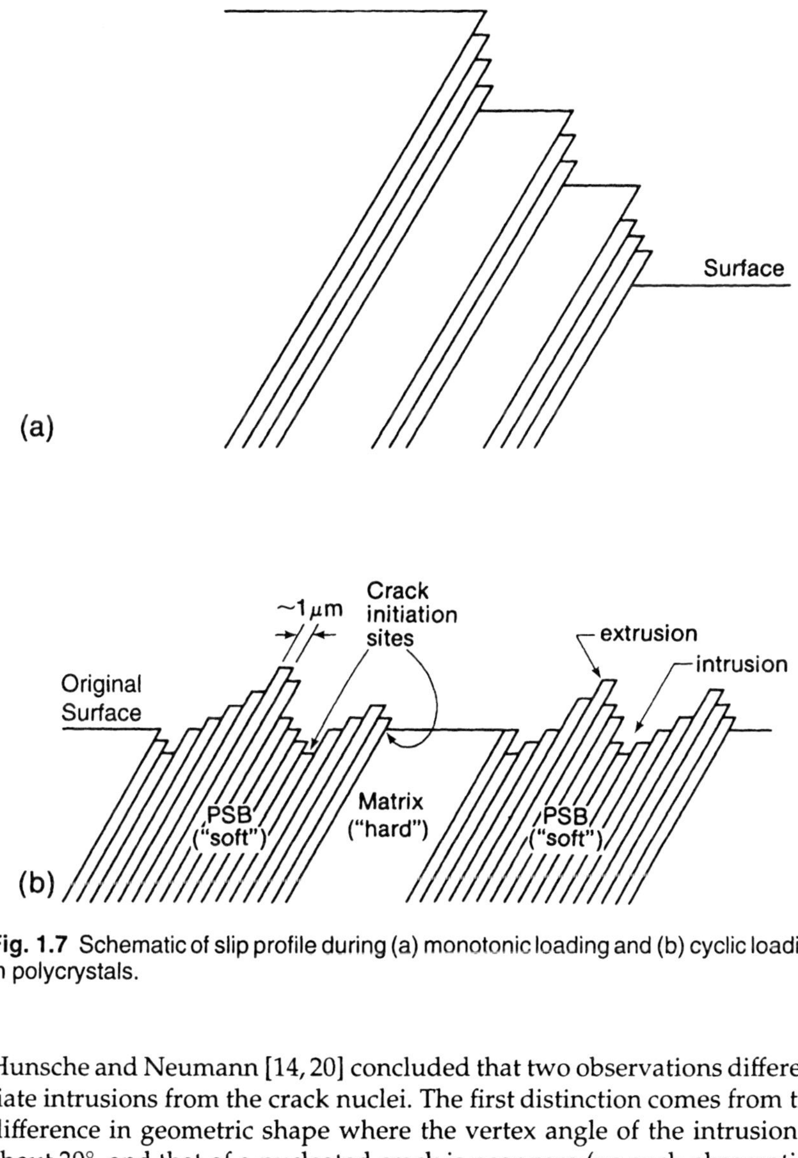

Fig. 1.7 Schematic of slip profile during (a) monotonic loading and (b) cyclic loading in polycrystals.

Hunsche and Neumann [14, 20] concluded that two observations differentiate intrusions from the crack nuclei. The first distinction comes from the difference in geometric shape where the vertex angle of the intrusion is about 30°, and that of a nucleated crack is near zero (no such observation was made by Ma and Laird [16]). The second observation is that the formation of intrusions is almost independent of the environment whereas crack formation is a strongly environmentally-dependent process. Basinski and Basinski [15] have observed crack formation anywhere within PSB material as well as near the interface. The most frequently observed crack sites were at very narrow PSBs and at isolated 'intrusions'. Dickson *et al.* [17] have also observed microcracks along very thin, ribbonlike extrusions. At present there appears to be some conflicting data regarding details of crack nucleation sites, however, the important point to keep in mind is that the cracks originate within the PSBs, including the borders.

It has been suggested that the mechanism of crack nucleation and growth is probably that proposed by Thompson *et al.* [19] which is depicted in Fig. 1.8. In this model a monolayer of oxygen or an oxide layer form rapidly on the freshly formed slip step exposed in air during a tension cycle. In the compression part of the cycle, the slip reverse takes place on the same plane, and some of the freshly absorbed oxygen is drawn into the body of the crystal in the form of dissolved atoms. During the next cycle this process is repeated and more oxygen is supplied to the crystal which could then diffuse along the slip band. The chemisorption of oxygen in and near the slip band causes weakening in the PSBs of the crystal eventually leading to cracking.

It is worth noting that the oxygen (or another aggressive medium) here plays an active role in assisting early crack nucleation and growth, and is not the cause of crack initiation. The environment merely affects the degree of slip irreversibly, thus an aggressive environment leads to rougher surface topology compared to an inert medium. In a vacuum environment crack nucleation may be retarded by up to two orders of magnitude (Mughrabi [12] contends that the effect of environment is negligible when the length of an initiated crack is defined to be about 10^{-4} μm. It is the rate of growth of these microcracks which is influenced by the environment). Thus the fundamental question which still remains to be answered is: what is the mechanism of crack nucleation?

Several mechanisms have been proposed in recent years [12, 21, 22] to supplement earlier models [13, 23]. The consensus at present appears to be that crack initiation may take place either at the slip steps between the emerging extrusions and the matrix, at the micro-notches near the outer

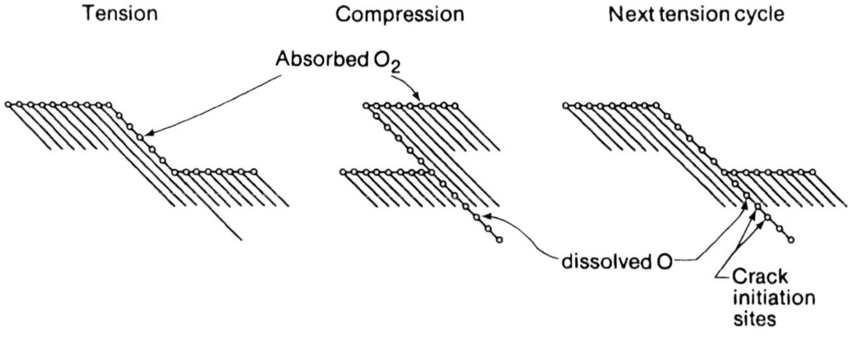

Fig. 1.8 Mechanism of environmentally-assisted stage I crack growth, after [19].

edges, or at the intrusion sites. All of these locations are sites of high stress/strain concentration. Across the PSB–matrix interface there is a high strain gradient due to a difference in compliance of the two parts (soft PSBs and harder matrix). The deformation compatibility requirement at the interface will result in high shear stresses along the interface leading to stage I cracks. The role of the PSB–matrix interface as a dominant site for crack initiation has been emphasized by many investigators [12, 16, 20–22].

The above conclusions are drawn from the experimental observations of single crystal materials, mostly copper and aluminium. As mentioned earlier, in the case of polycrystals, the net extrusion effect is much smaller and fatigue cracks generally occur in a transcrystalline mode in the PSBs. Crack initiation at sites where the PSBs impinge on the grain boundaries have also been observed, leading to intercrystalline crack initiation. This situation occurs when the active Burgers vector has a large component parallel to the specimen surface which prevents the emergence of extrusions along the PSB–matrix interface. The dislocations in this case pile up against the grain boundary causing a high strain concentration zone. A more common cause of crack initiation along the grain boundaries is attack by an aggressive or an embrittling environment or the effect of elevated temperature (sections 1.3.4 and 1.3.5).

Figure 1.9 shows various crack initiation sites, namely (i) transgranular, (ii) intergranular, and (iii) surface inclusion or pore. The first two are PSB-induced surface cracks, and the latter type is observed in some commercial alloys. There are also internal defects such as (iv) inclusions, (v) grain boundary voids and (vi) triple point grain boundary intersections which may be sources of crack initiation. The latter two are commonly observed at elevated temperatures.

1.3.3 CRACK PROPAGATION

Once cracks have initiated they may grow as a result of further cyclic deformation. In polycrystals grain boundaries often act as a barrier for micro-crack growth. On the other hand, in high temperature or corrosive environments grain boundaries are sources of crack initiation and the initiated crack may grow along the grain boundary.

In a uniaxial loading, Fig. 1.10, the maximum shear stress is on the plane whose normal makes an angle of 45° with the applied stress direction. Thus, micro-cracks (crystallographic cracks) may initiate along the active shear slip bands of approximate maximum shear stress in the suitably oriented crystals at about a 45° angle. With further cyclic deformation a few of the initiated cracks will first grow along the slip bands and then a dominant crack gradually emerges and changes direction to become perpendicular to the loading axis (Fig. 1.10). This transition from the active slip direction (stage I) to the non-crystallographic plane (perpendicular to

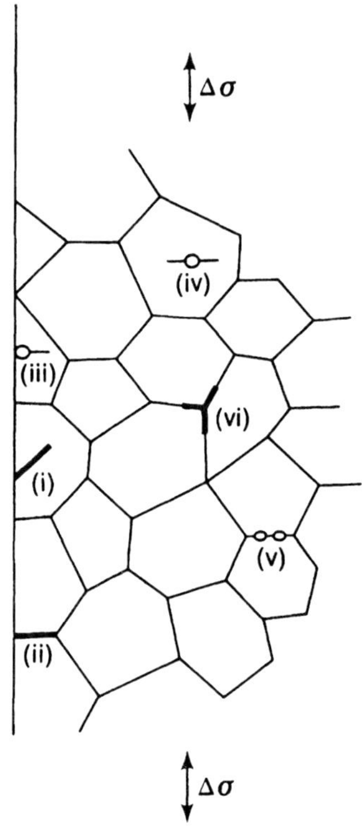

Fig. 1.9 Various sites for crack initiation in polycrystals (see text).

the stress axis) (stage II), occurs over a few grains. The length of the stage I crack depends on the material and loading amplitude, and is generally of the order of 3 to 4 grain size. The length of the stage I crack increases with an increase in the average grain size of the material and the decrease of the stress or strain amplitude. The stage II crack growth is often called the continuum crack propagation (mode I) and will be dealt with in Chapter 7. Figure 1.10 is a schematic drawing of stage I and stage II crack growth in a polycrystalline material. One of the surface grains is cracked along the active slip plane which subsequently results in cracking of the second grain along its active slip plane. The more constrained interior grains could have more than one slip plane activation, eventually leading to mode I (stage II) cracking.

It should be noted that crack growth under cyclic loading is considerably affected by proximity to the grain boundary. The so-called 'anomalous' small crack growth which presently cannot be predicted by using a nominal fracture mechanics parameter, is mainly due to the effect of

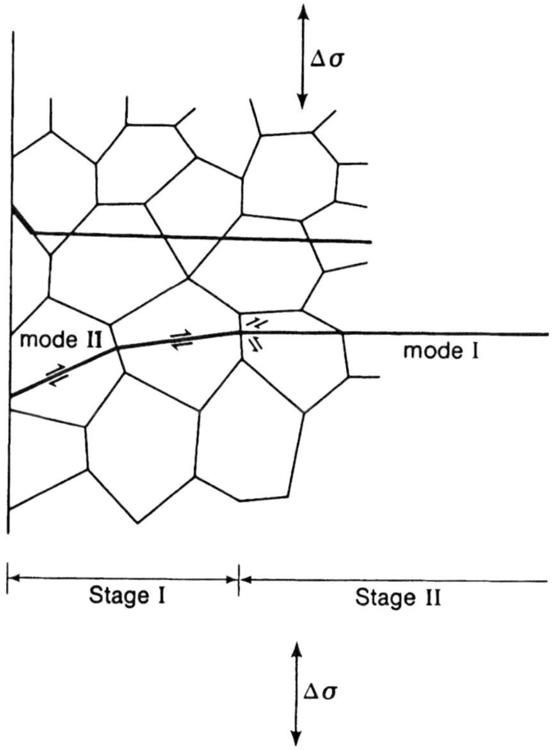

Fig. 1.10 Transgranular and intergranular crack initiation and transgranular crack growth, stage I and stage II crack growth.

microstructural barriers such as grain boundaries or banded microstructural features in metals and alloys. A retardation in the crack growth occurs because of the reduction in the crack driving force which has been referred to as crack tip shielding. This is the case in the breakdown of correlation with nominal fracture mechanics parameters. The growth of small cracks will be discussed in Chapter 9.

A further change in crack plane direction may occur from the flat plane (stage II) to an inclined plane whose normal is about 45° to the applied stress direction (Fig. 1.11). The latter plane is not parallel to the stage I (mode II) slip plane, it is generally an out-of-plane shear surface (mode III) often called a 'shear lip'. The crack growth direction gradually turns and grows along the slanted surface. This change in direction has been attributed to the change of constraint condition in front of the crack from a plane strain to a plane stress. The environment also has a strong influence on the mode change, i.e. an aggressive environment generally delays the

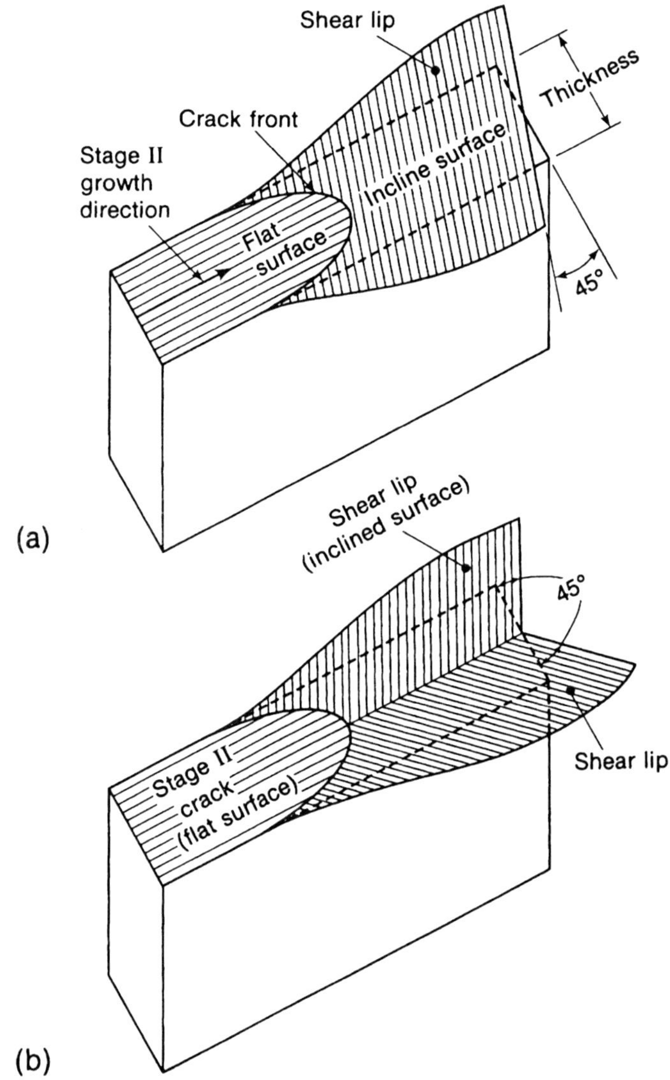

Fig. 1.11 Change in crack growth direction from the flat (stage II) to inclined surface: (a) 45° single shear lip; (b) double shear lip planes.

transition. Note that in the case of double shear planes the crack direction remains the same (Fig. 1.11(b)). Other types of transitions, e.g. transition from the mode I crack to mode II (in-plane shear) crack (kinked cracks), have also been observed [24].

At this stage it suffices to mention that the stage I fatigue crack is generally controlled by resolved shear stress along the slip plane whereas that of the stage II crack is controlled by the normal stress component. The crack path in turn is influenced by the material microstructure and environment–microstructure interaction. The transgranular fatigue crack growth depicted in Fig. 1.10 requires much less energy than the micro-void coalescence associated with the ductile fracture.

A crack phase diagram is shown in Fig. 1.12, which portrays three fatigue crack growth phases: short crack growth (SG), near threshold long crack growth (LT) and intermediate long crack growth (LG). For each phase the corresponding crack tip cyclic plasticity pattern is also depicted [25]. Short cracks and near threshold long cracks propagate along the cyclic slip bands. The crack growth rate for each phase, da/dN, is shown on the top right hand side inset. It can be seen that the growth rate of a short crack (SG) is much higher than that of a long crack under the same applied ΔK (stress intensity range). (Stress intensity factor range, ΔK is a measure of the crack driving force in a linear fracture mechanics context, $\Delta K = Y\,\Delta\sigma a^{1/2}$ where Y is a geometric parameter, see Chapter 7 for further explanation.)

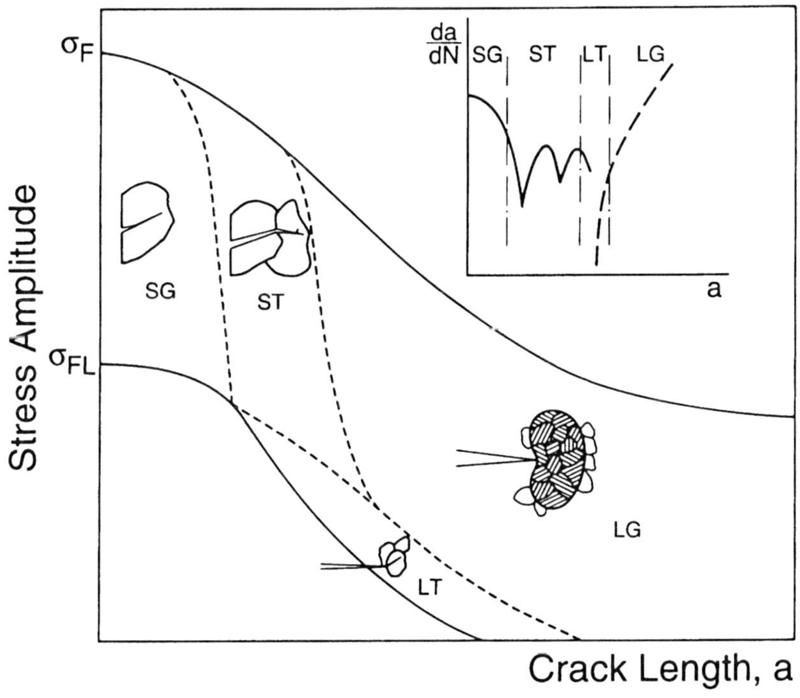

Fig. 1.12 A crack phase diagram showing three main growth phases: short crack growth (SG), near threshold long crack growth (LT) and intermediate long crack growth (LG) [25]. Reproduced with permission from The Society of Materials Science. The transition from short to long crack is depicted by (ST). The crack tip plasticity pattern and crack growth rate for each phase is also shown.

Once a short crack grows longer, it either branches into a secondary slip band or it meets the microstructural barriers. In either case, the rate of crack growth decreases and may even cease to propagate. This is a transition stage (ST) of the microstructure-sensitive short crack growth into the long crack growth regime. The crack tip cyclic plasticity plays a key role in the microstructure-sensitive growth of short cracks (see [25] where further explanation is provided).

It can be seen from the inset to Fig. 1.12 that there are essentially four crack growth regimes. The first two, concerning the short crack growth (SG and ST), will be discussed in Chapter 9, whereas the behaviour of the long (macro) cracks (LT and LG) is the topic of Chapter 7.

1.3.4 EFFECT OF TEMPERATURE

At high temperatures, $T \geqslant 0.5\, T_m$ (where T_m is the melting temperature), most metals under cyclic loading show damage in the form of grain boundary voids and wedge cracks (Fig. 1.9). The latter requires grain boundary sliding which results in a geometric incompatibility at 'triple points' and hence a stress concentration. This type of creep damage is generally observed at high load levels and lower temperatures. The effect of environment is also enhanced by the rise in temperature affecting the surface layer and initiating cracks in grain boundaries.

A cavity may nucleate as a result of either grain boundary sliding or matrix slip impingement, and its growth is sustained by either stress-assisted diffusion or further grain boundary sliding, or a combination of the two. This mechanism requires a rather high tensile stress to maintain stability and growth. Other factors such as migration of fatigue-generated vacancies, segregation of impurities and diffusion and development of internal gas pressure also contribute to the maintenance of cavity stability [18]. Cavity nuclei have been observed in the initial microstructure which may have been produced during the fabrication process [26].

It then follows that to resist a cavity formation, the grain boundary particles should not be too large to create high stress concentration regions nor too small to be ineffective in resisting grain boundary sliding. Austenitic steels are a good example of the above requirement.

During a cyclic deformation, the cavity stability requires a critical time period so that it is not sintered out in a compressive cycle. If the rate of loading is below the critical rate (or frequency), then the nucleated cavities usually grow along grain boundaries which are generally perpendicular to the maximum principal stress direction.

Void growth mechanism and coalescences of voids is a topic under intense investigation. A review article by Argon [27] discusses some of the issues. A schematic intergranular crack growth by micro-void coalescence is shown in Fig. 1.13.

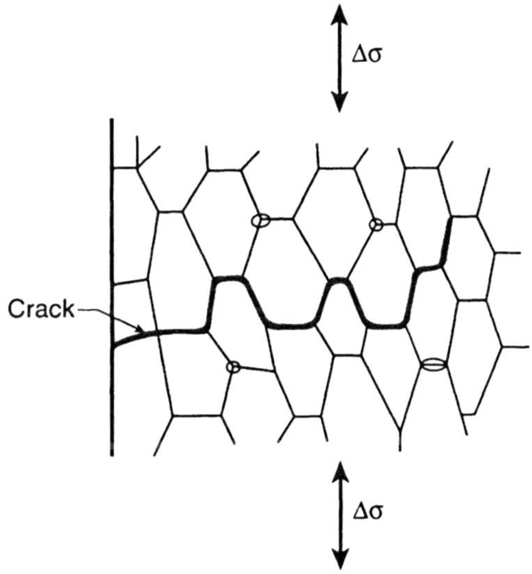

Fig. 1.13 Schematic of intergranular crack growth through micro-void coalescence.

An important design consideration in the high-temperature environment is the interaction between creep and fatigue loading. This will be discussed in Chapter 3.

1.3.5 EFFECT OF OXIDATION

The effect of oxygen on crack initiation at ambient temperature has been discussed (section 1.3.2). The influence of oxygen is more pronounced as the temperature is increased. Coffin [28] has reviewed the effect of oxide films on crack initiation and propagation. A suggested mechanism is the embrittlement of grain boundaries due to oxygen penetration. This embrittlement has been explained in terms of pinning of the grain boundaries due to the existence of oxides and small cavities. During cyclic loading at high temperatures, the lack of grain boundary mobility may lead to intergranular cracking. The oxidation processes have been found to be highly localized at crack nucleation sites and along the fracture path [28].

There is an interaction between cyclic loading and environment. The protective brittle oxide film is repeatedly ruptured by the localized slip action of cyclic deformation. The exposed fresh metal surface is then locally oxidized, acting as a notch leading to further strain/stress concentration, and eventual crack initiation.

Experiments have shown that symmetric cycling in a high temperature and high vacuum environment produces crack initiation at the slip band extrusions (transgranular cracks) as described earlier. However, unsymmetric wave forms and an air environment lead to quite different crack initiation sites mostly in the grain boundaries (intergranular cracks). The effect of oxygen on fatigue life will be discussed in Chapter 3.

1.4 FACTORS AFFECTING FATIGUE LIFE

Factors which affect fatigue life may be grouped into the following categories: microstructure of the material (grain size, texture); processing (deformation history, manufacture); load spectrum (sign, magnitude, rate, history); environment (temperature, corrosive medium); and geometry of a component (surface finish, notches, welds, connections, thickness). Some of these factors are interrelated as shown in Fig. 1.14. We have already discussed the effect of a few of these factors, e.g. temperature, environment, microstructure. Here we will endeavour to give a general guideline as to their influence on fatigue life. At the outset it has to be emphasized

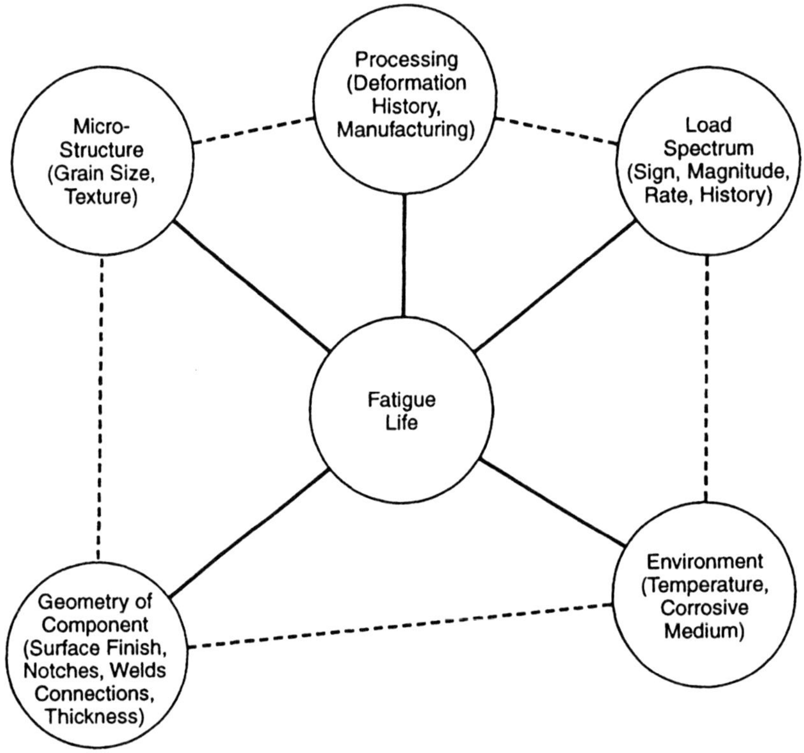

Fig. 1.14 Factors affecting fatigue life of a component.

that the degree of influence of each of the above-mentioned factors will vary for different materials.

1.4.1 EFFECT OF MICROSTRUCTURE

Grain structure, average grain size and microstructural flaws have a strong influence on fatigue life. For example, a material with large grain size would exhibit a lower fatigue limit in comparison to one with a smaller grain size when cycled at ambient temperature. However, at high temperatures (in the creep range of metals) the coarse-grained materials appear to exhibit better fatigue properties.

Fatigue damage in a metallic material is caused by plastic (irreversible) deformation which is generally produced by the motion of dislocations. The dislocation motion is affected by the precipitates, impurities, grain boundaries, etc. In addition, the cyclic deformation influences the dislocation substructure and density, eventually forming a stable structure (Chapter 2). The initial dislocation substructure may be quite different depending upon the prior processing, e.g. annealing or work hardening.

Phase transformation may also take place as a result of cyclic loading which could affect the fatigue life. For example, austenite–martensite transformation is observed in steels which results in a reduced fatigue life [23].

1.4.2 EFFECT OF PROCESSING TECHNIQUES

Processing techniques such as forging, rolling and extrusion produce directional properties in materials due to the grain orientation. Fatigue life is generally enhanced in the oriented direction and is lower in the transverse direction. The material processing may also produce microscopic defects such as cavities (dissolved gases). These defects act as stress raisers and could be potential sites for crack initiation.

Heat treatment, case hardening, cold and hot working, surface coating, plating, cladding, etc. can all influence fatigue life. For example, the manufacturing process could produce residual stresses. (The residual stresses are self-equilibrium stresses which exist in a material without application of external loads.) Compressive residual stresses at the external surfaces generally enhance fatigue life, while tensile ones are detrimental. Processes such as shot peening, cold rolling and static preloading are employed to induce compressive residual stresses.

Components are generally heat treated after manufacturing to enhance some properties, e.g. to reduce the residual stresses. In steels heat treatment causes surface decarburization, thus decreasing the surface toughness and causing micro-cracks. Processes such as forming, drawing, forging, extrusion, rolling, machining, punching, etc. produce rough surfaces in addition to the changes in microstructure and residual stress values. The rough surface produced in these processes generally causes a decrease in the fatigue life of the component (section 1.4.5).

1.4.3 ENVIRONMENTAL FACTORS

A corrosive environment causes degradation of the material, the most common visible effect being pitting or surface roughness (etching). These notch-like regions act as stress raisers and are generally the sites of crack nucleation. The most important locations are those where fatigue slip activity is taking place. A slight difference between the electrochemical potential of the outside and inside of the slip band can promote the corrosion process.

It was mentioned earlier that cyclic loading enhances the corrosion effect by continually breaking the protective oxide layer and exposing fresh surfaces, thus extending the corrosion effect from the surface into the interior of the metal. A corrosive environment thus shortens the crack nucleation stage. Once cracks are initiated, subsequent crack propagation may be enhanced by a corrosive environment. For example, corrosion may inhibit crack rewelding [14] and thus increase the crack growth rate. Embrittlement of the crack front material due to the diffusion of hydrogen has been observed in a corrosive environment [29]. On the other hand, for some materials crack tip blunting due to corrosion may occur which could cause a reduced rate of crack propagation. It can be said without loss of generality that an aqueous environment shortens the crack nucleation stage.

We have already indicated that an increase in temperature reduces the fatigue resistance of most metals. Increased temperature generally causes an increase in the crack propagation rate. Temperature, environment and cyclic loading have a synergistic effect. These effects are further discussed in Chapters 2, 3 and 7.

1.4.4 EFFECT OF LOAD SPECTRUM

The history, magnitude, sign and rate of loading influence fatigue life. It was pointed out in section 1.2 that the actual service loading may be a very complex one. The loading history could be a superposition of deterministic and stochastic loads. In most cases the loading is multiaxial, e.g. a combined axial, bending and torsional action on a slender member which results in a multiaxial state of stress that varies across and along the component. The effect of multiaxial loading is to reduce fatigue life in comparison to simple uniaxial loading, except in the case of pure torsional loading (Chapters 4 and 5).

In section 1.2 we have indicated how complex loading can be approximated into a sequence of alternating and mean loads. A technique to

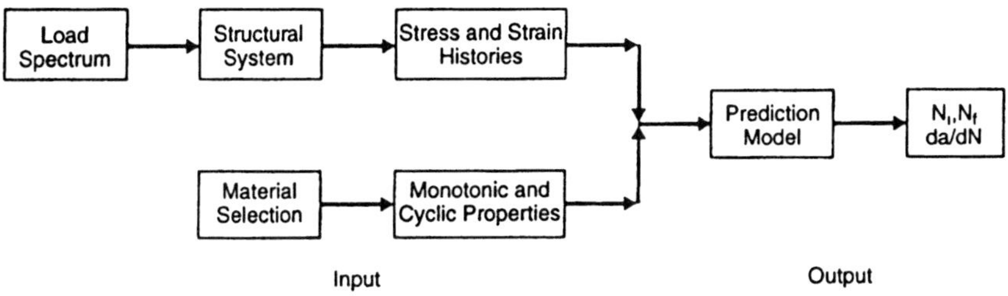

Fig. 1.15 The required sequence (flow chart) for fatigue life prediction.

construct the most damaging and least damaging loading histories from a compact storage of representative service loads has also been put forward by Pompetzki and Topper [30]. Irrespective of the method used to present the service load, a stress analysis will generally be required to convert these load spectra to the stress and strain histories within the component. The resulting stress and strain histories are then used to predict the component life. Figure 1.15 shows the sequence of required operations.

The outcome of the stress analysis would be stress or strain histories which could be partitioned into sequences of alternating stress (or strain) $\Delta\sigma/2 = (\sigma_{max} + \sigma_{min})/2$ with or without mean values $\sigma_m = (\sigma_{max} - \sigma_{min})/2$ (with similar expressions for the strain). Fatigue life decreases with the increase in magnitude of the alternating stress (or strain) (Fig. 1.1). The effect of the positive (tensile) mean stress is to reduce fatigue life whereas a negative (compressive) mean stress may increase it. The influence of the mean stress (or strain) is more significant in the low strain, high cycle fatigue regime of a material. In a high strain regime, especially in the presence of elevated temperature, a mean stress could cause ratcheting which may result in a catastrophic failure. History or sequence of loading is also an important factor in the process of accumulation of damage. The above-mentioned factors will be discussed further in Chapter 3.

1.4.5 EFFECT OF GEOMETRY

Surface finish has a significant influence on crack initiation. We have seen in section 1.3 that smooth surfaces become rough as a result of slip along the crystal planes. Pre-existing surface scratches or micro-notches could cause an early strain localization and thus reduce the crack initiation life. Geometric discontinuities in a component such as holes, notches and joints are sources of stress (or strain) concentration and are often the sites of crack initiation. In these locations stress and strain are much greater than the nominal applied stresses (or strains) and high stress and strain gradients exist in the vicinity of these geometric discontinuities.

The fatigue life of a notched component is generally less than that of an unnotched one. While the surface roughness of a component could be improved by polishing, notches cannot be avoided in most industrial components because of the functional requirements or connection details. Notch effects will be studied in Chapter 8.

In summary, fatigue is a complicated process which is affected by a number of factors, some of which are difficult to quantify. The complexity of the fatigue process is compounded by the fact that some of the factors discussed in this section are interrelated and cannot be easily isolated for a quantitative evaluation. This is partly the reason why there is such a vast literature on the subject, and intensive research is being carried out to delineate and quantify the effect of some of the factors discussed here.

1.5 FATIGUE DESIGN METHODOLOGY

Design criteria for the prevention of fatigue failure have evolved from the concept of almost infinite life to the present-day damage tolerance philosophy. As mentioned earlier, in the Introduction, in the indefinite life design methodology one wishes to ensure that the **maximum** stress remains below the material's fatigue (endurance) stress limit. If a material does not exhibit a definite fatigue limit, then the stress level corresponding to a predetermined number of cycles, say 10^8 or 10^9, may be chosen from the stress versus life (S–N) curve as the limiting value. Components of rotating machinery or engine valves, etc. are designed by this method.

1.5.1 SAFE-LIFE DESIGN METHOD

Designing for a finite life is sometimes termed the '*safe-life*' criterion. For example, in pressure vessel design codes, e.g. [31], the general membrane stress is held below the lesser of $\frac{2}{3}$ yield stress or $\frac{1}{3}$ of the ultimate strength in order to avoid general yielding. However, local yielding in the highly stressed regions is permitted for the local membrane and bending stresses up to a stress level corresponding to twice the yield strain (for a strain-hardening material the local stress will exceed the yield strength of the material). The limit for the peak stresses is obtained through a fatigue analysis as follows.

The procedure is based on the maximum shear stress theory of failure and consists of finding the amplitude of the maximum shear stress cycle. To facilitate the calculations, a measure called stress intensity (twice the maximum shear stress) is used in place of the shear stress itself. (This measure is not to be confused with the stress intensity factor, K, in linear fracture mechanics, which has dimensions of stress times the square root of length, i.e. MPa $m^{1/2}$.)

At each point of the vessel, at any given time, there are three principal stresses σ_1, σ_2 and σ_3, and three stress differences:

$$\begin{aligned} S_{12} &= \sigma_1 - \sigma_2, \\ S_{23} &= \sigma_2 - \sigma_3, \\ S_{31} &= \sigma_3 - \sigma_1. \end{aligned} \tag{1.2}$$

The extremes of the range through which each stress difference fluctuates are determined, and the absolute magnitude of this range for each stress difference is noted. The alternating stress intensity, S_{alt}, is the largest of the amplitudes of the three stress differences, i.e.

$$S_{alt} = \tfrac{1}{2} \max \left[|S_{12}|, |S_{23}|, |S_{31}| \right]. \tag{1.3}$$

This stress is then compared with the ASME design fatigue curve which is shown in Fig. 1.16 [32]. The ASME design fatigue curve was obtained from the fatigue data of En-25, A-225 and A-302 low alloy steels. The best fit line of these data is shown in Fig. 1.16 as a dotted line. In addition, the data for the ASTM A-516 Gr. 70 pressure vessel steel are also plotted in the figure [33]. It can be seen that the ASME design curve is obtained from the reference data by an almost uniform reduction of stress through a safety factor of about 3.5–4 (Fig. 1.16).

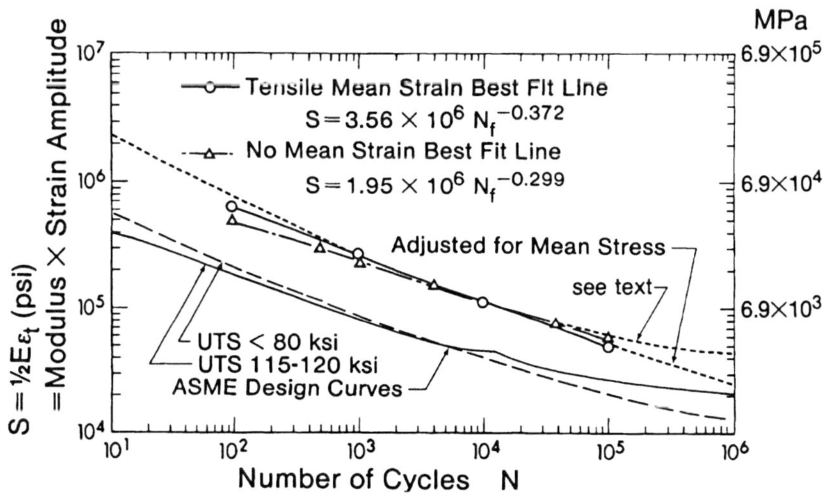

Fig. 1.16 ASME design fatigue curve for carbon low alloy series 4XX, high alloy steels and high tensile steels for temperatures not exceeding 700°F, and experimental data. Reproduced with permission from ASME from [33].

It should be noted that the ordinate in Fig. 1.16 is converted to a 'linear' stress by multiplying the strain amplitude by the elastic modulus. In other words, the ASME code permits a simple elastic analysis and converts plastic strains to an 'equivalent' elastic stress.

1.5.2 FAIL-SAFE DESIGN METHOD

This method of design was originally developed by the designers of aircraft structures. A fatigue crack is allowed to grow, but the component is designed in such a way that the presence of a crack will not lead to failure before it is detected. Generally crack arresters are built in several positions and/or multiple load (stress) paths are designed to achieve a fail-safe design. In other words this concept is based on designing alternative load-carrying members such that failure of one could be tolerated by load redistribution to the remaining members. A good discussion of this philosophy and some illustrative examples are given by Osgood [34].

The underlying assumption in the case of the 'safe-life' design concept is that the fatigue life of the component (structure) can be predicted under the expected loading conditions. This may not be achievable in the case of complicated structural assemblies or circumstances where service loads may be of a random nature containing occasional overloads.

In a fail-safe structural assembly the consequence of a failure is a small increase in displacement, and the integrity of the structure is ensured through the crack arresters. It should be pointed out that not all structures can be designed economically by a fail-safe concept, e.g. one-piece parts.

1.5.3 DAMAGE TOLERANCE DESIGN METHOD

In this design method one uses fracture mechanics to ascertain the extent of crack growth. Periodic inspections are planned to detect cracks of a certain size. An important consideration in this design methodology is the selection of materials with a high fracture toughness.

Periodic inspection serves two purposes. It is a very important aspect of preventive maintenance, and it does, in some cases, make an important contribution to safety. Periodic inspection for safety has for a long time been an integral part of the licensing process for conventional boiler and pressure vessels, but is a rather crude way of looking for gross effects such as massive corrosion or tube blockage.

In aviation, periodic inspection was at first considered in the context of a pre-set safe life. This was to be carried out at intervals determined by a program of testing and of major overhauls which could be modified in time by evidence coming from the inspection itself.

For nuclear reactors, the hostile radiation environment presents serious problems for re-inspection, requiring specially-developed equipment and a longer time to carry out the inspection.

The problem of establishing the frequency of re-inspection in practical terms has been a concern of the aviation industry since its inception. However, to survive in aviation it is necessary to learn to operate on the brink of a catastrophe, and consequently the safety margins are small and very dependent on re-inspection procedures. In nuclear matters in general the margins should be much greater and there need be less reliance on re-inspection to maintain a satisfactory safety level throughout the life of the plant.

A method has been developed [35] which provides for a rational basis for periodic inspection based on growth of cracks, taking into account variability of various parameters.

1.6 PROBABILISTIC APPROACH

It is generally agreed that there exists some degree of uncertainty in most phenomena encountered or observed in engineering practice. In mechanical systems, the loads (applied or transformed), the material strength to resist them and defect sizes are the most common variables which exhibit scatter. In recent years there has been a considerable effort to deal with such situations by employing the theory of probability. Today, even legal codes such as the Canadian National Building Code [36] have a probabilistic background.

As an example, let us consider a component in service subjected to repeated loads from the service–load spectrum. There will be a period during which the process of load application leads to the formation of macroscopic cracks. The time (or number of cycles) up to this stage will be termed service life to 'initial' failure. The cracks will propagate as the life continues with a progressive reduction in the residual strength of the component. When a crack length reaches a critical length, a_F, crack propagation will accelerate and component fracture takes place after very few load applications.

Figure 1.17 shows a population of components varying from weak ones to those strong in resisting crack initiation and propagation. We note in the figure three levels of crack length: microscopic, macroscopic and critical (unstable). In Fig. 1.17, the threshold crack length, a_0, may be envisaged as the crack level below which there is no reduction in the strength of the component.

At any given service life, N_s, there will be a variation in crack length in the population. Alternatively a specific crack length will be obtained at various service lifetimes in the population (Fig. 1.17). These variations can be represented in a statistical form, i.e. as probability density functions [35]. This issue will be discussed further in Chapter 10.

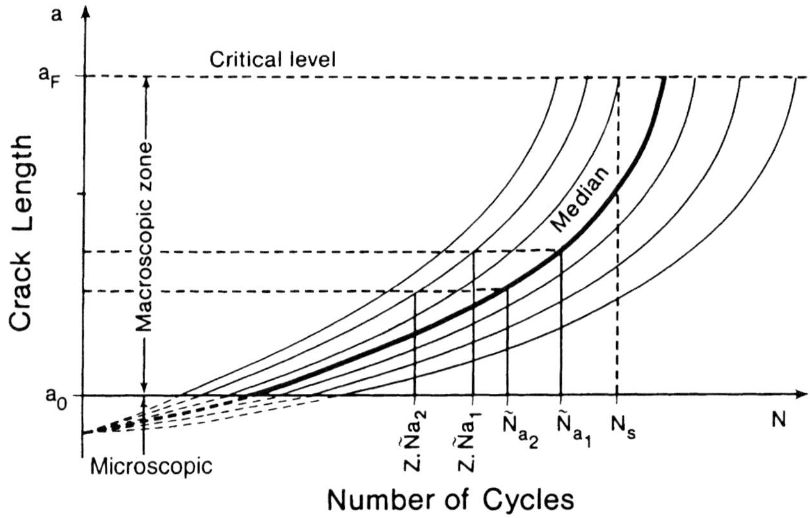

Fig. 1.17 Evolution of a fatigue crack during service life of a population of similar components. Reprinted with kind permission from [35].

REFERENCES

1. Wetzel, R.M. (ed.) (1977) *Fatigue Under Complex Loading: Analyses and Experiments*, Society of Automotive Engineers, Warrendale, PA.
2. Haugen, E.B. (1980) *Probabilistic Mechanical Design*, John Wiley & Sons, New York.
3. Crandall, S.H. and Mark, W.D. (1963) *Random Vibration in Mechanical Systems*, Academic Press, New York.
4. Lowak, H., De Jonge, J.B., Franz, J. and Schütz, D. (1979) *MINI TWIST: A Shortened Version of TWIST. LBF-TB-146*, Laboratorium für Betriebsfestigkeit, Amsterdam.
5. Schijve, J. (1963) The analysis of random load time histories with relation to fatigue tests and life calculation, in *Fatigue of Aircraft Structures* (eds W. Barrois and E.L. Ripley), Macmillan, New York, pp. 115–49.
6. Fuchs, H.O. *et al.* (1977) Shortcuts in cumulative damage analysis, in *Fatigue Under Complex Loading*, Society of Automotive Engineers, Warrendale, PA, pp. 145–62.
7. Conle, A. and Topper, T.H. (1982) Fatigue Service Histories: Techniques for Data Collection and History Reconstruction. *SAE Technical Paper 820093.*
8. American Society for Testing and Materials (1992) Standard Practice for Cycle Counting in Fatigue Analysis. *ASTM Standard E1049*, Vol. 03.01.
9. Matsuishi, M. and Endo, T. (1968) *Fatigue of Metals Subjected to Varying Stress.* Paper presented to Japan Society of Mechanical Engineers, Fukuoka, Japan, March 1968. See also: Endo, T. *et al.* (1974) *Damage Evaluation of Metals for Random or Varying Loading.* Proceedings of the 1974 Symposium on Mechanical Behaviour of Materials I, The Society of Materials Science, Kyoto, Japan, pp. 371–80.

10. Landgraf, R.W., Morrow, J. and Endo, T. (1961) Determination of the cyclic stress–strain curve. *J. Mater.*, **4**, 176–88.
11. Murakami, Y. (1992) *The Rainflow Method in Fatigue*, Butterworth Heinemann, Oxford.
12. Mughrabi, H. (1986) Cyclic deformation and fatigue: some current problems, in *Strength of Metals and Alloys* (*Proc. ICSMA 7*), vol. 3 (eds H.J. McQueen *et al.*), Pergamon Press, Oxford, pp. 1917–42.
13. Laird, C. and Duquette, D.J. (1972) Fatigue crack initiation, in *Corrosion Fatigue – Chemistry, Mechanics and Microstructure* (eds O.J. Devereux *et al.*), National Association of Corrosion Engineers, Houston, TX, pp. 88–117.
14. Hunsche, A. and Neumann, P. (1988) Crack nucleation in persistent slipbands, in *Basic Questions in Fatigue*, Vol. I, ASTM STP 924 (eds J.T. Fong and R.J. Fields), American Society for Testing and Materials, Philadelphia, PA, pp. 26–38.
15. Basinski, Z.S. and Basinski, S.J. (1985) Low amplitude fatigue of copper single crystals – Parts II and III. *Acta Metall.*, **33**, 1307–27.
16. Ma, B.T. and Laird, C. (1984) Overview of fatigue behaviour in copper single crystals. I. Surface morphology and stage I crack initiation sites for tests at constant strain amplitude. *Acta Metall.*, **37**, 325–36.
17. Dickson, J.I. *et al.* (1993) *Extrusion–Intrusion Formation in CU and 70 CU-302n.* Fatigue 93, Proc. 5th Int. Conf. on Fatigue and Fatigue Thresholds (Montreal) (eds J.P. Bailon and J.I. Dickson), EMAS, West Midlands, UK, Vol. III, pp. 1883–92.
18. Plumbridge, W.J. (1987) Metallography of high temperature fatigue, in *High Temperature Fatigue – Properties and Predictions* (ed. R.P. Skelton), Elsevier Applied Science, London, pp. 177–228.
19. Thompson, N., Wadsworth, N.J. and Louat, N. (1956) The origin of fatigue fracture in copper. *Phil. Mag.*, **1**, 113–26.
20. Hunsche, A. and Neumann, P. (1986) Quantitative measurement of persistent slip band profiles and crack initiation. *Acta Metall.*, **34**, 207–17.
21. Brown, L.M. and Ogin, S.L. (1985) Role of internal stresses in the nucleation of fatigue cracks, in *Fundamentals of Deformation and Fracture*, Eshelby Memorial Symposium (Sheffield) (eds B.A. Bilby *et al.*), Cambridge University Press, Cambridge, pp. 501–28.
22. Polák, J., Obrtlik, K. and Lisutin, P. (1988) Mechanisms of fatigue crack initiation, in *Basic Mechanisms in Fatigue of Metals* (eds P. Lukás and J. Polák), Elsevier, Amsterdam, pp. 101–9.
23. Klesnil, M. and Lukáš, P. (1980) *Fatigue of Metallic Materials*, Elsevier, Amsterdam.
24. Ewalds, H.L. and Wanhill, R.J.H. (1986) *Fracture Mechanics, Part V, Mechanisms of Fracture in Actual Materials*, Edward Arnold, London.
25. Ellyin, F. and Li, C. (1995) The role of cyclic plasticity in crystallographic crack growth retardation. *Mater. Sci. Res. Int.*, **1**, 137–43.
26. Swindeman, R.W. and Brinkman, C.R. (1988) What creep cavitation studies are needed to develop stainless steels resistant to time-dependent fatigue?, in *Basic Questions in Fatigue*, Vol. II, ASTM STP 924 (eds R.P. Wei and R.P. Gangloff), American Society for Testing and Materials, Philadelphia, PA, pp. 5–26.
27. Argon, A.S. (1982) Mechanisms and mechanics of fracture in creep alloys, in *Recent Advances in Creep and Fracture of Engineering Materials and Structures* (eds B. Wilshire and D.R.J. Owens), Swansea, Pineridge Press, pp. 1–52.

28. Coffin, L.F. (1980) Damage processes in time-dependent fatigue, in *Creep–Fatigue–Environment Interaction* (eds P.M. Pelloux and N.S. Stoloff), The Metallurgical Society of AIME, Warrendale, PA, pp. 1–23.
29. Bricknell, R.H. and Woadford, D.A. (1981) The embrittlement of nickel following high temperature air exposure. *Metall. Trans.* A, **12**, 425–33).
30. Pompetzki, M.A. and Topper, T.H. (1993) The least and most damaging histories which can be constructed from a specific from-to matrix, in *Fatigue Design*, ESIS16 (eds J. Solin *et al.*), Mechanical Engineering Publications, London, pp. 19–35.
31. American Society of Mechanical Engineers (1994) ASME Boiler and Pressure Vessel Code, Section VIII, Division 2, Alternative Rules, ASME, New York.
32. American Society of Mechanical Engineers (1969) Criteria of ASME Boiler and Pressure Vessel Code for Design by Analysis in the Sections III and VIII, Division 2, ASME, New York.
33. Ellyin, F. (1985) Effect of tensile-mean-strain on plastic strain energy and cyclic response, *J. Engng. Mater. Technol., Trans. ASME*, **107**, pp. 119–25.
34. Osgood, C.C. (1982) *Fatigue Design*, 2nd edn, Pergamon Press, Oxford.
35. Ellyin, F. (1985) A strategy for periodic inspection based on defect growth. *Theor. Appl. Fract. Mech.*, **4**, Elsevier Science – NL, Amsterdam, 83–96.
36. National Research Council of Canada (1985) National Building Code of Canada, NRCC No. 23174, Ottawa.

Cyclic stress–strain response

2

2.1 INTRODUCTION

This chapter is mainly concerned with the specification of stress–strain relationships under monotonic and cyclic loading. Various factors such as rate of loading, environment and temperature affect the stress–strain relationship. It will be seen that the cyclic stress–strain curve differs from that of the monotonic curve in several aspects. First, there is a transition accompanied by either strain softening or hardening with respect to the monotonic curve. A stable (or saturated) state is subsequently reached. The latter state is a manifestation of a stable micro-substructure. However, the transient response depends on the test control mode, i.e. deformation (strain) or load (stress) control, and prior history of deformation.

Microstructural changes taking place as a result of cyclic loading will be described throughout this chapter. Plastic strain is accomplished through dislocation motion, and materials can be grouped according to their slip characteristics. An equilibrium dislocation structure is observed at the stable (saturated) state. Dislocation motion and slip characteristics are influenced by factors such as temperature, loading rate and environment.

This chapter attempts to describe the stress–strain behaviour of metals from both macro- and microscopic viewpoints. To model the stable macroscopic response, explicit mathematical stress–strain relations are developed with a few material constants. The major thrust of this chapter is the description of behaviour under simple uniaxial stress states.

2.2 MONOTONIC BEHAVIOUR UNDER TENSION OR COMPRESSION

The stress–strain relationship of materials in multiaxial stress states has generally been an extension of the observations made in a uniaxial tension (or compression) test. A study of the material response under simple tension (or compression) is therefore an important step in the process of understanding the mechanics of material response.

When a smooth specimen of a circular cross-section with a uniform gauge length L_0 is subjected to a slowly increasing load P, we can measure the change in length, ΔL, of the uniform section with an extensometer. Instead of plotting load versus the corresponding displacement, it is more useful to choose quantities which render the results independent from the specimen geometry. If the applied load is properly aligned with the specimen axis, there will be an almost homogeneous strain state within the uniform gauge length, L_0. In this case, the best measures are:

$$e = \frac{\Delta L}{L_0} = \text{axial strain}; \quad s = \frac{P}{A_0} = \text{axial stress}. \tag{2.1}$$

The uniaxial stress–strain curve for a typical strain-hardening material is shown in Fig. 2.1. For small values of the applied load, the response is linear, and the process is reversible between 0 and A in Fig. 2.1, point A being the limit of the elastic response. As the load is increased beyond this limit, permanent deformations set in with the resulting plastic strain. At this stage if the load is reduced, the unloading response is initially

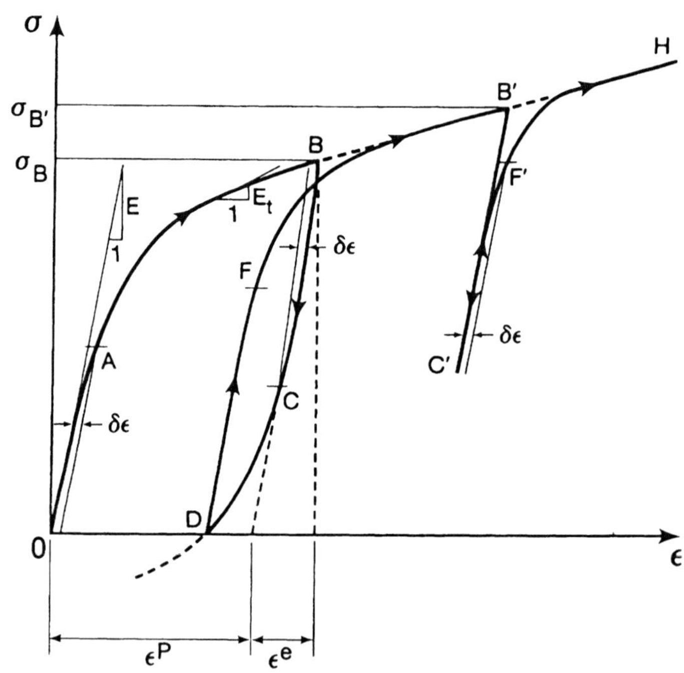

Fig. 2.1 Schematic diagram of a stress–strain curve and definition of offset yield stress for loading and unloading, after [41].

a straight line BC, which is essentially parallel to 0A, and it then deviates from the linear response, entering an irreversible domain, CD. Upon reloading we observe a linear portion DF, and a nonlinear part FB′. Two important observations are: (a) following reloading the curve does not pass through the unloading point B; (b) following the unloading path, the material yields at point C, at which the stress level is higher than the originally compressive yield stress level. The latter observation is generally known as the Bauschinger effect [1] and it indicates that after initial loading over elastic limit, the material property becomes anisotropic, particularly in the yield stress point (flow stress). This phenomenon is attributed to the existence of 'internal back stresses' which facilitate dislocation motion during the load reversal. It should be noted that most of the inelastic deformation of crystalline materials takes place by the glide and climb motion of dislocations. The kinetics of the motion of dislocations in a slip plane will depend on the current state of the obstacles. When imposing a reversed deformation, the mobile segment of dislocation is decomposed, not requiring the penetration of many obstacles, thereby producing an easily reversed plastic deformation.

The loop traced by the unloading and reloading, BCDFB′, is called a hysteresis loop, and is indicative of the absorbed plastic strain energy during the process. The majority of the irrecoverable energy is converted into heat, and internal damping absorbs a part of it.

The axial strain defined by (2.1) relates the change in length to the original length L_0. This is generally called the **engineering strain**. We may choose an alternative definition whereby the strain is related to the current length of the gauge length. In this approach, the total strain is regarded as the sum of the increments of strain, i.e.

$$\varepsilon = \sum \frac{\delta L}{L}, \tag{2.2}$$

where δL is the increment in elongation and L is the current length. At the limit,

$$\varepsilon = \int_{L_0}^{L} \frac{dL}{L} = \ln\frac{L}{L_0} = \ln(1+e), \tag{2.3}$$

where L is the current length of the gauge length. The strain defined by (2.3) is called the **true strain** or **logarithmic strain**. When a metal is deformed plastically, one observes that the material volume remains almost constant. More accurately, the change in volume during plastic deformation, at low and moderate temperatures, is negligible. (The volume increase is of the order of 0.05% to 0.5%.) For the tensile specimen described earlier, this is expressed as

$$A_0 L_0 \approx AL. \tag{2.4}$$

Substituting the above into (2.3) then leads to

$$\varepsilon = \ln\left(\frac{A_0}{A}\right) = 2\ln\left(\frac{D_0}{D}\right), \tag{2.5}$$

where D_0 and D now refer to the original and current diameters of the specimen, respectively.

In a similar manner, we can define the **true stress** by

$$\sigma_{\text{true}} = \frac{P}{A} = s\,(1+e), \tag{2.6}$$

where the last step is obtained by using (2.1) and (2.4).

In comparing (2.3) and (2.6) we note that when the engineering strain is small compared to unity, then the numerical difference between the true and engineering stress and strain are negligible. For example, when the engineering strain is 0.04, the true strain will be 0.039, a difference of about 2%. For the same deformation, the difference between the true stress and engineering stress would be 4%. Thus, the true stress versus true strain curve will always be above that of the engineering one; however, up to ~4–5% strain, the difference is negligible.

Figure 2.2 shows the stress–strain curve of a low alloy carbon steel (ASTM A-516 Gr. 70), for the entire range of the deformation process until fracture [2]. In the plastic regime three stages of responses are apparent. First we observe a high slope or rapid strain-hardening region, followed by an almost constant slope or constant strain-hardening regime. The latter is then followed by the stage three high strain-hardening region associated

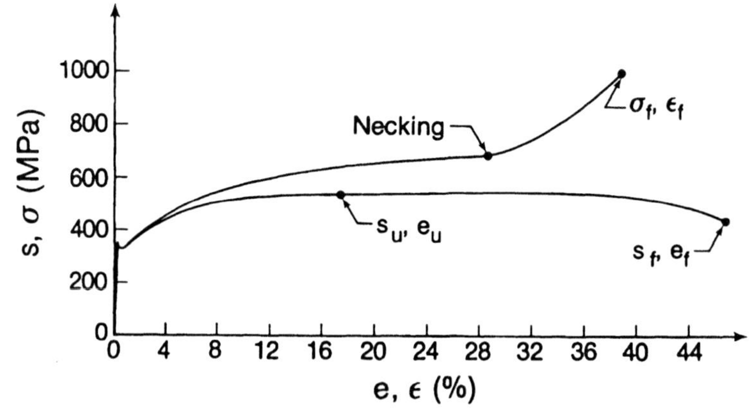

Fig. 2.2 Monotonic stress–strain curve of ASTM A-516 Gr. 70 low alloy carbon steel, after [2].

with the tensile instability. This type of behaviour appears to be characteristic of most polycrystalline metals.

From the theoretical viewpoint, it is preferable to use the true stress and true strain to define the state of stress and strain in a material. This definition also makes the tension and compression curves coincide. It is also consistent with the dislocation theory where the process is better described by an incremental concept. However, from a practical point of view, the distinction is not noticeable except for moderate to large strains. (In metal forming or in theories based on strain accumulation, it is imperative that the true strain be used.)

In this and in the following chapters, unless otherwise stated, we deal with small deformations, and therefore no attempt will be made to emphasize the nature of strain definition, although the incremental strain definition is implied.

The monotonic stress–strain curve 0ABB'H in Fig. 2.1 and part of Fig. 2.2 can be approximated by the Ramberg–Osgood relationship [3], written in its non-dimensional form:

$$\frac{\varepsilon}{\varepsilon_0} = \frac{\sigma}{\sigma_0} + \alpha\left(\frac{\sigma}{\sigma_0}\right)^{1/n}, \tag{2.7}$$

where $\varepsilon_0 = \sigma_0/E$, σ_0 is the yield stress and E is the Young's modulus of the material. The coefficient α and hardening exponent, n, are determined from the experimental data by a best fit procedure. Relation (2.7) can also be written as

$$\varepsilon = \varepsilon^{e} + \varepsilon^{p} = \frac{\sigma}{E} + \left(\frac{\sigma}{K}\right)^{1/n}, \tag{2.8}$$

where K is a coefficient with dimensions of stress. In a similar manner, the material constants K and n are found from the test data by a best fit procedure. From (2.7) and (2.8) we obtain the following relationship between the coefficients:

$$\alpha = \frac{(\sigma_0/K)^{1/n}}{(\sigma_0/E)}. \tag{2.9}$$

The approximation involved by fitting a stress–strain curve through (2.7) or (2.8) may not be accurate enough in certain cases. In these instances, the stress–strain curve is divided into sections and each part is fitted with an appropriate function, or numerical values at discrete points are provided in a function table and interpolation functions are used for in between points.

In describing the material stress–strain curve (Fig. 2.1) we implicitly assumed that the response was rate insensitive. A study of dislocation structure will indicate that the rate of strain hardening depends on the

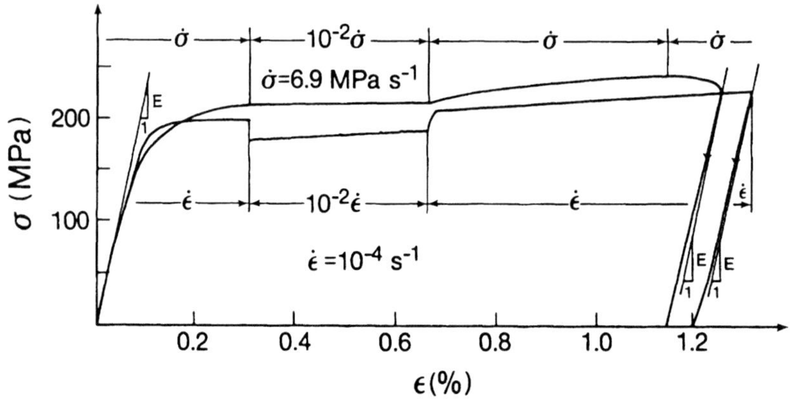

Fig. 2.3 The effect of different strain-rates and the rate control mode on the stress–strain curve of AISI type 304 stainless steel. Reprinted with kind permission from [5].

current obstacle state of the material (glide resistance or hardness). In addition, the rate of strain hardening would depend on the strain-rate. (An interested reader is encouraged to consult, for example, Argon [4] for further description of dislocation models).

In the case of rate-sensitive material (e.g. austenitic stainless steels and titanium alloys) the stress–strain curve will depend on the chosen control variable (stress or strain) and its rate of application. Figure 2.3 shows the stress–strain curves of an AISI type 304 stainless steel at room temperature under load (stress) and displacement (strain) controlled conditions [5]. We observe that the material response to the change in the rate and direction is instantaneous for the strain-controlled test, whereas in the case of the stress-controlled test, the response is gradual. Also, the transition from the elastic to the inelastic deformation is more gradual for the stress-controlled test. The rate effect is more pronounced with increasing temperature (section 2.8). However, at room temperature and small strain-rates (quasi-static loading), the rate effect may be neglected. The rate effect is further discussed in section 2.10.

2.3 MATERIAL RESPONSE TO CYCLIC DEFORMATION OR LOADING – TRANSIENT BEHAVIOUR

It was indicated in the previous section that the material response deviates from that of the monotonic, once a reversed plastic deformation takes place (Bauschinger effect). When a smooth solid specimen is subjected to one of the loading programs shown in Fig. 2.4, the response will differ depending upon the mode of controlled variable. The essential characteristics include cyclic hardening or softening, cyclic mean stress

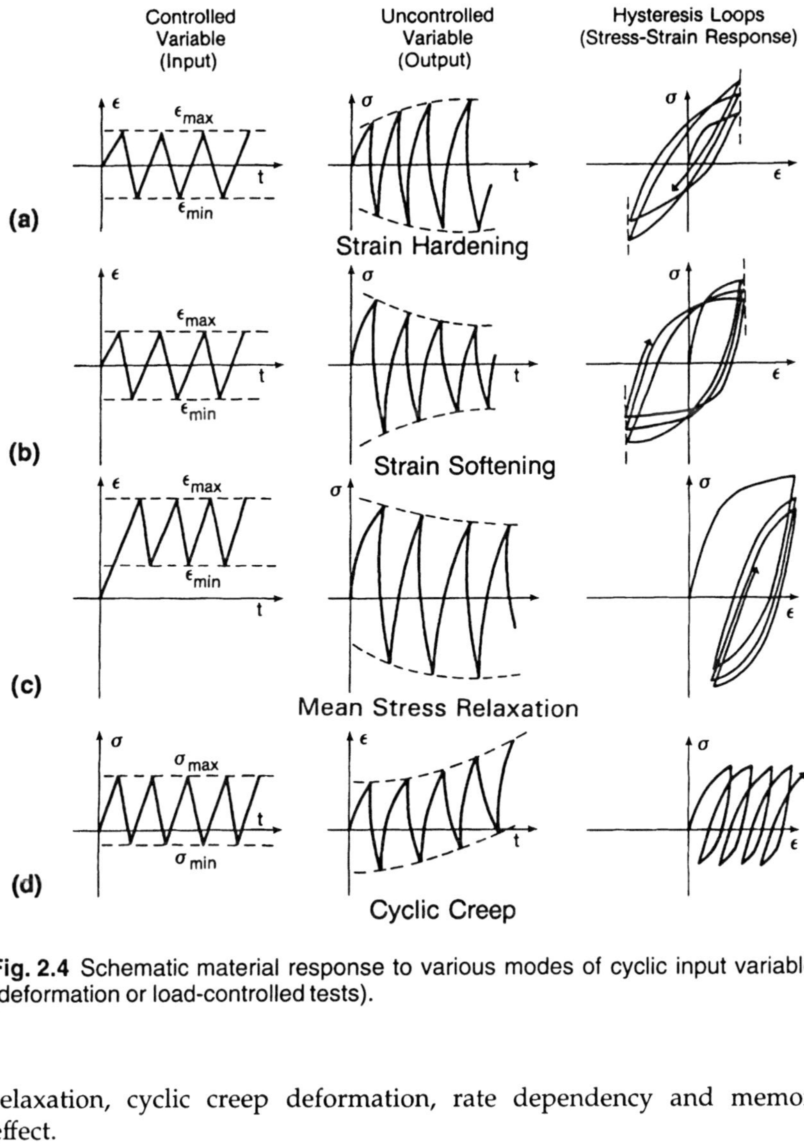

Fig. 2.4 Schematic material response to various modes of cyclic input variables (deformation or load-controlled tests).

relaxation, cyclic creep deformation, rate dependency and memory effect.

In the case of fully-reversed strain-controlled tests, the material response initially varies with the number of cycles and later stabilizes. Figure 2.5 shows the stress–strain response of a pearlitic–ferritic steel (ASTM A-516 Gr. 70) indicating a strain-softening phenomenon, i.e. the uncontrolled stress decreases with the increasing number of cycles until a stable (saturated) state is achieved. It should be noted that at higher strain ranges ($\Delta\varepsilon/2 > 0.4\%$) this steel strain hardens, i.e. the uncontrolled stress increases with the increasing number of cycles until it reaches a stable state

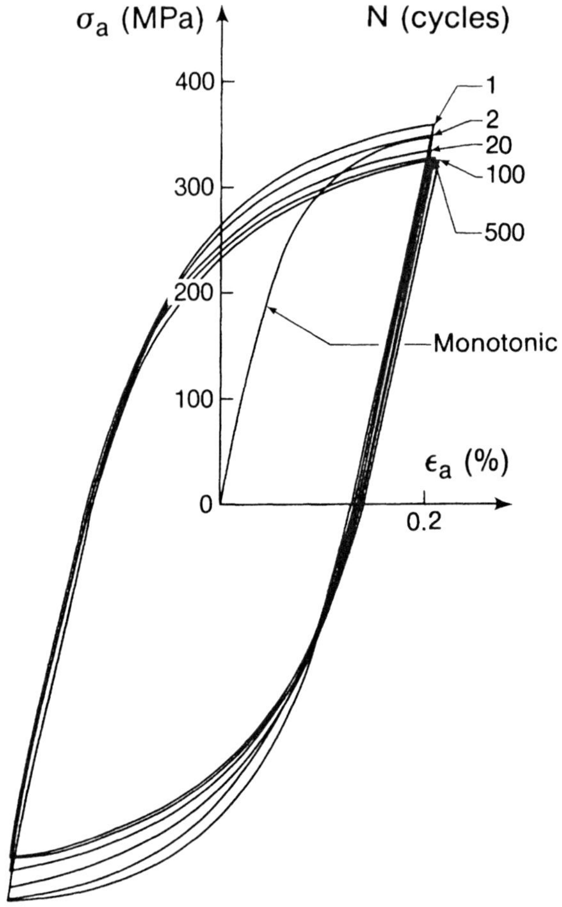

Fig. 2.5 Recorded stress–strain response of carbon steel A-516 Gr. 70 under fully-reversed strain-controlled conditions, indicating strain softening at this strain amplitude, $\Delta\varepsilon/2 = \pm 0.2\%$.

[2]. This type of behaviour is commonly observed for the low alloy carbon steels.

As mentioned in the previous section, during cyclic deformation the mobile segment of dislocation is reduced, leading to the formation of a different obstacle structure. The resistance to dislocation glide is generally increased, thus requiring higher stresses to continue the deformation. The cyclic hardening rate is smaller at low strain amplitudes than at high ones.

At this stage it is useful to define some of the terms used earlier in conjuction with Fig. 2.4. To define the magnitude of the cyclic strain or stress, we generally require two values, e.g. the maximum and minimum values as shown in Fig. 2.4. In addition, the form of the cycle (wave slope) has to be defined, i.e. variation in time. Any other two parameters involving the above-defined maxima and minima can also be used to describe the magnitude. In the fatigue literature the following parameters are often used:

Strain range: $\Delta\varepsilon = \varepsilon_{max} - \varepsilon_{min}$; Strain amplitude: $\varepsilon_a = \frac{1}{2}(\varepsilon_{max} - \varepsilon_{min})$ (2.10a)

Stress range: $\Delta\sigma = \sigma_{max} - \sigma_{min}$; Stress amplitude: $\sigma_a = \frac{1}{2}(\sigma_{max} - \sigma_{min})$ (2.10b)

Mean strain: $\varepsilon_m = \frac{1}{2}(\varepsilon_{max} + \varepsilon_{min})$; Strain ratio: $R_\varepsilon = \varepsilon_{min}/\varepsilon_{max}$ (2.10c)

Mean stress: $\sigma_m = \frac{1}{2}(\sigma_{max} + \sigma_{min})$; Stress ratio: $R_\sigma = \sigma_{min}/\sigma_{max}$. (2.10d)

Any two combinations of the stress (strain) range (or amplitude); mean stress (strain); and stress (strain) ratio may be used to define the magnitude of the cyclic input parameters. Note that when mean strain is zero, then $\varepsilon_a = \Delta\varepsilon/2$ and similarly when $\sigma_m = 0$, $\sigma_a = \Delta\sigma/2$.

The mean stress relaxation phenomenon under strain-controlled conditions is shown in Fig. 2.6 for the previously-mentioned pearlitic–ferritic

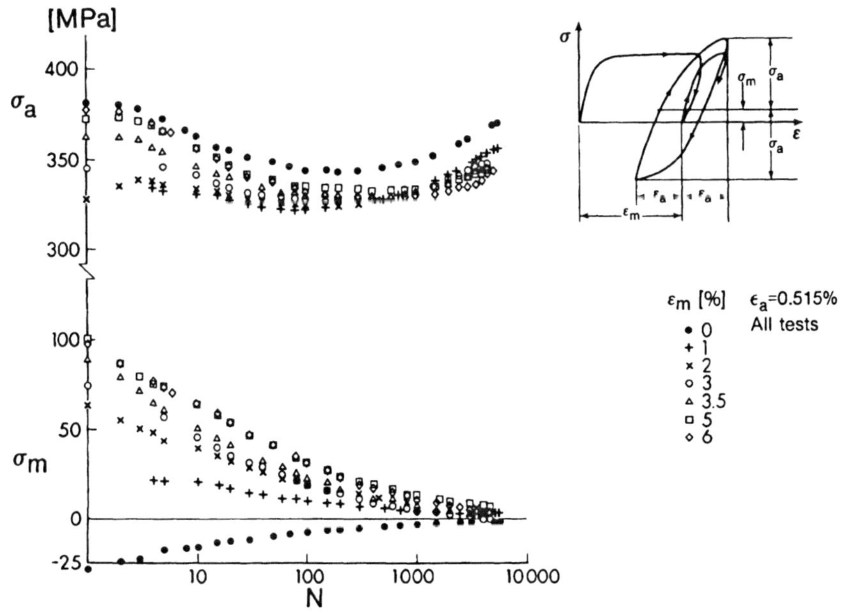

Fig. 2.6 Mean stress relaxation as a function of the number of cycles for fully-reversed strain-controlled tests of A-516 Gr. 70 carbon steel with a prior mean strain. Reproduced with kind permission from ASME from [6].

steel [6]. As shown in the inset at the top right corner, the specimens were first subjected to a prescribed mean strain and then cycled in a strain-controlled mode with a constant strain amplitude. The uncontrolled stress amplitude, σ_a, and the mean stress, σ_m, are plotted against the number of cycles, N. It can be seen that the mean stress relaxes quite noticeably during the softening phase, and after about 1000 cycles it reaches about 10% of its initial value. The rate of mean stress relaxation, $d\sigma_m/dN$, is greater for the higher values of mean strain, ε_m. In these tests the range of the cyclic strain was kept constant. In general the rate of mean stress relaxation is a function of the mean strain as well as the plastic strain range, $\Delta\varepsilon^p$.

For fully-reversed stress-controlled loading, one observes strain hardening or softening similar to the strain-controlled condition. The plastic strain decreases with the number of cycles and reaches a stable state for strain-hardenung metals. The reverse occurs for strain-softening materials.

When a mean stress is present, the response is rather complicated. For example, in the case of tensile mean stress, creep is observed under stress-controlled conditions (Fig. 2.4(d)). The ratcheting strain (cyclic creep strain) increases with the increasing number of cycles. Attainment of a stable cyclic response depends on the magnitude of the mean stress, stress amplitude and temperature. A stable response results in a shake-down (elastic) behaviour. Figure 2.7 shows the increase in cyclic creep strain with the number of cycles for both positive and negative mean stress values. In these tests the ratio of the mean stress to the stress range varied from ± 2–3.5%, and no stable cyclic response was observed. On the other hand, the plastic strain range, $\Delta\varepsilon^p = \Delta_B - \Delta_A$ (the width of hysteresis loops) reached a stable value for the smaller stress ranges (Figs 2.8 and 2.9). In the case of higher stress ranges, strain hardening follows the initial softening phase. The rate of strain hardening increases with the increase in mean and range of stress. However, it should be noted that in the load-controlled tests and in the presence of high creep strains, the true stress varies during the test. Therefore, it is essential that the effect of the creep strain be accounted for in the modelling of the stress-controlled cyclic condition.

The effect of prestrain on the cyclic response is shown in Fig. 2.10 for the fully-reversed stress-controlled condition [7]. The cyclic creep strain, ε_c, depends on the prestrain, ε^p_{pre}, and stress range, $\Delta\sigma$. For the tensile prestrain the creep strain decreases, i.e. $d(\varepsilon_c)/dN < 0$, while for the compressive prestrain the rate of creep strain is positive ($d(\varepsilon_c)/dN > 0$) but the magnitude of the creep strain per cycle decreases with the increasing number of cycles. In both cases, there is a softening of the material and the amount of stable creep strain is less than the prestrain value. For the range of experimental data reported in [7], the stable creep strain (ε_c) was about $\frac{1}{3}$–$\frac{1}{2}$ of ε^p_{pre}. This observation is indicative of a partial fading of prior deformation memory as the cycling progresses.

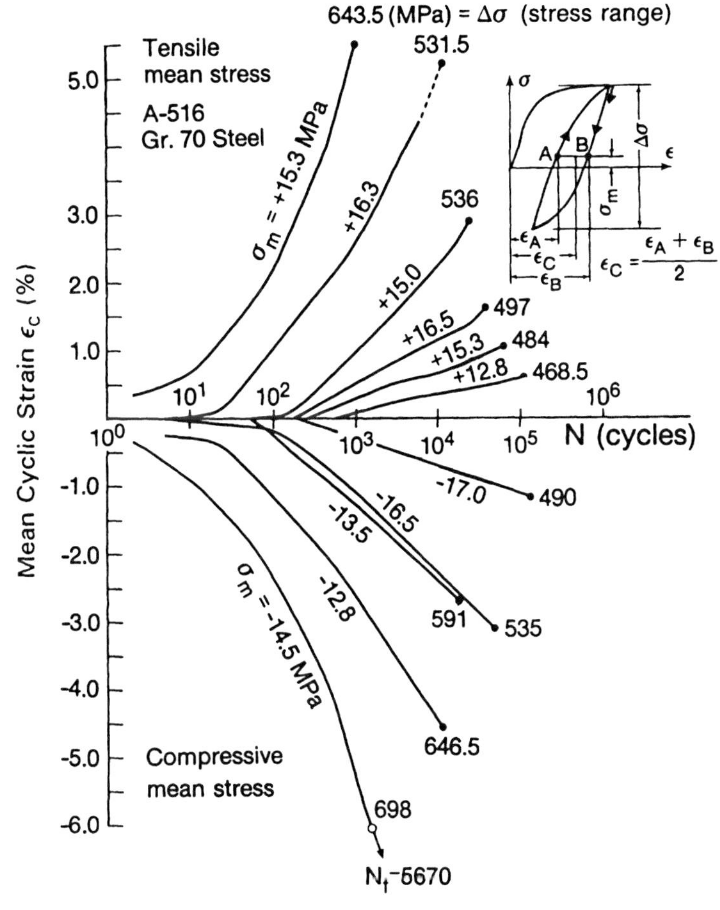

Fig. 2.7 Increase in cyclic creep strain with the number of cycles for various stress range and mean stress values. Each stress range and mean stress value is indicated on the corresponding curve.

It can be seen that for the types of loading shown in Fig. 2.4, the transient response is varied, and it is doubtful that this behaviour can be represented by a simple model containing a few variables. We will further discuss transient behaviour in Chapter 6, and in the meantime will concentrate on stable cyclic behaviour in the remaining sections of this chapter.

2.4 STABLE CYCLIC RESPONSE

In the majority of cyclic loading cases, the material response reaches a steady-state regime, or the variation from cycle to cycle becomes so small

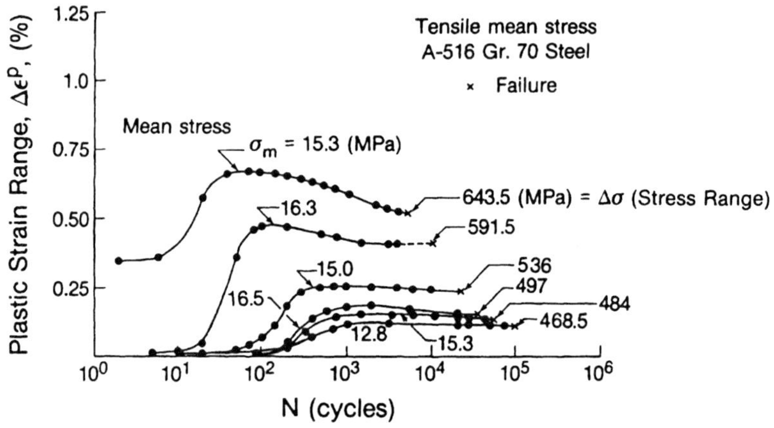

Fig. 2.8 Variation of plastic strain range with number of cycles for a range of positive mean stress and stress range values in stress-controlled tests.

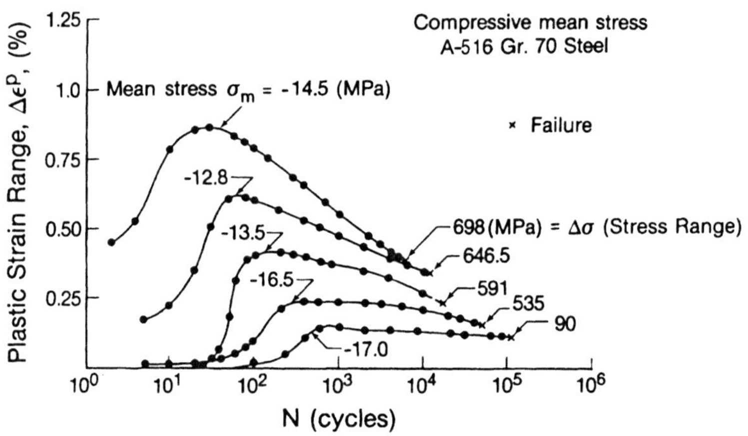

Fig. 2.9 Variation of plastic strain range with number of cycles for a range of negative mean stress and stress range values in stress-controlled tests.

that the difference is not appreciable. For example, Fig. 2.11 shows the maximum (tensile) and minimum (compressive) stresses plotted against the number of cycles, N, for the deformation-controlled tests with the strain varying from 0 to 2 ε_a with a tensile mean value of $\varepsilon_m = \varepsilon_a$. In the early life, the tensile (positive) stress decreases due to the mean stress relaxation as shown in Fig. 2.6. The compressive (negative) stress remains constant or decreases slightly at early life, and then increases by a small amount due to the strain-hardening characteristic of the metal. It can be seen from Fig.

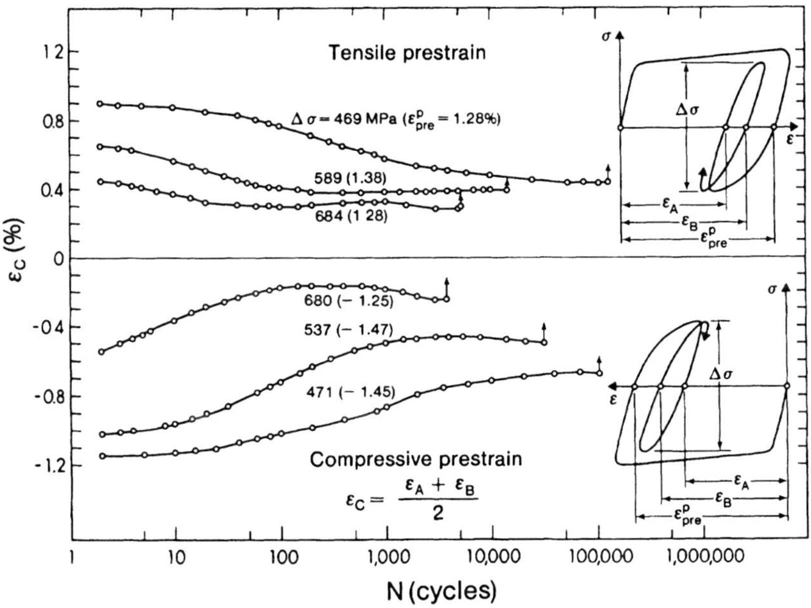

Fig. 2.10 Variation of cyclic creep strain with number of cycles for various values of stress range and plastic prestrain of low alloy carbon steel A-516 Gr. 70 under stress-controlled mode. Reproduced with permission from ASME from [7].

2.11 that the stable (saturated) behaviour is achieved after relatively short life, and that the majority of the life is spent in a stable cyclic state. Similar results for fully-reversed strain-controlled tests have been reported [8] for the same pearlitic–ferritic steel. From these results we may conclude that the saturation stress in the low alloy carbon steel (ASTM A-516 Gr. 70) is relatively insensitive to this type of prior strain history (mean strain). The above observations regarding the effect of prior prestrain have to be viewed within the limits of the uniaxial tests performed where the strain path is fixed.

2.4.1 MICROSTRUCTURAL CHANGES DURING CYCLIC LOADING

Feltner and Laird [9] have reported similar results to those above for copper and pure aluminium. They concluded that face-centred cubic (FCC) materials with a wavy slip character (high stacking fault energy), e.g. pure copper and aluminium, would display a unique stable (saturation) cyclic stress–strain curve. The saturation stress level is determined only by the plastic strain amplitude and test temperature. By contrast, in materials with a planar slip mode, e.g. brass and aluminium alloys (low

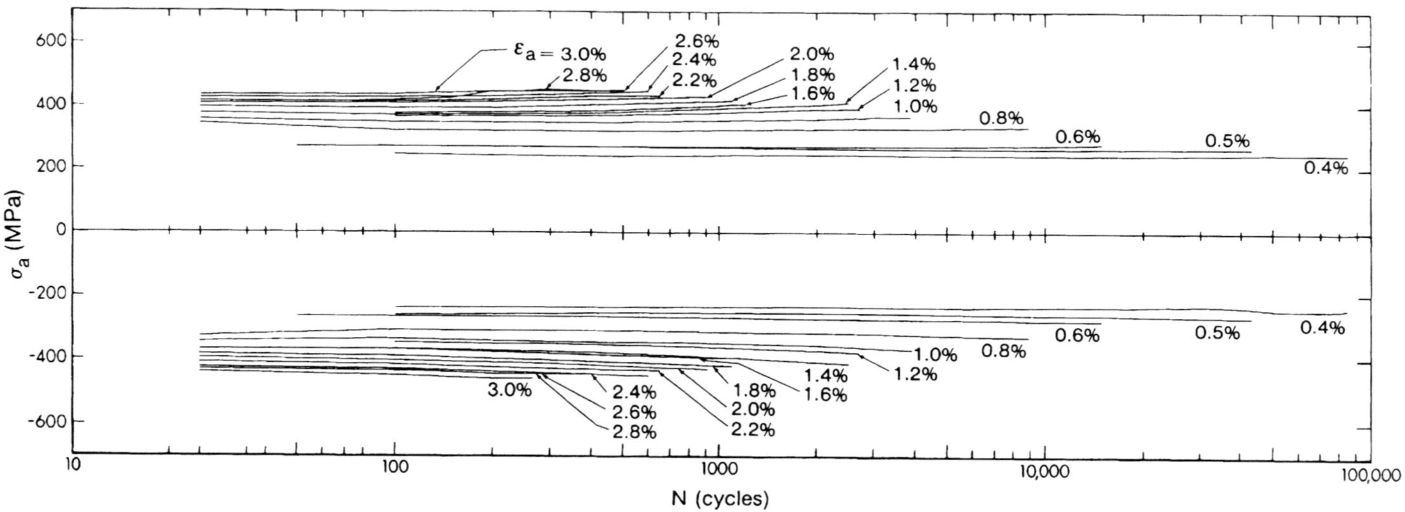

Fig. 2.11 Variation of the maximum (tensile) and minimum (compressive) stresses with number of cycles for various values of strain amplitude under deformation control (with $\varepsilon_m = \varepsilon_a$) of A-516 Gr. 70 carbon steel. Reproduced with permission from ASME from [6].

stacking fault energy), the saturation stress level would depend on the previous strain history. In general, for most materials one would expect some sensitivity with respect to the path and prior history of loading (Chapter 6).

The slip character is a measure of the degree of dislocation motion during plastic deformation. A planar slip occurs when the dislocations tend to stay in a planar array with a difficult cross slip, whereas a wavy slip is observed when the dislocations are uniformly distributed in the matrix, i.e. a cross slip is easy.

The stable (saturated) stress level has been related to an equilibrium dislocation structure which can accommodate the imposed plastic strain without further hardening [10–13]. Typical dislocation structures range from dipolar and multipolar configurations at low strain amplitudes to cell structures at the higher amplitudes. During a transition period the cyclic dislocation structure changes from veins and walls formed in monotonic deformation to multiple bundles and walls with significantly higher dislocation densities [14]. Cell structures develop when multiple slip systems are operative.

Figure 2.12 shows a map of typical dislocation structures of FCC materials as a function of the number of cycles to failure, N_f, as well as the applied plastic strain amplitude and the slip character of the material [9, 10]. For example, it seems that for a material with wavy slip, the dislocation structure could range from those typical of single slip, namely dipoles, dipole bundles, persistent slip bands (PSBs) to those characteristic of multiple slip, e.g. cell structures, depending on the applied plastic strain range.

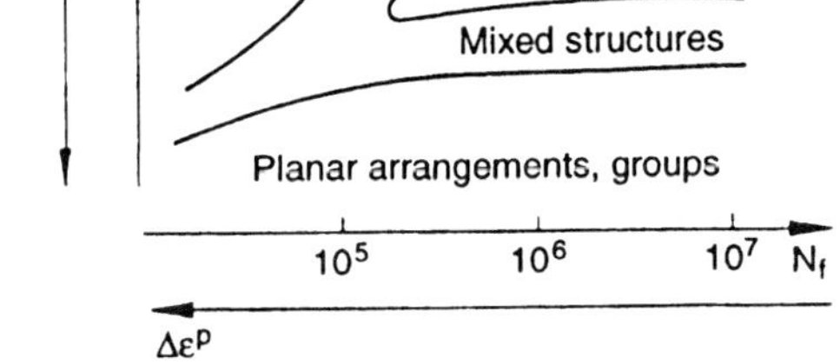

Fig. 2.12 A map of dislocation structures formed in FCC metals and alloys, depending on the slip characteristics and number of cycles to failure, from [10].

Figure 2.13 shows the dislocation structure of a commercially pure polycrystal copper reported by Schubert and Rie [12]. They observed that the dislocation cells started to form at low strain amplitudes ($\Delta\varepsilon/2 = \pm 0.2\%$), Fig. 2.13(a). At higher strain amplitudes ($\Delta\varepsilon/2 = \pm 0.7\%$), Fig. 2.13(b), the cells are more distinct and smaller in size. Similar results have been reported in [9, 11, 13]. The average cell size (or distance between the walls) increases with a decrease in the plastic strain amplitude. The cell size (or distance between the walls) increases with increase in temperature. As seen in Fig. 2.13, the cells contain very few dislocations, whereas the cell walls contain large numbers of tangled dislocations and defect structures.

2.4.2 DETERMINATION OF THE CYCLIC CURVE

The cyclic saturation regime, being associated with the 'constant' microstructure [12–14], provides the opportunity to examine the effect of strain-rate and temperature changes from such a state [15]. Departure from the steady-state regime during extended cycling is related either to the secondary hardening (mostly at low strain amplitudes) or to the gradual damage accumulation and cyclic strain localization in persistent slip bands (PSBs) near the surface [10, 15]. (The fatigue damage mechanism, which was described in Chapter 1, will be further discussed in Chapter 3.)

The variation in stress during fully-reversed strain cycling for four metals used in aircraft structures have been given in [6], where it was reported that the cyclic rate of change of stress amplitude, $d\sigma/dN$, was a minimum after 20–50% of the fatigue life. For aluminium alloys, 2024-T4 and 7075-T6, the material initially strain hardened and a stable cyclic response was evident soon after. The SAE 4340 steel (quenched and tempered at 1000 °F) and titanium alloy Ti-8Al-1Mo-1V exhibited strain softening.

The general trend of the monotonic and cyclic stress–strain response of a metal with different strain histories is shown in Fig. 2.14 [9]. In a deformation-controlled mode, the stress required to maintain the imposed strain levels may decrease (or increase) with increased number of cycles. The material is then said to cyclically strain soften (or strain harden). The rate of strain hardening (or strain softening) depends upon the cyclic strain amplitude, temperature and prior deformation history. The effect of the history of deformation is more pronounced in planar slip materials. These materials are characterized by low stacking fault energy and difficult cross slip. In non-strain aging predeformed metals softening occurs so as to decrease the internal energy due to relaxation or recovery of crystal distortion. An indication of attainment of stable cyclic conditions under load-controlled tests is the closure of the hysteresis loop during a cycle.

A convenient method of describing the stable material response is through the cyclic stress–strain curve. Generally a cyclic stress–strain curve is obtained by connecting the tips of stable hysteresis loops for

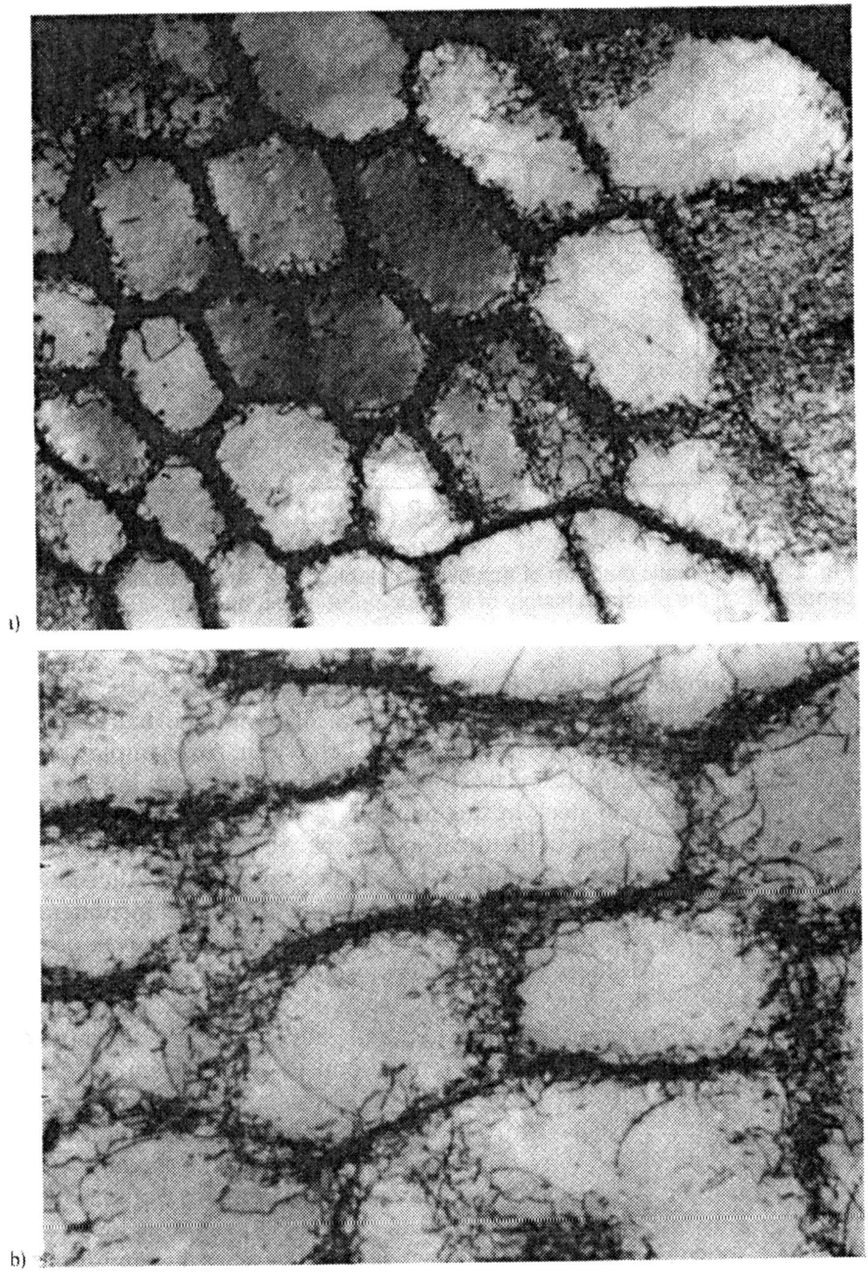

Fig. 2.13 Dislocation structure in a commercially pure polycrystal copper at (a) $\Delta\varepsilon/2 = \pm 0.2\%$, $2 \times 20\,000$ zone axis $(0\,\bar{1}\,1)$; (b) $\Delta\varepsilon/2 = \pm 0.7\%$, $2 \times 10\,000$ zone axis $(\bar{1}\,\bar{2}\,3)$ (courtesy of Dr. Schubert and Prof. Rie of Technical University of Braunschweig).

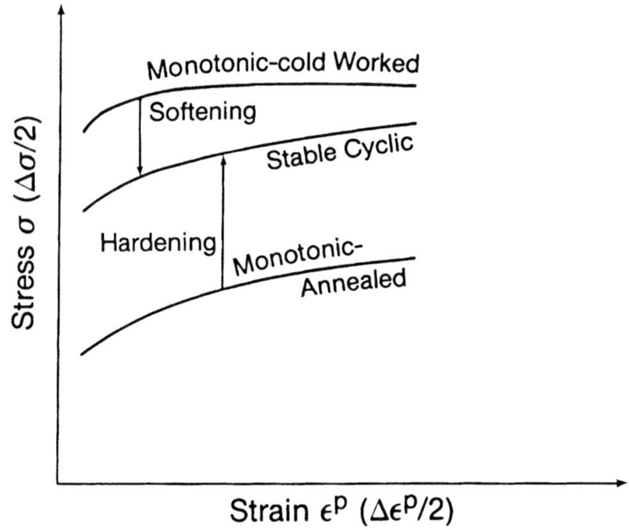

Fig. 2.14 Schematic diagram of achieving a stable cyclic stress–strain curve depending upon the prestrain history of the monotonic curve, after [9].

different strain amplitudes of fully-reversed strain-controlled tests. For the purpose of uniformity, the hysteresis loops at about the half-life are normally chosen as the representative stable behaviour. An example of this method of determining the cyclic stress–strain curve is shown in Fig. 2.15 for a low alloy carbon steel. In this method, generally termed the single step test (SST), a number of identical specimens are used and tested until failure occurs; thus the procedure is rather time-consuming. Alternative procedures are outlined in [17]; they are multiple (MST) and incremental (IST) step tests, monotonic tension after cyclic straining, and analysis of the individual hysteresis loop. Table 2.1 gives a summary of the most popular methods used to determine the cyclic stress–strain curve. It should be noted that the plastic response is, in general, path and history dependent. Therefore, one should expect different responses to the strain histories shown in Table 2.1. If we define the stress–strain curve obtained by the single step tests (SST) as a norm, then the curves constructed through other test methods should be viewed as an approximation to the above.

In the multiple step test (MST) one takes advantage of the observation made by Bairstow, as early as 1910 [18], that stable hysteresis loops are obtained soon after a sudden change in the cyclic strain amplitude. Generally, a sequence of low-to-high strain amplitudes is used. To avoid a mean-stress effect, a high-to-low strain sequence would be preferable. However, strain amplitudes which are too high may cause the initiation of cracks which could grow under lower amplitudes, thus resulting in lower stress

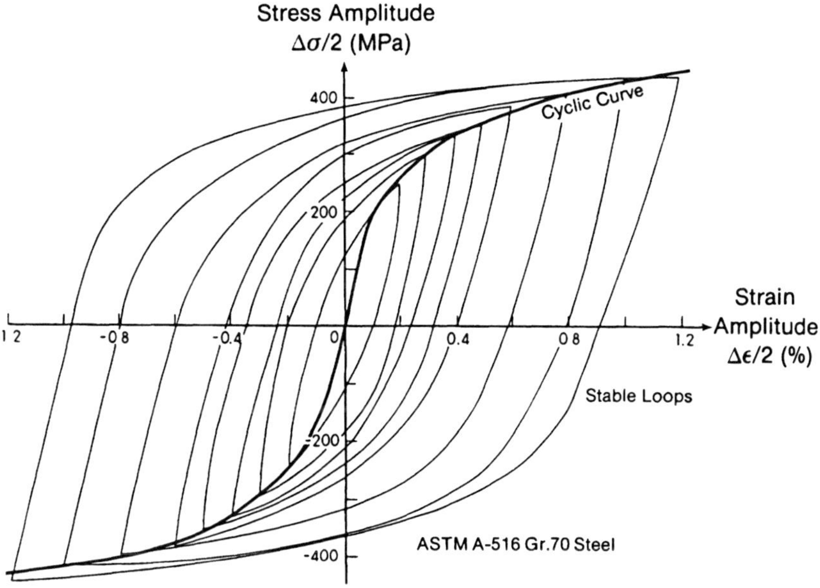

Fig. 2.15 Construction of a cyclic curve – actual data for ASTM A-516 Gr. 70 carbon steel.

Table 2.1 Methods for determining the cyclic stress–strain curve

Test method	Abbreviation	Schematic of strain cycles
Single step test	SST	
Multiple step test (I)	MST (I)	
Multiple step test (D)	MST (D)	
Incremental step test	IST	

values for the same strain range. Also, a cell structure obtained under a high strain amplitude will influence the stress level at lower strain amplitudes.

Blocks of increasing and decreasing strain amplitudes are used to determine stable cyclic stress–strain curves by incremental step tests (IST).

After application of a few blocks, the material reaches a cyclically stable state, and the locus of tips of the loop in the increasing sequence of the strain amplitude produces a cyclic curve. By using one or two specimens, the cyclic curve is quickly determined. However, an element of judgement is involved in this method with regard to determining when a stable material response is achieved. A comparison between the incremental method and that of the several stable hysteresis loops at half-life has been reported [8], Fig. 2.16. The data obtained by the incremental method are

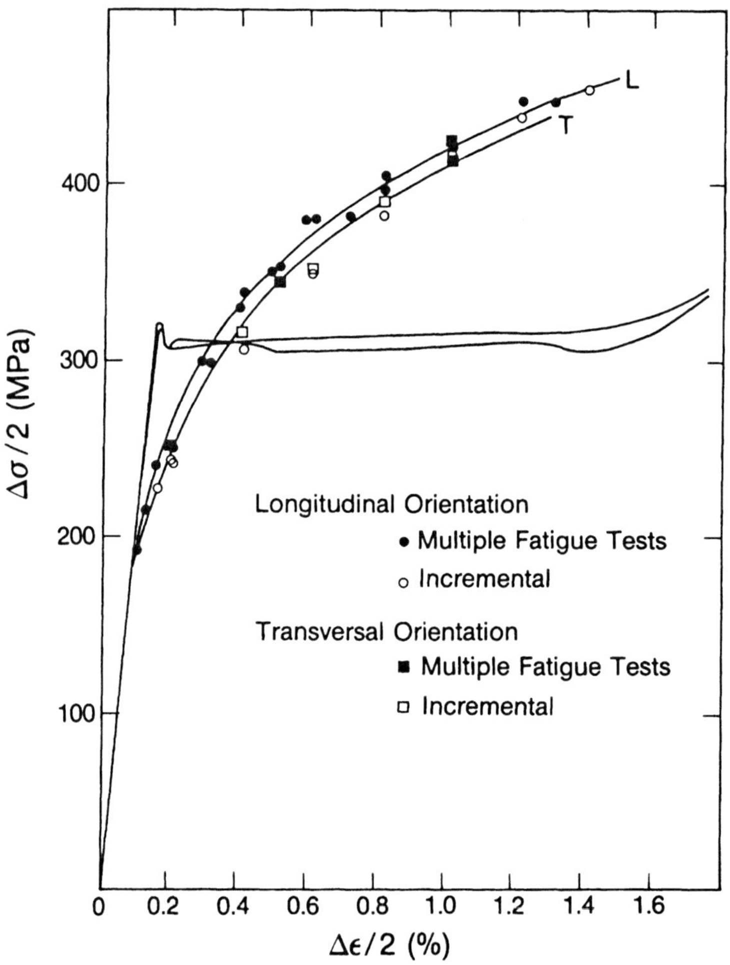

Fig. 2.16 Comparison between cyclic stress–strain curves obtained by multiple fatigue tests (Fig. 2.15) and the incremental method (IST) for A-516 Gr. 70 carbon steel. Also shown are the effects of specimen orientation and comparison with the monotonic curve. Reprinted with kind permission from [8].

generally below those of the half-life hysteresis loops, and similar trends are observed in [12, 16].

In the above-mentioned test methods, a material will experience different plastic strain histories prior to the measurement of the saturated stress. Schubert and Rie [12] and Pickard and Knott [13] found significant differences between different tests for polycrystalline copper and copper aluminium alloys. The dislocation cell structures are formed at a relatively early stage of cycling and are a function of the maximum plastic strain amplitude (Fig. 16 in [13]). From this observation we can explain the differences in results obtained by the various tests listed in Table 2.1. For example, in the IST method the dislocation structure formed would be that corresponding to an 'effective' strain, close to the maximum strain amplitude, e.g. Fig. 8 in [10]. The flow stress measured at lower strain levels would be influenced by the dislocation structure thus formed. Although in the MST (I) method the dislocation structure would correspond to the appropriate strain level of the SST, the results will be influenced by the mean stress generated during step changes. From the above discussion, it is therefore clear that the cyclic stress–strain curves obtained by MST, IST, etc. are of an approximate nature and the amount of approximation depends upon the slip characteristics of the material.

The monotonic tension curve, after achievement of the stable cyclic response by either of the methods described earlier, can produce a reasonable approximation of the cyclic stress–strain curve under certain circumstances. Another approximate way of determining the cyclic curve is by dividing the upper branch of a stable hysteresis loop by a factor of 2. Neither of these two short-cut methods are widely used, and there are obvious disadvantages in employing them.

2.4.3 MATHEMATICAL DESCRIPTION OF THE STRESS–STRAIN RELATIONSHIP

The cyclic stress–strain curve (Fig. 2.16) can be represented by

$$\frac{\Delta\varepsilon}{2} = \frac{\Delta\sigma}{2E} + \left(\frac{\Delta\sigma}{2K'}\right)^{1/n'} \tag{2.11}$$

where K' is the cyclic strength coefficient and n' is the cyclic strain-hardening coefficient. Equation (2.11) is similar to (2.8) except that the strain and stress are replaced by the strain and stress amplitudes, $\Delta\varepsilon/2$, $\Delta\sigma/2$, and the material constants are denoted by the primed letters. This notation will be used hereafter.

The cyclic hardening coefficient, n', lies in a narrow range for metals and alloys, generally between 0.05 and 0.25. A comparison between monotonic and cyclic behaviour of a number of materials leads to the following quantitative conclusion regarding cyclic softening and hardening characteristics [17]. Materials with a high monotonic exponent will cyclically harden while those with a low exponent ($n < 0.1$) may cyclically soften. A mixed or stable cyclic behaviour can be expected at intermediate values of n. The cyclic strain-hardening exponent, n', is about 0.15 for materials with a wavy slip mode, whereas for those with a planar slip mode, n' is about $\frac{1}{2}$–$\frac{2}{3}$ of this value [9]. In general, fully annealed metals cyclically harden, and those heavily cold worked will cyclically soften (Fig. 2.14).

Based on a series of tests, Smith *et al.* [19] were able to propose an empirical relation by which the cyclic softening and hardening characteristics can be predicted on the basis of the monotonic properties. It was observed that metals with a ratio of ultimate strength to yield strength (s_u/σ_y) greater than 1.4 cyclically hardened, and those with a ratio less than 1.2 cyclically softened. For those between 1.2 and 1.4 a mixed mode could be expected. Figure 2.16 shows the monotonic and stable cyclic curves of a pearlitic–ferritic steel [8]. Up to about 0.4% strain, the material cyclically softens and beyond this strain level, it cyclically hardens. For this steel $s_u/\sigma_y = 534/325 = 1.64$ [2].

The cyclic hardening and softening mechanisms have been discussed by Feltner and Laird [9]. Abdel-Raouf *et al.* [11] have shown that for a relatively pure iron, the saturation stress is inversely proportional to the cell size. For metals with a wavy slip mode such as aluminium, Al–Mg alloys, copper, pure iron, and dilute iron–carbon alloys, the stable (saturation) cyclic stress is

$$\frac{\Delta\sigma}{2} = f\left(\frac{\Delta\varepsilon^{p}}{2}, \dot{\varepsilon}, T\right), \tag{2.12}$$

where $\dot{\varepsilon} = d\varepsilon/dt$ is the strain-rate and T is the temperature. At a constant strain-rate and temperature, the functional form of (2.12) is given by (2.11), i.e.

$$\frac{\Delta\sigma}{2} = K'\left(\frac{\Delta\varepsilon^{p}}{2}\right)^{n'}. \tag{2.13}$$

Noting that the number of cycles to saturation, N_s, decreases with increasing plastic strain amplitude, a linear relation was found in [11] relating $(\Delta\varepsilon^{p})^{1/2}$ with $\log N_s$, i.e.

$$(\Delta\varepsilon^{p})^{1/2} + A \log N_s = B, \tag{2.14a}$$

where A and B are constants. From (2.14a) the number of cycles to saturation is given by

$$N_s = C \exp[-(\Delta\varepsilon^{p})^{1/2}/A], \tag{2.14b}$$

where the constant $C = \exp(B/A)$. The number of cycles to saturation decreases with the increasing plastic strain in an exponential fashion.

For a discussion of microstructural change during cyclic loading, the reader is referred to [9–15, 20, 21] where further references to other works are cited.

At this stage it is useful to mention that the applied stress in (2.12) is composed of two terms, namely an internal stress, σ_I, associated with strain hardening and recovery, and an effective stress component σ_e, which is responsible for slip, i.e.

$$\sigma = \sigma_\mathrm{I} + \sigma_\mathrm{e}\,. \tag{2.15}$$

As the deformation proceeds, the internal stress builds up in response to strain hardening and reaches a steady-state value at the saturated state.

2.5 ANALYSIS OF HYSTERESIS LOOPS

A number of phenomena in cyclic behaviour can be deduced from the shape of the hysteresis loop, and thus a study of it, and its description in an analytical form, is important. While the cyclic stress–strain curve describes the relationship between the stable stress and strain amplitudes, it cannot, in general, describe the hysteresis loop branches. Masing [22] originally suggested that either branch of the hysteresis loop was geometrically similar and that they could be obtained from the monotonic curve when scaled by a factor of 2. This assumption has been modified by replacing the monotonic by the cyclic stress–strain curve. Then a material is said to exhibit a 'Masing-type' behaviour when the hysteresis loop branches (ascending and descending ones), and can be described by the cyclic stress–strain curve magnified by a factor of 2. For this type of material, when the compressive tips of stable hysteresis loops of various strain amplitudes are transferred to a common origin the maximum compressive stress – the upper branch would form a unique curve. Materials whose behaviour differs from the above description are termed 'non-Masing type'. Figure 2.17 shows a non-Masing behaviour, which appears to be the norm for carbon steels.

In a study of hysteresis loop shapes for a range of metals and alloys [23], it was observed that Masing behaviour was associated with the structure in which no dislocation cells were able to form. For this group of metals and alloys the deformation is structurally controlled and is independent of matrix properties, which is characteristic of low stacking fault energy materials. On the other hand, non-Masing behaviour was observed in the group of metals in which cyclic deformation is controlled by the matrix properties, and dislocation cells are formed at relatively low strain ranges, typical of high stacking fault energy materials. For some materials, e.g. ferritic–pearlitic steels, a Masing type of behaviour was observed at low strain range, of the order of 0.0036, whilst at higher strain ranges the response was of a non-Masing type. Christ and Mughrabi [10] contend that

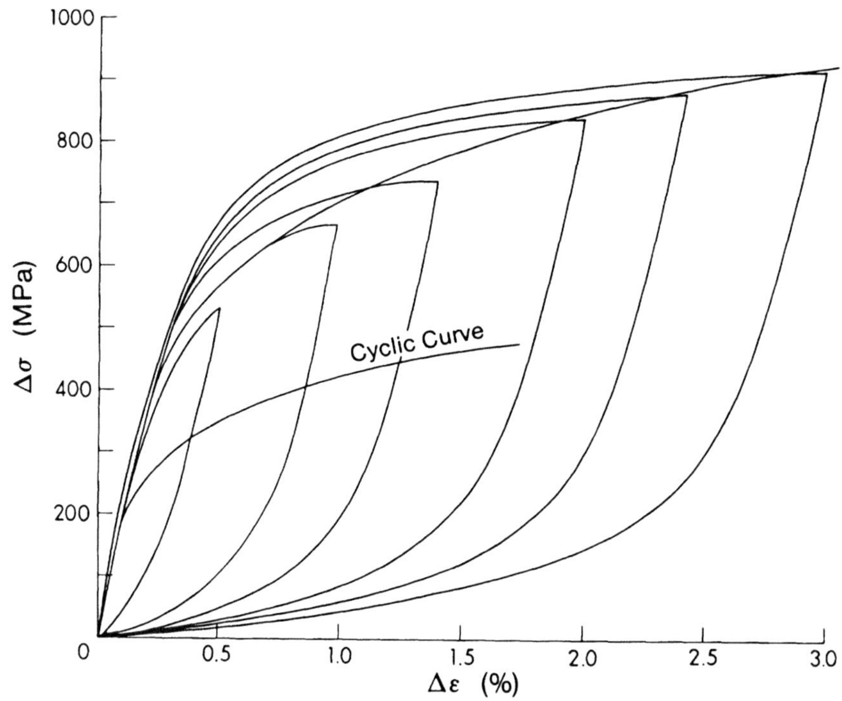

Fig. 2.17 Stable hysteresis loops and a best fit curve of peak stress and strain ranges. Superposed on this figure is the stable cyclic curve of Fig. 2.15. The A-516 Gr. 70 steel indicates a non-Masing behaviour. Reprinted with kind permission from [6].

in the case of single-phase polycrystalline materials a wavy slip behaviour is a prerequisite for Masing-type behaviour.

Studies of the above type are useful in attempting to classify materials with regard to Masing or non-Masing types of behaviour, based on their slip modes and dislocation structures. However, a mathematical description of the branches of hysteresis loops is required, so that properties such as hysteretic energy, etc. can be calculated. In the case of Masing-type material, as stated earlier, the two ascending and descending branches of the hysteresis loop are similar and when the compressive tips of various stable hysteresis loops are made to coincide, the ascending (upper) branches form a unique curve (Fig. 2.18). The equation of the 'Masing curve' is then obtained from (2.11) when magnified by a factor of 2, i.e.

$$\Delta\varepsilon = \frac{\Delta\sigma}{E} + 2\left(\frac{\Delta\sigma}{2K'}\right)^{1/n'}. \tag{2.16}$$

It should be noted that in the case of a Masing-type material (2.16), the exponent n' is the ratio of the complementary plastic strain energy range

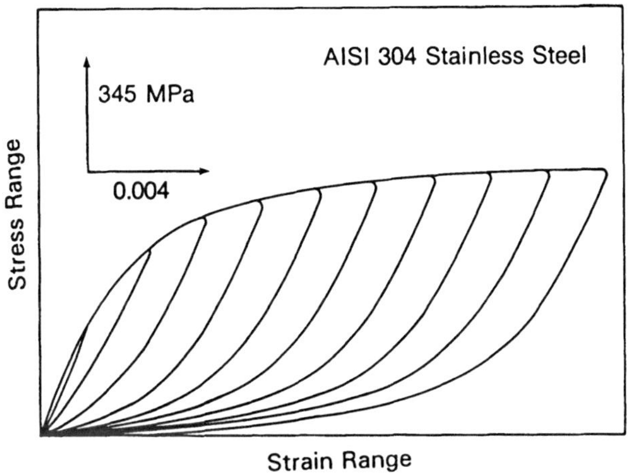

Fig. 2.18 Masing behaviour of AISI type 304 stainless steel, after [23].

$\Delta W^{\rm p}(\Delta\sigma)$ to the plastic strain energy range, $\Delta W^{\rm p}(\Delta\varepsilon^{\rm p})$, i.e.

$$n' = \frac{\Delta W^{\rm p}(\Delta\sigma)}{\Delta W^{\rm p}(\Delta\varepsilon^{\rm p})} = \frac{\int_0^{\Delta\sigma} \Delta\varepsilon^{\rm p}\ {\rm d}(\Delta\sigma)}{\int_0^{\Delta\varepsilon^{\rm p}} \Delta\sigma\ {\rm d}(\Delta\varepsilon^{\rm p})}. \tag{2.17}$$

For non-Masing behaviour, Jhansale and Topper [24] observed that the cyclic yield stress under tension and compression varied with the strain amplitude for a wide range of structural metals, however the nonlinear portion of the hysteresis loops did not change appreciably. Matching the upper branches of the hysteresis loops through translation along the linear portion, they defined a 'skeleton curve' with a minimum proportional range. Lefebvre and Ellyin [8] and Ellyin and Kujawski [7] expanded the above concept and provided a general expression for the cyclic stress–strain curve applicable to both Masing and non-Masing types of materials.

2.6 DESCRIPTION OF THE MASTER CURVE

A master curve is defined as a unique curve which envelops the ascending (loading) branches of hysteresis loops. It is obtained from matching the upper branches of the hysteresis loops for different strain amplitudes, by translating each loop along the linear response portion. Figure 2.19 shows the manner in which a master curve is obtained, and a coordinate system with its origin based at the smallest linear portion, 0*. On the same figure the hysteresis loop with a strain range of 1.8% is partitioned by drawing

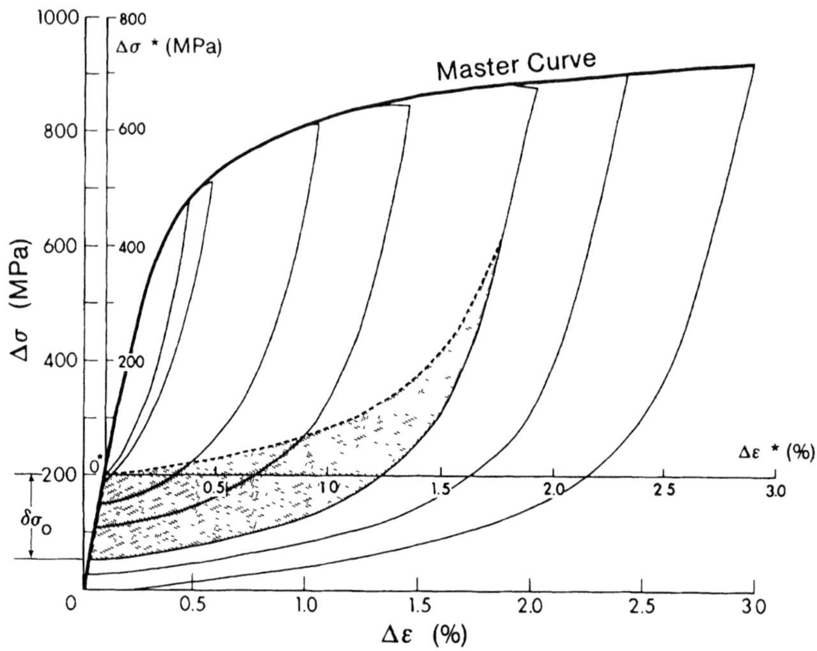

Fig. 2.19 Construction of a master curve using the stable hysteresis loops of Fig. 2.17 for ASTM A-516 Gr. 70 carbon steel. Reproduced with permission from ASME from [6].

a symmetric lower branch in a dotted line. The cross-hatched area is a measure of deviation from the Masing description (2.16). The larger the strain amplitude, the greater is the deviation from the Masing-type description.

The equation for the master curve with the origin at 0*, corresponding to the hysteresis loop with the minimum proportion range, can be written as

$$\Delta\varepsilon^* = \frac{\Delta\sigma^*}{E} + 2\left(\frac{\Delta\sigma^*}{2K^*}\right)^{1/n^*}, \tag{2.18}$$

where the asterisk superscript indicates that the quality is measured with respect to the coordinate at the origin 0*. K^* and n^* are found from fitting (2.18) to the experimental data. The relationship between the two coordinate systems shown in Fig. 2.19 is:

$$\Delta\sigma = \Delta\sigma^* + \delta\sigma_0\,, \tag{2.19a}$$

$$\Delta\varepsilon^{\mathrm{p}} = \Delta\varepsilon^{*\mathrm{p}}\,, \tag{2.19b}$$

$$\Delta\varepsilon^{\mathrm{e}} = \Delta\varepsilon^{*\mathrm{e}} + \frac{\delta\sigma_0}{E}\,, \tag{2.19c}$$

where $\delta\sigma_0$ is the increase in the proportional stress limit, a measure of cyclic expansion of the elastic range. $\delta\sigma_0$ can be determined from (2.18) and (2.19), i.e.

$$\delta\sigma_0 = \Delta\sigma - \Delta\sigma^* = \Delta\sigma - 2K^*\left(\frac{\Delta\varepsilon^{\mathrm{p}}}{2}\right)^{n^*}. \qquad (2.20)$$

If the cyclic stress–strain curve is known, then for a given strain (or stress) amplitude, the corresponding stress amplitude, $\Delta\sigma/2$ (or strain amplitude, $\Delta\varepsilon/2$) can be calculated from (2.16). Furthermore, knowing the strength coefficient, K^*, and the strain-hardening exponent, n^*, of the master curve, we can then determine $\delta\sigma_0$ from (2.20) or (2.19c).

The master curve (Fig. 2.19) can be constructed from a minimum of two specimens using the single step test (SST) method. One specimen is cycled at a strain amplitude slightly higher than the monotonic yield strain, i.e. $\Delta\varepsilon/2 \approx \varepsilon_y$. The other specimen is cycled at the highest strain level for which the specimen results are valid. The hysteresis loops at the half-life are then matched as shown in Fig. 2.20. These two hysteresis loops could also be used to construct the cyclic stress–strain curve as depicted in Fig. 2.15. Alternatively, a single specimen can be used to obtain the master and cyclic curves through the multiple step test (MST (I)) method. Of course, as described in the previous section, the latter curves are an approximation of the curves constructed by the SST method.

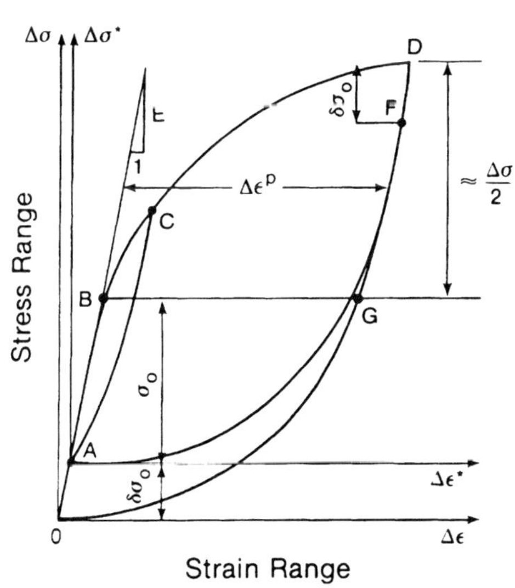

Fig. 2.20 Description of the linear range 0AB and master curve coordinates with the origin at A, and various definitions.

In certain cases, fitting (2.18) to the experimental data may not give accurate results for the total range. In these circumstances, we may construct a function table ($\Delta\sigma^*$ vs. $\Delta\varepsilon^*$) and interpolate between the prescribed points in the table.

An approximate value of $\delta\sigma_0$ can be directly calculated from the cyclic stress–strain curve (Fig. 2.16). Denoting the limit of proportionality by σ_0, the difference between the stable cyclic stress amplitude, $\Delta\sigma/2$, and σ_0, is the increase in the proportional range, Fig. 2.20 [8], i.e.

$$\delta\sigma_0 \simeq \left(\frac{\Delta\sigma}{2}\right) - \sigma_0 \,. \tag{2.21}$$

Note that the degree of the above approximation deteriorates as the strain range increases.

The master curve (2.18) also describes the transient hysteresis loops sufficiently well. For example, Fig. 2.21 shows hysteresis loops for a few stress-controlled tests [7]. Each loop is identified by two numbers, the first one indicating the specimen number and the second one the number of

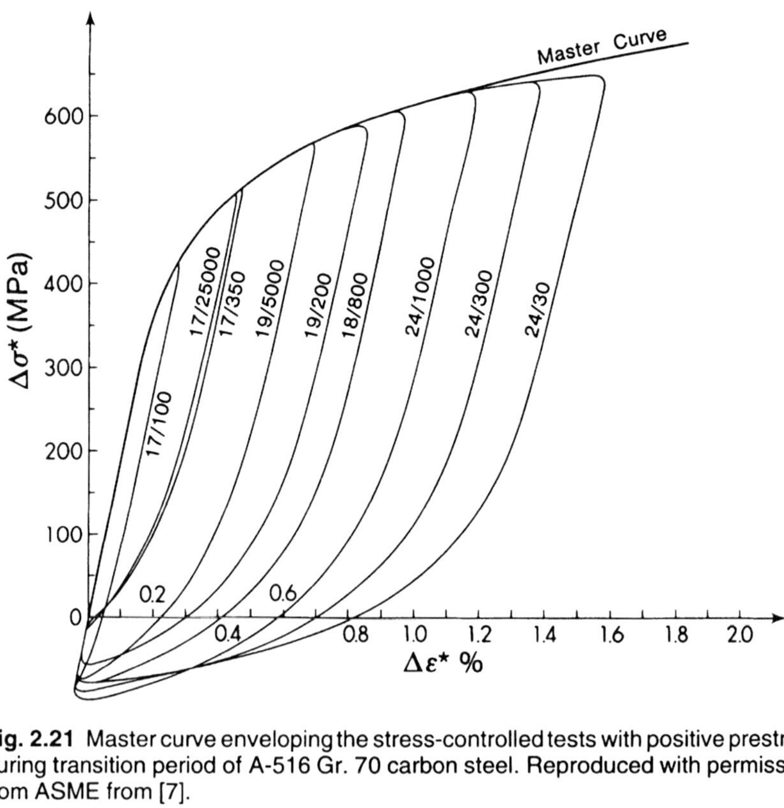

Fig. 2.21 Master curve enveloping the stress-controlled tests with positive prestrain during transition period of A-516 Gr. 70 carbon steel. Reproduced with permission from ASME from [7].

cycles. For large values of the stress range, $\Delta\sigma$ (specimen No. 24), the master curve overestimates the slope of the ascending branch during the initial stage of hardening. However, as the number of cycles increases, the ascending branches tend toward the master curve. Similar results were observed for the strain-controlled tests of annealed commercially pure copper with a high rate of strain hardening, Fig. 2.22 [12]. This type of transient behaviour (under stress or strain control) can be explained in terms of the formation of the dislocation structure. As described earlier, a stable dislocation structure is formed when a saturated cyclic state is achieved. In the interim, the hardening rate increases with the decrease in the mobile dislocation density. A master curve is the locus of the stable

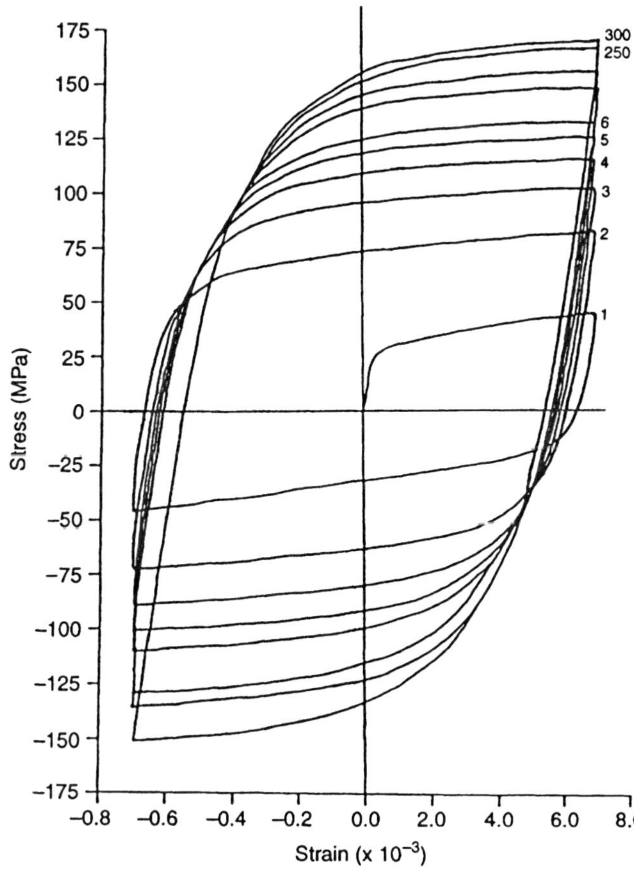

Fig. 2.22 Pronounced strain hardening of a commercially pure copper (courtesy of Dr. Schubert and Prof. Rie). See Chapter 6 for various stable cyclic loops (Figs 6.18–6.20).

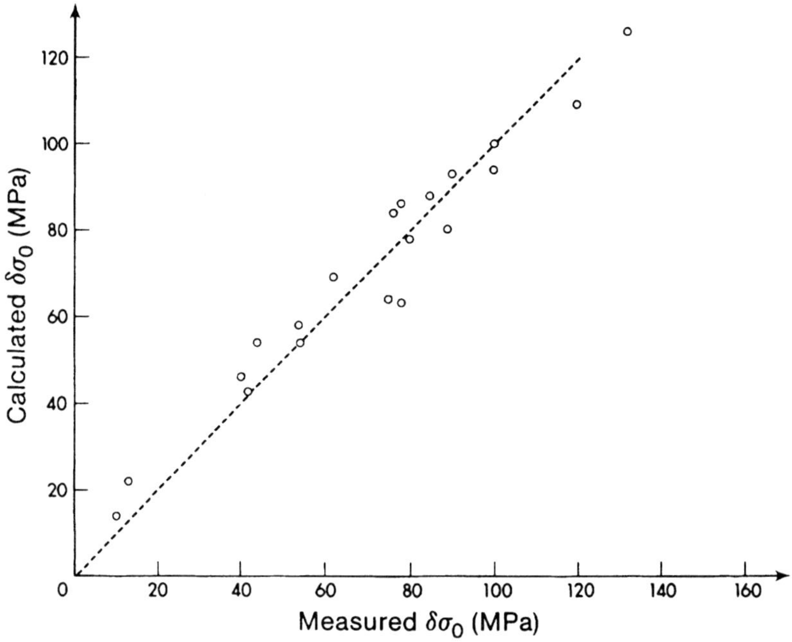

Fig. 2.23 Comparison between calculated and measured increases in the proportional stress limit of stress-controlled tests of A-516 Gr. 70 carbon steel. Reproduced with permission from ASME from [7].

(saturated) flow stresses, and all interim (transient) states will have hardening rates lower than that of the master curve.

A comparison between the measured and calculated $\delta\sigma_0$ (2.20) for stress-controlled tests is shown in Fig. 2.23. The results obtained for strain-controlled tests are in better agreement [7].

From the foregoing discussion it is clear that (2.18) and (2.20) constitute a general cyclic stress–strain relation under uniaxial stress conditions. A Masing-type material is a particular case of (2.18) and can be described by setting $\delta\sigma_0 = 0$ in (2.20), in which case, $\Delta\sigma^* = \Delta\sigma$ and $n^* = n'$, $\Delta\varepsilon^* = \Delta\varepsilon$ and (2.16) is then recovered.

2.7 SLOPE OF THE STRESS–STRAIN CURVE DURING LOAD REVERSAL

It is generally assumed that the slope of the stress–strain curve immediately after load reversal is equal to that at the start of the loading cycle. While this may be reasonably so for monotonic loading (Fig. 2.1), the elastic modulus changes during cyclic loading [25]. The effective elastic unloading modulus, E', is defined as the slope of the 'linear' portion of the stable hysteresis loop upon stress or strain reversal from the maximum load. This modulus depends on the cyclic characteristic of the material. For example, for the ferritic–pearlitic steel, ASTM A-516 Gr. 70, E' decreases with cyclic softening and increases with cyclic hardening of the material (Fig. 1 in [25]).

From the stable hysteresis loops of fully-reversed strain-controlled tests, one can obtain the elastic strain range, $\Delta\varepsilon^e = \Delta\sigma / E'$ and the plastic strain range, $\Delta\varepsilon^p = \Delta\varepsilon^t - \Delta\varepsilon^e$, where $\Delta\varepsilon^t$ is the total strain range. Figure 2.24 shows a plot of the elastic and plastic strain amplitudes ($\varepsilon_a^e = \Delta\varepsilon^e/2$, $\varepsilon_a^p = \Delta\varepsilon^p/2$) versus the stress amplitude ($\sigma_a = \Delta\sigma/2$) on a log–log scale for these test results. A power law relationship, similar to that of the plastic strain amplitude, appears to fit the data, i.e.

$$\frac{\Delta\varepsilon^e}{2} = \left(\frac{\Delta\sigma}{2K_e}\right)^{1/n^e}. \tag{2.22}$$

The prediction of the linear relationship, with $E = 204\,000$ MPa, is also shown in the figure by a dashed line. The specific values of the coefficients in (2.22) are given in the figure. Attempts have been made to relate the change in the elastic modulus with cyclic damage, however, caution should be exercised in interpreting such results [25].

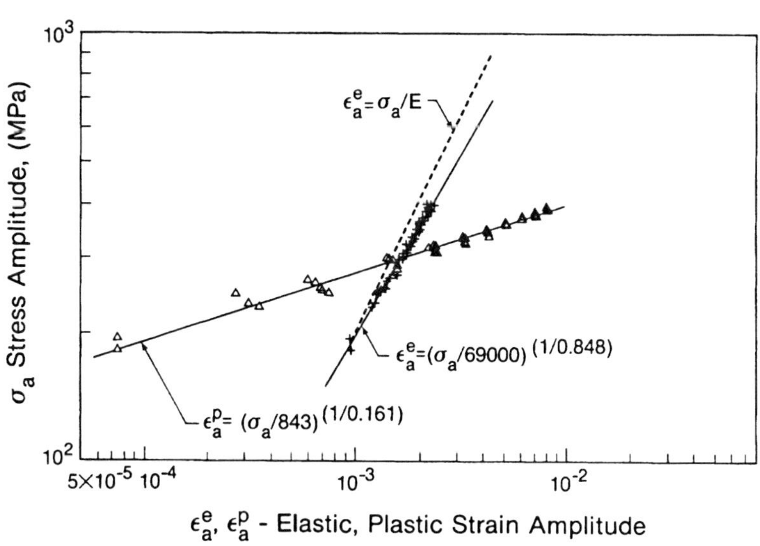

Fig. 2.24 Relationship between the cyclic stress amplitude and elastic and plastic strain amplitudes, indicating a nonlinear behaviour for A-516 Gr. 70 carbon steel in both elastic and plastic regimes. Reprinted with kind permission from [25].

2.8 EFFECT OF TEMPERATURE ON THE CYCLIC STRESS–STRAIN RELATIONSHIP

In the case of monotonic stress–strain curves, the effect of increased temperature is to decrease the stress required to achieve a given plastic strain. The monotonic stress–strain curves at higher temperatures follow essentially the same trend as those at room temperature ($\sim 20°$ C) and are more or less parallel to one another in the plastic range, e.g. Fig. 2.25 from [26].

In a similar manner, temperature affects the cyclic stress–strain curve of a material, and the response will vary with the material type and microstructure. For example, while Fig. 2.25 shows that the cyclic stress–strain curves remain parallel for various temperatures, there is also evidence that for materials which exhibit a three-stage behaviour, the increase in temperature results in a change of the slope of the cyclic stress–strain curve [27].

Referring to (2.15), the effective stress which is responsible for the plastic deformation is given by

$$\sigma_e = \sigma - \sigma_I . \tag{2.23}$$

From Fig. 2.25 it follows that for a given applied stress, the internal stress, σ_I, is temperature dependent and it should generally decrease with the increase in temperature. At higher temperatures most materials become time dependent; consequently, the internal stress is both temperature and strain-rate dependent, i.e.

$$\sigma_I = \sigma_I(\varepsilon, \dot{\varepsilon}, T) . \tag{2.24}$$

There is ample experimental evidence supporting the above functional relationship, e.g. Polák *et al.* [28, 29].

Figure 2.26 depicts a stable hysteresis loop under a load-controlled condition in an elevated temperature environment. Upon unloading at point A, the strain tends to increase in spite of the decreasing stress until point B, where the tangent is equal to the elastic modulus. This behaviour has been associated with the creep of the material, and the resulting stress is defined as the effective stress, σ_e (Krempl [30] and later Asada *et al.* [31] have defined it as an 'overstress'). However, as we have seen in the previous section, the tangent modulus immediately upon unloading is not a linear one at room temperature. Thus the observed nonlinearity at the descending branch AB in Fig. 2.26 is not entirely due to the creep of the material, it also depends on the cyclic characteristic of the material.

In a stress-controlled experiment, one would expect that the effective stress, σ_e, would also have a functional relationship similar to that expressed by (2.24). In a totally strain-controlled experiment some materials, e.g. $2\frac{1}{4}$ Cr-1 Mo steel, show a weak dependence of the effective stress σ_e on temperature and strain rate [28, 31]. However, σ_e in general would have

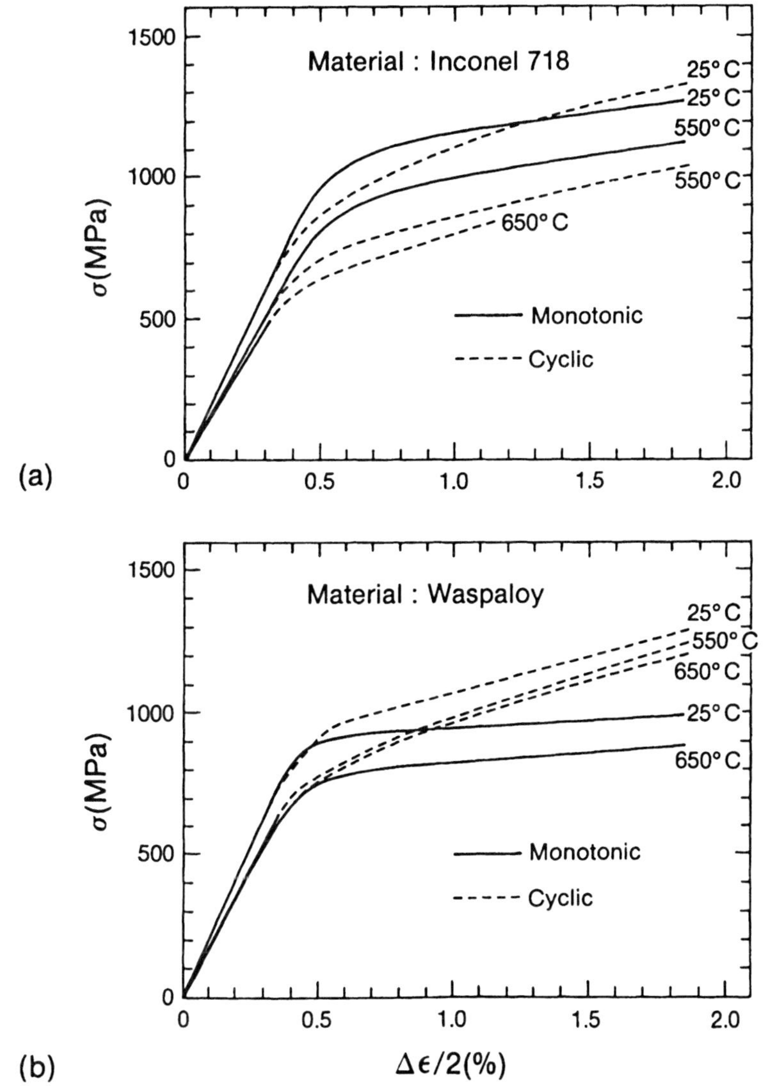

Fig. 2.25 Monotonic and cyclic stress–strain curves of (a) Inconel 718 alloy and (b) Waspaloy at various temperatures, from [26].

the following functional relationship,

$$\sigma_e = \sigma_e(\varepsilon, \dot{\varepsilon}, T), \tag{2.25}$$

in which dependence on $\dot{\varepsilon}$ and T is not too strong.

At high temperatures, the elastic modulus, E, decreases with an increase in temperature. The cyclic strength coefficient K in (2.8) also decreases with

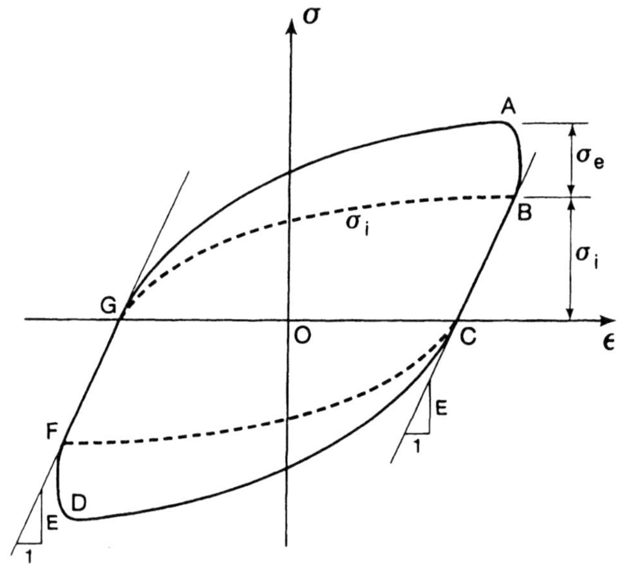

Fig. 2.26 Schematic representation of a stable hysteresis loop under stress-controlled conditions and definition of effective and internal stress components.

an increase in temperature, although contrary trends have been noted for some metals [32]. The rate of hardening or softening is also affected by the temperature, which may result in the variation of cyclic exponent n' (2.11) with temperature.

The cyclic action enhances the diffusion process of creep due to the acceleration of activation processes. The extent of this action depends on the material and rate of loading. The effect of environment is also more pronounced at higher temperatures.

When a certain stress or strain is held constant at an elevated temperature it is generally known as the hold- or dwell-time. In a stress-controlled test this leads to an increased strain (creep strain). Stress relaxation occurs when the test is conducted under strain-controlled conditions.

Changes in microstructure resulting from the increase in temperature are manifested in changes in slip characteristics, dislocation cross-slipping and climbing, grain boundary alteration and formation of cavities. For example, Inconel 718 alloy which exhibits cyclic strain softening due to a temperature increase (Fig. 2.25) has a planar slip character. On the other hand, a wavy slip character is noted for a Waspaloy which exhibits strain hardening at elevated temperatures (Fig. 2.25). Since the stacking fault energy is an increasing function of temperature, at elevated temperatures $T > 0.5\ T_m$, where T_m is the melting temperature, most metals exhibit a wavy slip characteristic. Furthermore, an increasing temperature causes thermal activation, resulting in the dislocation dispersion throughout the matrix via cross-slip and climbing motions. This process is a time-dependent function, hence the internal stress σ_1 (2.23) should be a time-dependent function the history of total dislocations.

Cyclic deformation accelerates dislocation migration. In section 2.3 we have noted that the dislocation structure may initially consist of tangles, and due to cyclic deformation, dislocations migrate to form wall structures and eventually dislocation cells [15] (see also [33] for a description of dislocation evolution at high temperatures). Thermal activation enhances this process. For example, it has been shown that cell size in the austenitic steels increases with increasing temperature [34]. Since the stress is inversely proportional to the cell size, it therefore follows that the stress will decrease with increasing temperature as observed in Fig. 2.25.

The grain boundaries tend to weaken in comparison to the matrix as temperature increases. This is mainly due to the dislocation climb and the formation of cavities which result from vacancy clustering, dislocation pile-up and grain boundary sliding. For further discussion regarding the effect of temperature on dislocation structure, the reader is encouraged to consult [10, 15, 21, 26, 27, 33, 35], where further references to other works are listed.

Another microstructural change due to elevated temperature is alloy instability. At high temperatures alloying components are transformed to various solid solutions (aging and precipitation reactions). The diffusion of interstitial particles and the formation and growth of precipitates are enhanced by the cyclic action at high temperatures, which results in dynamic strain aging (negative strain-rate effect).

In summary, the most important effect of increased temperature is the prevalence of time-dependent behaviour (creep, relaxation) and microstructural changes as discussed above.

2.9 EFFECT OF ENVIRONMENT ON THE STABLE CYCLIC STRESS–STRAIN RELATIONSHIP

The effect of environment would be to influence the surface or subsurface layers of an uncracked specimen through chemical or physical interaction. This process would then be observed in the macroscopic scale through a change in plastic deformation. In a high vacuum environment there is no chemical reaction and a cyclic stress–strain curve obtained in this environment can be used as the reference. It has been found [36, 37] that the bulk macroscopic stress–strain curves are very slightly affected when tests are carried out in a vacuum or air environment. It should be noted, however, that the damage accumulates at the surface layer and the rate of damage accumulation depends on the environment, as will be seen in the next chapter.

Table 2.2 Cyclic properties of ASTM A-516 Gr. 70 steel [37]

Properties	Vacuum	Air
E (MPa)	204 000	204 000
K' (MPa)	1203	1067
n'	0.218	0.193
K^* (MPa)	616.9	725.6
n^*	0.129	0.152

Stable cyclic hysteresis loops of a ferritic–pearlitic steel tested in a 0.1 mPa vacuum and in air at room temperature, were used to obtain cyclic stress–strain and master curves [37]. Table 2.2 gives the cyclic coefficients (2.11) and master curve coefficients (2.18) obtained through a least squares technique.

A plot of the two sets of curves using the data in Table 2.2 shows that the effect of environment on the stable stress–strain curve is not significant for the range of experimental data from which they were derived, $\Delta\varepsilon \leqslant 3\%$.

The same trend has been observed for a copper polycrystal where the cyclic stress–strain response as well as the dislocation structure showed no influence by the environment (air vs. vacuum) at comparable numbers of cycles [38].

The effect of environment is enhanced by a temperature increase and cyclic action, which allow for rapid diffusion of oxides and other gases into the metal. Oxidation results in the weakening and embrittlement of the grain boundaries. This process is generally a surface phenomenon and,

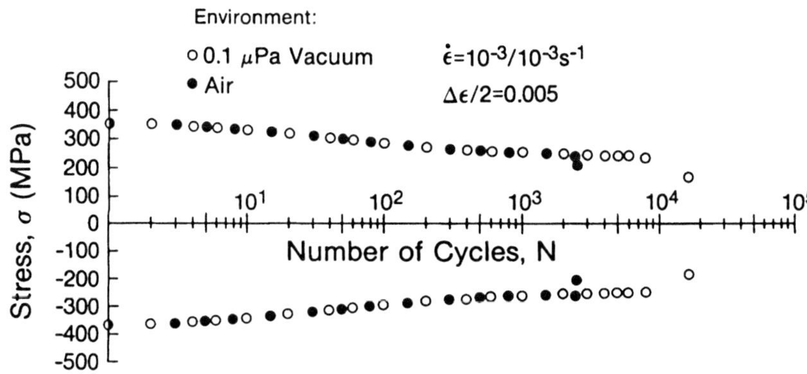

Fig. 2.27 Variation of the maximum and minimum stress values with number of cycles for $2\frac{1}{4}$ Cr-1 Mo steel at 550°C in high vacuum and air environments (courtesy of Prof. Y. Asada, Tokyo University).

although it does not appreciably affect the bulk response of the material (Fig. 2.27), it has a significant effect on the initiation and propagation of cracks. With increased temperature, the crack propagation path is changed from the transgranular to the intergranular. Intergranular crack propagation is faster than that of trangranular, and thus the fatigue life is shortened (Fig. 2.27). This will be discussed further in Chapters 3 and 7.

2.10 EFFECT OF RATE OF LOADING ON THE STABLE CYCLIC RESPONSE

In section 2.2 we briefly discussed the effect of rate of loading on the monotonic stress–strain curve. The same trend is observed in the case of the stable cyclic stress–strain curve, i.e. an increase in the loading rate results in an increase in the proportional (elastic) limit and a change in the slope of the elastic–plastic transition part of the stress–strain curve. After the transition zone, the curves for various rates are almost parallel. If the stress-rate, $\dot{\sigma}$, and the strain-rate, $\dot{\varepsilon}$, are changed by the same ratio, the effect of $\dot{\varepsilon}$ on the stress–strain curve is more pronounced than that of $\dot{\sigma}$, and the transition part of the stress-controlled curve is more gradual than that of the strain-controlled one.

The spacing of the stress–strain curve is highly nonlinear with respect to changes in the loading rate, i.e. while the loading rate may vary by several orders of magnitude, the change in stress would be less than an order of magnitude. The effects of change in the loading rate are similar to that of temperature, except that the loading rate effects are much smaller.

Figure 2.28 shows the stress range versus the strain range ($\Delta\sigma$ vs. $\Delta\varepsilon$) of the stable hysteresis loop of a $2\frac{1}{4}$ Cr-1 Mo steel tested in a vacuum (100 μPa at 550°C under total strain control, $\Delta\varepsilon - 1\%$). The effect of a change of two orders of magnitude in the strain rate is depicted in the figure. The rate of unloading and compressive loading was kept the same for all three tests ($\dot{\varepsilon} = 1 \times 10^{-3}\,s^{-1}$).

It should be pointed out that certain materials, e.g. low alloy carbon steels and low carbon steels, show a negative strain-rate effect, i.e. at a certain temperature range, the stress decreases with increased strain-rate (dynamic strain-aging), e.g. [39]. The critical temperature range is about 200–400°C for low carbon and low alloy steels and about 400–600° C for austenitic steels.

2.11 CYCLIC STRESS–STRAIN RELATIONSHIP FOR MULTIAXIAL STRESS STATES – PROPORTIONAL LOADING PATHS

In this section we will generalize the master curve concept to multiaxial stress states. Cyclic hardening (or softening) is characterized by the

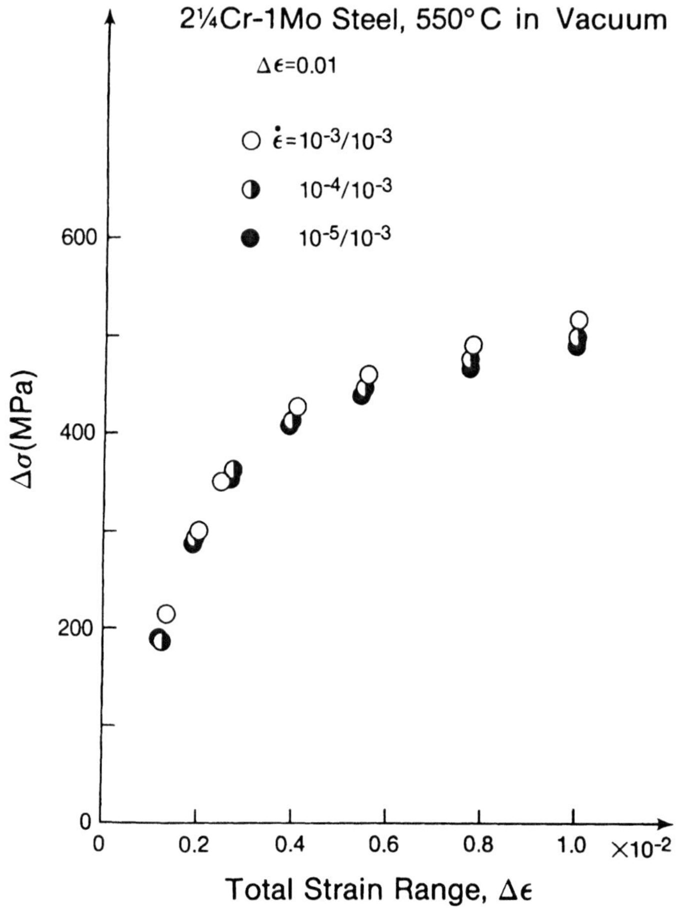

Fig. 2.28 Effect of strain-rate on the stable cyclic strain response of $2\frac{1}{4}$Cr-1 Mo steel at 550°C in a vacuum environment (courtesy of Prof. Y. Asada, Tokyo University).

expansion (or contraction) of the initial yield surface. For example, in experiments conducted by Ellyin and Neale [40], it was observed that the shape of the steady-state cyclic yield surface is independent of the stress ratio, but depends on the stress amplitude (Fig. 2.29). (Experimental determination of the yield surfaces requires special attention [41] and will be briefly discussed in Chapter 6.) The motion of the stabilized yield surface is then in the direction of loading when the loading path is proportional. The above observation is a simple extension of the uniaxial stress–strain curve (cf. Figs 2.1 and 2.15) into multi-dimensional stress space. Results of the strain-controlled tests (Fig. 2.30) indicate that the cyclic steady-state yield surface is independent of the strain ratio, but is a function of the strain

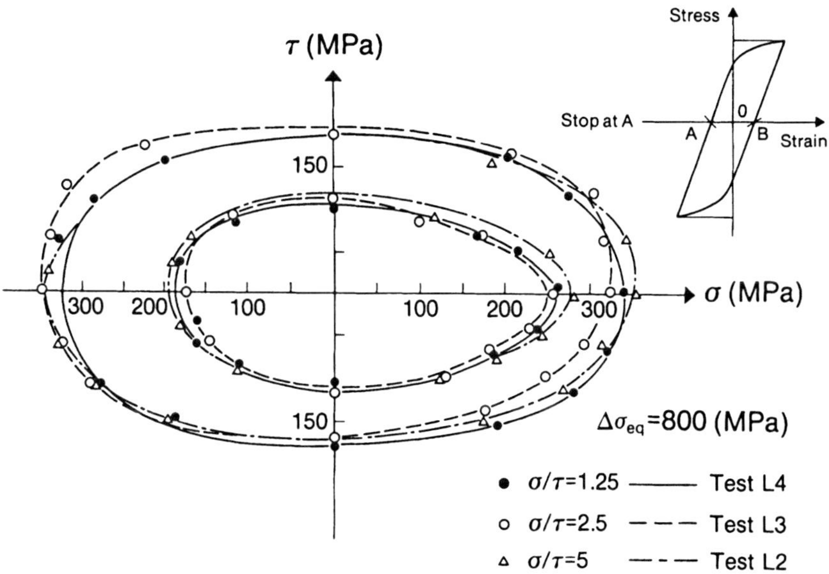

Fig. 2.29 Monotonic and cyclic steady-state yield surfaces for load-controlled tests of 2024-T351 aluminium alloy under combined torsional and axial load cycling, $\Delta\sigma_{eq} = (\Delta\sigma^2 + 3\Delta\tau^2)^{1/2}$. Reproduced with permission from ASME from [40].

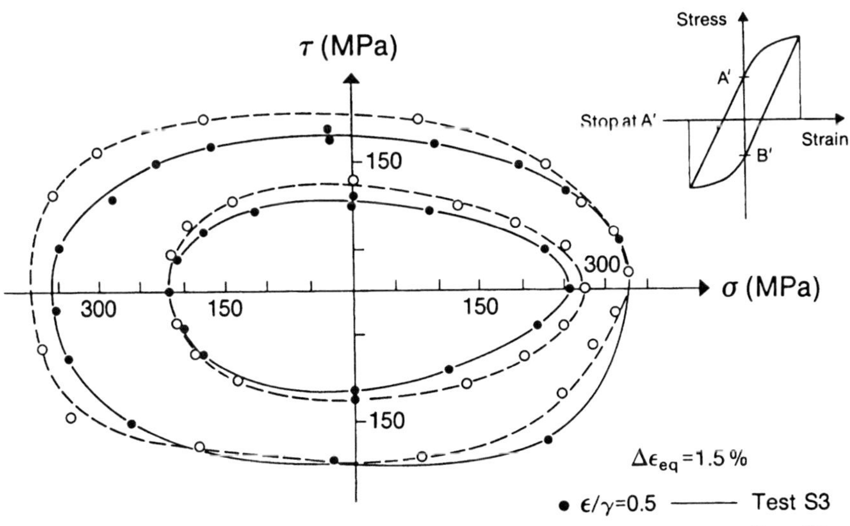

Fig. 2.30 Monotonic and cyclic steady-state yield surfaces for deformation-controlled tests of 2024-T351 aluminium alloy under combined shear and axial strains, $\Delta\varepsilon_{eq} = (\Delta\varepsilon^2 + \frac{1}{3}\Delta\gamma^2)^{1/2}$. Reproduced with permission from ASME from [40].

amplitude. A comparison of Figs 2.29 and 2.30 suggests a one-to-one mapping of the equivalent stress space to that of the equivalent strain space. These quantities will be defined shortly.

Let us now consider a different approach. In the theory of plasticity, plastic strains, ε_{ij}^{p}, can be directly related to final deviatoric stresses, s_{ij}, when the loading path is proportional [42], i.e.

$$\varepsilon_{ij}^{p} = F(\sigma_{ij}) s_{ij}, \quad (i, j = 1, 2, 3) \tag{2.26}$$

where

$$s_{ij} = \sigma_{ij} - \tfrac{1}{3}\sigma_{kk}\delta_{ij} \tag{2.27}$$

is the deviatoric stress, and δ_{ij} is the Kronecker delta, equal to unity when $i = j$, otherwise zero. A summation is implied when an index is repeated, i.e. $\sigma_{kk} = \sigma_{11} + \sigma_{22} + \sigma_{33}$. In the equivalent stress-equivalent strain space (σ_{eq} vs. ε_{eq}), we note that the relationship is almost identical to that of the uniaxial stress–strain curve. There is ample experimental evidence to support the above contention for both monotonic and cyclic loading under proportional paths, for example [43, 44]. Thus when (2.26) is written in terms of the cyclic equivalent stress and strain, its form would be similar to that of the plastic part of (2.18), i.e.

$$\Delta\varepsilon_{eq}^{p} = \frac{1}{K^*}\left(\frac{\Delta\sigma_{eq}^*}{2K^*}\right)^{(1-n^*)/n^*} \Delta\sigma_{eq}^*, \tag{2.28}$$

where

$$(\Delta\varepsilon_{eq}^{p})^2 = \tfrac{2}{3}\Delta\varepsilon_{ij}^{p}\Delta\varepsilon_{ij}^{p}, \tag{2.29}$$

$$(\Delta\sigma_{eq}^*)^2 = \tfrac{3}{2}\Delta s_{ij}^*\Delta s_{ij}^*, \quad (\Delta s_{ij}^* = s_{ij}^{*\max} - s_{ij}^{*\min}) \tag{2.30}$$

are von Mises-type equivalent plastic strain and stress ranges, respectively. The origin of the equivalent stress range, $\Delta\sigma_{eq}^*$, is placed at the minimum elastic limit (Fig. 2.19) thus,

$$\Delta\sigma_{eq}^* = \Delta\sigma_{eq} - (\delta\sigma_0)_{eq}. \tag{2.31}$$

Multiplying (2.26) by itself and using (2.29) and (2.30) and comparing with (2.28) we obtain the components of the plastic strain range, $\Delta\varepsilon_{ij}^{p}$, in terms of the deviatoric stress range, Δs_{ij}^*, i.e.

$$\Delta\varepsilon_{ij}^{p} = 3\frac{(\Delta\sigma_{eq}^*)^{(1-n^*)/n^*}}{(2K^*)^{1/n^*}}\Delta s_{ij}^*. \tag{2.32}$$

The elastic components of the strain range are obtained from the generalized Hooke's law [45]

$$\Delta\varepsilon_{ij}^{e} = \frac{1+\nu}{E}\Delta\sigma_{ij} - \frac{\nu}{E}\Delta\sigma_{kk}\delta_{ij}. \tag{2.33}$$

The total strain range is then obtained from adding (2.32) and (2.33) which results in

$$\Delta\varepsilon_{ij} = \frac{1+\nu}{E}\Delta\sigma_{ij} - \frac{\nu}{E}\Delta\sigma_{kk}\delta_{ij} + 3(2K^*)^{-1/n^*}(\Delta\sigma^*_{eq})^{(1-n^*)/n^*}\Delta s^*_{ij}. \quad (2.34)$$

The above equation relates the components of the stable cyclic strain range to those of the stress and deviatoric stress ranges. Apart from the elastic constants – Poisson's ratio, ν and Young's modulus, E – (2.34) contains two other material constants: the cyclic strength coefficient, K^*, and the master curve strain-hardening exponent, n^*. These and the increase in the elastic limit, $(\delta\sigma_0)_{eq}$, can be determined from the uniaxial test data. Thus, all the material constants in (2.34) can be determined from simple uniaxial tests, as described in the master curve section. A minimum of two tests will be required (Fig. 2.20).

A comparison of the prediction of (2.34) with the experimental results is reported in [46].

The multiaxial constitutive relationship for the general case of transient non-proportional and out-of-phase loading will be discussed in Chapter 6.

REFERENCES

1. Bauschinger, J. (1886) On the change of the position of elastic limit of iron and steel under cyclic variation of stress. *Mitt. Mech.–Tech. Lab., Munich*, **13** (1).
2. Lefebvre, D., Ellyin, F. and Nguyen, Qoc T. (1982) *Low Cycle Fatigue of A 516 Gr. 70 Steel*, Proc. 1982 Joint Conf. on Experimental Mechanics, Society for Experimental Stress Analysis, Hawaii, May 1982, Brookfield Center, CT, Part I, pp. 128–35.
3. Ramberg, W. and Osgood, W.R. (1943) Description of Stress-Strain Curves by Three Parameters. *NACA TN* 402, National Advisory Committee for Aeronautics.
4. Argon, A. (ed.) (1975) *Constitutive Equations in Plasticity*, MIT Press, Cambridge, MA, Ch. 1, pp. 1–22 and Ch. 3 by U.F. Kocks, pp. 81–115.
5. Kujawski, D., Kallianpur, V. and Krempl, E. (1980) An experimental study of uniaxial creep, cyclic creep and relaxation of AISI 304 stainless steel at room temperature. *J. Mech. Phys. Solids*, **28**, Elsevier Science Ltd, Oxford, 129–48.
6. Ellyin, F. (1985) Effect of tensile-mean-strain on plastic strain energy and cyclic response. *J. Engng. Mater. Technol., Trans. ASME*, **107**, 119–25.
7. Ellyin, F. and Kujawski, D. (1984) Plastic strain energy in fatigue failure. *J. Pressure Vessel Technol., Trans. ASME*, **106**, 342–7.
8. Lefebvre, D. and Ellyin, F. (1984) Cyclic response and inelastic strain energy in low cycle fatigue. *Int. J. Fatigue*, **6**, Elsevier Science Ltd, Oxford, 9–15.
9. Feltner, C.E. and Laird, C. (1967) Cyclic stress-strain response of F.C.C. metals and alloys. I. Phenomenological experiments and II. Dislocation structure and mechanisms. *Acta Metall.*, **15**, 1621–53.

10. Christ, H. and Mughrabi, H.J. (1992) Microstructure and fatigue, in *Low Cycle Fatigue and Elasto-Plastic Behaviour of Materials* (eds K.-T. Rie *et al.*), Elsevier, London, pp. 56–69.
11. Abdel-Raouf, H., Benham, P.P. and Plumtree, A. (1970) Mechanical behaviour and substructure of strain-cycled iron. *Can. Metall. Quart.*, **10**, 87–95.
12. Schubert, R. and Rie, K.-T. (1988) Cyclic hardening, yield surface and dislocation arrangement at LCF (in German). *Material Wissenschaft u. Werkstofftechnik*, **19** (11), pp. 376–83.
13. Pickard, A.C. and Knott, J.F. (1988) Effects of testing method on cyclic hardening behaviour in face-centered-cubic alloys, in *Low Cycle Fatigue*, ASTM STP 942 (eds H.D. Solomon *et al.*), American Society for Testing and Materials, Philadelphia, PA, pp. 58–76.
14. Grosskreutz, J.C. and Mughrabi, H. (1975) Description of the work-hardened structure at low temperature in cyclic deformation, in *Constitutive Equations in Plasticity* (ed. A.S. Argon), MIT Press, Cambridge, MA, pp. 251–326.
15. Mughrabi, H. (1986) Cyclic deformation and fatigue: some current problems, in *Strength of Metals and Alloys* (Proc. ICSMA 7), vol. 3 (eds H.J. McQueen *et al.*), Pergamon Press, Oxford, pp. 1917–42.
16. Endo, T. and Morrow, J.D. (1969) Cyclic stress-strain and fatigue behaviour of representative aircraft metals. *J. Mater.*, **4**, 159–75.
17. Landgraf, R.W., Morrow, J.D. and Endo, T. (1969) Determination of the cyclic stress-strain curve. *J. Mater.*, **4**, 176–88.
18. Bairstow, L. (1910) The elastic limits of iron and steel under cyclic variation of stress. *Trans. R. Soc. London*, A **210**, 35–55.
19. Smith, R.W., Hirschberg, M.H. and Manson, S.S. (1963) Fatigue Behaviour of Materials Under Strain Cycling in Low and Intermediate Life Range, NACA TN D-1574, National Advisory Committee for Aeronautics.
20. Klesnil, M. and Lukáš, P. (1980) *Fatigue of Metallic Materials*, Materials Science Monographs, vol. 7, Elsevier, London.
21. Plumtree, A. (1987) Correlation between microstructure and cyclic behaviour, in *Low Cycle Fatigue and Elasto-Plastic Behaviour of Materials* (ed. K.-T. Rie), Elsevier, London, pp. 19–30.
22. Masing, G. (1926) *Eigenspannungen und Verfestigung beim Messing*. Proc. 2nd Int. Congress of Applied Mechanics, Zurich, Orell Füssliverlag, pp. 332–5.
23. Abdel-Raouf, H., Topper, T.H. and Plummtree, A. (1977) Cyclic plasticity and Masing behaviour in metals and alloys, in *Advances in Research on the Strength and Fracture of Materials*, Proc. ICF 4, Waterloo, Canada, vol. 2B, part IV (ed. D.M.R. Taplin), Pergamon Press, pp. 1207–15.
24. Jhansale, H.R. and Topper, T.H. (1973) Engineering analysis of the inelastic stress response of a structural metal under variable cyclic strains, in *Experimentation and Failure Prediction*, ASTM STP 519, American Society for Testing and Materials, Philadelphia, PA, pp. 246–70.
25. Kujawski, D. and Ellyin, F. (1987) The effect of cyclic loading on the slope of the stress-strain immediately upon load reversal. *Res. Mechanica*, **22**, Elsevier Science Ltd, Oxford, 295–9.
26. Pineau, A. (1983) High temperature fatigue behaviour of engineering materials in relation to microstructure, in *Fatigue in High Temperature* (ed. R.P. Skelton), Elsevier, London, pp. 305–64.
27. Fritzemeier, L.G. and Tien, J.K. (1988) The cyclic stress-strain behaviour of

nickel-base superalloys – Part I. Polycrystals, and Part II. Single crystals. *Acta Metall.*, **36**, 275–90.

28. Polák, J., Klesnil, M. and Helesic, J. (1986) Cyclic stress-strain response of $2\frac{1}{4}$Cr-1 Mo steel at elevated temperature. *Fatigue Fract. Engng. Mater. Struct.*, **9**, 185–94.
29. Polák, J. and Helesic, J. (1987) Cyclic stress in polycrystalline copper, in *Low Cycle Fatigue and Elasto-Plastic Behaviour of Materials* (ed. K.-T. Rie), Elsevier, London, pp. 37–42.
30. Krempl, E. (1979) Viscoplasticity based on total strain. *J. Engng. Mater. Technol., Trans. ASME*, **101**, pp. 380–6.
31. Asada, Y. *et al.* (1987) Creep-fatigue behaviour of 304 stainless steel and $2\frac{1}{4}$Cr-1 Mo steel based on the overstress concept, in *Advances in Analysis and Life Assessment for Pressure Vessel and Piping*, PVP 129 (eds S.J. Chang *et al.*), American Society of Mechanical Engineers, New York, pp. 93–9.
32. Skelton, R.P. (1987) Cyclic stress-strain properties during high strain fatigue, in *High-Temperature Fatigue – Properties and Predictions* (ed. R.P. Skelton), Elsevier, London, pp. 27–112.
33. Plumbridge, W.J. (1987) Metallography of high temperature fatigue, in *High-Temperature Fatigue – Properties and Predictions* (ed. R.P. Skelton), Elsevier, London, pp. 177–228.
34. Challenger, K.D. and Moteff, J. (1973) Correlation of substructure with the elevated temperature low cycle fatigue of AISI 304 and 316 stainless steels, in *Fatigue at Elevated Temperatures*, ASTM STP 529, American Society for Testing Materials, Philadelphia, PA, pp. 68–74.
35. Eifler, D. and Macherauch, E. (1990) Microstructure and cyclic deformation behaviour of plain carbon and low-alloyed steels. *Int. J. Fatigue*, **12**, 165–74.
36. Verkin, B.I. and Grinberg, N.M. (1979) The effect of vacuum on the fatigue behaviour of metals and alloys, *Mater. Sci. Engng.*, **41**, 149–81.
37. Ellyin, F. and Li, B. (1993) Crack initiation and fatigue lives in air and vacuum environments, *Mater. Sci. Engng.*, **A171**, 105–13.
38. Christ, H.-J., Mughrabi, H. and Wittig-Link, C. (1988) Cyclic deformation behaviour, microstructure and fatigue crack initiation of copper polycrystals fatigued in air and in vacuum, in *Basic Mechanisms in Fatigue of Metals* (eds P. Lukáš and J. Polák), Elsevier, Amsterdam, pp. 83–92.
39. Abdel-Raouf, H., Plumtree, A. and Topper, T.H. (1973) Effects of temperature and deformation rate on cyclic strength and fracture of low-carbon steel, in *Cyclic Stress-Strain Behaviour–Analysis, Experimentation and Failure Prediction*, ASTM STP 519, American Society for Testing and Materials, Philadelphia, PA, pp. 28–57.
40. Ellyin, F. and Neale, K.W. (1979) Effect of cyclic loading on the yield surface. *J. Pressure Vessel Technol., Trans. ASME*, **101**, 54–63.
41. Ellyin, F. and Grass, J.-P. (1975) Détermination expérimentale de surfaces d'ecoulement plastique du titane 50A, Premiere partie, Instrumentation. *Trans. CSME*, **3**, 156–62.
42. Katchanov, L.M. (1971) *Foundation of the Theory of Plasticity*, North-Holland, Amsterdam.
43. Ellyin, F. (1978) *Behaviour of Titanium 50A under Multiaxial Stress States.* Proc. V Interamerican Conference on Material Technology, Sao Paulo, Brazil, American Society of Mechanical Engineers, New York, pp. 361–8.

44. Tanaka, E., Murakami, S. and Oaka, M. (1985) Effect of plastic strain amplitudes on non-proportional cyclic plasticity. *Acta Mech.*, **47**., 167–82.
45. Drucker, D.C. (1967) *Introduction to Mechanics of Deformable Solids*, McGraw-Hill, New York.
46. Ellyin, F. and Golos, K. (1987) A constitutive relation for multiaxial stress states under cyclic conditions, in *Constitutive Laws for Engineering Materials – Theory and Application*, vol. 1 (eds C.S. Desai *et al.*), Elsevier, New York, pp. 523–9.

Phenomenological approach to fatigue life prediction under uniaxial loading

3

3.1 INTRODUCTION

Fatigue failure of metals due to repeated loads has been recognized for over a hundred years. In general, this phenomenon has attracted attention due to the advances in technology which mean that progressively more and more is demanded from machine parts in terms of speeds of operation and loads to be sustained. The required safety and/or reliability of modern engineering structures has also generated interest in this subject. Since Wöhler's 1871 work on fatigue limit, the field of fatigue has been subdivided into a number of specific topics such as high- and low-cycle fatigue, accumulation of damage, fatigue of notched members, and the initiation, propagation and arrest of fatigue cracks. Each of these areas has developed on the basis of available experimental results. Studies have ranged from the microscopic dislocation mechanism to phenomenological material behaviour to full-scale structural testing and analyses.

The majority of investigations into the fatigue process have been concerned with the phenomenological aspect of fatigue failure, i.e. estimating the total life of components under a specified loading condition. In this type of approach no distinction is usually made between the crack initiation and propagation phases. Therefore, a considerable amount of effort has been expended in defining a suitable empirical damage parameter, ψ, to obtain the best correlation between the damage parameter and life to failure (number of cycles N_f, or reversals $2N_f$). For instance, quantities such as stress, strain, plastic work, strain energy density, or some function of these have been used as the damage parameter. The word 'failure' in this context and throughout this chapter is used in a generic sense, i.e. it could encompass a range from life to crack initiation to life to complete separation of a section. Ideally separate criteria are required for the crack initiation phase and that of propagation. However, the demarcation between crack initiation and propagation stages is not easily defined. Instead the life to achieve a crack of a certain length, e.g. of the order of a millimetre, is taken to represent the 'fatigue life' of a smooth specimen. This is also sometimes referred to as the life to crack initiation in the case of large components.

The choice of a crack of the order of a millimetre is related to the ability to detect cracks by non-destructive techniques. The size of the crack is also important in the sense that continuum fracture mechanics can be applied to the crack propagation phase beyond the 'initiation life'. Short cracks of the order of grain size behave in a different fashion and their growth cannot be predicted by the nominal fracture mechanics parameters. The microstructure (orientation of adjacent grains), grain boundaries and other microstructural barriers have a strong influence on the growth of short cracks. Thus, a crack should be long enough (more than a few grains) before continuum theories would apply. However, a precise definition of a threshold crack beyond that defined above is not necessary since 'life curves' (damage parameter vs. number of cycles to failure) have similar shapes for different *initiated* macro-cracks (this will be demonstrated in section 3.5.4). From an engineering point of view, the demarcation between the life to initiate a dominant macro-crack and the subsequent crack propagation is when the latter phase is a stable one.

3.1.1 GENERAL APPROACH

In the case of one-parameter damage representation, the life to initiate a macro-crack, which is generally termed the fatigue failure criterion, can be written in the form of

$$\psi = q(t)\,, \tag{3.1}$$

where ψ is a damage parameter and t is the total life in terms of cycles N_f or reversals $2N_f$ to 'failure'. The relation (3.1) is a decreasing function of t. If the material exhibits a fatigue (endurance) limit, then for $t \to \infty$, $q(t) \to \psi_0$, thus a possible form of $q(t)$ is

$$q(t) \sim \kappa(t)^{\alpha} + \psi_0\,, \tag{3.2}$$

with $\alpha < 0$ and $\kappa > 0$, having dimensions of ψ which will be specified later on. By setting $t = 2N_f$, (3.1) can be written as

$$\psi = \kappa(2N_f)^{\alpha} + \psi_0\,. \tag{3.3}$$

The above relationship is shown schematically by a full line in Fig. 3.1 on a log–log scale, with a smooth transition at the region of intersection of rectilinear segments of the curve. The point of intersection of the linearized segments determines the theoretical limiting life, $2N_L$ (Fig. 3.1). For those cases where the fatigue (endurance) limit does not exist, e.g. high strength

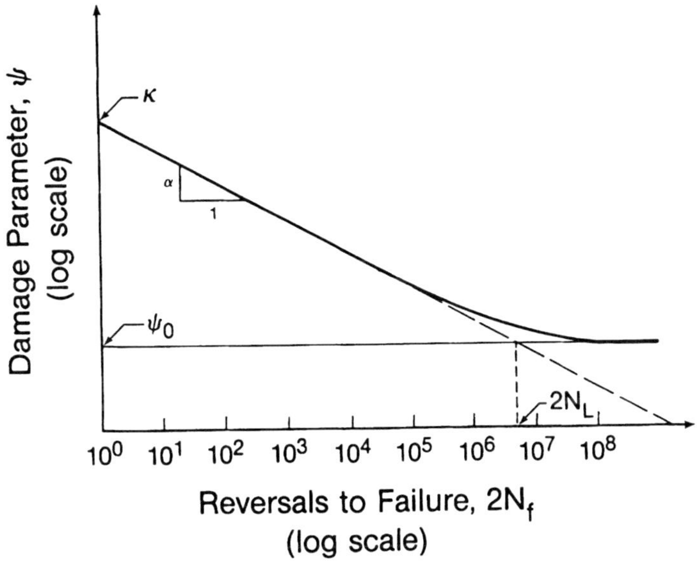

Fig. 3.1 A schematic plot of a fatigue life curve in terms of damage parameter versus number of reversals to failure.

steels or aluminium alloys, we set $\psi_0 = 0$ in (3.3) to obtain

$$\psi = \kappa (2N_f)^{\alpha} . \tag{3.4}$$

The relation (3.4), shown by a dashed line in Fig. 3.1, is applicable for very long lives, say $2N_f > 10^6$. The rectilinear segments of the curve can be used in those instances where there is a sharp knee at the transition region. The coefficient κ, with dimensions of ψ, and the exponent α of (3.3) are evaluated by a best fit technique from experimental data. In the past, experimental methods used to obtain the data were used to correlate the fatigue life, $2N_f$ (or N_f), of a smooth specimen with either the stress or strain amplitude. These 'intrinsic' (S–N) curves, popular with designers, generally do not convey the complete damage process which takes place in a material. In the S–N approach, the interrelation between stress and strain (path dependency) and the fatigue damage process is usually overlooked [1]. On the microscopic level the cyclic inelastic deformation is related to the movement of dislocations, and the cyclic stress is related to the resistance to motion. Because of the irreversibility of the micro-plastic deformation, in each cycle of loading there is a dissipation of strain energy. Therefore, the idea of relating the fatigue life to the plastic work or total strain energy during a load cycle has been proposed as an alternative approach. The significance of the energy approach is in its ability to unify microscopic and macroscopic testing data, and to formulate a multiaxial failure criteria.

3.2 STRESS-BASED APPROACH

The very first systematic study of metal fatigue using the stress-based approach was by Wöhler on rotating axles [2]. Since Wöhler's 1871 work, fatigue data have been presented in the form of *S–N* curves, where *S* is the engineering stress amplitude and *N* is the number of cycles to failure, on a semi-log or log–log scale.

If the true stress amplitude $\sigma_a = \Delta\sigma/2$ is used as the damage parameter instead of engineering stress, *s*, the entire plot of $\log(\Delta\sigma/2)$ versus $\log(2N_f)$ is given by [1]

$$\sigma_a = \Delta\sigma/2 = \sigma_f'(2N_f)^b, \tag{3.5}$$

where σ_f' and b are the fatigue strength coefficient and the fatigue strength exponent. A similar relation (in terms of N_f) was proposed by Basquin as early as 1910 [3]. The relation (3.5) is a straight line on a log–log scale, as illustrated in Fig. 3.2(a). The so-called Basquin's coefficients, σ_f' and b, are evaluated by a best fit technique from the experimental data. The coefficient σ_f' is taken as the intercept of the plot at $2N_f = 1$ and the exponent b is

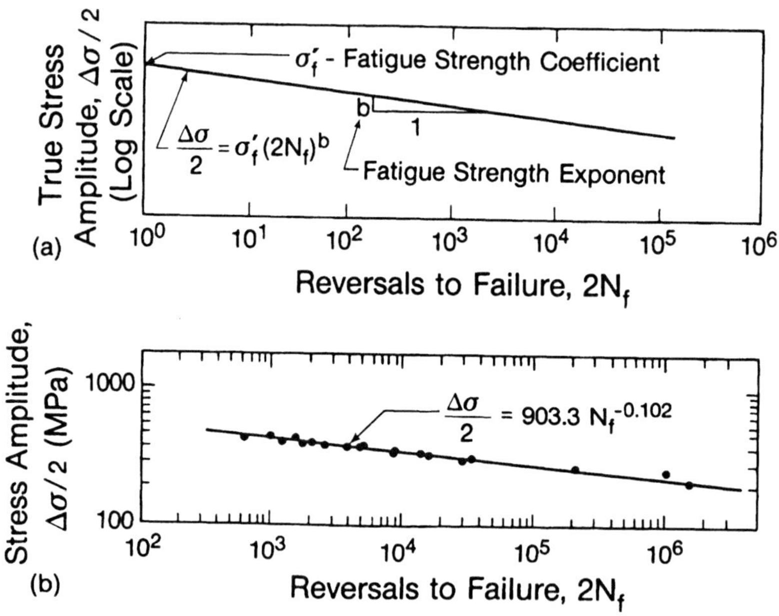

Fig. 3.2 Stress amplitude versus number of reversals to failure. (a) schematic plot and the corresponding equation; (b) experimental data, ASTM A-516 Gr. 70 steel.

the slope of the plot. Figure 3.2(b) shows the stress amplitude–fatigue life behaviour of ASTM A-516 Gr. 70 low alloy carbon steel [4].

The fatigue strength coefficient σ_f' may be estimated from the monotonic tension test by setting $\sigma_f' \approx \sigma_f$ or, for steels, to about 500 BHN hardness, as $\sigma_f' \approx s_u + 350$ MPa. Fatigue strength exponent b varies from -0.05 to -0.15 and for most metals has an average value of -0.085. Morrow [1] has shown through an energy argument that b is related to the cyclic strain hardening exponent n' as follows

$$b = \frac{-n'}{1 + 5n'}. \tag{3.6}$$

This correlation is compared in Fig. 3.3 with the representative data for different types of steels with $n' < 0.2$ [5]. No general agreement is observed although some sort of central tendency may be detectable with a wide dispersion.

The presence of a fatigue limit in certain irons and steels may be accounted for by introducing a fatigue strength limit σ_{FL}. For most metals the

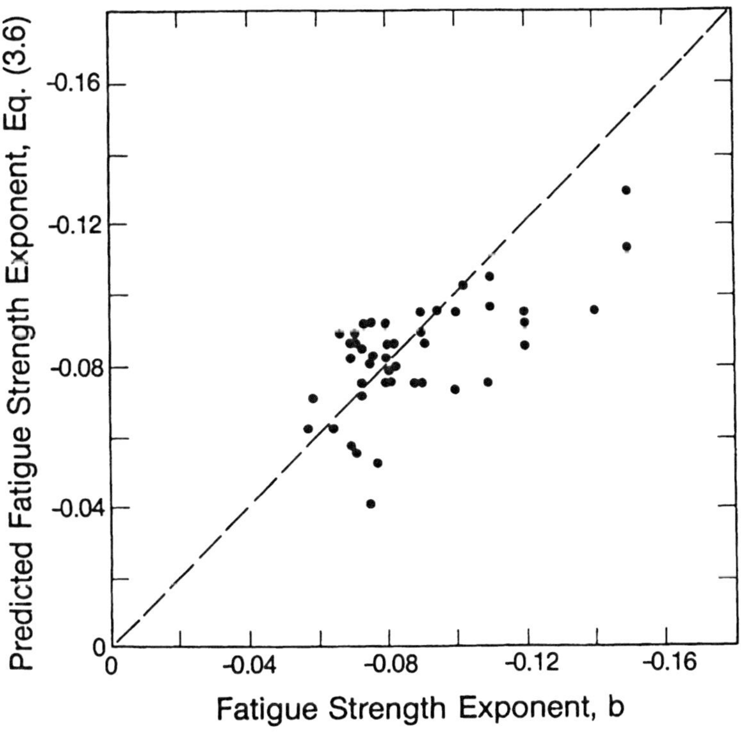

Fig. 3.3 Correlation between (3.6) and experimental data for various steels.

fatigue (endurance) limit varies from $\frac{1}{3}$–$\frac{1}{2}$ s_u depending on the number of reversals. In the absence of available data, $\sigma_{FL} \simeq \frac{1}{3} s_u$ at 10^8 reversals and $\sigma_{FL} \simeq \frac{1}{2} s_u$ at 2×10^6 reversals may be used as approximate values for most metals.

The mean-stress effect can be accounted for in (3.5) as follows [1]

$$\frac{\Delta\sigma}{2} = (\sigma_f' - \sigma_m)(2N_f)^b, \tag{3.7}$$

where σ_m is the mean stress. The effect of the tensile mean stress is thus equivalent to a reduction of the fatigue strength coefficient, and a compressive mean stress has the reverse effect. Equation (3.7) can also be used to predict the effect of residual stresses on the fatigue life, provided they do not relax during the life. An alternative approach was proposed by Smith, Watson and Topper [6]. They showed that experimental data with a mean stress fall into a scatter band, if the damage parameter is expressed as $(\frac{1}{2}\sigma_{max}\,\Delta\varepsilon\, E)^{1/2}$.

In the previous chapter it was shown that a combination of mean stress and stress amplitude may result in a cyclic-dependent creep (ratcheting) strain which can lead to premature creep–fatigue fracture. The effect of ratcheting strain on the fatigue life will be discussed later, in section 3.9. It should be noted that the stress-based approach is mostly applicable to the high-cycle fatigue regime where the strains are essentially elastic. Note also that in low-cycle fatigue, the magnitude of the stress used as an input, or predicted from the *S–N* curve, is an 'equivalent' elastic stress, not the real one. It is worthwhile mentioning again that the fatigue damage process is controlled by the plastic deformation in both micro- and macro-scales.

3.3 STRAIN-BASED APPROACH

Strain-based approaches essentially use the observed trend of the stress approach, i.e. a linear log–log relation between elastic or plastic strain amplitude and fatigue life is employed. Since the elastic component of strain in the uniaxial stress state is $\Delta\varepsilon^e/2 = \Delta\sigma/2E$, one can divide (3.5) by E to obtain

$$\frac{\Delta\varepsilon^e}{2} = \frac{\sigma_f'}{E}(2N_f)^b. \tag{3.8}$$

In the case of high-cycle fatigue (where the bulk plastic strain range, $\Delta\varepsilon^p \approx 0$) both relations, i.e. (3.5) and (3.8), are equivalent. For low-cycle fatigue $\Delta\varepsilon^p > \Delta\varepsilon^e$, and in 1954, Coffin and Manson independently established that the plastic strain amplitude versus life data could be linearized on a log–log coordinate [7, 8]. The Coffin–Manson equation can be written in the following form, adopting Morrow's notation [1]

$$\frac{\Delta \varepsilon^{p}}{2} = \varepsilon_{f}' (2N_{f})^{c}, \tag{3.9}$$

where ε_f' is the fatigue ductility coefficient which corresponds to the plastic strain at one reversal ($2N_f = 1$), and c is the fatigue ductility exponent. The latter is the slope of the plot of (3.9) on the log–log scale. Equation (3.9) is schematically illustrated in Fig. 3.4.

By adding the elastic and plastic components of strain amplitude, (3.8) and (3.9) respectively, we obtain the fatigue resistance relationship in terms of the total cyclic strain amplitude, i.e.

$$\frac{\Delta \varepsilon}{2} = \frac{\Delta \varepsilon^{e}}{2} + \frac{\Delta \varepsilon^{p}}{2} = \frac{\sigma_f'}{E} (2N_f)^{b} + \varepsilon_f' (2N_f)^{c}. \tag{3.10}$$

Figure 3.5 shows the variation of the elastic, plastic and total strain amplitudes with the number of reversals to failure for the ASTM A-516 Gr. 70 steel [4]. Attempts have been made to relate the fatigue ductility coefficient ε_f' with the monotonic fracture ductility ε_f. No general agreement has been found, and a variation ranging from 0.35 ε_f to $\sim \varepsilon_f$ has been observed [9]. Thus, approximating ε_f' by setting $\varepsilon_f' = \varepsilon_f$ is not recommended.

The fatigue ductility exponent c varies from -0.4 to -0.8 with a mean value of about -0.6 for the majority of steels. A similar relation to that of (3.6) was proposed between c and n' through Morrow's energy consideration [1]

$$c = -\frac{1}{1 + 5n'}. \tag{3.11}$$

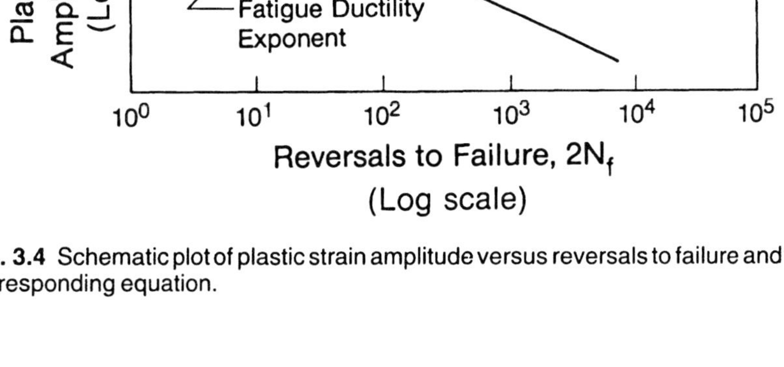

Fig. 3.4 Schematic plot of plastic strain amplitude versus reversals to failure and the corresponding equation.

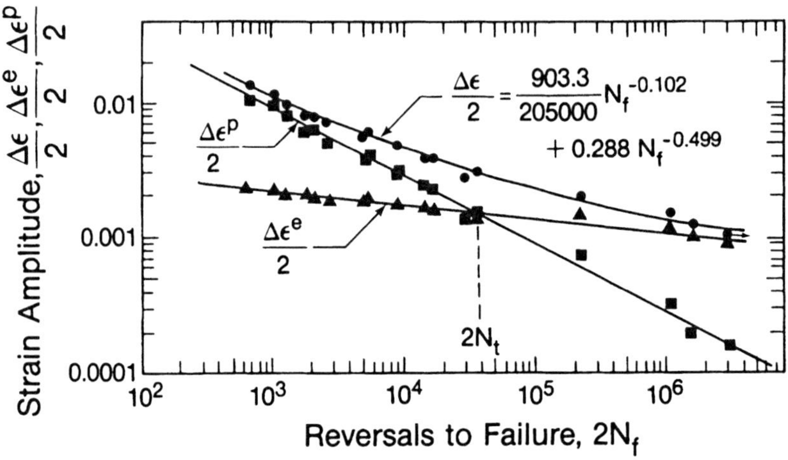

Fig. 3.5 Elastic, plastic and total strain amplitudes versus reversals to failure of a low alloy carbon steel (ASTM A-516 Gr. 70).

This correlation is compared in Fig. 3.6 with representative data for different types of steels with $n' \leqslant 0.2$ [5]. The correlation is rather weak and its usage, in general, is not recommended.

The mean-strain effect in low-cycle fatigue can be accounted for by modifying the Coffin–Manson relation(see the discussion by Libertiny in [10]) as follows

$$\frac{\Delta\varepsilon^p}{2} = (\varepsilon_f' - \varepsilon_m)\,(2N_f)^c\,, \tag{3.12a}$$

where ε_m is the mean strain. The above relation can be rewritten as

$$\frac{\Delta\varepsilon^p}{2} + \varepsilon_m\,(2N_f)^c = \varepsilon_f'(2N_f)^c\,. \tag{3.12b}$$

Assuming that ε_m is of the order of the plastic strain amplitude, and setting $c = -0.6$, the second term on the LHS of (3.12b) is rather small in comparison with $\Delta\varepsilon^p/2$, and can be neglected for $2N_f > 4000$ ($(2N_f)^c = (4000)^{-0.6} = 0.007$). In the case when the mean strain is of the order of the fatigue ductility coefficient, ε_f', the required prestrain would be likely to initiate cracks, and thus the fatigue life would be reduced. Therefore, it can be concluded that for moderate mean strains, i.e. $\varepsilon_m/\Delta\varepsilon^p \leqslant 1$, the low-cycle fatigue life will not be appreciably affected by the introduction of prestrain. Consequently, the fully-reversed strain test results ((3.9) or (3.10)) could be used to correlate experimental data. Ellyin [11] has investigated the effect of tensile mean strain (up to 6%) on the fatigue life of ASTM A-516 Gr. 70

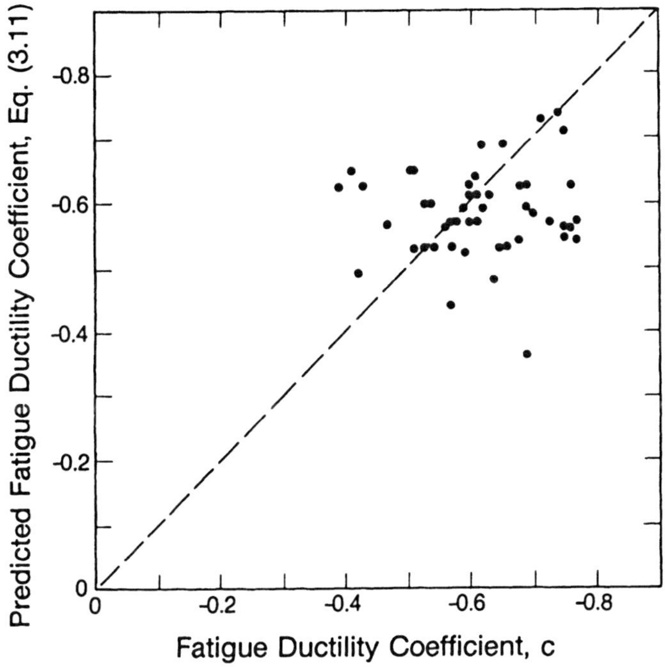

Fig. 3.6 Correlation between (3.11) and experimental data for various steels.

steel. The experimental results and comparison with the best fit curves with zero mean strain are shown in Fig. 3.7. It can be seen that the effect of the mean strain is very small and is negligible for $N_f > 1000$.

The plastic strain-based approach was essentially introduced for thermal and low-cycle fatigue. Progress in this area accelerated once the closed-loop testing facility became widely available, mostly in the early sixties. Subsequently, the totally strain-based approach became very popular, and is generally applicable to the fatigue resistance of notched elements. At highly stressed regions near a notch, plastic deformation is controlled by the surrounding elastic matrix; thus, a totally strain-controlled test is a fair representation of the condition experienced by the material at regions deformed beyond the elastic limit.

3.4 ENERGY-BASED APPROACH

On the microscopic level, the irreversible nature of the micro-plastic deformation caused by each cycle of loading is associated with the dissipation of strain energy. The dissipated strain energy per cycle (or a portion

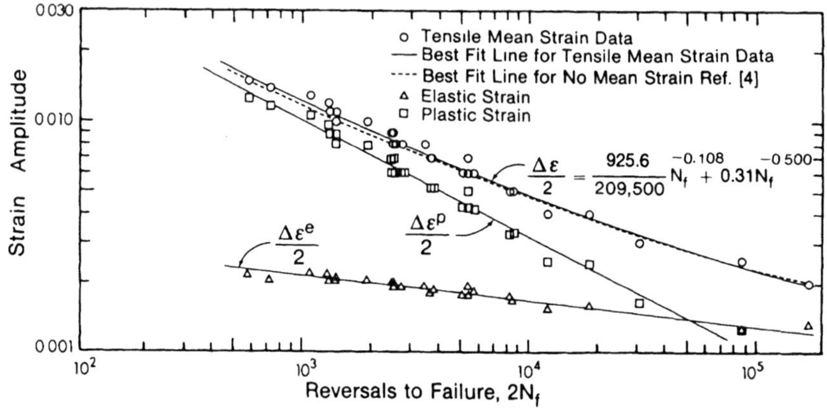

Fig. 3.7 Elastic, plastic and total strain amplitudes versus number of reversals to failure for tensile mean stress tests of ASTM A-516 Gr. 70 low alloy carbon steel, after [11] with permission from ASME.

of it) may be regarded as a main contributor to the fatigue damage process taking place in each cycle.

The relevance of the hysteresis energy to the fatigue phenomenon was recognized as early as 1910 by Bairstow [12] who qualitatively described the role of hysteresis energy in the fatigue process. The plastic strain energy per cycle ΔW^p (area of the hysteresis loop) has the advantage of being almost constant during the life, under strain-controlled conditions [13–15, 21, 23]. The Palmgren–Miner hypothesis of cumulative damage in fatigue [16] was stated as: 'The phenomenon of cumulative damage under repeated load [is] related to the network absorbed by a specimen . Enomoto [17] assumed that fatigue fracture occurs when the energy absorbed in each cycle, that is in excess of a non-damaging amount, reaches a critical value. Martin [18] considered that only the energy associated with strain hardening is damaging in a fatigue sense.

Energy-based criteria can be classified into two categories depending upon which of the following hypotheses is used:

- The total absorbed energy to fracture (fatigue toughness) is constant and is independent of the number of cycles to failure [16, 19, 20].
- The total absorbed energy to fracture depends on the number of reversals (cycles) to failure [1, 4, 11, 13–15, 21–24].

Halford [23] compiled over 1200 individual tests on metals in which energy to fracture was measured or calculated from the available data. He then concluded that the total energy to fracture increases with life, and a slope of $\frac{1}{3}$ is seen to fit the central tendency of all data points. Therefore, we will limit our analysis to the approaches in which fatigue toughness increases with increasing life.

The thermodynamic basis of the energy method can be summarized as follows. When a material element is loaded by an external agency, part of the supplied energy is dissipated into heat, dQ, and the remaining part is stored in the material. The stored energy is generally composed of two parts, a recoverable (elastic) part, $dW_{\text{recoverable}}$, and an irrecoverable part, $dW_{\text{irrecoverable}}$, which causes plastic deformation and damage, i.e.

$$dW_{\text{irrecoverable}} = dW_{\text{supplied}} - (dQ + dW_{\text{recoverable}}) \, . \tag{3.13a}$$

The LHS of (3.13a) is the energy required to cause permanent inelastic deformation and damage in the material,

$$dW_{\text{irrecoverable}} = dW_{\text{inelastic}} + dW_{\text{damage}} \, . \tag{3.13b}$$

Because of measurement difficulties with regard to heat loss, $-dQ$, it is generally assumed that

$$dW_{\text{damage}} \propto dW_{\text{irrecoverable}} \, . \tag{3.13c}$$

The hysteresis loop is taken to represent the irrecoverable energy part. Alternatively, the energy associated with damage can be related to the supplied energy

$$dW_{\text{damage}} \propto dW_{\text{supplied}} \, . \tag{3.13d}$$

3.4.1 HYSTERESIS ENERGY APPROACH

During cyclic loading, energy is dissipated because of plastic deformation. The plastic strain energy, ΔW^{p} absorbed in a cycle is the area of the hysteresis loop (area 0ABC0 in Fig. 3.8) and for a Masing material is given by [1]

$$\Delta W^{\text{p}} = \frac{1-n'}{1+n'} \Delta\sigma \Delta\varepsilon^{\text{p}} \, , \tag{3.14}$$

where $\Delta\varepsilon^{\text{p}}$ is the plastic strain range, $\Delta\sigma$ is the stress range, and n' is the cyclic strain-hardening exponent. For a non-Masing material, the prediction of (3.14) may not be accurate [4, 11, 15]. The procedure for calculating the plastic strain energy for a non-Masing material is outlined in [4, 15]. Figure 3.9 shows a hysteresis loop (ABCA) for the 'minimum' proportional limit, $\Delta\sigma/2 \simeq \sigma_0$, and another (0ABCDF0) with a bigger proportional limit. The equation of the 'master' curve ABCD, where the origin A corresponds to the lower tip of the minimum proportional limit hysteresis loop is given by (2.18).

The hysteresis loop ABCDFA (Fig. 3.9) corresponding to Masing behaviour, is obtained by translating the lower branch DF0 by a quantity equal to

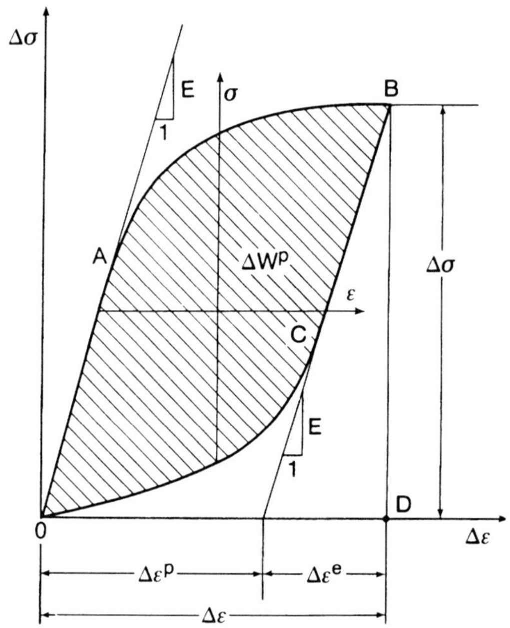

Fig. 3.8 Stable cyclic hysteresis loop for a Masing material and the corresponding nomenclature.

the increase in the proportional stress limit, i.e. $\delta\sigma_0$. The plastic strain energy of the (Masing) loop ABCDFA would be

$$\Delta W_M^p = \frac{1-n^*}{1+n^*}\,\Delta\sigma^*\,\Delta\varepsilon^p = \frac{1-n^*}{1+n^*}\,(\Delta\sigma - \delta\sigma_0)\,\Delta\varepsilon^p\,, \tag{3.15}$$

where n^* is the strain-hardening exponent of the master curve. The cross-hatched area 0AF0 in Fig. 3.9 represents the increase in the plastic strain energy per cycle due to the non-Masing behaviour. It can be evaluated in the following manner. For a coordinate system with the origin at D, the plastic strain energy for the loading from point D to A is given by

$$W_1^p = \int_0^{\Delta\varepsilon^p} \sigma^* \mathrm{d}\varepsilon^p\,. \tag{3.16}$$

If the coordinate system is translated to point F, the plastic strain energy for the loading from point F to 0 is also given by (3.16). For loading DF0, we have

$$W_2^p = \int_0^{\Delta\varepsilon^p} \sigma^* \mathrm{d}\varepsilon^p + \delta\sigma_0 \Delta\varepsilon^p\,. \tag{3.17}$$

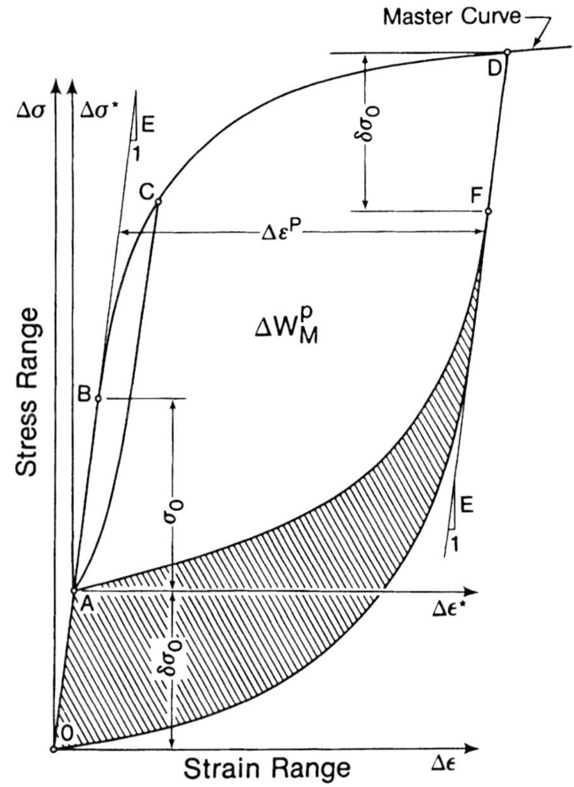

Fig. 3.9 Description of hysteresis loop for a material which does not follow a Masing hypothesis and the corresponding nomenclature (see text).

Therefore, the area 0AF0 (cross-hatched in Fig. 3.9) is equal to

$$(0AF0)_{area} = W_2^p - W_1^p = \delta\sigma_0 \Delta\varepsilon^p . \tag{3.18}$$

The plastic strain energy of the hysteresis loop 0ABCDF0 is then given by

$$\Delta W^p = \frac{1-n^*}{1+n^*}(\Delta\sigma - \delta\sigma_0)\,\Delta\varepsilon^p + \delta\sigma_0\Delta\varepsilon^p . \tag{3.19}$$

Equation (3.19) can be written in the following form:

$$\Delta W^p = \frac{1-n^*}{1+n^*}\Delta\sigma\Delta\varepsilon^p + \frac{2n^*}{1+n^*}\delta\sigma_0\Delta\varepsilon^p . \tag{3.20}$$

The increase in the proportional stress limit $\delta\sigma_0$, which describes the cyclic hardening or softening, can be easily obtained when the cyclic and master curve equations are known. The parameter $\delta\sigma_0$ (Fig. 3.9) is given by ((2.20) in Chapter 2)

$$\delta\sigma_0 = \Delta\sigma - \Delta\sigma^* = \Delta\sigma - 2K^*\left(\frac{\Delta\varepsilon^p}{2}\right)^{n^*}. \tag{3.21}$$

The plastic strain energy of the hysteresis loop measured experimentally [15] and calculated by (3.20) are shown in Figs 3.10 and 3.11, for strain- and stress-controlled tests, respectively. It can be seen from Figs 3.10 and 3.11 that the agreement is fairly good.

A series of ten dual-phase steels (ferrite–martensite microstructure) containing approximately 1.2% Mn and 0.65% Si (by weight) and varying carbon contents were investigated by Mediratta, Ramaswamy and Rama Rao [25]. They found that all dual-phase steels exhibited non-Masing material behaviour. The measured average strain energy per cycle, ΔW^p, was compared with a number of proposed methods and they concluded that only '... the equation developed by Lefebvre and Ellyin [26] (3.20) for

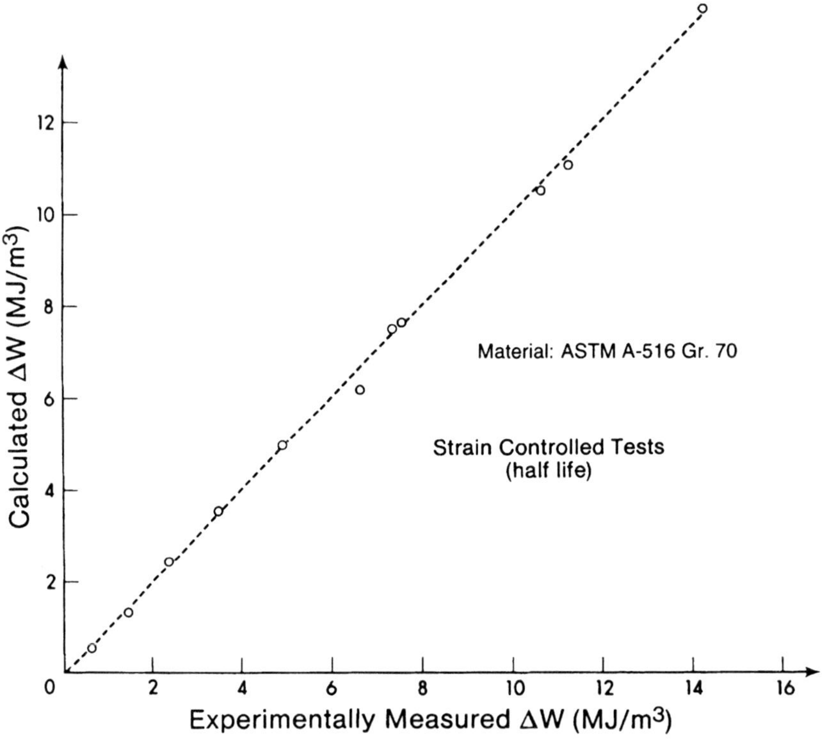

Fig. 3.10 Calculated versus experimentally-measured plastic strain energy per cycle for strain-controlled tests of ASTM A-516 Gr. 70 steel. Reproduced with permission from ASME from [15].

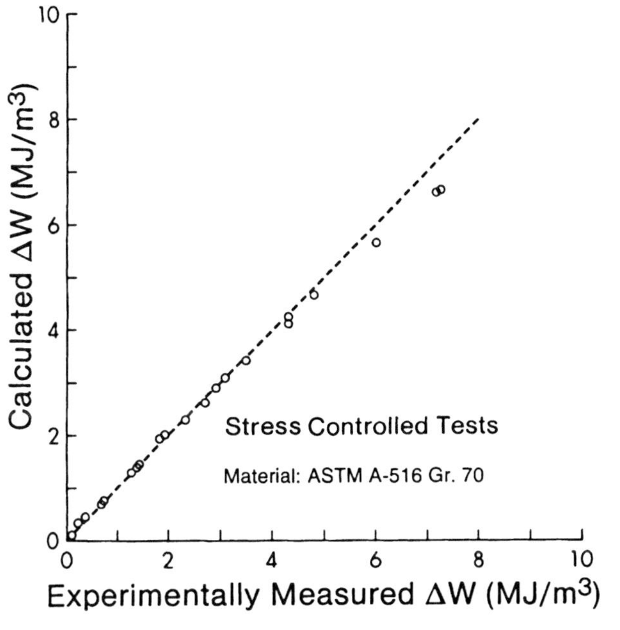

Fig. 3.11 Calculated versus experimentally-measured plastic strain energy per cycle for stress-controlled tests of ASTM A-516 Gr. 70 steel. Reproduced with permission from ASME from [15].

non-Masing materials provides energy values comparable to the measured ones'.

Equations (3.20) and (3.21) constitute a general relationship for calculating the plastic strain energy (the hysteresis loop area) under uniaxial stress conditions. Note that in the Masing material description, the elastic limit, σ_0, of the stable hysteresis loops for different strain ranges remains constant, and thus $\delta\sigma_0 = 0$. For the Masing material the standard relation (3.14) used to calculate the hysteresis energy can be obtained from (3.20) by setting $\delta\sigma_0 = 0$ and $n^* = n'$.

3.4.2 PLASTIC STRAIN ENERGY APPROACH

A major part of the hysteresis energy (3.20) is dissipated into heat and vibration, and only the remaining part causes damage in the form of slip along the crystallographic planes and dislocation movements. However, as the strain range $\Delta\varepsilon$ decreases the plastic strain range $\Delta\varepsilon^p$, and the corresponding plastic strain energy density ΔW^p, also decrease. Therefore, it becomes more difficult to measure a bulk plastic strain at small values of $\Delta\varepsilon$. The significance of the plastic strain in high-cycle fatigue was investigated by Lukáš and Klesnil [27]. They were able to show the applicability of the strain-based approach to the high-cycle fatigue region. In a similar fashion, the steady-state hysteresis loop can be used as a suitable damage parameter to formulate a proper fatigue criterion for the entire range of the fatigue life.

The hysteresis energy, ΔW^p, can be substituted into the LHS of (3.3) so that the fatigue failure criterion becomes

$$\Delta W^p = \kappa_p(2N_f)^{\alpha_p} + \Delta W_0^p, \tag{3.22}$$

where $\kappa_p > 0$, $\alpha_p < 0$ and the superscript p refers to the plastic strain. When $2N_f \to \infty, \Delta W^p \to \Delta W_0^p$, therefore ΔW_0^p in (3.22) is the hysteresis energy of the material fatigue (endurance) limit. The other two material constants, κ_p and α_p, may be determined from the best fit to experimental data. For most metals the hysteresis energy at the fatigue limit, ΔW_0^p, is very small, and for lives up to $2N_f < 5 \times 10^5$, it can be omitted. Figure 3.12 is a plot of the measured ΔW^p versus $2N_f$ for a variety of tests for the ASTM A-516 Gr. 70 steel [24]. It is clear from this figure that ΔW^p has the attributes of a proper damage parameter for the investigated life range.

To predict fatigue life from (3.22), in general, three material constants have to be determined. Neglecting ΔW_0^p, (3.22) plots as a straight line on the log ΔW^p vs. log $(2N_f)$ scale (Fig. 3.12) and α_p is the slope of this line and log κ_p is the intercept $2N_f = 1$ on the vertical axis. The LHS of (3.22) is given by (3.20) and can be calculated, if the loading condition (stress or strain amplitude) is specified, and the master and cyclic curves are known. As

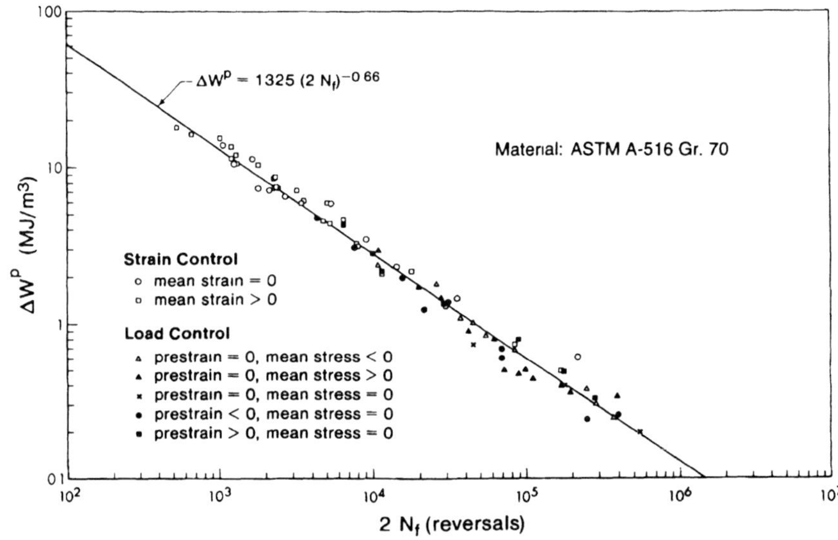

Fig. 3.12 Absorbed plastic strain energy density per cycle, ΔW^p, versus number of reversals to failure of a low alloy carbon steel (ASTM A-516 Gr. 70).

mentioned earlier, for lives close to the fatigue limit of the material, the corresponding ΔW^p approaches a very small value and is therefore difficult to measure accurately. To overcome this, the sum of the plastic and elastic strain energies can be used to advantage.

3.4.3 TOTAL STRAIN ENERGY APPROACH

In the previous section it was pointed out that in the high-cycle fatigue regime, as the strain range $\Delta\varepsilon$ decreases, $\Delta\varepsilon^p \rightarrow \Delta\varepsilon^p_{\text{fatigue limit}}$ and the corresponding cyclic plastic strain energy, $\Delta W^p \rightarrow \Delta W^p_0$. The value of ΔW^p_0 for most steels is in the range of 10^{-4} to 5×10^{-2} MJ m^{-3}, with a mean of $\sim 1.5 \times 10^{-2}$ MJ m^{-3} [28]. It is obvious that in this case, the elastic strain energy can be used as an appropriate substitute parameter in fatigue analysis; thus, using the thermodynamic basis (3.13d) instead of (3.13c).

Ellyin and Kujawski [24] have used the strain energy range per reversal, ΔW, which includes both the elastic and plastic strain energies, to unify low- and high-cycle fatigue. The expression for calculating the strain energy range, the area 0ABD0 in Fig. 3.8, can be written as [26]

$$\Delta W = \tfrac{1}{2}\Delta W^p + \tfrac{1}{2}\Delta\sigma\Delta\varepsilon\,, \tag{3.23}$$

where ΔW^p is the hysteresis energy given by (3.20). Substituting ΔW into the LHS of (3.3), the fatigue failure criterion in terms of the total strain energy range becomes

$$\Delta W = \kappa(2N_f)^\alpha + \Delta W_0\,, \tag{3.24}$$

where $\kappa > 0$ and $\alpha < 0$. When $2N_f \rightarrow \infty$, $\Delta W \rightarrow \Delta W_0$, therefore

$$\Delta W_0 = \left(\Delta W^p_0 + \frac{\Delta\sigma^2}{2E}\right)_{\text{fatigue limit}} \approx \left.\frac{\Delta\sigma^2}{2E}\right|_{\text{fatigue limit}} \tag{3.25}$$

i.e. constant ΔW_0 in (3.24) is approximately equal to the elastic energy range at a material's fatigue (endurance) limit. The other two material constants κ and α may be determined from the best fit to experimental data. The total strain energy range per reversal, ΔW, at the half-life of A-516 Gr. 70 steel is given in terms of the number of reversals to failure in Fig. 3.13. The unifying nature of the total strain energy range approach is evident from examining this figure. It can be seen that the exponents α_p and α have similar values of -0.66 and -0.65, respectively. The value of $\Delta W_0 = 0.5$ MJ m^{-3} gives a fatigue (endurance) stress range value $\Delta\sigma_{\text{lim}} = 453$ MPa, and a fatigue limit amplitude of $\sigma_{FL} = 226$ MPa.

These two approaches (using ΔW^p or ΔW as a damage parameter) have certain limitations regarding their range of applicability. They are suitable only for fully- or almost fully-reversed tests, because they are not sensitive to the mean stress. To alleviate this shortcoming, Golos and Ellyin [29, 30]

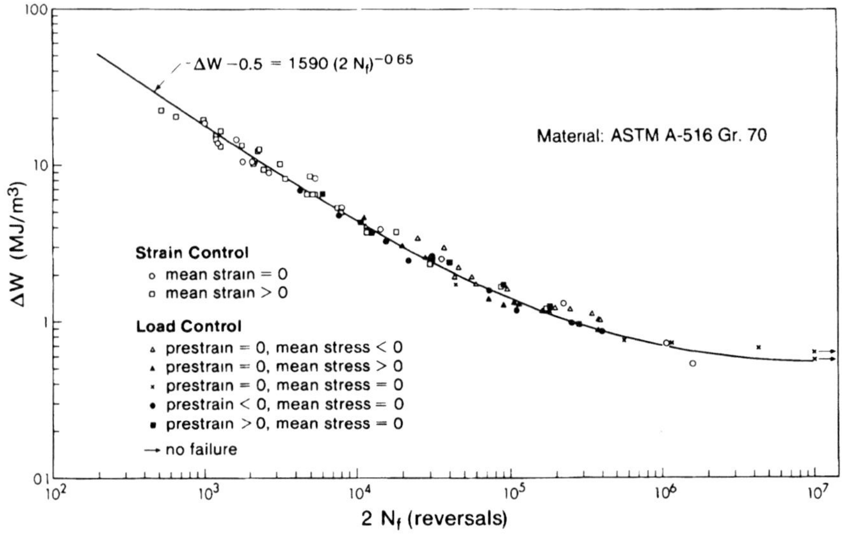

Fig. 3.13 Strain energy density range per reversal, ΔW, versus number of reversals of a low alloy carbon steel (ASTM A-516 Gr. 70).

have introduced another version of the total strain energy, ΔW^{t}. In this approach, the sum of the hysteresis energy, ΔW^{p}, and an elastic energy associated with the tensile stress, $\Delta W^{\mathrm{e+}}$, is used, i.e.

$$\Delta W^{\mathrm{t}} = \Delta W^{\mathrm{p}} + \Delta W^{\mathrm{e+}} \,. \tag{3.26}$$

For a cyclic loading with $\sigma_{\min} \leqslant 0$, the tensile elastic strain energy density per cycle, $\Delta W^{\mathrm{e+}}$ (Fig. 3.14), is determined by

$$\Delta W^{\mathrm{e+}} = \frac{1}{2E}\left(\frac{\Delta\sigma}{2} + \sigma_{\mathrm{m}}\right)^2 = \frac{\sigma_{\max}^2}{2E} \,. \tag{3.27}$$

Substituting from (3.20) and (3.27) into (3.26) the 'total' strain energy density per cycle, ΔW^{t}, for $\sigma_{\min} \leqslant 0$, is given by

$$\Delta W^{\mathrm{t}} = \frac{1-n^*}{1+n^*}\,\Delta\sigma\Delta\varepsilon^{\mathrm{p}} + \frac{2n^*}{1+n^*}\,\delta\sigma_0\Delta\varepsilon^{\mathrm{p}} + \frac{\sigma_{\max}^2}{2E} \,. \tag{3.28}$$

Introducing ΔW^{t} to the LHS of (3.3) we have

$$\Delta W^{\mathrm{t}} = \kappa_{\mathrm{t}}(2N_{\mathrm{f}})^{\alpha_{\mathrm{t}}} + \Delta W_0^{\mathrm{t}} \,, \tag{3.29}$$

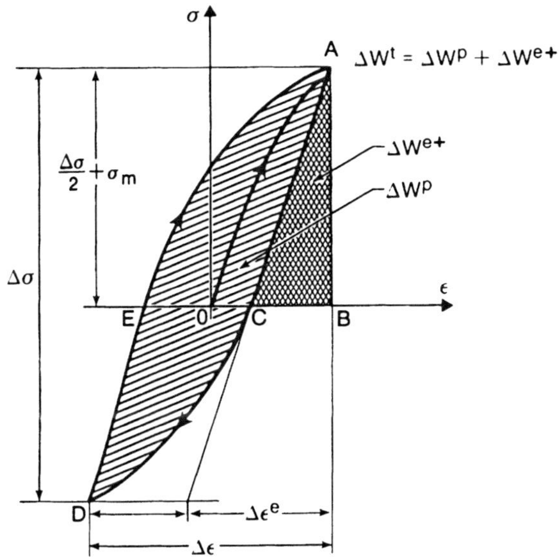

Fig. 3.14 Hysteresis loop and the definition of positive elastic strain energy.

where $\kappa_t > 0$ and $\alpha_t < 0$. When $2N_f \rightarrow \infty$, $\Delta W^t \rightarrow \Delta W^{e+}_{lim} \approx \Delta W^t_0$, thus

$$\Delta W^t_0 \approx \left(\frac{\sigma^2_{max}}{2E}\right)_{\text{fatigue limit}} \quad \text{for } \sigma_{min} \leqslant 0 \,. \tag{3.30}$$

The constant ΔW^t_0 is the tensile elastic energy at the material fatigue (endurance) limit. The other two material constants, κ_t and α_t, may be determined from the best fit to experimental data. Figure 3.15 is a plot of the measured ΔW^t versus $2N_f$ for a variety of tests of the ASTM A-516 Gr. 70 steel. It can be seen from the figure that ΔW^t is a very good correlating parameter for the life range reported therein. All the results fall in a narrow scatter band. The exponent $\alpha_t = -0.65$ and the value of ΔW^t_0 is 0.1 which corresponds to a fatigue limit stress amplitude of $\sigma_{FL} = 202$ MPa. This is in very close agreement with the experimentally determined value of $\sigma_{FL} = 205$ MPa reported in [26]. The relationship between ΔW_0 given by (3.25) and ΔW^t_0 given by (3.30) for a fully-reversed loading is $\Delta W_0 \approx 4\Delta W^t_0$. The values obtained from the fitted curves in Figs 3.13 and 3.15 differ slightly from the aforementioned geometric relationship. This is due to the choice at which ΔW_0 was calculated, e.g. in Fig. 3.13 it was taken at $2N_f = 10^7$, whereas in Fig. 3.15 it was taken at $2N_f = 10^8$.

From Figs 3.12, 3.13 and 3.15 we can see that the exponent α is not sensitive to the choice of the measure of strain energy, (3.20), (3.23) and

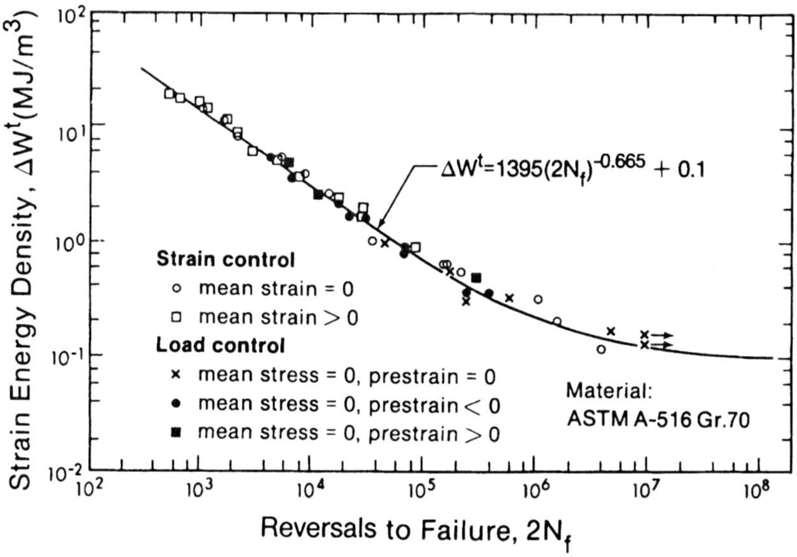

Fig. 3.15 Total strain energy density, ΔW^t, versus number of reversals of a low alloy carbon steel. Reproduced with permission from ASME from [30].

(3.26). That is, the damaging part of the strain energy, ΔW^p, $\Delta W - \Delta W_0$, or $\Delta W^t - \Delta W_0^t$, when divided by $(2N_f)^\alpha$, is a constant, equal to the intercept energy at $2N_f = 1$. This proportionality constant then depends on the choice of the damage parameter, i.e.

$$\frac{\Delta W^p}{\kappa_p} = \frac{\Delta W - \Delta W_0}{\kappa} = \frac{\Delta W^t - \Delta W_0^t}{\kappa_t} = (2N_f)^\alpha . \tag{3.31}$$

Equation (3.31) indicates that the non-dimensional damaging part of energy plotted against number of reversals to failure is a **unique** straight line on a log–log scale, irrespective of the energy form used.

The effect of mean stress included in (3.28) is restricted to those situations whereby the cyclic loading is deformation controlled or the mean stress to stress amplitude ratio is small ($\sigma_m/\sigma_a < 0.1$) in a stress-controlled mode. In the low-cycle regime when the cyclic loading is applied under the load-controlled mode, considerable creep (ratcheting) strain may result. In the latter case a different approach would be required (section 3.9).

3.4.4 NOTE ON THE MEAN-STRESS EFFECT

It should be noted that the experimental data in Figs 3.11, 3.12 and 3.15 include tests with a positive mean strain (stress) under strain control, and both positive and negative mean stress under stress-controlled conditions. All three variations of the energy criterion are generally in good agreement with the experimental data, however, the quality of agreement improves as we proceed from Fig. 3.11 to 3.12 to 3.15. The first two criteria, the plastic strain energy (3.22), and the total strain energy (3.24), do not contain the mean-stress effect in an explicit form and as such are insensitive to it.

We should emphasize here that the mean-strain (stress) effect is of second order in the high-strain regime under strain control due to the mean stress relaxation with increased cycles (Fig. 2.6 in Chapter 2, and Fig. 3.7). It then follows that any major mean-stress effect is confined to the low-strain regime where the elastic strain component is larger than the plastic one, and the hysteresis loop area is small (see also the discussion at the end of section 3.3). In such cases, the energy criterion (3.29) would be expected to correlate better with the experimental data containing a mean stress or mean strain loading.

For tests under stress control, the situation is more complicated. In the high-strain regime, a mean stress will produce a cyclic creep (ratcheting) strain which may lead to failure due to geometric instability (buckling) or creep rupture. The fatigue failure criteria discussed in sections 3.2 to 3.5, which include explicit expressions for the mean stress (or strain), are expressed by (3.7), (3.12) and (3.29). They apply only to those situations where the creep strain-rate is small and the hysteresis loop is a closed one (Fig. 2.10). The effects of mean stress and ratcheting strain will be further discussed in section 3.9.

3.5 CUMULATIVE DAMAGE

Cumulative damage, which accounts for the previous damage history and loading order effects, can be formulated through a phenomenological approach. In this case one defines a damage variable which essentially controls the fatigue process [31]. A deterministic approach will be used to describe the damage accumulation process, and in the presence of the observed scatter of data, any variable is to be interpreted with respect to its central tendency. A comparison will be made later between the various proposed models to show their relationship with respect to the general framework, which is presented first.

3.5.1 DESCRIPTION OF THE CONCEPT

Let us consider a material which is pre-damaged with its state of damage denoted by D. This state of damage could have been arrived at by various means, e.g. by fatigue, creep, machining, etc. (for a virgin material $D = 0$). The pre-damaged material is then subjected to some form of alternating loading for which a controlling damage parameter, ψ, is kept constant. The controlling damage parameter can be, for example, the range of stress, strain, or input of energy per cycle of loading. It has been stated earlier that the fatigue damage phenomenon is a highly nonlinear process. In general, it encompasses the development of persistent slip bands (PSBs), the initiation of micro-cracks and subsequent propagation to the final (sudden) fracture.

The increment of damage, dD, is a function of the actual state of damage, D, the fatigue process defined by the controlling damage parameter ψ, the previous history of pre-damage, h, and the increment of cycles, dn, and some material properties, p (e.g. $\varepsilon_f', \sigma_f', \sigma_y', \sigma_{FL}$, etc.). We will assume that during the loading process the temperature and environment are not altered. In this case, the general form of the damage process can be expressed as

$$\mathrm{d}D = F(D, \psi, h, \mathrm{d}n; p)\,, \tag{3.32a}$$

where D, ψ, h and dn are variables and p is a parameter. The variable h, the previous history of pre-damage, can be taken as the number of cycles, n, which has to be applied to the virgin material to produce the same magnitude of damage, D. Of course, this is to be achieved through the same fatigue damage process defined by ψ.

Therefore, (3.32a) can be written as

$$\mathrm{d}D = F(D, \psi, n, \mathrm{d}n; p)\,. \tag{3.32b}$$

Let us assume that the relative increase of damage, dD/D, is a linear function of the relative increase of cycles, dn/n, i.e.

$$\frac{\mathrm{d}D}{D} = f(\psi; p)\,\frac{\mathrm{d}n}{n}\,. \tag{3.33}$$

Noting that the damage process is irreversible, it follows that $f(\psi; p) \geqslant 0$ for any ψ and $p = \text{constant}$. Damage at any stage is then given by

$$\int \frac{\mathrm{d}D}{D} = \int f(\psi; p)\,\frac{\mathrm{d}n}{n}\,, \tag{3.34}$$

and since $f(\psi; p) = \text{constant}$ when $\psi = \text{constant}$ (we assume that macro-mechanical properties p are not affected by the damage process), then from (3.34) we get

$$\ln D = f(\psi; p) \ln n + C\,, \tag{3.35}$$

where C is a constant of integration. Noting that for $n = N_f$, $D = D_f$, then

$$C = \ln D_f - f(\psi; p) \ln N_f\,. \tag{3.36}$$

Substituting (3.36) into (3.35) and rearranging, we obtain

$$\frac{D}{D_f} = \left[\frac{n}{N_f}\right]^{f(\psi, p)}, \tag{3.37}$$

where the ratio $d = D/D_f$ is dimensionless damage, with $d = 0$ for the virgin material and $d = 1$ at failure. The measure of damage D can be taken arbitrarily, but the following condition should be satisfied: $D = 0$ for virgin material and $D = D_f = \text{constant}$ at failure for any fatigue process.

If for $\psi_1 > \psi_2$ the rate of damage, d, is higher in the early stage of life, then ψ_2 is less severe than ψ_1. Therefore, from (3.37) it is evident that $f(\psi; p)$ is not an increasing function of ψ, i.e.

$$f(\psi_1; p) \leqslant f(\psi_2; p) \quad \text{if} \quad \psi_1 > \psi_2 \,. \tag{3.38}$$

Based on the above analysis, the damage curves for different ψs are shown schematically in Fig. 3.16, from which the concept of the accumulation of damage can be deduced. For example, if n_1 cycles are applied at the ψ_1 level, where N_{f1} is the number of cycles to failure, then the level of damage accumulated will be represented by point A. Changing to a second loading level represented by ψ_2 with a mean life to failure N_{f2}, point A moves to A′ at the same damage level, $d_A = d_{A'}$, and cycling can proceed for n_2 cycles as damage accumulates from $d = d_{A'}$ to $d = 1$ at failure. Thus knowing the damage curves for different ψ values, it is easy to calculate n_2/N_{f2} once n_1/N_{f1} is specified, and N_{f1} and N_{f2} are known. Here the subscripts $(i = 1, 2, \ldots)$ refer to the sequence of the applied level of ψ, i.e. 1 is the first and 2 is the second level, etc. Since $d_A = d_{A'}$, we have

$$\left(\frac{n_1}{N_{f1}}\right)^{f(\psi_1, p)} = \left(1 - \frac{n_2}{N_{f2}}\right)^{f(\psi_2, p)}, \tag{3.39}$$

thus, n_2/N_{f2} can be calculated from (3.39) as

$$\frac{n_2}{N_{f2}} = 1 - \left(\frac{n_1}{N_{f1}}\right)^{f(\psi_1, p)/f(\psi_2, p)}. \tag{3.40}$$

From (3.40) it can be seen that if $f(\psi_1; p) = f(\psi_2; p)$ the linear damage rule $\sum_{i=1}^{2} n_i/N_{fi} = 1$ will be obtained for any two-level loading. Thus, loading order becomes unimportant during the fatigue process if $f(\psi_i; p) = \text{constant}$ for any ψ_i. In this case all damage lines (3.37) are reduced to a single damage curve in the normalized damage (d) versus life (n/N_f) plot (Fig. 3.16). The shape of the single damage curve depends only on the value of $f(\psi; p) = \text{constant}$. When $f(\psi; p) = 1$, for all values of ψ the damage line is a straight line (shown by the dashed line in Fig. 3.16).

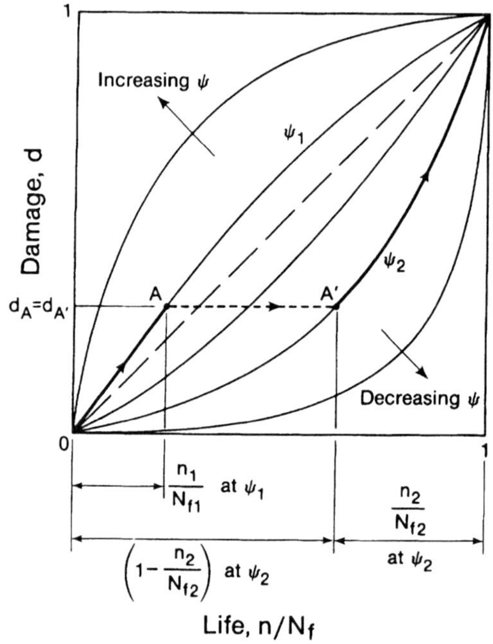

Fig. 3.16 Schematic of normalized damage curves. Reprinted by permission from [31].

3.5.2 MULTI-LEVEL CYCLIC LOADING

The method described in the previous section can easily be generalized to multi-level loading blocks. The extension to multi-level loading will be demonstrated through a three-level cyclic loading as shown in Fig. 3.17. Let the first block of loading be applied for n_1 cycles with the fatigue damage parameter ψ_1 where N_{f1} is the number of cycles to failure at ψ_1. Subsequently, the second block is applied for n_2 cycles with N_{f2} being the number of cycles to failure at ψ_2. The third block with a controlling damage parameter ψ_3, is applied until failure takes place. We wish to determine the number of cycles n_3 associated with the third level of loading to failure. Let us consider the first two loading blocks 1 and 2. Following the procedure outlined in section 3.5.1, we first have to determine the equivalent number of cycles n_2^* applied at the level ψ_2 which will cause the same amount of damage as the first stage of loading with ψ_1, i.e. $d_A = d_{A'}$ in Fig. 3.17. Using (3.37) we can write

$$\left(\frac{n_1}{N_{f1}}\right)^{f(\psi_1,\,p)} = \left(\frac{n_2^*}{N_{f2}}\right)^{f(\psi_2;\,p)}. \tag{3.41}$$

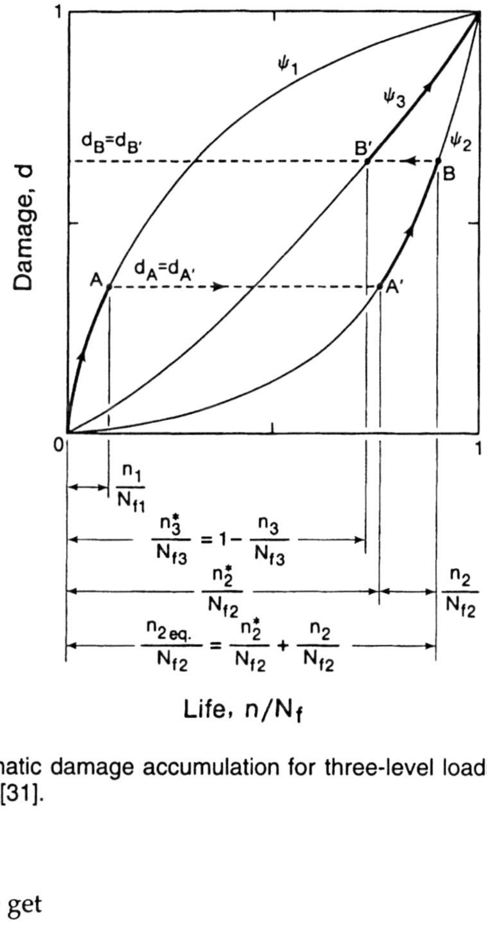

Fig. 3.17 Schematic damage accumulation for three-level loading. Reprinted by permission from [31].

From (3.41) we get

$$\frac{n_2^*}{N_{f2}} = \left(\frac{n_1}{N_{f1}}\right)^{f(\psi_1, p)/f(\psi_2, p)}. \tag{3.42}$$

The total number of equivalent cycles, n_{2eq}, at the end of the second level of loading is then given by

$$\frac{n_{2eq}}{N_{f2}} = \frac{n_2^*}{N_{f2}} + \frac{n_2}{N_{f2}}. \tag{3.43}$$

The damage d_B (Fig. 3.17) at the end of the second loading block is

$$d_B = \left[\frac{n_{2eq}}{N_{f2}}\right]^{f(\psi_2, p)}. \tag{3.44}$$

We can now treat the equivalent number of cycles (n_{2eq} at ψ_2 and n_3 at ψ_3) as a two-stage loading, i.e.

$$\left[\frac{n_{2eq}}{N_{f2}}\right]^{f(\psi_2,p)/f(\psi_3,p)} + \frac{n_3}{N_{f3}} = 1 . \tag{3.45}$$

Substituting (3.42) and (3.43) into (3.45) we obtain the following damage criterion for the three-stage loading

$$\left[\left(\frac{n_1}{N_{f1}}\right)^{f(\psi_1,p)/f(\psi_2,p)} + \frac{n_2}{N_{f2}}\right]^{f(\psi_2,p)/f(\psi_3,p)} + \frac{n_3}{N_{f3}} = 1 . \tag{3.46}$$

In the case of i-stage loading, the analysis outlined earlier can be followed in a step-by-step manner until the total accumulated damage equals unity. In general, for the i-step of loading, we have

$$\left\{\left[\left(\frac{n_1}{N_{f1}}\right)^{f(\psi_1,p)/f(\psi_2,p)} + \frac{n_2}{N_{f2}}\right]^{f(\psi_2,p)/f(\psi_3,p)} + \cdots + \frac{n_{i-1}}{N_{fi-1}}\right\}^{f(\psi_{i-1},p)/f(\psi_i,p)} + \frac{n_i}{N_{fi}} = 1 . \tag{3.47}$$

It should be noted that in the case where test conditions, e.g. temperature and environment are changing during the fatigue process, this has to be taken into account by introducing an appropriate additional variable into the damage function f. The change in the material parameter, p, which may occur as a result of the fatigue cycling, e.g. change in the fatigue limit, is generally a gradual (continuous) process. However, for the sake of convenience, any change may be divided into a number of stages which coincide with the end of each loading block.

Note that, as stated earlier, function $f(\psi;p)$ is not an increasing function of ψ, i.e. for $\psi_1 > \psi_2$, $f(\psi_1;p) <$ (or $\leqslant$) $f(\psi_2;p)$. Therefore, (3.37) states that damage is accumulated at a faster pace in the earlier stage of cyclic loading with ψ_1, than the one with ψ_2. On the other hand, damage 'production' is more rapid at the later stages of a test with ψ_2 than that with ψ_1. The fatigue process with ψ_1 can be related to the low-cycle fatigue where fatigue cracks initiate early in the life. Conversely, the test with ψ_2 can be related to the high-cycle fatigue where the crack initiation occupies a majority of the total life and fast crack propagation takes place at the end of the life. Similar observations have been made by Hunter and Fricke [32] that up to 'slip saturation' the amount of damage increases rapidly, whereas beyond this stage, the rate of increase of damage is much slower. With reference to Fig. 3.16, this observation is associated with the function $f(\psi;p)$ being bounded between 0 and 1.

The described concept of fatigue damage accumulation states that the relative damage d, rather than the absolute magnitude of damage D, account for the sequence effects in a multi-level loading test [33]. The relative damage, d, herein encompasses the damage which takes place at the atomic, micro- and macro-levels.

3.5.3 SPECIFICATION OF THE DAMAGE FUNCTION WITH REFERENCE TO OTHER APPROACHES

The essential step in the specification of the fatigue damage accumulation concept defined earlier is the identification of the function $f(\psi;p)$. In general, a particular form of the damage function can be deduced from the experimental data, or a postulate can be put forward. For example, the simplest damage function $f(\psi;p)=1$ was postulated by Palmgren [34] and Miner [16], which gives a linear damage accumulation law

$$\sum \frac{n_i}{N_{fi}} = 1 \,, \tag{3.48}$$

where i denotes the number of loading levels.

As discussed earlier, the linear damage accumulation rule can also be obtained from (3.47) when the function $f(\psi;p)$ = constant, which results in a single curve for d vs. n/N_f (Fig. 3.16).

A stress-dependent theory was proposed by Marco and Starkey [35]. The fatigue damage process was described by a power law relationship of the form

$$D = \left(\frac{n}{N_f}\right)^{x_\sigma} , \tag{3.49}$$

where the exponent x_σ is a variable quantity whose magnitude is dependent on the applied stress condition. Comparing (3.37) with (3.49) it can be seen that $f(\psi;p) = x_\sigma$. The values of x_σ can be determined from two-step fatigue tests where the cumulative damage law is in the form of

$$\left(\frac{n_1}{N_{f1}}\right)^{x_{\sigma 1}/x_{\sigma 2}} + \frac{n_2}{N_{f2}} = 1 \,. \tag{3.50}$$

Let us now consider some experimental observations on fatigue curves representing different degrees of pre-damage. Bennett [33] published in 1946 the results of fatigue tests on specimens cycled at two successive levels (Fig. 3.18). Line PA is the semi-log S–N curve of the virgin material. Specimens prestressed at ± 372 MPa (± 54 ksi) for 33% of the mean life were then tested at other stress levels to failure. This resulted in the remaining S–N curve PB with a higher slope than PA of the virgin material. In the case when 90% of the life was applied at the same stress level (± 372 MPa) the remaining S–N curve PC with the highest slope was obtained. All three S–N curves converge to a single point P, with the coordinate ± 413.4 MPa (± 60 ksi) at about 20 000 cycles. The concept of convergence of the

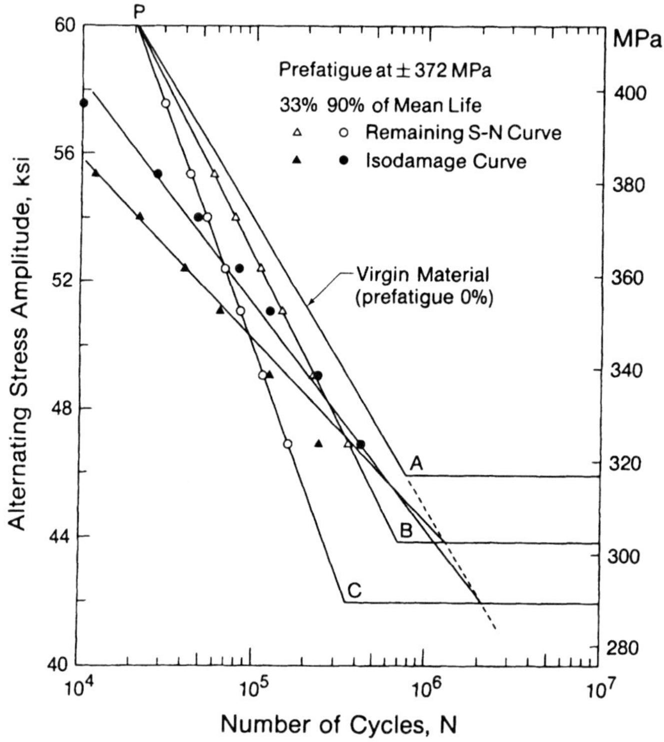

Fig. 3.18 Fatigue curves for virgin and prefatigued specimens of X4130 steel, from [33].

remaining *S–N* curves was an essential feature of the cumulative fatigue damage approach proposed by Manson *et al.* [36] assuming a rotation of the *S–N* curve. Knowing the convergence point, P (assumed a material constant) and subtracting the prestress cycles from the life of the virgin material at the prestress level, the remaining *S–N* curve is obtained by passing a straight line through this point and P, the point of rotation. Once the rotation point is known (it can be estimated as the point in the *S–N* curve at a stress equal to the ultimate tensile strength [36]) the effect of multi-level loading can be determined by successive step-by-step application of the two-level loading procedure. It is evident from Fig. 3.18 that the 'apparent' fatigue limit of the remaining *S–N* curves decreases with the increasing pre-damage.

More recently Subramanyan [37] in 1976 and Hashin and Rotem [38] in 1978 proposed an alternative approach of the convergence concept of isodamage lines at the fatigue limit (knee point) of the *S–N* curve of the virgin material. Srivatsavan and Subramanyan [39] later modified the method by taking into account a reduction in the fatigue limit of the material due to overstressing. The concept of a set of isodamage curves is based on the following observations [37]. It is assumed that the zone of fatigue damage has an upper bound which is the *S–N* curve of the virgin material with $D_f = \tan\theta_f$ (Fig. 3.19). The lower bound is the horizontal line through the fatigue (endurance) limit $D = 0$, shown by a dashed line in Fig. 3.19. The number of cycles required to reach a particular state of damage, $D = \tan\theta$, at different stress levels, fall on a smooth curve which tends to be almost linear. These lines also appear to converge in the neighbourhood of the fatigue limit of the material [32, 33, 40].

Now suppose that a specimen is prestressed for n_1 cycles at the stress level σ_1 for which failure will occur at N_1. A line joining point (n_1, σ_1) and the fatigue limit (knee point), represents an isodamage line (Fig. 3.19). The slope, $\tan\theta$, of this line is a measure of damage, D_1. The ratio $\tan\theta/\tan\theta_f$ is the normalized damage d where $d = 0$ for the virgin material and $d = 1$ represents fatigue failure, i.e. the *S–N* curve of the material. The specimen, after the application of n_1 cycles at σ_1, is further cycled at the σ_2 level until failure occurs. It is necessary to determine the number of cycles n_2 at the σ_2 level. By moving from point 1 to point 2 on the isodamage curve $D_1 = D_2$ we can easily find the equivalent number of cycles n_{21} at σ_2 for which $D_2 = D_1$, and $n_2 = N_2 - n_{21}$ (Fig. 3.19). Thus, by subtracting the remaining number of cycles (open symbols on the remaining *S–N* curves in Fig. 3.18) from the total life for a particular stress level, we can calculate the number of cycles (filled symbols in Fig. 3.18) required to reach a particular state of

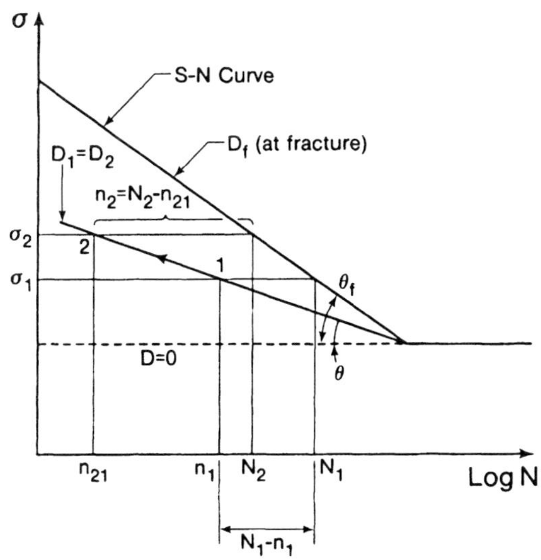

Fig. 3.19 Schematic of isodamage curves. Reproduced with permission from [31].

damage – isodamage points. It can be seen that the isodamage points at different stress levels, tend to fall on straight lines which converge at the 'apparent' fatigue limit on the *S–N* curve of the virgin material. Figure 3.18 shows that the remaining *S–N* curves can be transformed to a set of isodamage curves passing through the 'apparent' (reduced) fatigue limit.

Transforming extensive data on the SAE 4130 steel of Manson *et al.* [36] (shown in inset to Fig. 3.20) in the manner described above, we obtain the isodamage curves depicted in Fig. 3.20. On the semi-log scale, in general, *S–N* curves may not be linear for the full range of low- to high-cycle

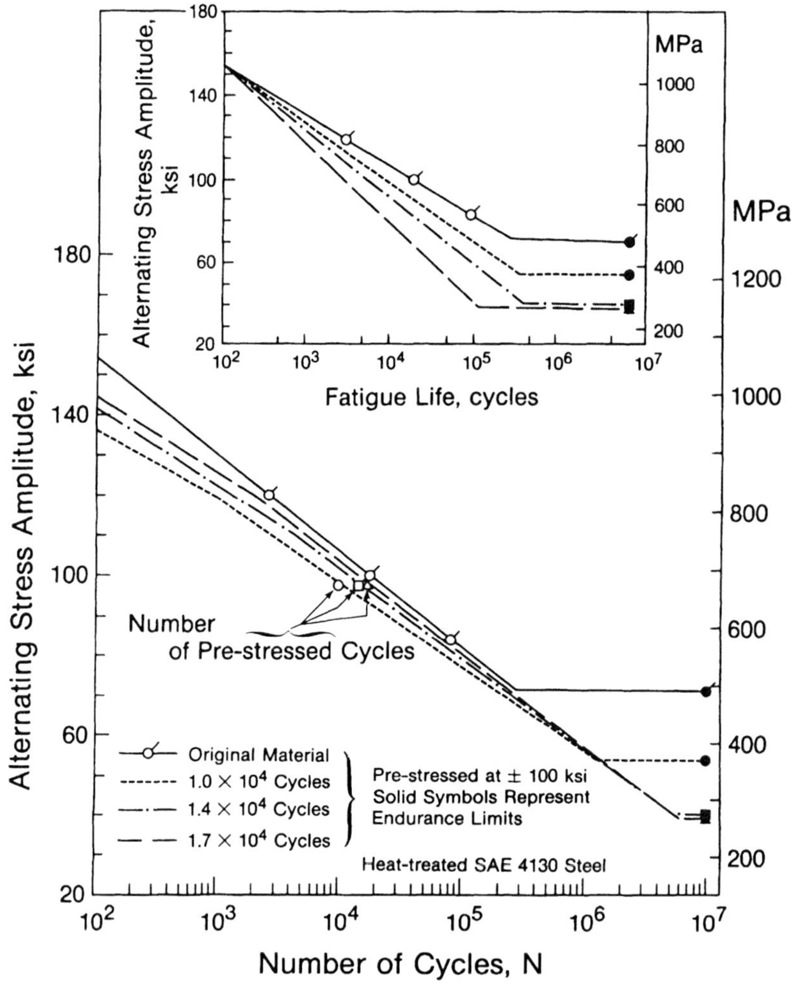

Fig. 3.20 Fatigue curves for virgin and prefatigued specimens. Reproduced with permission from [31].

regimes. Isodamage curves can also deviate from linearity on a semi-log scale. From Figs 3.18 and 3.20 it is evident that the isodamage curves do not converge to the fatigue (endurance) limit of the virgin material. This type of isodamage curve, with varying apparent fatigue limit, were recently used by Kujawski and Ellyin [41] to evaluate the accumulation of damage resulting from the application of cyclic stress (or strain) sequences of varying amplitude. The damage variable was based on the plastic strain energy ($\psi = \Delta W^p$ where ΔW^p is the hysteresis energy per cycle) as a function of cyclic life in the form of [15]

$$\Delta W^p = \kappa_p (N_f)^{\alpha_p}, \tag{3.51}$$

where κ_p is the energy value at which the straight line with slope α_p of the life curve intersects the energy axis. This concept includes both the crack initiation and propagation stages and is based on the following assumptions:

- For the crack initiation stage, the material has a fatigue limit defined by the plastic strain energy ΔW_e^p or $\Delta\varepsilon_e^p\,\Delta\sigma_e$, and the life N_e.
- For the crack propagation stage, the material has an 'apparent' (reduced) fatigue limit defined by ΔW_e^{p*} or $\Delta\varepsilon_e^{p*}\,\Delta\sigma_e^*$ and life N_e^*, where N_e^* is taken from the extrapolation of the original log ΔW^p vs. log N_f curve to the 'apparent' fatigue limit ΔW_e^{p*}. The apparent stress range, $\Delta\sigma_e^*$, can be obtained from the fracture mechanics approach, and $\Delta\varepsilon_e^{p*}$ from the cyclic stress–strain relation, once $\Delta\sigma_e^*$ is known. Alternatively, the apparent fatigue limit in terms of ΔW_e^{p*} can be obtained from a two-step test, provided the ΔW^p vs. N_f curve is known.
- The life curve and isodamage curves are assumed to be straight lines on a log ΔW^p vs. log N_f plot and they all converge to the fatigue limit, N_e, for the crack initiation stage and to N_e^* for the crack propagation stage.
- The transition between crack initiation and propagation stages can be expressed in terms of the critical damage curve, which can be associated with the French curve [42].

The advantage of the plastic strain energy in comparison to stress as a fatigue damage variable, ψ, is that ΔW^p (the area of the hysteresis loop) is almost constant during the life under strain-controlled tests. Furthermore, the interrelation between the strain and stress (path dependency of fatigue damage) can be taken into account in the energy approach. Also, a straight line plot of log ΔW^p vs. log N_f(or $2N_f$) reflects more closely the trend of experimental data in low- and high-cycle fatigue [15], Fig 3.12. However, there are some limitations to the ΔW^p implementation, namely its insensitivity with respect to the hydrostatic pressure and the difficulty of its measurement near the fatigue (endurance) limit, which was discussed in section 3.4.3. Note that at the fatigue limit $\Delta\varepsilon^p \rightarrow \Delta\varepsilon_e^p$, where $\Delta\varepsilon_e^p$ is of the order of 10^{-5} to 5×10^{-4} for most metals.

To overcome the above limitation, as pointed out earlier, attempts have been made to use either the range of total strain energy, $\Delta W = \Delta W^p + \Delta W^e$ [24], or the total strain energy density as a sum of ΔW^p and the tensile part of the elastic strain energy, $\Delta W^t = \Delta W^p + \Delta W^{e+}$ [29, 30], to correlate fatigue lives. All the approaches based on isodamage curves give the cumulative damage law in the form of (3.40) and (3.47) for two- and multi-level tests, respectively.

The form of the damage function $f(\psi; p)$ depends on the chosen ψ and p, and also on the representation of the life curve on a semi-log or log–log coordinate system. Various forms of $f(\psi; p)$ deduced from earlier investigations are summarized in Table 3.1.

It should be noted that the double linear damage rule (DLDR) proposed by Manson [44] can, in principle, be obtained by an appropriate approximation of the damage curves (Fig. 3.16) by two straight segments and using the linear damage law for each segment separately. However, this linearization is not unique and would involve an element of judgement.

Comparison of the different approaches with the experimental data can be found in [29, 30, 35–39, 41]. As an example, Miller and Zachariah's [45] two-level test data for the torsional cyclic straining of thin-walled steel tubes are shown in Fig. 3.21. The dotted curves represent the double cumulative damage rule developed in [46] in terms of crack initiation and propagation. The Hashin and Rotem approach used by Hashin and Laird [47] is shown by dashed lines. The approach based on the plastic strain energy [41] is depicted by full lines in this figure.

The two- and multi-level test results of a pressure vessel steel, A-516 Gr. 70, reported by Golos and Ellyin [30] are shown in Fig. 3.22 and summarized in Table 3.2, respectively. In this approach the total strain energy density ΔW^t (3.28) was used as the controlling damage variable, ψ. It can be seen that the correlation with the experiments is fairly good.

3.5.4 DETERMINATION OF THE CRITICAL DAMAGE CURVE

In formulating damage accumulation in terms of the cyclic strain energy, a distinction was made between the crack initiation and propagation phases. In section 3.5.3 we indicated that the transition between crack initiation and propagation stages can be expressed in terms of a critical damage curve. Such a curve essentially represents the demarcation between the macroscopically undamaged and damaged states when cycled at a given strain energy (or other damge parameter) level. In other words, the cracks, if initiated as a result of cyclic loading, have not reached a size whereby the macro-mechanical and cyclic properties are affected.

To establish a critical damage curve, we proceed as follows. A specimen is cycled at a predetermined total strain energy density, ΔW_i^t (or any other damage parameter), for n_i cycles. The level of the strain energy is then

Table 3.1 Various definitions of damage variables, the required material property and the corresponding damage function

Total life representation (S–N curve)	Controlling damage variable, ψ	Material parameter, p	Damage function $f(\psi; p)$	References
$\sigma_a - \log N_f$	σ_a	σ_e	$\frac{1}{(\sigma_a - \sigma_e)}$	Subramanyan, 1976 [37]
$\log \sigma_a - \log N_f$	σ_a	σ_e	$\frac{1}{\log(\sigma_a/\sigma_e)}$	Hashin and Rotem, 1978 [38]
$\sigma_a - \log N_f$	σ_a	σ_e^*	$\frac{1}{(\sigma_a - \sigma_e^*)}$	Srivatsavan and Subramanyan, 1978 [39]
$\log \Delta W^p - \log N_f$	ΔW^p	ΔW_e^{p*}	$\frac{1}{\log(\Delta W^p/\Delta W_e^{p*})}$	Kujawski and Ellyin, 1984 [41]
$\log \Delta W^t - \log N_f$	ΔW^t	ΔW_e^+ (or $\Delta W_e^{+\prime}$)	$\frac{1}{\log(\Delta W^+/\Delta W_e^+)}$ or $\frac{1}{\log(\Delta W^+/\Delta W_e^{+\prime})}$	Golos and Ellyin, 1987, 1988 [29, 30]
	Not specified. (Damage curves based on analogy to early crack growth)		$(N_f)^{0.4}$	Manson and Halford, 1981 [43]

where:
σ_a = stress amplitude;
σ_e = stress amplitude at fatigue limit of a virgin material;
σ_e^* = stress amplitude at 'apparent' fatigue limit;
ΔW^p = hysteresis energy (area of hysteresis loop);
ΔW_e^{p*} = hysteresis energy at 'apparent' fatigue limit;
ΔW^t = total strain energy density; $\Delta W^p + \sigma_{max}^2/2E$ (for $\sigma_{min} \leqslant 0$);
ΔW_e^+ = tension part of elastic strain energy at fatigue limit used for the crack initiation stage;
$\Delta W_e^{+\prime}$ = constant $< \Delta W_e^+$ – an equivalent fatigue limit for the crack propagation stage.

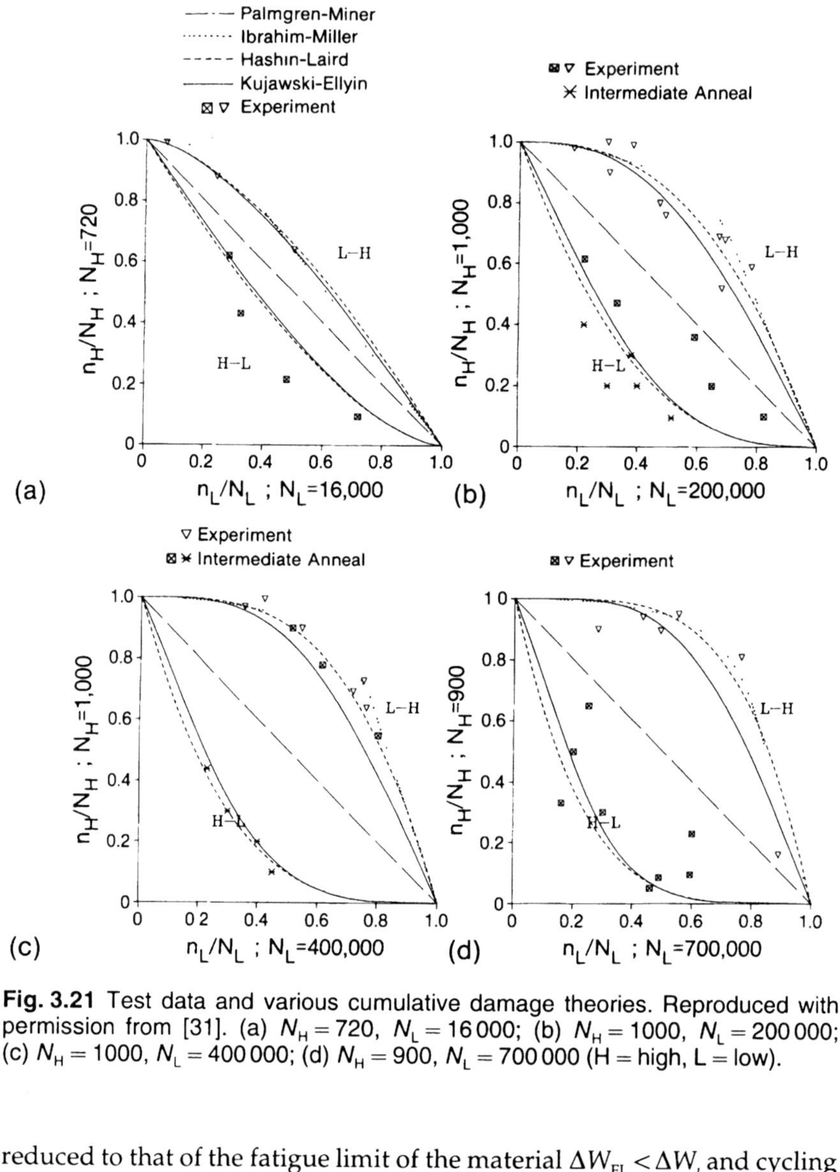

Fig. 3.21 Test data and various cumulative damage theories. Reproduced with permission from [31]. (a) $N_H = 720$, $N_L = 16\,000$; (b) $N_H = 1000$, $N_L = 200\,000$; (c) $N_H = 1000$, $N_L = 400\,000$; (d) $N_H = 900$, $N_L = 700\,000$ (H = high, L = low).

reduced to that of the fatigue limit of the material $\Delta W_{FL} < \Delta W_i$ and cycling is continued as shown in the inset to Fig. 3.23. If the specimen does not fail, say after 10^6 to 10^7 cycles, then the initial level of cumulative cycling would not have produced a crack which could propagate at the fatigue limit. This is because short and long cracks will both propagate at stress levels below a material's fatigue limit (Chapters 7 and 9). We continue the process by subjecting a similar specimen to the same level of energy, ΔW_i^t, but increase the initial number of cycles from n_i to n_j where $n_j > n_i$ and follow

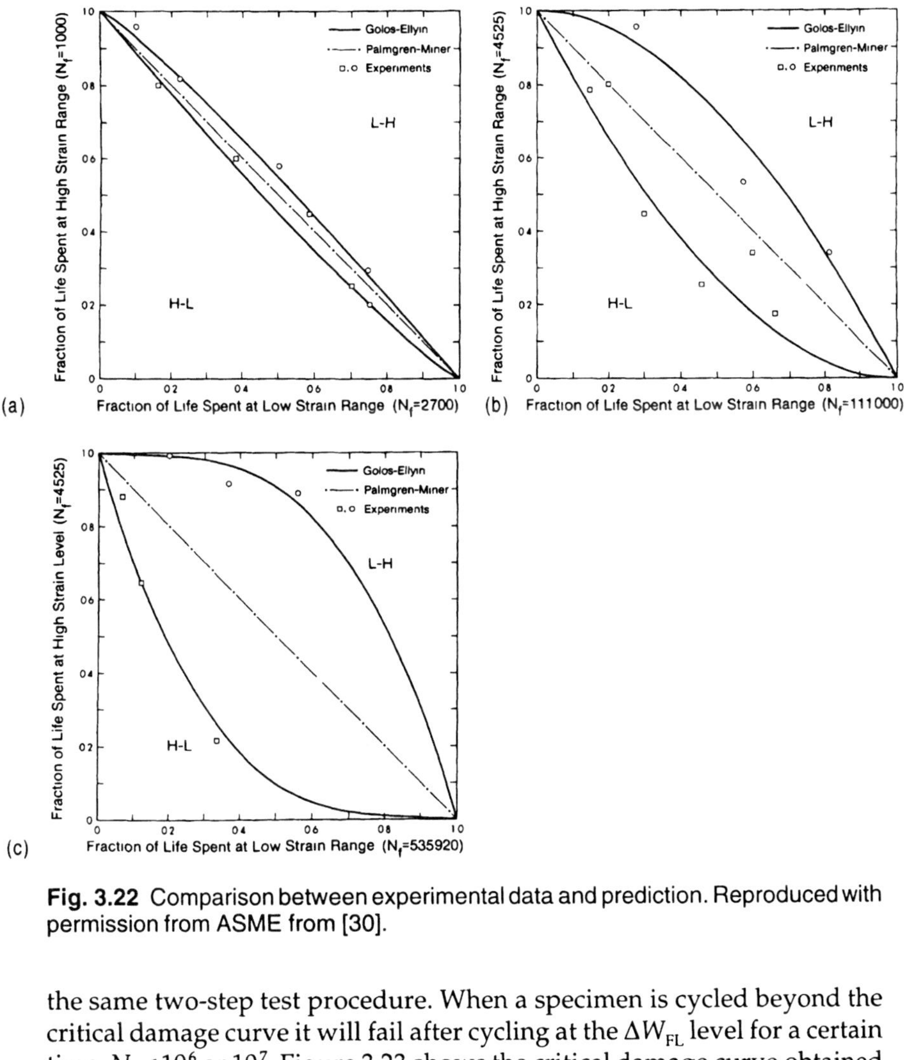

Fig. 3.22 Comparison between experimental data and prediction. Reproduced with permission from ASME from [30].

the same two-step test procedure. When a specimen is cycled beyond the critical damage curve it will fail after cycling at the ΔW_{FL} level for a certain time, $N_f < 10^6$ or 10^7. Figure 3.23 shows the critical damage curve obtained in this manner. In this figure the filled circles correspond to specimens which failed during cycling at the ΔW^t_{FL} level at less than 2×10^6 cycles. The open circles correspond to specimens which did not fail after 2×10^6 cycles. For the sake of comparison, the best fit line of the life curve (Fig. 3.15) is drawn in the figure.

All tests reported here were conducted at an average strain rate $\dot{\varepsilon} = 5 \times 10^{-3}\,s^{-1}$, and subjected to fully-reversed strain-controlled sinusoidal load. It can be seen that at the high strain (low-cycle fatigue) regime about 10–15% of the total life is spent in the critical crack initiation phase, whereas in the low strain (high-cycle) regime the critical crack initiation

Table 3.2 Multi-level cumulative damage strain-controlled fatigue test results of A-516 Gr. 70 carbon steel [30]

Stage 1		Stage 2		Stage 3		Stage 4		Predicted value of cycle ratio at the last stage	Sequence of loading
Strain amplitude $\Delta\varepsilon_1/2$ (%)	Cycle ratio n_1/N_{f2}	Strain amplitude $\Delta\varepsilon_2/2$ (%)	Cycle ratio n_2/N_{f2}	Strain amplitude $\Delta\varepsilon_3/2$ (%)	Cycle ratio n_3/N_{f3}	Strain amplitude $\Delta\varepsilon_4/2$ (%)	Cycle ratio n_4/N_{f4}		
0.800	0.50	0.600	0.30	0.200	0.07			0.10	H–L
0.800	0.50	0.200	0.18	0.600	0.30			0.16	H–L–H
0.200	0.18	0.600	0.50	0.800	0.62			0.51	L–H
0.200	0.18	0.800	0.62	0.600	0.54			0.33	L–H–L
0.800	0.30	0.600	0.19	0.400	0.21	0.200	0.10	0.12	H–L
0.200	0.09	0.800	0.10	0.400	0.54	0.600	0.59	0.32	L–H–M
0.800	0.10	0.200	0.50	0.600	0.10	0.400	0.39	0.13	H–L–M
0.200	0.27	0.400	0.27	0.600	0.37	0.800	0.44	0.34	L–H

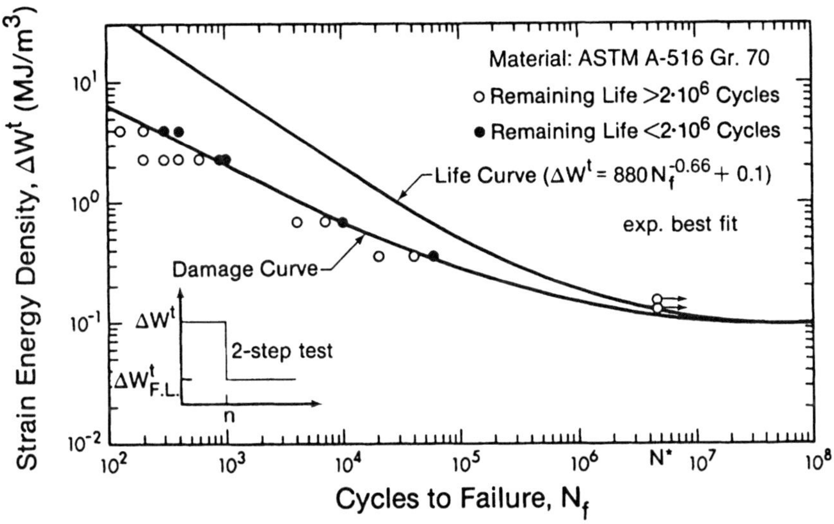

Fig. 3.23 Determination of the critical damage curve and its relation to the life curve. Reproduced with permission from [30].

phase occupies a major part of the total life [30]. Note that a relation of the form (3.29) can also describe the critical damage curve, although obviously the material constants κ_t and α_t would be different for the two curves representing the critical damage and life curves.

It is worthwhile mentioning again that in the phenomenological approaches described in sections 3.2 to 3.4, the general form of the fatigue life equations is independent of the definition of the failure (or crack size); the fitted coefficients are, on the other hand, dependent on the adopted definition.

3.5.5 NOTE ON THE DAMAGE CONTROLLING VARIABLE, ψ

One of the current research topics in fatigue failure is the identification of a suitable damage variable, ψ, which can be used to describe the fatigue damage process. This will consequently unify low- and high-cycle fatigue in a way consistent with the nonlinear fracture mechanics theory. Once this damage variable ψ is adequately described then the effect of loading sequence can be determined in a manner described in this section. From our previous discussion and based on experimental observation, the energy approach appears to be the most promising one. Plasticity, which is responsible for the dissipation of strain energy, is independent of the hydrostatic (mean) stress state whereas the damage process is highly dependent upon it. Therefore, any fatigue loading path (cycle in terms of stress) has to be considered in the stress space with respect to the hydrostatic stress components. When the hydrostatic stresses are positive, the damage is accumulated at a higher rate due to its contribution to the opening of the defects (cracks) with the increased energy input, whereas for compressive hydrostatic stresses, the defect (crack) will tend to close with the applied energy input, and consequently the rate of damage will be smaller.

Let us consider a loading path ABCDA as shown in Fig. 3.24, in the principal stress space. Points A and C represent the intersection of the loading path with the Π-plane. The Π-plane is a plane normal to the hydrostatic pressure axis with a unit normal $\underset{\sim}{n}$: $(1/\sqrt{3}, 1/\sqrt{3}, 1/\sqrt{3})$ passing through the origin of the principal axes. The points of intersection A and C divide the path into two segments ABC and CDA for which the hydrostatic pressure is positive and negative, respectively. Therefore, based on the above arguments of the opening and closing nature of defects (cracks), for the same increase of input energy, the rate of damage will be higher for the ABC path than that of CDA. Therefore any damage

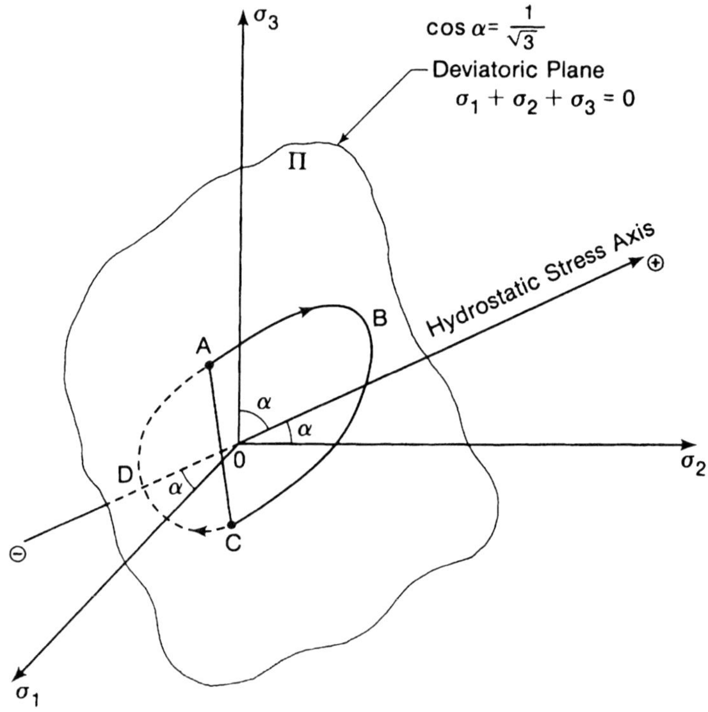

Fig. 3.24 A cyclic loading path in the principal stress space in relation to the hydrostatic stress. Reproduced with permission from [31].

parameter ψ should reflect this dependency on the hydrostatic pressure. Consequently, when the stress path is above the Π-plane, it will be more damaging than that below the plane, for the same amount of input energy.

Based on the above argument, it is easily understood why fully-reversed biaxial loading is more damaging than uniaxial which in turn is more damaging than pure shear, for the same amount of input energy per half cycle (Chapter 4). Likewise, the effect of the mean stress and negligible effect of the mean strain in low-cycle fatigue in the uniaxial stress state can be explained by the above-mentioned hydrostatic stress effect. In the case of the mean plastic strain (in low-cycle fatigue), the mean stress relaxes and consequently the hydrostatic stress component is reduced with respect to the Π-plane.

3.6 TIME-DEPENDENT FATIGUE

At elevated temperatures most metals show some form of time dependency. The fatigue resistance of metals is generally affected by a temperature rise and shorter lives are observed in comparison to the time-independent fatigue where the influence of time and loading rate are negligible. The interest in recent years in fatigue resistance at high temperatures stems generally from the development of efficient engines for aerospace applications, nuclear reactor components – especially those for fast breeder reactors – and steam turbines in power plants.

Alloys have been developed primarily to give high creep strength under static loading at high temperatures. Early tests to assess the fatigue resistance of high-temperature alloys were conducted in air. Fortunately, it was found that alloys which have good resistance to creep deformation generally exhibit acceptable fatigue strength up to a certain limiting temperature. For example, ferritic steels with molybdenum, vanadium or chromium alloying elements are more suitable for use at temperatures below 600°C while austenitic steels with higher chromium content have better performance at temperatures above 600°C.

Testing in air at high temperatures introduces an additional environmental effect (oxidation) superimposed on those of creep and fatigue. Thus, to eliminate this environmental effect, fatigue tests are carried out at high vacuum [49, 50]. For the purpose of illustration, Fig. 3.25, adapted from [50], shows that for the continuous **symmetric** waves, rate dependency is observed when testing in air, while the strain-rate influence is almost absent in a high vacuum (0.1 μPa). There is ample experimental evidence showing the same trend, e.g. review papers [51–56], where references to other works are listed. Thus, a high vacuum environment ($\leqslant 10\ \mu$Pa) provides a medium whereby the creep–fatigue interaction can be investigated. In such an environment any manifestation of time dependency will be due to creep.

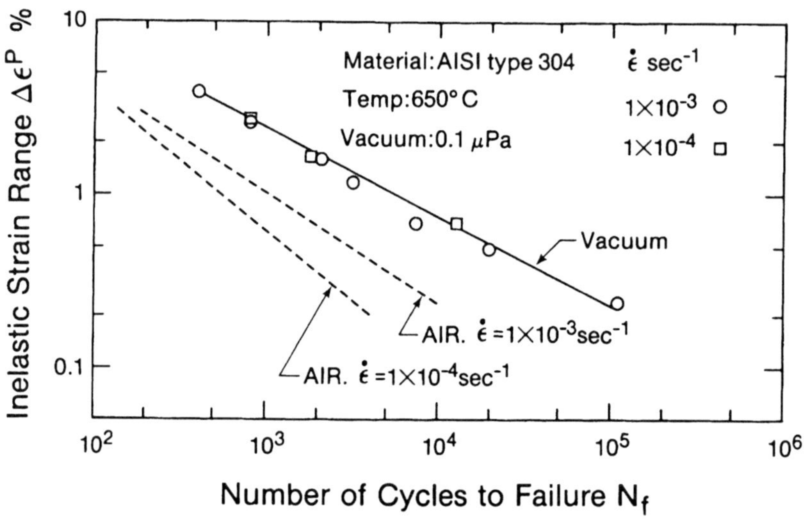

Fig. 3.25 Symmetric continuous tests of stainless steel AISI type 304 conducted in a high vacuum and air for two different strain rates. Reprinted with kind permission from [70].

To determine the effect of temperature on fatigue behaviour, tests are conducted using a variety of waveshapes [51]. Some typical waveshapes are shown in Fig. 3.26. As indicated earlier, the effects of test frequency cycling are not significant in a high vacuum environment for symmetric wave forms. (Frequency measured in Hz (cycles per second) is not a good measure for time-dependent fatigue, strain-rate (s^{-1}) is a better measure). Therefore, one would expect any difference in life between symmetric and asymmetric cycles to be caused by the creep behaviour of the material. There is a large amount of experimental data which indicate that a reduction in fatigue life occurs when the **total time** spent under tensile stress exceeds that of compressive stress [49–55].

3.6.1 EFFECT OF THE WAVE FORM

Let us now examine the wave shape effect on the fatigue life, excluding for the time being the environmental effects, i.e. tests performed under high vacuum. The effect of the strain-rate can be investigated by comparing tests performed under balanced (symmetric) and unbalanced (unequal forward and reversed strain-rates) wave forms at the same temperature. Experimental results indicate that fatigue life is reduced in slow–fast cycles and is affected to a lesser degree in fast–slow cycles in comparison to the symmetric cycle. Of particular interest is that in the case of the slow–fast wave shape, the fractured surface is intergranular, and cavity damage is

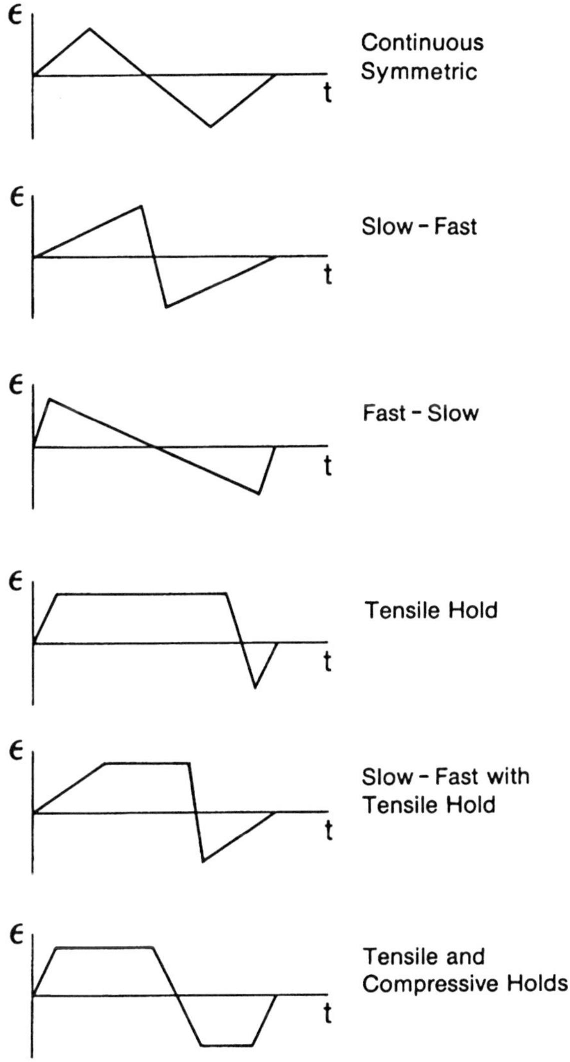

Fig. 3.26 Various strain wave forms at which tests are performed to assess creep–fatigue interaction.

observed in the interior of the material. However, in the equal rate or fast–slow wave shapes, the fracture path is generally transgranular, e.g. [44, 49–51].

To investigate the effect of hold-time, tests are generally conducted with hold-times imposed at either the tension or compression peak of each cycle (Fig. 3.26). A reduction in the fatigue life is observed for the hold-time under tension and the cracked surface is intergranular due to the creep damage sustained during the tensile hold period. A recovery is observed when a compressive peak strain is superimposed on a tensile hold experiment. Coffin [52] contends that in a high vacuum environment creep damage by cavity growth occurs only when the time spent under tensile stress (slow–fast or tension hold-time) exceeds that of the compressive stress.

3.6.2 LIFE PREDICTION METHODS

Life prediction methods for high-temperature fatigue can be grouped into five main categories, namely: linear summation of time and cycle [57, 58]; modifications of low-temperature fatigue relationships [49, 59]; ductility exhaustion [60]; strain range partitioning [61]; and continuous damage parameter [62–67].

The linear summation rule of time and cycle leads to conservative estimates of life and has been widely used because it is the procedure which is advocated by the ASME Boiler and Pressure Vessel Code [68].

The frequency-modified fatigue failure approach, and more recent extensions termed frequency separation methods, have been largely developed by Coffin and co-workers [51]. In the earlier versions, i.e. frequency-modified fatigue criteria, it was assumed that the period of the cycle, but not the wave form, affects the creep–fatigue life. However, as discussed in section 3.6.1, the wave form has a significant effect on the fatigue life at high temperatures, and thus, in the newer versions of the method the frequency was substituted by the tensile loading frequency at first, and a more involved stress range dependent criterion [51].

The strain range partitioning method was developed on the basis of experimental data to establish a strain range–life relationship by Manson and co-workers [61]. Four sets of cyclic life experiments are required to specify the creep–fatigue criterion. The inelastic strain range is partitioned into completely reversed plasticity, $\Delta\varepsilon_{pp}$, completely reversed creep, $\Delta\varepsilon_{cc}$, mixed tensile plasticity reversed by compressive creep, $\Delta\varepsilon_{pc}$, or the tensile creep reversed by the compressive plasticity, $\Delta\varepsilon_{cp}$. A given hysteresis loop is broken into the pp, cc and pc (or cp) strain range components, and using an approximate linear damage rule, the life prediction is made based on the lives of the constituent strain range components. This method predicts two saturation bounds, an upper bound at high strain-rates, pp, and a lower bound at lower frequencies. The later, as well as the basic testing method, have been questioned, e.g. Coffin [51].

More recent methods have been based on defining a damage function to account for the material degradation. The pioneering work of Kachanov [62] has had a major influence in treating creep damage. The continuum damage approach to the creep–fatigue damage process has been reviewed by Murakami [55]. In principle, an incremental damage parameter is introduced as a function of stress and stress increment and its evolution with the number of cycles is postulated to fit the experimental data. Examples of this type of approach can be found in Majumdar and Maiya [63, 64], Lemaitre and Plumtree [65], Saitoh and Krempl [66] and Asada *et al.* [67], among others.

Table 3.3 summarizes some of the above-mentioned creep–fatigue failure criteria. They have also been compared with the experimental data in [56]. A large deviation between the predictions and bench mark experiments was noted. This work thus pointed out the need for further improvements in predictive models.

3.7 A MECHANISM-BASED DAMAGE FUNCTION FOR TIME-DEPENDENT FATIGUE

In a situation involving mechanical and/or thermal cycling (creep–fatigue) we assume that the damage is comprised of two functions, namely a rate-independent (fatigue) damage component ϕ_{i} and a rate-dependent (creep) damage component ϕ_{d}, i.e.

$$D = D(\phi_{\mathrm{i}}, \phi_{\mathrm{d}}) \,. \tag{3.52}$$

The rate of damage accumulation per cycle is then

$$\frac{\mathrm{d}D}{\mathrm{d}N} = \frac{\partial D}{\partial \phi_{\mathrm{i}}} \frac{\mathrm{d}\phi_{\mathrm{i}}}{\mathrm{d}N} + \frac{\partial D}{\partial \phi_{\mathrm{d}}} \frac{\mathrm{d}\phi_{\mathrm{d}}}{\mathrm{d}N} \,. \tag{3.53}$$

In the previous sections we have shown that the damage controlling function in the case of rate-independent ('pure') fatigue is

$$\frac{\mathrm{d}\phi_{\mathrm{i}}}{\mathrm{d}N} \sim \Delta W_{\mathrm{i}} \,, \tag{3.54}$$

where ΔW_{i} refers to the damaging part of the strain energy (3.31). In the case of the high strain creep–fatigue loading where $\Delta W^{\mathrm{p}} \gg \Delta W^{\mathrm{e}+}$, ΔW_{i} may be interpreted as the area of the hysteresis loop for continuous symmetric fast–fast strain-controlled cycling. However, from (3.31) we have

$$\left(\frac{\Delta W_{\mathrm{i}}}{\kappa_{\mathrm{i}}}\right) = (2N_{\mathrm{fi}})^{\alpha} \,, \tag{3.55a}$$

or

$$2N_{\mathrm{fi}} = \left(\frac{\Delta W_{\mathrm{i}}}{\kappa_{\mathrm{i}}}\right)^{1/\alpha} \,. \tag{3.55b}$$

For the sake of convenience, we adopt the notation:

$$2N_{\mathrm{fi}} = 2N_{\mathrm{fo}} \quad \text{and} \quad -1/\alpha = \alpha' \,. \tag{3.56}$$

Table 3.3 Various proposed creep-fatigue criteria

Source	Creep–fatigue criterion	Remarks
Linear summation of time and cycle Robinson [57], Taira [58]	$\frac{n}{N_f} + \int \frac{dt}{t_R} = 1$	N_f = fatigue life without creep t_R = creep rupture time.
Frequency separation Coffin [49]	$N_f = \left(\frac{C_2}{\Delta\varepsilon^{in}}\right)^{1/\beta}\left(\frac{\nu_t}{2}\right)^{1-k}\left(\frac{\nu_c}{\nu_t}\right)^{c}$ frequency $\frac{1}{\nu} = \frac{1}{\nu_t} + \frac{1}{\nu_c}$	ν_t and ν_c are tensile and compressive frequencies; C_2, β, k and c are material constants.
Strain range partition Manson *et al.* [61]	$\frac{1}{N_f} = \frac{F_{pp}}{N_{pp}} + \frac{F_{cc}}{N_{cc}} + \frac{F_{cp}}{N_{cp}}\left(\text{or } \frac{F_{pc}}{N_{pc}}\right)$ $N_{ij} = c_{ij}(\Delta\varepsilon^{in})^{\alpha_{ij}},\ F_{ij} = \frac{\Delta\varepsilon_{ij}}{\Delta\varepsilon^{in}}$ $\Delta\varepsilon^{in} = \Sigma\,\Delta\varepsilon_{ij}\,(i, j, = \mathrm{p,c})$	Subscripts p and c refer to plastic and creep. First letter in a mixed mode refers to the tensile portion of a cycle, and second to the compressive part.
Damage rate Majumdar and Maiya [63]	$N_f = \frac{[-1 + (1 + 2D_c/D_t)^{1/2}]}{D_c}$ where $D_c = \int_{cycle} \left\{\begin{matrix} A_g \\ -A_g \end{matrix}\right\} \lvert\varepsilon^{in}\rvert^{m} \lvert\dot{\varepsilon}^{in}\rvert^{k_c}\, dt$ $D_t = \int_{cycle} \left\{\begin{matrix} C_t \\ C_c \end{matrix}\right\} \lvert\varepsilon^{in}\rvert^{m} \lvert\dot{\varepsilon}^{in}\rvert^{k}\, dt$	A_g and C_t are used for the tensile stress and $-A_g$ and C_c for the compressive stress. $\dot{\varepsilon}^{in}$ is the inelastic strain rate. A_g, C_t, C_c, m, k_c and k are material parameters.
Damage rate Lemaitre and Plumtree [65]	$\frac{dD}{dN} = \frac{dD_f}{dN} + \frac{dD_c}{dN}$	σ_{max} and σ_{avg} are the maximum and average stress during one cycle. Other

	$\frac{\mathrm{d}D_f}{\mathrm{d}N}=\left\{\frac{\sigma_{max}-\sigma_{avg}}{B(\sigma_{avg})(1-D)}\right\}^{\beta}(1-D)^{-p}$ $\frac{\mathrm{d}D_c}{\mathrm{d}N}=\left\{\frac{\lvert\sigma\rvert}{A^{\pm}(1-D)}\right\}^{\gamma}(1-D)^{-q}$	parameters have to be found from suitable experiments.
Incremental life Saitoh and Krempl [66]	$D_f+D_c=1$ where $\dot{D}_f=H(\dot{\varepsilon}^{in})\lvert\dot{\varepsilon}^{in}\rvert^{n_f}\left\lvert\frac{\sigma}{\sigma_f}\right\rvert^{m_f}$ $\dot{D}_c=\mathrm{sgn}(\sigma)\lvert\dot{\varepsilon}^{in}\rvert^{n_c}\left\lvert\frac{\sigma}{\sigma_c}\right\rvert^{m_f}$	$H(\dot{\varepsilon}^{in})=\begin{cases}1, \dot{\varepsilon}^{in}\geqslant 0\\ 0, \dot{\varepsilon}^{in}<0\end{cases}$ $\mathrm{sgn}(\sigma)=\begin{cases}1, \sigma\geqslant 0\\ -1, \sigma<0\end{cases}$ n_f, m_f, σ_f, n_c, m_c and σ_c are material constants.
Damage rate Asada *et al.* [67]	$N_f(C_1D_I^{n_1}+C_2D_D^{n_2})=1$ where $D_I=\int_{cycle}\sigma_e\mathrm{d}\varepsilon^{p}$ $D_D=\int_{cycle}\sigma_e\mathrm{d}t$	σ_e is the overstress, and C_1, n_1, C_2 and n_2 are material parameters.

Equation (3.55b) can then be written as

$$\frac{1}{2N_{\mathrm{fo}}} = \left(\frac{\Delta W_{\mathrm{I}}}{\kappa_{\mathrm{I}}}\right)^{\alpha'} \quad \alpha' > 0. \tag{3.57}$$

In the case of rate-dependent fatigue, the failure mechanism involves: (a) nucleation of the cavities; and (b) growth of the cavities.

Rice [69] has investigated the growth of a single void in a cavitated grain boundary facet. In this model the cavities are assumed to be uniformly distributed over the facet with an average spacing of $2b$ and a radius a (Fig. 3.27(a)). The diffusion along the void surface is assumed to be fast

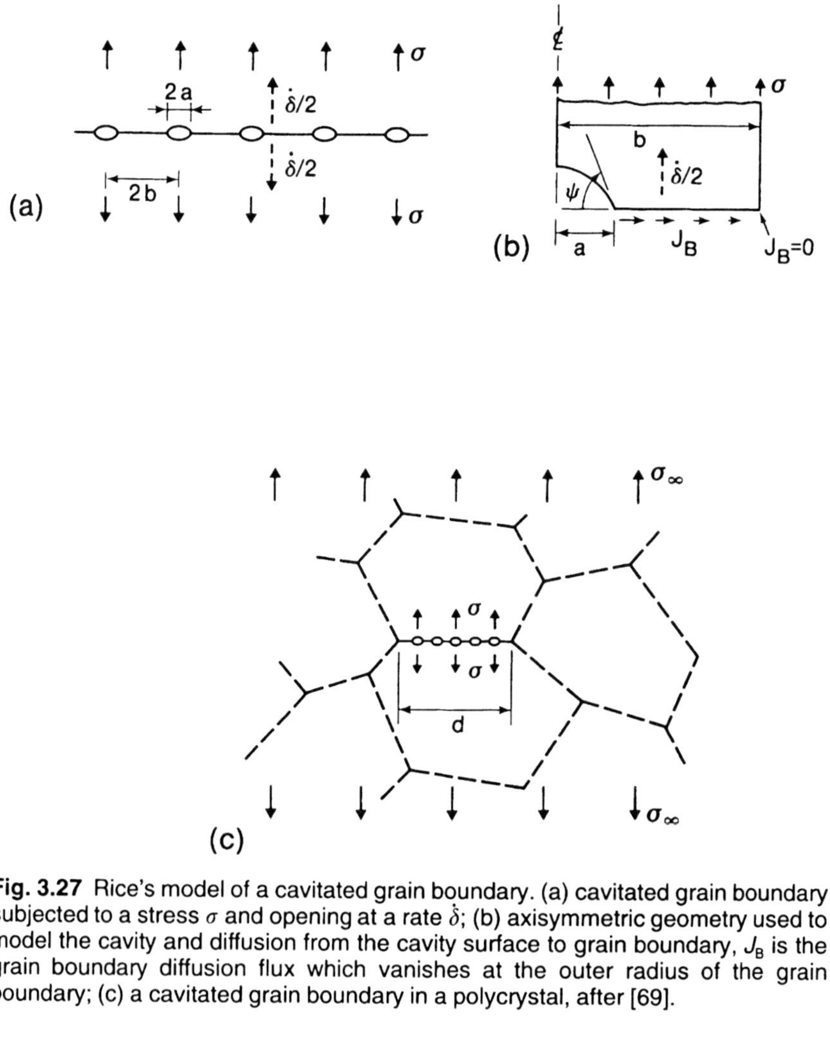

Fig. 3.27 Rice's model of a cavitated grain boundary. (a) cavitated grain boundary subjected to a stress σ and opening at a rate $\dot{\delta}$; (b) axisymmetric geometry used to model the cavity and diffusion from the cavity surface to grain boundary, J_B is the grain boundary diffusion flux which vanishes at the outer radius of the grain boundary; (c) a cavitated grain boundary in a polycrystal, after [69].

compared to that along the grain boundary so that a quasi-equilibrium spherical cap with an aperture angle ψ is maintained (a typical value of ψ is about 70°).

The remotely applied uniaxial stress is denoted by σ_∞ which causes a remote strain-rate of $\dot{\varepsilon}_\infty$. The average opening rate of the grain facet (Fig. 3.27) is given by

$$\dot{\delta} = \lambda d\left(1 - \frac{\sigma}{\sigma_\infty}\right)\dot{\varepsilon}_\infty , \tag{3.58}$$

where λ is a dimensionless factor and d is the grain facet diameter. The stress σ in (3.58) is the tension on cavity surfaces restraining them from opening. Note that if $\sigma = \sigma_\infty$, then there would not be a cavity growth, i.e. $\dot{\delta} = 0$. The rate of cavity growth is thus controlled by the difference between the applied remote stress, σ_∞, and the local stress constraining the cavity, σ, i.e. $(\sigma_\infty - \sigma)$. Thus, (3.58) could be written as

$$\dot{\delta} = k_{\text{constraint}}\,\dot{\varepsilon}_\infty . \tag{3.59}$$

The rate of work done to open a cavitated facet is proportional to $(\sigma_\infty - \sigma)\dot{\delta}$ i.e.

$$\dot{W} \sim (\sigma_\infty - \sigma)\dot{\delta}. \tag{3.60}$$

Now returning to the cyclic loading, the rate-dependent part of the work done in a cycle will be proportional to

$$\Delta W_{\text{d}} \sim \int_{\text{cycle}} \sigma\dot{\varepsilon}\,\text{d}t - \int_{\text{cycle}} \sigma_{\text{ref}}\,\dot{\varepsilon}\,\text{d}t , \tag{3.61}$$

where σ_{ref} is the stress associated with a symmetric continuous high strain-rate regime where no rate dependency is observed (Fig. 3.25), and σ is the applied remote stress in a slower strain-rate regime [70]. To illustrate the above, let us compare two deformation-controlled tests. The first one is a continuous symmetric loading at high strain-rate, $\dot{\varepsilon} = \dot{\varepsilon}_{\text{ref}}$, herein termed fast–fast and shown by symbol FF. The second test is performed with a slower strain-rate $\dot{\varepsilon} < \dot{\varepsilon}_{\text{ref}}$ in the tension going leg and $\dot{\varepsilon} = \dot{\varepsilon}_{\text{ref}}$ in the compression leg (reversed loading) and will be termed slow–fast (SF). The remote strain and stress versus time (ε_∞ vs. t, σ_∞ vs. t) plots are shown in Fig. 3.28. For a material which does not exhibit a rate effect, the two sets of curves will coincide. Therefore, any difference between them is due to the rate dependency of the material and the rate-dependent part of the energy is given by (3.61).

When comparing the corresponding hysteresis loop energies, the time scale is eliminated and we have the integrated part of each energy on the RHS of (3.61). Thus, to obtain the energy due to the rate effect we have to make a transformation of coordinates, and match the reversed loading

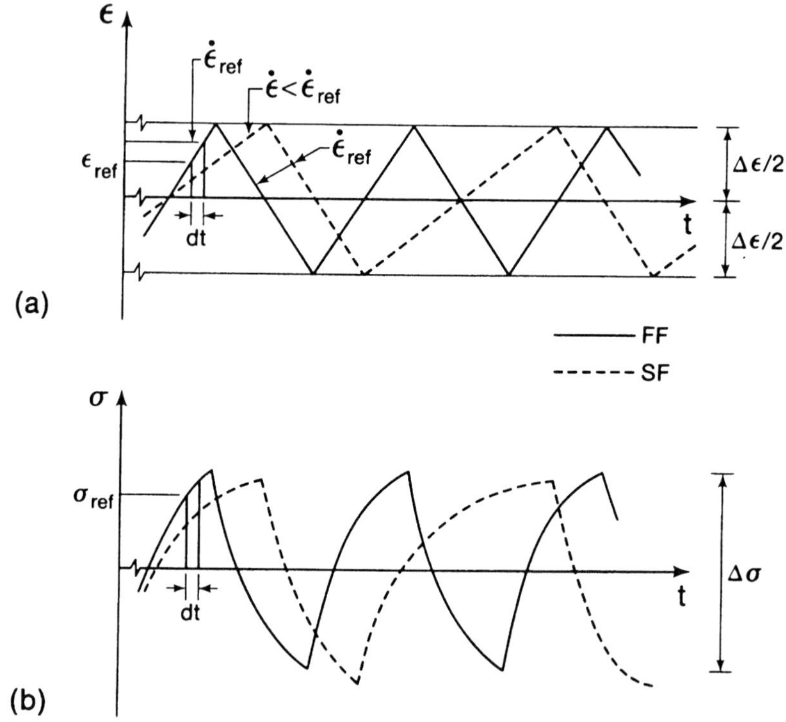

Fig. 3.28 (a) A program of strain-controlled continuous symmetric fast–fast tests and a continuous asymmetric slow–fast test. (b) Schematic of the stress response. Reproduced with kind permission from [70].

branches with the same strain-rate $\dot{\varepsilon} = \dot{\varepsilon}_{ref}$. In this manner, the area due to the rate effect can be determined. Figure 3.29 shows two hysteresis loops, one FF with $\dot{\varepsilon}_{ref} = 1 \times 10^{-3}\ s^{-1}$ and the other SF $(10^{-4}/10^{-3})s^{-1}$, superimposed on each other. Since the cavity growth occurs during the positive (tensile) part of the stress, the damaging part of the rate-dependent strain energy is given by

$$\Delta W_d^+ = \int_{cycle} (\sigma - \sigma_{ref}) H(\sigma - \sigma_{ref})\, d\varepsilon\,, \tag{3.62}$$

where $H(\sigma - \sigma_{ref})$ is the Heaviside function, equal to unity when $\sigma - \sigma_{ref} \geqslant 0$, otherwise nil.

Returning to (3.53), we then have

$$\frac{d\phi_d}{dN} \sim \Delta W_d^+\,, \tag{3.63}$$

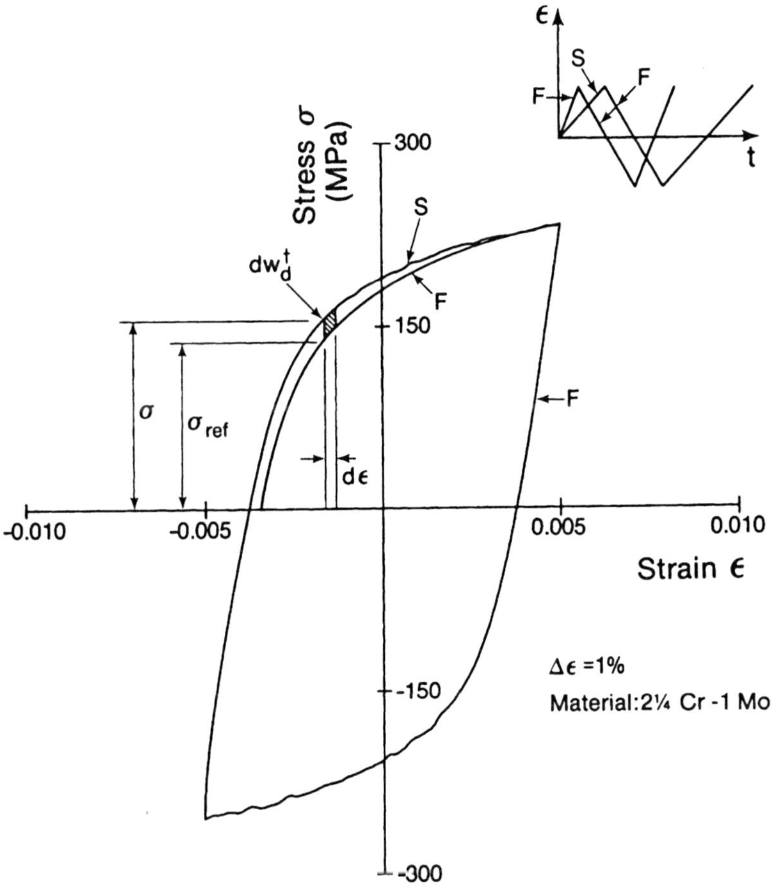

Fig. 3.29 Matching the unloading branches of two fast–fast and slow–fast tests to obtain the rate-dependent effects. Reproduced with kind permission from [70].

and in a manner similar to the rate-independent part, we can write

$$\left(\frac{\Delta W_d^+}{\kappa_d}\right) = (2N_{fd})^\beta \quad \beta < 0\,, \tag{3.64a}$$

or

$$\frac{1}{2N_{fd}} = \left(\frac{\Delta W_d^+}{\kappa_d}\right)^{\beta'} \quad \beta' = \frac{-1}{\beta} > 0\,. \tag{3.64b}$$

Substituting from (3.63), (3.64(b)), (3.54) and (3.57) into (3.53) we get

$$A(\Delta W_i)^{\alpha'} + B(\Delta W_d^+)^{\beta'} = \frac{1}{2N_f}, \tag{3.65a}$$

or

$$\frac{1}{N_{fo}} + \frac{1}{N_{fd}} = \frac{1}{N_f}. \tag{3.65b}$$

In deriving (3.65), an implicit assumption was that

$$\frac{\partial D}{\partial \phi_1} = C_1 \quad \text{and} \quad \frac{\partial D}{\partial \phi_d} = C_d, \tag{3.66}$$

where C_1 and C_d are constants. The first condition for the rate-independent part has been verified experimentally (section 3.4). The second condition will be confirmed through correlation with experiments in the following section.

At this stage it would be instructive to see whether the derived relationship for the time-dependent fatigue (3.65) satisfies the two limiting cases, namely 'pure fatigue' (rate-independent) and creep (rate-dependent) conditions. In the case of the rate-independent response, $\sigma = \sigma_{ref}$ for any $\dot{\varepsilon}$ and thus $\Delta W_d^+ = 0$, and we get

$$A(\Delta W_1)^{\alpha'} = \frac{1}{2N_f}, \tag{3.67a}$$

or

$$\frac{\Delta W_1}{\kappa} = (2N_f)^{\alpha}, \tag{3.67b}$$

which is (3.31) with $\alpha = -1/\alpha'$ and $\kappa = (A)^{1/\alpha'}$.

In the case of 'pure creep', $\sigma =$ constant, $\sigma_{ref} = 0$, thus $\Delta W_1 \equiv 0$ and from (3.61) we see that the energy required for creep rupture depends on the applied stress level and strain-rate $\dot{\varepsilon}$. We could thus conclude that the time-dependent fatigue (creep–fatigue) failure criterion (3.65) possesses proper damage parameters derived from observation in both micro- and macro-levels.

3.7.1 CORRELATION WITH EXPERIMENTAL DATA

Results of experiments performed in a 0.1 μPa vacuum environment at 550°C on a $2\frac{1}{4}$Cr-1 Mo steel are reported in [70]. The stable hysteresis loops of each continuous symmetric fast–fast test were analysed to measure the rate-independent plastic strain energy density, ΔW_1^p. This energy is plotted versus the inverse of the numbers of cycles to failure, $1/N_{fo}$, on a log–log scale, in Fig. 3.30. The coefficients of (3.57) are then determined through a best fit regression technique and are indicated in Fig. 3.30.

The rate-dependent plastic strain energy density, ΔW_d^{+p}, is determined for the unsymmetric wave forms as described earlier, and is depicted in Fig. 3.29. The energy thus obtained is plotted in Fig. 3.31 against the inverse

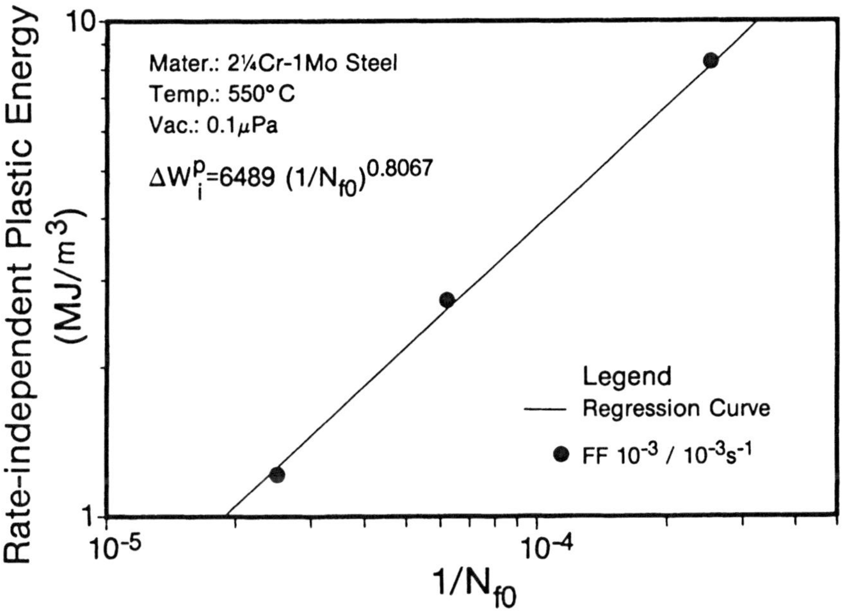

Fig. 3.30 Relationship between rate-independent plastic strain energy density and inverse of cycles to failure. Reprinted with kind permission from [70].

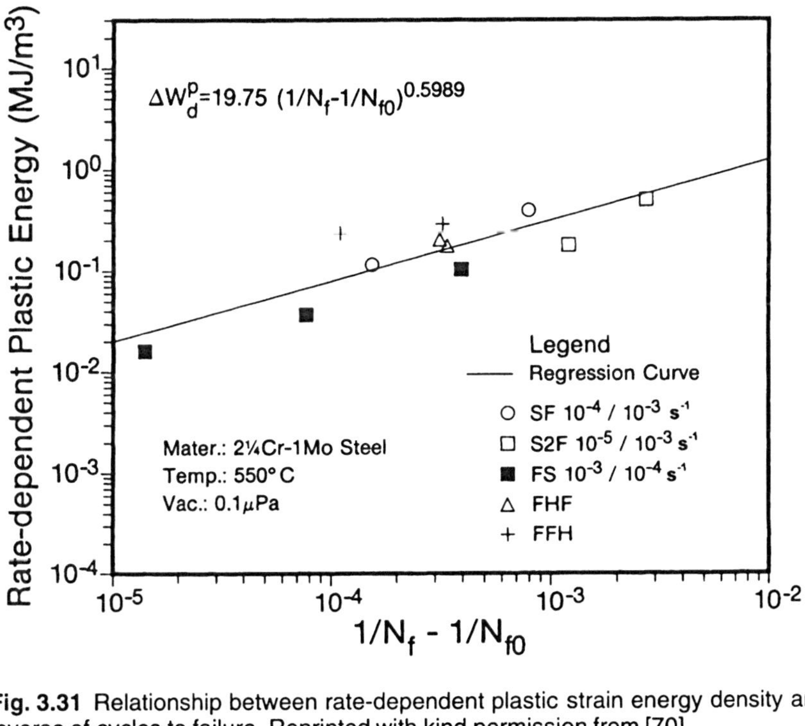

Fig. 3.31 Relationship between rate-dependent plastic strain energy density and inverse of cycles to failure. Reprinted with kind permission from [70].

of the remaining life, $(1/N_f - 1/N_{fo})$ on a log–log scale. The constants of (3.64a) are then evaluated through a best fit regression line of the data (Fig. 3.31). It can be seen that (3.64) predicts fairly well the trend of experimental results, despite the fact that the planimeter used to measure the hysteresis areas had an accuracy of $10^{-2}\,\mathrm{MJ\,m^{-3}}$, thus rendering the determination of ΔW^{+p} of the fast–slow (FS) tests somewhat less accurate.

With the material constants thus evaluated, the predicted life from the creep–fatigue criterion (3.65a) can be compared with the experimentally-observed life as shown in Fig. 3.32. It can be seen that most of the results fall within the standard scatter band of fatigue data, except for a couple of tests. Bearing in mind the difficulty of performing elevated temperature tests in a high vacuum, and conversion of diametral strain control to an equivalent axial strain for the reported experiments we can conclude that

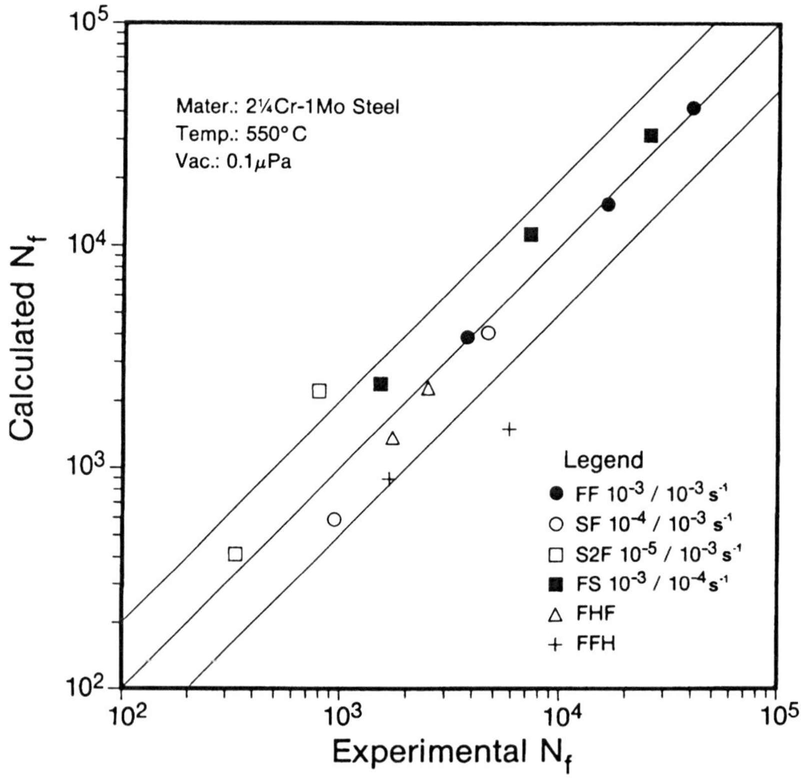

Fig. 3.32 Comparison between calculated and observed cycles to failure in the creep–fatigue regime. Reprinted with kind permission from [70].

the theory described in this section for time-dependent fatigue is capable of predicting the creep–fatigue phenomenon fairly well.

3.8 EFFECT OF ENVIRONMENT ON CRACK INITIATION AND FATIGUE LIFE

It was mentioned earlier that fatigue failure is a multi-stage process which involves plastic strain accumulation resulting in crack initiation and growth of cracks. The demarcation between the crack initiation phase and that of propagation has generally been based on an arbitrarily defined crack length, ranging from a grain diameter to as high as 1 mm. The underlying process in the microscopic scale is formation of dislocations, initially at the surface or subsurface layers, and later at the front of initiated cracks. At the macroscopic scale this is manifested by a measurable, irreversible plastic strain. If an environment is to have an effect on the fatigue life, it must influence either surface or subsurface layers of an uncracked specimen, or the crack front material, through chemical or physical interaction. This process would then be observed in the macroscopic scale through a change in plastic deformation.

In Chapter 1 we briefly discussed the effect of environment on crack initiation. Figure 1.8, reproduced here as Fig. 3.33, shows that during the tensile part of a cycle, the fresh slip surface chemically interacts with the environment, e.g. absorbing O_2 from the moist laboratory air. In the compressive part of the cycle, slip reversal becomes difficult compared to a vacuum or an inert atmosphere. Dissolved molecular oxygen within a crystal (or other embrittling chemical agents) reduce slip reversibility thus causing early crack initiation.

It was pointed out in Chapter 2 that the bulk macroscopic stress–strain curves are very slightly affected when tests are carried out in a vacuum or

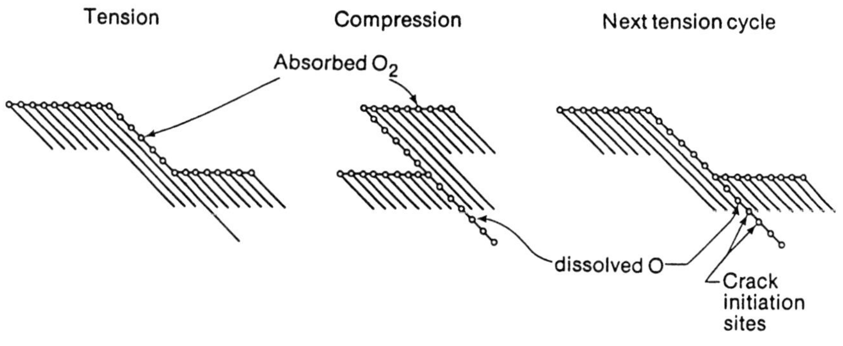

Fig. 3.33 Environmentally-assisted stage I crack growth mechanism, after [19] in Chapter 1.

air environment. However, damage accumulates at the surface or subsurface layers, and the rate of damage accumulation depends on the environment.

Numerous studies have been conducted to examine the effect of environment on fatigue growth, e.g. [71–74] (see also section 7.11). Similarly, there have been various investigations into the effect of the environment, particularly at high temperatures, on the **total** fatigue life of smooth specimens, e.g. [75–78] among others. In these investigations, a comparison is generally made of the fatigue behaviour in a vacuum and air to show the effect of environmental reaction. While a clear trend is shown regarding the effect of environment, no **quantitative** life prediction has generally been made. In contrast, *S–N* fatigue results for tests conducted in a baseline air environment are compared to those in water environments of various chemistries in [79]. The objective of this study was to formulate recommendations to the ASME Boiler and Pressure Vessel Code for changes in fatigue design curves in a nuclear reactor coolant environment. Despite a large amount of collected data, the effect of the environment has not been clearly defined.

It was shown in section 3.5.4 (Fig. 3.23) that the fatigue life of a smooth specimen is mainly controlled by the initiation and subsequent propagation of small cracks. For example, the period to initiate a short crack may occupy from 20–80% of the total life of a smooth specimen, depending on the magnitude of the applied strain range [30].

In the following some recent experimental results will be presented, showing the effect of environment on both the crack initiation and propagation phases of smooth specimens. The material used in the investigation was a ferritic–pearlitic steel, ASTM A-516 Gr. 70, whose mechanical properties in both air and vacuum environments are listed in Table 2.2.

Two types of tests were carried out. The first was a series of two-step experiments to determine the critical damage curve, CDC, as described in section 3.5.4. Another series of single-step tests was performed to obtain the total fatigue life. Both series of tests were carried out in a 0.1 mPa vacuum at ambient temperature ($\approx 20°C$). All tests were conducted in a total strain-controlled mode with a sinusoidal wave of $R_\varepsilon = \varepsilon_{min}/\varepsilon_{max} = -1$ at an average strain-rate of $\dot{\varepsilon} = 5 \times 10^{-3}\,s^{-1}$. Fatigue failure was defined in terms of a drop of 10% of the peak tensile stress which was equivalent to a surface crack length of 3–4 mm [80].

Figure 3.34 shows the critical damage curve plotted in terms of imposed total strain range ($\Delta\varepsilon^t$) and number of reversals ($2N_i$) on a log–log scale. For the sake of comparison, similar curves obtained in an air environment (Fig. 3.23) are superimposed on Fig. 3.34. It can be seen that the initiation period, $2N_i$, in a vacuum is longer than that in air by an order of magnitude for this material. However, the total fatigue life, $2N_f$, is longer by a factor of about 2. It follows that for the same fatigue life, $2N_f$, that the

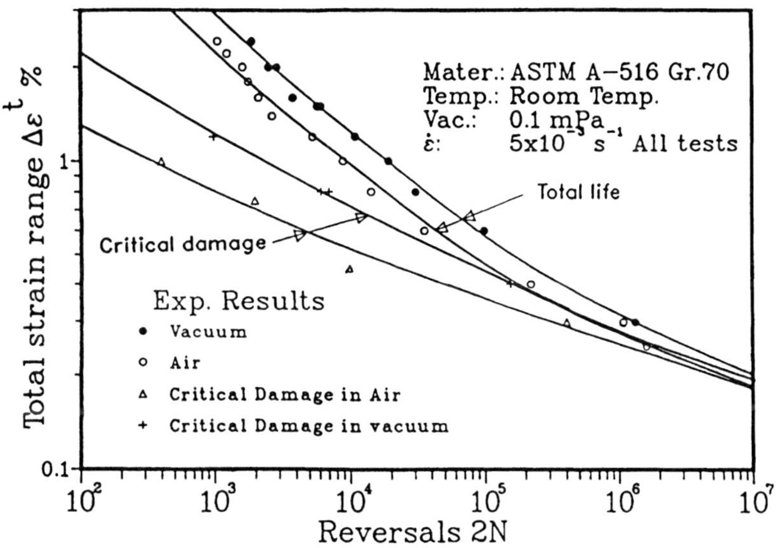

Fig. 3.34 A comparison of critical damage and fatigue life curves in vacuum and air environments, from [80].

crack propagation period, $2N_p$, is longer in air compared to that in a vacuum. These results suggest that initiation of short cracks and their rate of growth, da/dN, are accelerated in air compared to a vacuum. This illustrates the important role played by the thin oxide film in both initiation and propagation phases.

Similar qualitative results were obtained for the fatigue life of copper single crystals tested in high vacuum and air environments [81, 82]. Wang and Mughrabi [82] concluded that the reason for longer lives in a vacuum is mainly due to the reduced rate of early crack growth in Stage I and in the transition region from Stage I to Stage II. However, it should be noted that in polycrystals, the grain boundaries do slow down the growth rates, and qualitative results from monocrystals cannot be directly compared to those of polycrystals.

An analysis of the life curve data in Fig. 3.34 shows that the effect of environment is best observed with respect to the plastic strain. A decomposition of the imposed total strain into elastic and plastic components is shown in Fig. 3.35. The strain–life relationship (Basquin–Coffin–Manson (3.10)) was fitted to the experimental data. The coefficients and exponents of (3.10) for a vacuum environment are reported in Fig. 3.35, whereas those for air are reported in Fig. 3.5. It can be seen that the major difference between the two sets of data is a marked increase in the ductility coefficient

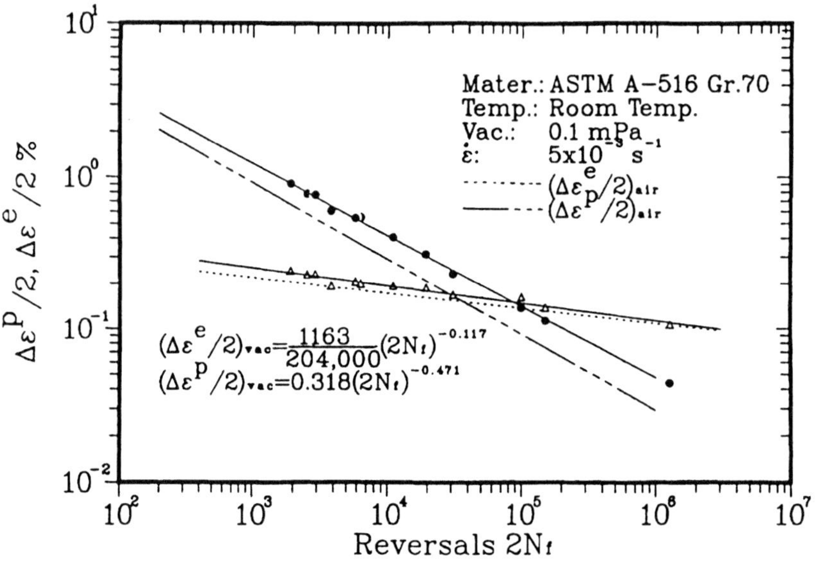

Fig. 3.35 Elastic and plastic components of the applied strain versus number of reversals to failure in air and vacuum environments, from [80].

ε_f' in a vacuum compared with air (0.318 vs. 0.288). Thus, the presence of oxygen decreases ductility and leads to shorter lives.

The results obtained in a 0.1 mPa vacuum may be considered as the data representing 'pure' (mechanical) fatigue damage. Creep damage is negligible since equal rates in tension and compression half cycles were applied. In the case of an air environment, the fatigue damage is composed of two components: one due to the 'pure' fatigue, and the other representing the environmental contribution. The environmental component is expected to depend on the time available for an environmental reaction. This suggests that the environmental effect should be small at very short lives and would increase with increasing lives. Analysis of data reported in [83] clearly supports the above conclusions, i.e. the slopes for crack initiation and total fatigue life curves in air are steeper than those in a vacuum.

3.8.1 CORRELATION WITH PLASTIC STRAIN ENERGY

In the above the results were presented in terms of the imposed strain range. While the correlation shown in Figs 3.34 and 3.35 is good, the extension of strain as a measure of damage to multiaxial states of stress has several drawbacks (Chapter 4). It was indicated earlier that the environment affects the plastic strain, and as a result damage could be related to the input plastic strain energy. Both crack initiation and total fatigue lives are correlated with the plastic strain energy, ΔW^{p}, in Figs 3.36 and 3.37, respectively. It can be seen that an equation of the form

$$\Delta W^{p} = \kappa_{p} N^{\alpha_{p}} \tag{3.68}$$

fits the data. The value of κ_p is independent of the environment while the slope α_p depends on the environment. Numerical values of κ_p and α_p for crack initiation and fatigue lives are reported in Figs 3.36 and 3.37. In Fig. 3.37, in addition to vacuum and air environments, results for a pure oxygen environment are also depicted. It should be noted that the presence of oxygen, rather than its concentration, is important.

For the sake of completeness, the fatigue life data have also been correlated with the total strain energy,

$$\Delta W^{t} = \kappa_{t}(N_{f})^{\alpha_{t}} + C. \tag{3.69}$$

It can be seen from Fig. 3.38 that again κ_t is independent of the environment while α_t and C are both influenced by the environment.

At high temperatures most metals exhibit rate dependency and there is some experimental evidence, e.g. [84] to suggest that the coefficients in (3.69) would have the following functional form

$$\alpha = \alpha(\mathrm{En}, T), \quad \kappa = \kappa(T) \quad \text{and} \quad C = C(\mathrm{En}, T) \tag{3.70}$$

where 'En' denotes a specific environment and T refers to temperature.

In summary, the crack initiation and fatigue lives correlate fairly well with both plastic and total strain energies (Figs 3.36–3.38). It is found that

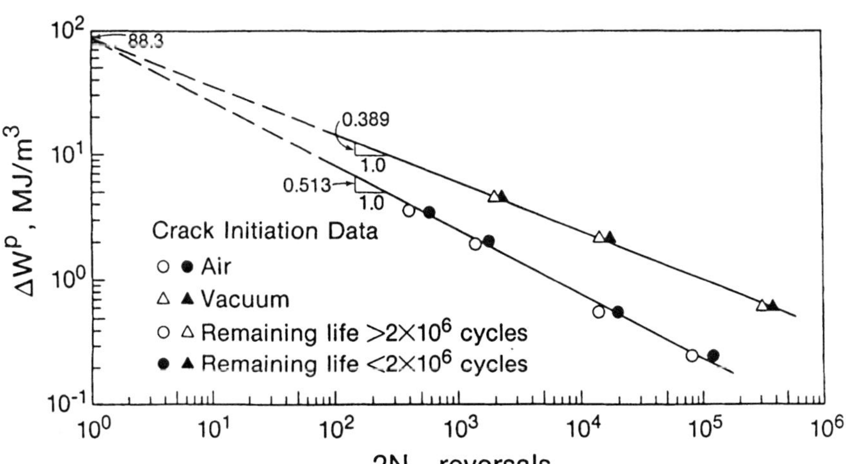

Fig. 3.36 Plastic strain energy per cycle versus number of reversals to crack initiation. Reprinted with permission from [83].

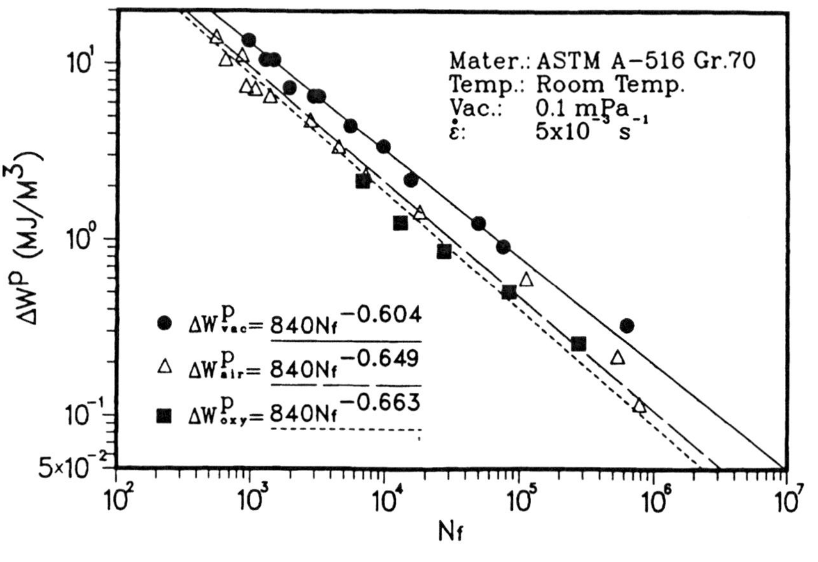

Fig. 3.37 Plastic strain energy per cycle versus number of cycles to fatigue failure in three environments: vacuum, air and pure oxygen.

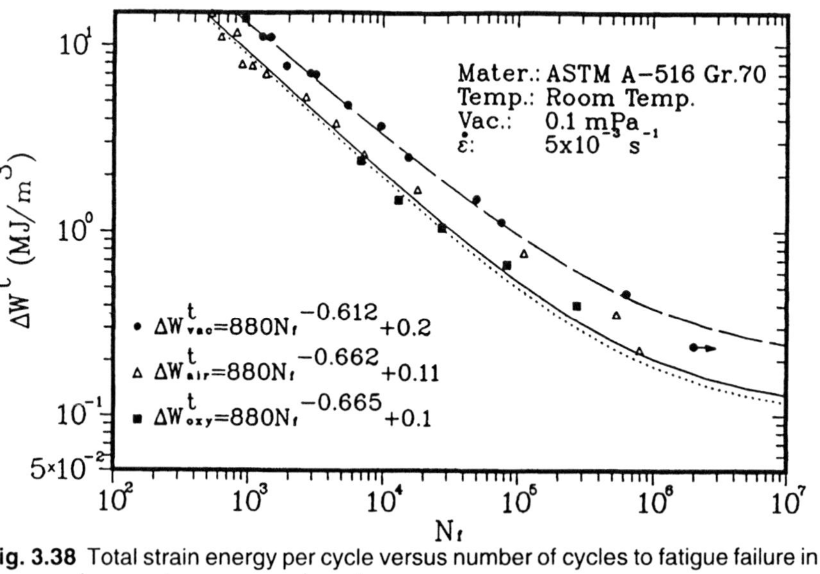

Fig. 3.38 Total strain energy per cycle versus number of cycles to fatigue failure in three environments: vacuum, air and pure oxygen.

the influence of the surrounding environment is reflected in the material constants α and C of (3.68) and (3.69), but not on κ.

It was mentioned earlier that the effect of environment is a time-dependent process. The results presented here were obtained at a fixed strain-rate and for a symmetric cyclic loading. To formulate a general environmentally-assisted fatigue failure criterion, one can proceed in a manner similar to that described in section 3.7. A time-dependent energy part, similar to (3.61), has to be evaluated for the environmental effect. From the survey of the literature, it would appear that a function of the strain-rate, $f(\dot{\varepsilon})$, has emerged as an appropriate damage parameter, e.g. Garud [85]. It is the specification of the functional form and comparison with the carefully conducted experiments in inert and corrosive environments which is the subject of intensive investigation at the present time.

3.9 EFFECT OF MEAN STRESS AND RATCHETING STRAIN ON FATIGUE LIFE

To investigate the effect of mean stress two types of tests are generally performed: strain-controlled cycling with a constant mean strain, e.g. [11, 86], or stress-controlled cycling with a constant mean stress, e.g. [87, 88]. If the bulk material response is elastic, as it generally is in the high-cycle regime, then the above two types of tests are equivalent and both can be used to study the mean-stress effect on the fatigue life. However, if the material response is elasto-plastic, which is the case in the low-cycle regime, the above types of tests will produce quite different results. In the strain-controlled case, the plastic deformation causes a mean stress relaxation in early life, as discussed in section 3.3 and depicted in Fig. 2.6 in Chapter 2. Therefore, the low-cycle fatigue life will not be appreciably affected by the introduction of a mean strain when compared to the fully-reversed cyclic tests (Fig. 3.7).

By contrast, in the stress-controlled mode, the plastic deformation causes cyclic creep (or ratcheting) strain. The accumulation of the ratcheting (cyclic creep) strain induces an additional damage which results in a shorter fatigue life when compared to a 'purely' mean stress test i.e. one with a negligible ratcheting strain. Thus, it follows that in the stress-controlled condition, both the mean stress and ratcheting strain effects have to be included in a fatigue damage theory. However, the ratcheting behaviour of materials is a complex phenomenon and depends on a number of factors, e.g. the magnitude of mean stress and stress amplitude, frequency (rate) of loading, load history, and microstructural characteristics, etc. The prediction of the ratcheting strain poses a challenge for most constitutive relations (Chapter 6). Perhaps this is the reason why most fatigue failure criteria dealing with the effect of mean stress do not contain a parameter which is directly related to the ratcheting strain.

Early attempts to deal with mean-stress effects were mainly based on the modification of the S–N curve, e.g. [1,6,89], among others. This type of approach has been discussed and compared with the experimental data by Kujawski and Ellyin [90].

The mechanism by which damage is accumulated due to the mean stress is different from that of the ratcheting (cyclic creep) strain. To quantify each component, we have to devise test procedures whereby the effect of the ratcheting strain can be decoupled from that of the mean stress. Such tests have been recently performed in the author's laboratory and are reported in [91]. All tests were conducted under stress-controlled conditions and the maximum stress, except for one test, was above the cyclic yield stress of the material. The ratio of the mean stress to the stress amplitude, σ_m/σ_a, was positive, ranging from 0.5 to 3. Detailed test results are given in [91], and they will be compared with the fatigue failure criterion to be outlined in the following section.

3.9.1 A FATIGUE CRITERION WITH MEAN-STRESS AND RATCHETING STRAIN EFFECTS

The energy criterion developed in section 3.4.2 was essentially for a fully-reversed loading condition or for a very small mean stress to stress amplitude ratio. The ΔW^{e+} in (3.27) which includes the mean stress value, would generally overestimate the mean-stress effect, if the magnitude of the mean stress is large. A study of the mean-stress effect on threshold conditions [90] indicates that the severity of the ΔW^e depends on the relative measure of the tensile mean stress to alternating stress, σ_m/σ_a. The fatigue damage parameter (3.29) for fully-reversed loading can be written as

$$\psi = \Delta W^p + \Delta W_a^e = \kappa_t (2N_f)^\alpha + C \tag{3.71}$$

where κ_t, α and C are material constants obtained from the fully-reversed fatigue tests, and the elastic energy ΔW_a^e is obtained from (3.27) by setting $\sigma_m = 0$, i.e.

$$\Delta W_a^e = \frac{\sigma_a^2}{2E}\,. \tag{3.72}$$

A generalization of (3.27) to include a normalized mean stress can be written in the form of

$$\psi_m = \Delta W^p + f\left(\frac{\sigma_m}{\sigma_a}\right) \Delta W_a^e = \kappa_t (2N_f)^\alpha + C \tag{3.73}$$

where

$$f(\sigma_m/\sigma_a) = 1 + \eta(\sigma_m/\sigma_a)\,. \tag{3.74}$$

Note that for $\sigma_m = 0$, (3.73) becomes identical to (3.27). Near the fatigue limit (threshold region), $\Delta W^p \ll \Delta W_a^e$, $(2N_f)^\alpha \to 0$, and (3.73) reduces to

$$\sigma_a = \sqrt{f} = (\sigma_a)_{-1} \tag{3.75}$$

where f is given by (3.74) and $(\sigma_a)_{-1}$ is the fatigue limit for the fully-reversed tests ($R = -1$). By setting $\eta = 1$ in (3.74), the relation (3.75) reduces to the well-known SWT parameter of [6]. Other particular cases of (3.75) are discussed in [90]. The coefficient $1 \geqslant \eta \geqslant 0$, which characterizes the material sensitivity to mean stress, is evaluated from a couple of tests with a mean stress near the threshold regime where there is no appreciable cyclic creep strain.

For those conditions where ΔW^p in (3.73) is not negligible, one can determine its value from (3.14) or (3.20). For a ferritic–pearlitic steel (ASTM A-516 Gr. 70) the coefficients on the RHS of (3.73) have been determined (Fig. 3.15), where $\kappa_t = 1395\,\mathrm{MJ\,m^{-3}}$, $\alpha = -0.665$ and $C = 0.1\,\mathrm{MJ\,m^{-3}}$. From a few test data, η in (3.74) is found to be equal to 0.665. Thus, (3.73) is now fully determined. A plot of ψ_m versus $2N_f$ is shown in Fig. 3.39, together with the test data from [91]. It can be seen that for the tests without appreciable ratcheting strains (open symbols), the data are very close to the curve predicted by (3.73). However, for the tests with appreciable ratcheting strain (filled symbols) the data points fall below the curve. With the

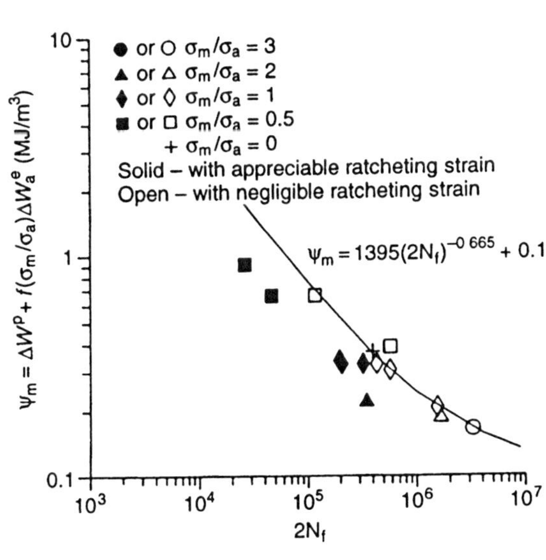

Fig. 3.39 Damage parameter with mean stress function versus number of reversals to failure (3.73), compared to the experimental data. Reproduced with kind permission from [91].

same mean applied stress and stress amplitude, the fatigue life can decrease considerably, depending on the amount of ratcheting strain during cyclic loading (Table 1 in [91]). The mean stress function, $f(\sigma_m/\sigma_a)$, which is related to the ratio of σ_m/σ_a (3.74), has no direct relationship to the ratcheting strain, and as such is unable to account for the ratcheting effect.

It is therefore clear that an additional parameter is needed to properly describe the effect of ratcheting on the fatigue life. Following the approach for time-dependent fatigue (section 3.7), we assume that the damage in a material is comprised of two functions, namely, a cyclic fatigue damage component, ϕ_f and a cyclic creep (ratcheting) damage component, ϕ_r, i.e.

$$D = D(\phi_f, \phi_r)\,. \tag{3.76}$$

The rate of damage accumulation per cycle is then

$$\frac{dD}{dN} = \frac{\partial D}{\partial \phi_f}\frac{d\phi_f}{dN} + \frac{\partial D}{\partial \phi_r}\frac{d\phi_r}{dN}\,. \tag{3.77}$$

The rate $d\phi_f/dN$ could be related to the damage parameter $(\psi_m - C)$, defined by (3.73), i.e.

$$\frac{d\phi_r}{dN} \sim (\psi_m - C) = \kappa_t(2N_{ff})^{\alpha} \tag{3.78}$$

where $2N_{ff}$ is the number of reversals to failure in the absence of ratcheting strain.

To account for the ratcheting effect, the following rate form (per cycle) parameter is defined:

$$\psi_r = \sigma_{max}\,\dot{\varepsilon}^{r}_{aver}\,, \tag{3.79}$$

where $\dot{\varepsilon}^{r}_{aver}$ is the average ratcheting strain per cycle, and σ_{max} is the maximum stress value during a cycle. In a manner similar to (3.73) we can write

$$\psi_r = \sigma_{max}\,\dot{\varepsilon}^{r}_{aver} = \kappa_r(2N_{fr})^{\beta_r} + C_r \tag{3.80}$$

where $2N_{fr}$ can be envisaged as a 'pure ratcheting' life, and κ_r, β_r and C_r are material constants related to the ratcheting effect. The constant C_r is a threshold value of ψ_r, i.e. the ratcheting effect could be neglected if $\psi_r \leqslant C_r$. The rate $d\phi_r/dN$ is then related to the parameter $(\psi_r - C_r)$ i.e.

$$\frac{d\phi_r}{dN} \sim (\psi_r - C_r)\,. \tag{3.81}$$

Assuming that the $\partial D/\partial\phi_f$ and $\partial D/\partial\phi_r$ in (3.77) are constants (section 3.7), we can obtain

$$\frac{1}{N_{ff}} + \frac{1}{N_{fr}} = \frac{1}{N_f}\,. \tag{3.82}$$

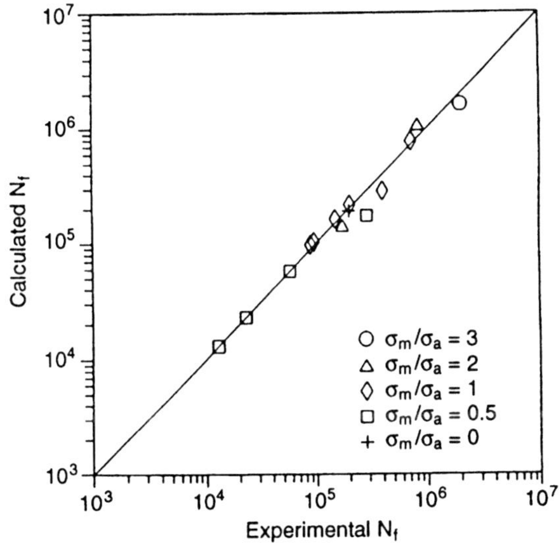

Fig. 3.40 Calculated number of cycles to failure from (3.82) compared to that from test results. Reproduced with kind permission from [91].

From (3.78), N_{ff} and then the difference $1/N_{fr} = 1/N_f - 1/N_{ff}$ can be obtained. Based on the representative test results, the best fit values of κ_r, β_r, C_r can then be determined. For the tests reported in [91], $\beta_r = -1.490$, $\kappa_r = 9194\,\mathrm{MJ\,m^{-3}}$, $C_r = 1.44 \times 10^{-5}\,\mathrm{MJ\,m^{-3}}$. The N_f calculated from (3.82) using (3.78) and (3.80) is compared with the experimental data in Fig. 3.40. It can be seen that the predicted results are in very good agreement with the test data which include the mean stress and ratcheting strain.

The energy-based fatigue failure criterion outlined in section 3.4 is thus extended to include the 'pure' mean-stress effect through the introduction of $f(\sigma_m/\sigma_a)$ in (3.73), and by the damage parameter $\psi_r = \sigma_{max}\,\dot{\varepsilon}^r_{aver}$ to account for the effect of ratcheting (cyclic creep) strain.

REFERENCES

1. Morrow, J.D. (1965) Cyclic plastic strain energy and fatigue of metals, in *Internal Friction, Damping and Cyclic Plasticity*, ASTM STP 378, American Society for Testing and Materials, Philadelphia, PA, pp. 45–84.
2. Wöhler, A. (1871) Tests to determine the forces acting on railway carriage axles and the capacity of resistance of the axles. *Engineering*, **11**.
3. Basquin, O.H. (1910) The exponential law of endurance tests. *Proc ASTM*, **10** (II) 625–30.
4. Lefebvre, D. and Ellyin, F. (1984) Cyclic response and inelastic strain energy in low-cycle fatigue. *Int. J. Fatigue*, **6** (1), pp. 9–15.

5. SAE Handbook (1978) Technical Report on Fatigue Properties, Section J-1099, Society of Automotive Engineers, Warrendale, PA.
6. Smith, K.N., Watson, P. and Topper, T.H. (1970) A stress–strain function for the fatigue of metals. *J. Mater.*, **5**, 767–78.
7. Manson, S.S. (1954) Behaviour of Materials Under Conditions of Thermal Stress, NACA TN-2933, National Advisory Committee for Aeronautics.
8. Coffin, L.F. (1954) A study of the effects of cyclic thermal stresses on a ductile metal. *Trans. ASME*, **76**, 931–50.
9. Landgraf, R.W. (1970) The resistance of metals to cyclic deformation, in *Achievement of High Fatigue Resistance in Metals and Alloys*, ASTM STP 467, American Society for Testing and Materials, Philadelphia, PA, pp. 3–36.
10. Dubuc, J. *et al.* (1970) Effect of mean stress and of mean strain in low-cycle fatigue of A-517 and A-201 steels. *J. Engng. Mater. Technol., Trans. ASME*, **92**, 35–52.
11. Ellyin, F. (1985) Effect of tensile-mean-strain on plastic strain energy and cyclic response. *J. Engng. Mater. Technol., Trans. ASME*, **107**, 119–25.
12. Bairstow, L. (1910) The elastic limits of iron and steel under cyclic variation of stress. *Phil. Trans. R. Soc. London* A, **210**, 35–55.
13. Behan, P.P. (1961–2) Axial load and strain-cycling fatigue of copper at low endurance. *J. Inst. Metals*, **89**, 328–38.
14. Behan, P.P. and Ford, H. (1961) Low endurance fatigue of a mild steel and an aluminum alloy. *J. Mech. Engng. Sci.*, **3**, 119–32.
15. Ellyin, F. and Kujawski, D. (1984) Plastic strain energy in fatigue failure. *J. Pressure Vessel Technol., Trans. ASME*, **106**, 342–7.
16. Miner, M.A. (1945) Cumulative damage in fatigue. *Trans. ASME*, **67**, A159–64.
17. Enomoto, N. (1955) On fatigue tests under progressive stress. *Proc. ASTM*, **55**, 903–13.
18. Martin, D.E. (1961) An energy criterion for low-cycle fatigue. *J. Basic Engng., Trans. ASME*, **83**, 565–71.
19. Feltner, C.E. and Morrow, J.D. (1961) Microplastic strain hysteresis energy as a criterion for fatigue failure. *J. Basic Engng., Trans. ASME*, **83**, 15–22.
20. Ivanova, V.S. (1963) *Fatigue Failure of Metals* (in Russian), Metallurgizdat, Moscow.
21. Gillemot, L.F. (1966) *Low-Cycle Fatigue by Constant Amplitude True Mean Stress.* Proceedings of the 1st International Conference on Fracture, 1965, Japan. The Japanese Society for Strength and Fracture of Materials, Tokyo, pp. 1461–77.
22. Hanstock, R.F. (1947) Damping capacity, strain hardening and fatigue. *Proc. Physical Soc.*, **59**, 275–87.
23. Halford, G.R. (1966) The energy required for fatigue. *J. Mater.*, **1**, 3–18.
24. Ellyin, F. and Kujawski, D. (1986) An energy-based fatigue failure criterion, in *Microstructure and Mechanical Behaviour of Materials*, Vol. II (eds H. Gu and J. He), EMAS, West Midlands, UK, pp. 541–600.
25. Mediratta, S.R., Ramaswamy, V. and Rama Rao, P. (1988) On the estimation of the cyclic plastic strain energy of dual-phase steels. *Int. J. Fatigue*, **10**, 13–19.
26. Lefebvre, D. and Ellyin, F. (1984) Cyclic response and inelastic strain energy in low-cycle fatigue. *Int. J. Fatigue*, **6**, 9–15.
27. Lukáš, P. and Klesnil, M. (1973) Cyclic stress–strain response and fatigue life of metals in low amplitude region. *Mater. Sci. Engng.*, **11**, 345–6.
28. Troshchenko, V.T. *et al.* (1985) *Cyclic Deformation and Fatigue of Metals* (in Russian), Naukova Dumka, Kiev.

29. Golos, K. and Ellyin, F. (1987) Generalization of cumulative damage criterion to multilevel cyclic loading. *Theor. Appl. Fract. Mech.*, **7**, 169–76.
30. Golos, K. and Ellyin, F. (1988) A total strain energy density theory for cumulative damage. *J. Pressure Vessel Technol., Trans. ASME*, **110**, 36–41.
31. Kujawski, D. and Ellyin, F. (1988) On the concept of cumulative fatigue damage. *Int. J. Fract.*, **37**, Kluwer Academic Publishers, Dordrecht, pp. 263–78.
32. Hunter, M.S. and Fricke, W.G. (1954) Metallographic aspects of fatigue behaviour of aluminum. *Proc. ASTM*, **54**, 717–36.
33. Bennett, J.A. (1946) A study of the effect of fatigue stressing on X4130 steel. *Proc. ASTM*, **46**, 693–771.
34. Palmgren, A. (1924) Endurance of ball bearing. *Z. Ver. Deutsch Ing.*, **68**, 339–41.
35. Marco, S.M. and Starkey, W.L. (1954) A concept of fatigue damage. *Trans. ASME*, **76**, 627–32.
36. Manson, S.S., Nachtigall, A.J. and Freche, J.C. (1961) A proposed new relation for cumulative fatigue damage in bending. *Proc. ASTM*, **61**, 679–92.
37. Subramanyan, S. (1976) A cumulative damage rule based on the knee point of the S–N curve. *J. Engng. Mater. Technol., Trans. ASME*, **98**, 316–22.
38. Hashin, Z. and Rotem, A. (1978) A cumulative damage theory of fatigue failure. *Mater. Sci. Engng.*, **34**, 147–60.
39. Srivatsavan, R. and Subramanyan, S. (1978) A cumulative damage rule based on successive reduction in fatigue limit. *J. Engng. Mater. Technol., Trans. ASME*, **100**, 212–14.
40. Kuno, T., Shimizu, H. and Yamada, K. (1969) *Micro-structural Aspects of Fatigue Behaviour of Rapid Heat Treated Steel.* Proc. 2nd Conf. On Fracture, April 1969, Brighton, UK (ed. P.L. Pratt): Chapman & Hall, pp. 630–42.
41. Kujawski, D. and Ellyin, F. (1984) A cumulative damage theory for fatigue crack initiation and propagation. *Int. J. Fatigue*, **6**, 83–8.
42. Lukáš, P. and Kunz, L. (1981) Influence of notches on high cycle fatigue life. *Mater. Sci. Engng.*, **47**, 93–8.
43. Manson, S.S. and Halford, G.R. (1981) Practical implementation of the double linear damage rule and damage curve approach for treating cumulative fatigue damage. *Int. J. Fract.*, **17**, 169–92.
44. Manson, S.S. (1966) Interfaces between fatigue, creep, and fracture. *Int. J. Fract.*, **2**, 327–63.
45. Miller, K.J. and Zachariah, K.P. (1977) Cumulative damage laws for fatigue crack initiation and Stage I propagation. *J. Strain Anal.*, **12**, 262–70.
46. Ibrahim, M.F.E. and Miller, K.J. (1980) Determination of fatigue crack initiation life. *Fatigue Engng. Mater. Struct.*, **2**, 351–60.
47. Hashin, Z. and Laird, C. (1980) Cumulative damage under two level cycling. *Fatigue Engng. Mater. Struct.*, **2**, 345–50.
48. Manson, S.S. and Halford, G.R. (1986) Re-examination of cumulative fatigue damage analysis – an engineering perspective. *Engng. Fract. Mech.*, **25**, 539–71.
49. Coffin, L.F. (1973) Fatigue at high temperatures, in *Fatigue at High Temperatures*, ASTM STP 520, American Society for Testing and Materials, Philadelphia, PA, pp. 5–43.
50. Morishita, M. and Asada, Y. (1984) Creep–fatigue interaction: data base analysis and application. *Nucl. Engng. Design*, **83**, 367–77.
51. Coffin, L.F. (1977) *Fatigue at High Temperature.* Fracture 1977, Proc. ICF 4, Part I (ed. D.M.R. Taplin). Pergamon Press, Waterloo, Canada, pp. 263–92.

52. Coffin, L.F. (1980) Damage processes in time-dependent fatigue, in *Creep–Fatigue–Environment Interaction* (eds P.M. Pelloux and N.S. Stoloff), The Metallurgical Society of AIME, Warrendale, PA, pp. 1–23.
53. Bui-Quoc, T. (1982) Recent developments of continuous damage approaches for the analysis of material behaviour under fatigue-creep loading, in *Inelastic Analysis and Life Prediction in Elevated Temperature Design*, PVP 59 (ed. G. Baylac), American Society of Mechanical Engineers, New York, pp. 211–26.
54. Ueda, M., Tanigawa, M. and Hara, Y. (1986) *Evaluation of Life Prediction Methods using Published Creep-Fatigue Data*. Proc. Int. Conf. on Creep, April 1986, Tokyo, Japan. Japan Society of Mechanical Engineers, pp. 397–402.
55. Murakami, S. (1987) Progress of continuum damage mechanics. *JSME Int. J.*, **30**, 701–10.
56. Inoue, T. *et al.* (1989) Fatigue–creep life prediction of $2\frac{1}{4}$ Cr-1 Mo steel by inelastic analysis. *Nucl. Engng. Design*, **114**, 311–21.
57. Robinson, E.L. (1952) Effect of temperature variation on the long-time rupture strength of steels. *Trans. ASME*, **74**, 777–81.
58. Taira, S. (1962) Lifetime of structures subjected to varying load and temperature, in *Creep in Structures* (ed. N.J. Hoff), Academic Press, New York, pp. 96–124.
59. Manson, S.S. (1966) Interfaces between fatigue, creep and fracture. *Int. J. Fract. Mech.*, **2**, 327–63.
60. Polhemus, J.F., Spaeth, C.E. and Vogel, W.H. (1973) Ductility exhaustion model for prediction of thermal fatigue and creep interaction, in *Fatigue at Elevated Temperatures*, ASTM STP 520, American Society for Testing and Materials, Philadelphia, PA, pp. 625–36.
61. Manson, S.S., Halford, G.R. and Hirschberg, M.H. (1971) Creep–fatigue analysis by strain-range partitioning, in *Design for Elevated Temperature Environment* (ed. S.Y. Zamrik), American Society of Mechanical Engineers, New York, pp. 12–24.
62. Kachanov, L.M. (1961) Rupture time under creep conditions, in *Problems of Continuum Mechanics*, SIAM, Philadelphia, PA, pp. 202–18.
63. Majumdar, S. and Maiya, P.S. (1980) A mechanistic model for time-dependent fatigue. *J. Engng. Mater. Technol., Trans. ASME*, **102**, 159–67.
64. Majumdar, S. (1984) Relationship of creep, creep–fatigue and cavitation damage in type 304 austenitic stainless steel. *J. Engng. Mater. Technol., Trans. ASME*, **111**, 123–31.
65. Lemaitre, J. and Plumtree, A. (1979) Application of damage concept to predict creep–fatigue failure. *J. Engng. Mater. Technol., Trans. ASME*, **101**, 284–92.
66. Saitoh, M. and Krempl, E. (1982) An incremental life prediction law for creep–fatigue interaction, in *Material Behaviour at Elevated Temperatures and Components Analysis*, PVP 10 (eds Y. Yamada *et al.*), American Society of Mechanical Engineers, New York, pp. 71–9.
67. Asada, Y. *et al.* (1987) Creep–fatigue damage evaluation of 304 stainless steel and $2\frac{1}{4}$ Cr-1 Mo steel based on overstress concept, in *Advances in Piping Analysis and Life Assessment for Pressure Vessels and Piping*, PVP 129 (eds S.J. Chang *et al.*), American Society of Mechanical Engineers, New York, pp. 93–9.
68. American Society of Mechnical Engineers (1994) ASME Boiler and Pressure Vessel Code, Section III, Rules for Construction of Nuclear Power Plant Components, ASME, New York.
69. Rice, J.R. (1981) Constraints on the diffusive cavitation of isolated grain boundary facets in creeping polycrystals. *Acta Metall.*, **29**, 675–81.

70. Ellyin, F. and Asada, Y. (1991) Time-dependent fatigue failure: the creep–fatigue interaction. *Int. J. Fatigue*, **13**, Elsevier Science Ltd, Oxford, pp. 157–64.
71. Verkin, B.I. and Grinberg, N.M. (1979) The effect of vacuum on the fatigue behaviour of metals and alloys. *Mater. Sci. Engng.*, **41**, 149–81.
72. Wei, R.P. and Simmons, G.W. (1981) Recent progress in understanding environmental assisted fatigue crack growth. *Int. J. Fract.*, **17**, 235–47.
73. Coffin, L.F. (1983) Overview of temperature and environment effect on fatigue of structural metals, in *Fatigue, Environment and Temperature Effects* (eds. J.J. Burke and V. Veiss), Plenum Press, New York, pp. 1–40.
74. Gangloff, R.P. (1990) *Corrosion Fatigue Crack Propagation*. Proc. Environment-Induced Cracking of Metals, NACE-10, Houston, Texas (eds R.P. Gangloff and M.B. Ives). National Association of Corrosion Engineers, pp. 55–109.
75. Marchionni, M., Ranucci, D. and Picco, E. (1982) *High-Temperature Low-Cycle Fatigue Behaviour of IN738LC Alloy in Air and Vacuum*. Proc. Conf. High Temperature Alloys for Gas Turbines, Dordrecht, Holland (ed. R. Brunetaund), pp. 791–804.
76. Liu, K.C. and Loring, C.M. Jr (1984) Low-cycle fatigue behaviour of oxygen-free high-conductivity copper at 300°C in high vacuum. *J. Nucl. Mater.*, **122, 123**, 783–8.
77. Earthman, J.C., Eggeler, G. and Ilschner, B. (1989) Deformation and damage process in a 12% Cr-Mo-V steel under high temperature low cycle fatigue condition in air and vacuum. *Mater. Sci. Engng.*, **A110**, 103–14.
78. Ranucci, D., Marchionni, D., Picco, E., Gherardi, F. and Caciorgna, O. (1983) *Effect of Strain Rate in Air and Vacuum on the High-Temperture Low-Cycle Fatigue Behaviour of Cast IN100 Alloy*. Proc. 6th Int. Conf. on Strength of Metals and Alloys, Melbourne, Australia, vol. 2 (ed. Grfkins). Pergamon Press, pp. 927–32.
79. Van Der Sluys, W.A. and Yukawa, S. (1995) Status of PVRC evaluation of LWR coolant environment effects on the S–N fatigue properties of pressure boundary materials, in *Fatigue and Crack Growth – Environmental Effects, Modeling Studies and Design Considerations*, PVP 306 (eds S. Yukawa, D.P. Jones and H.S. Mehta), American Society of Mechanical Engineers, New York, pp. 47–58.
80. Ellyin, F. and Li, B. (1993) Crack initiation and fatigue lives in air and vacuum environments. *Mater. Sci. Engng.*, **A171**, 105–13.
81. Wang, R., Mughrabi, H., McGovern, S. and Rapp, M. (1984) Fatigue of copper single crystals in vacuum and in air, I: persistent slip bands and dislocation microstructures. *Mater. Sci. Engng.*, **65**, 219–33.
82. Wang, R. and Mughrabi, H. (1984) Fatigue of copper single crystals in vacuum and in air, II: Fatigue crack propagation. *Mater. Sci. Engng.*, **65**, 235–43.
83. Kujawski, D. and Ellyin, F. (1992) Initiation and total fatigue life of a low alloy carbon steel in vacuum and air. *J. Testing Eval.*, **20**, ASTM, 391–5.
84. Li, P., Marchand, N.J. and Ilschner, B. (1989) Crack initiation mechanisms in low cycle fatigue of aluminum alloy. *Mater. Sci. Engng.*, **A119**, 41–50.
85. Garud, Y.S. (1991) Qualitative evaluation of environmentally assisted cracking: A survey of developments and application of modeling concepts. *J. Pressure Vessel Technol., Trans. ASME*, **113**, 1–9.
86. Koh, S.K. and Stephens, R. (1991) Mean stress effects on low cycle fatigue for a high strength steel. *Fatigue Fract. Engng. Mat. Struct.*, **14** (4), 413–28.
87. Wang, J. (1992) Low cycle fatigue and cycle dependent creep with continuum damage mechanics. *Int. J. Damage Mechanics*, **1**, 237–44.

88. Pokluda, J. and Stanek, P. (1978) Evaluation of cyclic creep for 12010, 14331 steels and effect of cyclic asymmetry on life in the range of time-dependent fatigue reliability (in Czech). *Kovové Materiály*, **5** (16), 583–99.
89. Manson, S.S. (1966) *Thermal stress and low cycle fatigue*, McGraw-Hill, New York.
90. Kujawski, D. and Ellyin, F. (1995) A unified approach to mean stress effect on fatigue threshold conditions. *Int. J. Fatigue*, **17**, 101–6.
91. Xia, Z., Kujawski, D. and Ellyin, F. (1996) Effect of mean stress and ratcheting strain on fatigue life of materials. *Int. J. Fatigue*, **18** (5), 333–41.

Fatigue failure under multiaxial states of stress 4

4.1 INTRODUCTION

Multiaxial stress states are encountered in the majority of mechanical/structural components, e.g. shafts, beams, pressure vessels, machine elements, etc. Indeed the very early investigations into fatigue were on shafts of rotating machines where bending and torsional couples were applied to simulate service loads [1]. The motivation for multiaxial fatigue investigations has been to produce design guidelines, preferably based on simple uniaxial stress data.

Thus, most of the effort in the past was directed towards refinement of empirical relationships, and less attention was paid to the fatigue mechanism and theoretical developments. Although investigations into yield theories and fatigue started at about the same time, their paths soon diverged. While the former progressed further towards development of inelastic constitutive relations (Chapter 6), research into fatigue was mostly confined to the elastic regime. Perhaps one of the major factors was the inability of theories of the strength of material to predict failure at load levels below a material's ultimate or yield strength. It was not until the accelerated development of linear fracture mechanics in the fifties [2] that efforts were directed towards describing the fatigue mechanism through crack growth. Even at present, there is no agreed theory describing fatigue failure by a constitutive-type approach. Later on in this chapter we will describe such a theory based on the cyclic strain energy. Here, as in the previous chapter, the word 'failure' is used in a general sense. It refers to attaining a macro-crack of a certain size.

In the multiaxial stress state, essentially we wish to describe a hypersurface which is the locus of the fatigue failure. For example, in the plane stress (biaxial) state we could represent the fatigue failure domain as a surface, schematically shown in Fig. 4.1. This surface is not generally symmetric with respect to the life axis, N. At low life, the base is large and it

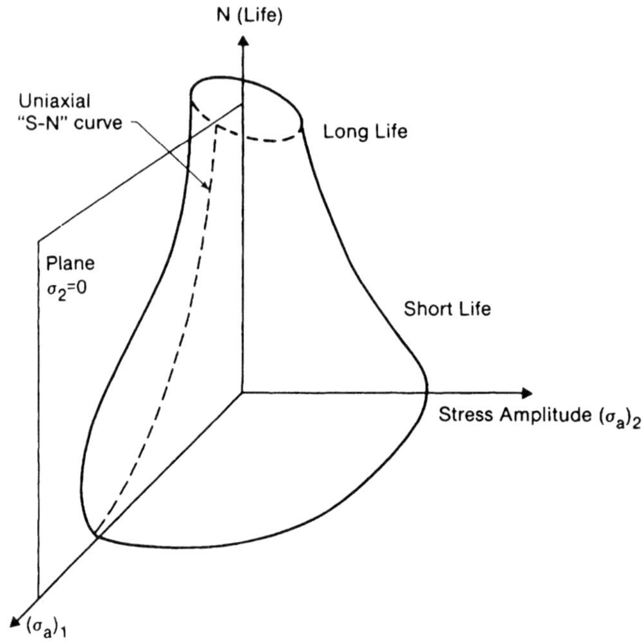

Fig. 4.1 Schematic representation of a failure surface in the biaxial stress state.

tapers to a smaller cross-section as the stress magnitude decreases (high life regime). Constant life contours may be obtained through the intersection with the $N =$ constant planes.

The constant life contours may change in shape, and attempts have been made to approximate them by various yield or failure theories, e.g. the von Mises or Tresca yield condition. This, however, is not satisfactory as will be explained later.

At this stage of development there are a number of approaches being used to predict multiaxial fatigue: modified yield theories; various combinations of yield theories and strain energy; strain energy per cycle, critical plane approach, etc. In the following some of the major proposals will be briefly reviewed. However, before embarking on the discussion of various approaches, it would be useful to define what we mean by a multiaxial stress state.

We already know that for the simplest stress state, i.e. uniaxial, the state of strain is three-dimensional, i.e. $\varepsilon_2 = \varepsilon_3 = -\nu\varepsilon_1$. Furthermore, for the same uniaxial stress condition, on planes inclined but not perpendicular or parallel to the loading direction, there are two stress components: shear and normal stresses. Thus, when we refer to a uniaxial loading condition, it

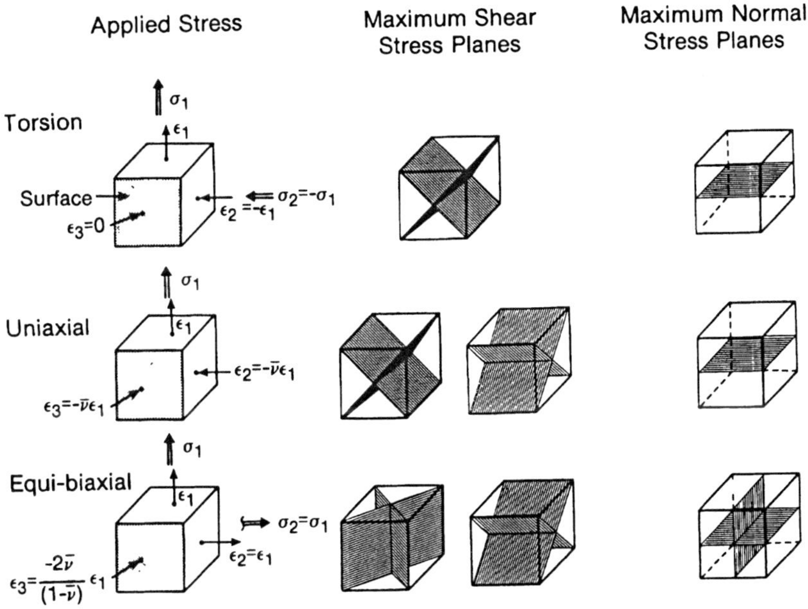

Fig. 4.2 An element near a surface subjected to three states of stress (torsion, uniaxial and equi-biaxial) and the corresponding strain states. The maximum shear stress planes and maximum normal stress planes for each stress state are shown in the second and third columns, respectively.

must be interpreted as an applied uniaxial principal stress, σ_1. Consequently, a multiaxial loading has to be identified with the existence of more than one principal stress component. Figure 4.2 shows a material element near a stress-free surface of a component subject to three stress states, namely torsion, uniaxial and equi-biaxial loading. We note that the uniaxial and equi-biaxial loading conditions are obtained from the torsional loading by adding a tensile stress equal to the maximum principal stress, σ_1, successively (see the first column in Fig. 4.2). Observe that the principal strains do not follow the same pattern, and an in-plane strain ratio $\rho = \varepsilon_2/\varepsilon_1$ varies from -1 to $-\bar{\nu}$ to $+1$, where $\bar{\nu}$ is a composite Poisson's ratio bounded between elastic and plastic values. The corresponding principal shear stress planes are shown in the second column and the maximum normal stress planes are depicted in the third column of Fig. 4.2. We will return to this figure later on, and at this stage it suffices to mention that fatigue life decreases as one proceeds from the torsion to the uniaxial and the equi-biaxial stress states shown in Fig. 4.2 for the same applied equivalent stress or strain.

4.2 PREVIOUS INVESTIGATIONS*

The study of multiaxial fatigue has followed a similar subdivision to that of uniaxial stress, i.e. multiaxial high- and low-cycle fatigue. The demarcation line is not clear except that in high-cycle fatigue the bulk material response is elastic, although it has been recognized that inelastic strains occur at some surface grains or localized regions where micro-cracks or defects are present. Biaxial testing equipment, mostly torsion–axial load systems, had been fully developed for operation in the elastic response domain. Therefore, early investigations were directed towards high-cycle, low strain regimes. The results of these investigations have been well-documented in most texts on fatigue, e.g. see [4] for a reasonably complete review of high-cycle multiaxial fatigue up to 1974.

It is not our intention to present here an exhaustive review of the multiaxial fatigue literature (such a review is given by Ellyin and Valaire [5]). Instead, we will attempt to classify the proposed fatigue failure criteria for the multiaxial stress states under three major headings and show the direction of recent developments.

As mentioned before, early investigators were mostly interested in providing design formulae and generally did not delve into the fatigue damage mechanism. However, as early as 1943, Guest [6] had pointed out that in the case of long life fatigue, damage started at points of localized plasticity, and theories used to predict the onset of plastic flow would also predict damage due to cyclic loading. Indeed numerous test results later confirmed that Tresca and von Mises yield conditions **do** reasonably predict multiaxial long life fatigue for ductile materials from uniaxial results. For brittle materials, the appropriate failure criteria are the maximum principal stress and the maximum principal strain.

A general criticism of the long life fatigue criteria based on the yield theories is their insensitivities to hydrostatic pressure. Tests have shown that fatigue life is decreased by the presence of hydrostatic tension, and is increased by hydrostatic compression. The effect of hydrostatic pressure is small but nevertheless definite, which renders pressure insensitive criteria unacceptable on theoretical grounds (see also discussion in section 3.5.5).

Let us now examine a general functional form of the fatigue failure criterion. The foregoing discussion suggests that such a criterion can be represented in the form of

$$f(I_1, I_2) = q(2N_f)\,, \tag{4.1}$$

where $I_1 = \sigma_1 + \sigma_2 + \sigma_3$ and $I_2 = \sigma_1\sigma_2 + \sigma_2\sigma_3 + \sigma_3\sigma_1$, are the first and second invariants of the stress tensor in the principal stress space, and N_f is the number of reversals to failure. An explicit form of the RHS of (4.1) can be

*This section and parts of section 4.3 are largely based on a review article by the author [3].

obtained by noting that for high-cycle fatigue, $2N_f \to \infty$ and $q(2N_f) \to C$, thus a possible form of $q(2N_f)$ is

$$q(2N_f) \sim K(2N_f)^{\alpha} + C\,, \tag{4.2}$$

with $\alpha < 0$ and $C > 0$.

4.2.1 STRESS-BASED CRITERIA

In the case of high-cycle fatigue, $2N_f \to \infty$, and (4.1) then becomes

$$f(I_1, I_2) = C\,. \tag{4.3}$$

We note that (4.3) is similar in form to the pressure sensitive yield surface functions in the theory of plasticity. In particular, when the effect of the hydrostatic pressure is neglected, the simplest form of (4.3) is then

$$J_2 = C_0\,, \tag{4.4}$$

where $J_2 = s_{ij}s_{ij}/2$ is the second invariant of the deviatoric stress

$$s_{ij} = \sigma_{ij} - \sigma_{kk}\delta_{ij}/3\,, \tag{4.5}$$

and $J_1 \equiv 0$. Condition (4.4) with $C_0 = \sigma_0^2/3$ is the von Mises yield condition, where σ_0 is the yield stress in simple tension. In the case of long life fatigue, σ_0 is replaced by a material's fatigue (endurance) limit or a failure stress amplitude, σ_{FL}.

Some of the widely-used stress-based criteria for fatigue failure are special cases of (4.4). For example, let us write (4.4) in its explicit form in terms of the stress components, σ_{ij}, i.e.

$$(\sigma_x - \sigma_y)^2 + (\sigma_y - \sigma_z)^2 + (\sigma_z - \sigma_x)^2 + 6(\tau_{xy}^2 + \tau_{yz}^2 + \tau_{zx}^2) = 2\sigma_{FL}^2\,, \tag{4.6}$$

and in the principal stress space, it has the form of

$$(\sigma_1 - \sigma_2)^2 + (\sigma_2 - \sigma_3)^2 + (\sigma_3 - \sigma_1)^2 = 2\sigma_{FL}^2\,. \tag{4.7}$$

Adopting the notation $\sigma_1 > \sigma_2 > \sigma_3$, the maximum shear stress is on the plane inclined at 45° with respect to the principal planes 1 and 3, and the magnitudes of the shear and normal stresses on this plane are

$$\tau_{max} = \frac{\sigma_1 - \sigma_3}{2} \quad \text{and} \quad \sigma_n = \frac{\sigma_1 + \sigma_3}{2}\,. \tag{4.8}$$

On the **maximum shear plane**, the von Mises criterion is given by

$$3\tau_{max}^2 + \sigma_n^2 = \sigma_{FL}^2\,, \tag{4.9}$$

which indicates a relationship between the maximum shear stress and the normal stress on the plane of the maximum shear stress. Equation (4.9) can be written in the general form

$$\tau_{max} + f(\sigma_n) = 0\,. \tag{4.10}$$

It can easily be shown that the failure criteria proposed by several investigatiors [7–10] are particular variations of (4.10). Fatigue failure criteria of the form (4.10) associated with the maximum shear stress plane have recently been termed 'critical plane' criteria.

The stress-based criteria discussed above are more suitable in predicting long life (high-cycle) fatigue failure [11, 12]. Table 4.1 gives a summary of some of the widely quoted criteria. As noted in Table 4.1, suitable adjustment of a coefficient reduces the majority of these criteria to either the Tresca or von Mises yield condition. Introduction of a second material constant [7–10] is intended to improve the correlation with the experimental data. However, the perceived improvement may be illusory, given the accuracy with which stresses were applied and measured in the experimental investigations. Therefore, a single material constant criterion would suffice and could be considered adequate.

4.2.2 STRAIN-BASED CRITERIA

Strain-based criteria are generally used in connection with low-cycle fatigue. At highly stressed regions, inelastic strains are constrained by the surrounding elastic matrix; therefore, a strain-controlled test is a fair representation of the conditions experienced by the material at regions stressed beyond the elastic limit. Progress in this area accelerated once the closed-loop testing facility became widely available, mostly in the early sixties. Earlier investigations were mostly concerned with the uniaxial stress condition. This activity culminated with the Coffin–Manson relationship [13, 14] (section 3.3).

The extension to the multiaxial stress state has been through definition of a suitable equivalent strain measure. The general relation can be written in the form of

$$\Delta\varepsilon_{\mathrm{eq}}^{\mathrm{p}} = k(2N_{\mathrm{f}})^{c}, \tag{4.11}$$

where Δ refers to the range during a cycle and k is a constant. Various equivalent strain measures which are used are: the octahedral shear strain [12, 15], the maximum shear strain [12, 16], the maximum normal strain [17], the modified von Mises strain [16, 18, 19], the maximum total strain [20] and the von Mises equivalent strain [21]. Some of the frequently used equivalent (or effective) strain measures are summarized in Table 4.2. It should be noted that there is only a small amount of reliable experimental data in this area, and caution should be exercised in applying these criteria in a design situation.

The equivalent strain is given by the following expression:

$$\varepsilon_{\mathrm{eq}} = \beta'[(\varepsilon_1 - \varepsilon_2)^2 + (\varepsilon_2 - \varepsilon_3)^2 + (\varepsilon_3 - \varepsilon_1)^2]^{1/2}. \tag{4.12}$$

Table 4.1 Stress-based fatigue failure criteria

Criterion	Relationship	Notes
Maximum principal stress	$\sigma_1 = \sigma_{FL}$	$\sigma_1 > \sigma_2 > \sigma_3$
Maximum shear stress (Tresca)	$\tau_{max} = \frac{1}{2}\max[\lvert\sigma_1 - \sigma_2\rvert, \lvert\sigma_2 - \sigma_3\rvert, \lvert\sigma_3 - \sigma_1\rvert] = \frac{1}{2}\sigma_{FL}$	Hydrostatic pressure insensitive
Strain energy density (SED)	$\sigma_1^2 + \sigma_2^2 + \sigma_3^2 - 2\nu(\sigma_1\sigma_2 + \sigma_1\sigma_3 + \sigma_2\sigma_3) = \sigma_{FL}^2$	
Distortion strain energy density (von Mises)	$(\sigma_1 - \sigma_2)^2 + (\sigma_2 - \sigma_3)^2 + (\sigma_3 - \sigma_1)^2 = 2\sigma_{FL}^2$	Hydrostatic pressure insensitive
Gough and Pollard [7]	$\left(\frac{\sigma}{\sigma_{FL}}\right)^2 + \left(\frac{\tau}{\tau_{FL}}\right)^2 = 1$	When $\sigma_{FL}/\tau_{FL} = 2$ (Tresca) $\sigma_{FL}/\tau_{FL} = \sqrt{3}$ (von Mises) $\sigma_{FL}/\tau_{FL} = [2(1+\nu)]^{1/2}$ (SED)
Stanfield [8]	$\tau_{max} + \lambda\sigma_n = \tau_{FL}$	For flexion $\lambda = (2\tau_{FL}/\sigma_{FL}) - 1$
Findley [9]	$(\sigma_n/\sigma_{FL})^{\sigma_{FL}/\tau_{FL}} + (\tau_{max}/\tau_{FL})^2 = 1$	When $\sigma_{FL}/\tau_{FL} = 2$ (Tresca)
McDiarmid [10]	$\tau_{max} + \lambda\sigma_n^n = \tau_{FL}$	$\lambda = \frac{(\tau_{FL} - \sigma_{FL}/2)}{(\sigma_{FL}/2)^n}$ When $\tau_{FL} = \sigma_{FL}/2$ (Tresca)
Equivalent stress	$\sigma_{eq} = \beta[(\sigma_1 - \sigma_2)^2 + (\sigma_2 - \sigma_3)^2 + (\sigma_3 - \sigma_1)^2]^{1/2} = \sigma_{FL}$	$\beta = 1/\sqrt{2}$ (von Mises) $\beta = \frac{1}{3}$ (octahedral shear stress) $\beta = 1/\sqrt{15}$ (RMS of shear on all planes through a point) Hydrostatic pressure insensitive

Table 4.2 Strain-based fatigue failure criteria

Criteria	Relationship	Notes
Equivalent plastic strain	$\Delta\varepsilon^p_{eq} = kN_f^c$	$\varepsilon^p_{eq} = \gamma^p_{oct}/\sqrt{2} = [(\varepsilon^p_1)^2 + (\varepsilon^p_2)^2 + (\varepsilon^p_3)^2]^{1/2}/\sqrt{3}$ (von Mises equivalent strain) Insensitive to hydrostatic pressure[a]
Maximum plastic shear strain	$\Delta\gamma^p_{max} = kN_f^c$	$\gamma^p_{max} = \max[\lvert\varepsilon^p_1 - \varepsilon^p_2\rvert, \lvert\varepsilon^p_2 - \varepsilon^p_3\rvert, \lvert\varepsilon^p_3 - \varepsilon^p_1\rvert]$ (Tresca equivalent strain) Insensitive to hydrostatic pressure
Maximum normal strain	$(\Delta\varepsilon_n)_{max} = kN_f^c$	$(\varepsilon_n)_{max} = \max[\lvert\varepsilon_1\rvert, \lvert\varepsilon_2\rvert, \lvert\varepsilon_3\rvert]$
Maximum total strain	$\Delta\varepsilon^t_{max} = kN_f^c$	$\varepsilon^t_{max} = \mathrm{Max}[\varepsilon_1^2 + \varepsilon_2^2 + \varepsilon_3^2]^{1/2}$ (when $\nu = \frac{1}{2}$, $\varepsilon^t_{max} = \sqrt{3}\gamma_{oct}/2$)
Brown and Miller [23]	$\gamma_{max} = g(\varepsilon_n)$	Represents constant life contours in γ_{max} vs. ε_n plane. g can be made to vary with life.
Ellyin and Valaire (AMPSTF) [24]	$N_f = N_{fo}a^{\bar{\varepsilon}}$	$\bar{\varepsilon} = \varepsilon^p_1 + \varepsilon^p_2$ for in-phase condition $\bar{\varepsilon} = \varepsilon^p_1$ for fully out-of-phase N_{fo} and a are determined from uniaxial tests.

[a] To account for the hydrostatic pressure effect, it has been proposed to make the 'material constants' k and c dependent on stress or strain, e.g. [17, 21, 22].

When $\beta' = \frac{2}{3}$, then $\varepsilon_{eq} = \gamma_{oct}$, the octahedral shearing strain. If the plastic components of strain are used, the constancy of volume results in

$$\gamma^p_{oct} = \sqrt{\frac{2}{3}}\,[(\varepsilon^p_1)^2 + (\varepsilon^p_2)^2 + (\varepsilon^p_3)^2]^{1/2}\,. \tag{4.13}$$

Similarly when $\beta' = \sqrt{2}/3$, we obtain the von Mises equivalent strain. In particular, the plastic component is given by setting $\nu = 0.5$, i.e.

$$\varepsilon^p_{eq} = (\tfrac{1}{3})^{1/2}\,[(\varepsilon^p_1)^2 + (\varepsilon^p_2)^2 + (\varepsilon^p_3)^2]^{1/2}\,. \tag{4.14}$$

Note that the difference between the octahedral plastic strain (4.13) and the equivalent von Mises plastic strain (4.14) is a numerical factor, i.e. $\varepsilon^p_{eq}/\gamma^p_{oct} = 1/\sqrt{2}$.

Sines and Ohgi [12] concluded that the plastic alternating octahedral shear strain $\Delta\gamma^p_{oct}$ (4.13), and the maximum alternating shear strain $\Delta\gamma^p_{max}$ (Table 4.2), correlate reasonably well with the experimental results.

In a manner similar to the stress-based criteria, we can find an explicit relation between the maximum shear strain γ_{max}, and the normal strain to the maximum shear direction, ε_n. When the principal strains are ordered, i.e. $\varepsilon_1 > \varepsilon_2 > \varepsilon_3$, then

$$\gamma_{max} = \varepsilon_1 - \varepsilon_3 \quad \text{and} \quad \varepsilon_n = \frac{\varepsilon_1 + \varepsilon_3}{2} . \tag{4.15}$$

Failure is assumed to take place when substituting into (4.12) from (4.15) and neglecting the effect of intermediate principal strain, ε_2. We then obtain

$$\gamma_{max}^2 + \frac{4}{3}\varepsilon_n^2 = \frac{2}{3}\left(\frac{\gamma_F}{\beta'}\right)^2 , \tag{4.16}$$

where γ_F is a failure strain. The above relation can be expressed in the general form

$$\gamma_{max} + g(\varepsilon_n) = 0 , \tag{4.17}$$

which is (4.10) written in terms of strains. Brown and Miller [23] have argued that in metals, cracks are initiated when crystallographic slip planes are activated by the shear strain on the **planes** of maximum shear strain. (Shear strains, when small, are the decrease in angle between a pair of lines which originally form a 90° angle. The maximum shear strain is specified by a direction in a plane, thus the word 'planes' of maximum shear strain is strictly ill defined.) The rate of crack growth is accelerated when the normal strain on the 'plane of maximum shear strain' is tensile. The same factors were thought to influence the rate of crack growth in Stage II when the crack advances by a process of slip and decohesion in the shear ears at the crack tip (Chapter 7). Therefore (4.17) would represent constant life contours in the γ_{max} vs. ε_n plane, the so-called 'Γ-plane'. This type of theory has also been termed 'critical plane' theory in recent years. Various forms of function g can be obtained by assuming a suitable failure criterion. For example, the octahedral shear strain criterion is obtained from (4.16) by substituting for $\beta' = \frac{1}{3}$, thus the functional relationship becomes

$$\gamma_{max}^2 + \tfrac{4}{3}\varepsilon_n^2 = 6\gamma_F^2 . \tag{4.18}$$

Makinde and Neale [25] have generalized (4.17) in a parametric form as

$$g(\theta, N_f) = k f_1(\theta) f_2(N_f) , \tag{4.19}$$

where g is a vector from the origin to a point on the life contour on the γ–ε_n plane, θ is the associated polar angle, N_f is the number of cycles to failure, k is a constant, and $f_1(\theta)$ and $f_2(N_f)$ are functions of θ and N_f, respectively. They were able to show that a number of previously proposed strain-based criteria are particular cases of (4.19).

Ellyin and Valaire [24] have proposed a criterion which is a variation of the maximum principal strain theory. It can be represented in the form of

$$N_f = N_{fo} a^{\bar{\varepsilon}} , \tag{4.20}$$

where $\bar{\varepsilon} = \max[|\varepsilon_1^p|, |\varepsilon_2^p|, |\varepsilon_3^p|]$ where ε_1^p, ε_2^p and ε_3^p are principal plastic strains. This criterion was termed 'absolute maximum plastic strain theory of failure' (AMPSTF). It states that similar fatigue lives will be achieved for biaxial and uniaxial conditions when the maximum plastic strains are equal. The experimental results [24] showed trends which could also have been fitted to either the plastic octahedral shear strain theory or the maximum plastic shear strain (Table 4.2). However, the AMPSTF was found to be a better fit, and it is also consistent with the crack growth orientation.

Socie [26] has reviewed the so-called critical plane approach, which attempts to connect fatigue crack initiation at the surface of materials with the maximum shear strain range and/or the maximum normal strains. He concludes that, 'Stresses play an important role and cannot be ignored'. Damage parameters based on energy contain both stress and strain, and therefore may be deemed more appropriate models for fatigue damage assessment.

4.2.3 ENERGY-BASED CRITERIA

The stress- or strain-based criteria discussed in the preceding sections do not account for the interaction between the stress and strain in a deformation process. They cannot, therefore, reflect the path dependence of the material response. Also, some of the proposed criteria do not possess the invariant properties with respect to the coordinate transformation – frame indifference properties of constitutive laws in continuum mechanics.

To overcome these shortcomings and with the realization that the plastic strain energy causes irreversible damage, energy-based criteria have been proposed. Ellyin and co-workers [27, 28] have formulated a fatigue failure theory by relating the cyclic plastic strain energy to that of the uniaxial plastic strain energy. The von Mises equivalence between the uniaxial and multiaxial condition was assumed, and the failure criteria of the form

$$\Delta\sigma_{eq}\,\Delta\varepsilon_{eq} = kN_f^{\alpha}, \tag{4.21a}$$

or

$$\Delta\sigma_{eq}\,\Delta\varepsilon_{eq}^{p} = k'N_f^{\alpha}, \tag{4.21b}$$

were derived. In the above relations $\Delta\sigma_{eq}$ and $\Delta\varepsilon_{eq}$ (or $\Delta\varepsilon_{eq}^{p}$) are the equivalent stress and strain ranges based on the von Mises criterion, and $k\,(k')$ and α are material parameters which can be either constant or depend on the multiaxial stress state. The predicted results were compared with the experimental data, and the agreement was found to be good [28].

A similar theory was also presented by Garud [29] where plastic work per cycle, W_c, was related to fatigue life, N_f, through a power law:

$$N_f = AW_c^{\gamma}, \tag{4.22}$$

where

$$W_c = \int_{\text{cycle}} \sigma_{ij} \, d\varepsilon_{ij}^{p},$$

and an incremental theory was used to relate stress and strain increments. In correlating the cyclic plastic strain energy with the experimental results an unexplained 'weighting factor' of $\frac{1}{2}$ was applied to the shear strain energy, thus obtaining a reduced plastic work. This is not justified in view of the results presented in [28], and more recent work [30].

4.3 A GENERAL APPROACH TO MULTIAXIAL FATIGUE

It has been shown in Chapter 3 that the strain energy per cycle, ΔW, is an appropriate damage parameter and correlates well with a wide range of experimental data. The salient feature of the approach is based on the premise that the damage caused in the material is related to the mechanical input energy. A major part of this energy is dissipated into heat and vibration, and the remaining part causes plastic deformation and subsequent crack initiation and propagation. In the early stages the energy is required for the dislocation movement along the crystallographic planes and formation of slip bands.

If cracks initiate in grains whose glide planes are oriented along the maximum shear stress planes, then we see from the second column of Fig. 4.2 that for torsional loading the initiated cracks will grow along the surface. In other words, the initiated cracks will be shallow. For uniaxial stress (second row of Fig. 4.2) the cracks may also grow into the material (across the thickness). In the case of equi-biaxial loading, the initiated cracks will grow across the thickness, i.e. the crack depth will be deeper compared to the other two cases. The difference between the uniaxial and equi-biaxial maximum shear planes is that in the equi-biaxial case both pairs of planes at 45° with 1–3 and 2–3 principal planes will initiate crack growth into the material (third row, second column of Fig. 4.2). However, in the case of uniaxial loading on the plane at 45° with the 2–3 principal stress planes, both shear and normal stress are zero. Similarly, the difference between the torsional and uniaxial maximum shear planes making a 45° angle with the 1–2 principal stress planes, is the absence of the normal stress in the case of torsional loading. Table 4.3 summarizes the states of stress on various maximum shear planes for the above three loading cases. In this table n_i $(i = 1, 2, 3)$ are the components of the unit normal to the maximum shear stress plane with respect to the principal stress axes, σ_1, σ_2, σ_3. Since the fatigue life decreases as one proceeds from torsion to equi-biaxial loading, it becomes evident that not only the maximum shear stress, but also the plane on which it acts with respect to the free surface, and also

Table 4.3 Directions of maximum shear stress planes and magnitudes of shear and normal stresses acting on them (Fig. 4.2)

Stress state	Maximum shear planes				
	n_1	n_2	n_3	$\lvert\tau\rvert$	$\lvert N\rvert$
Torsion	$\pm\sqrt{2}/2$	$\mp\sqrt{2}/2$	0	σ_1	0
Uniaxial	$\pm\sqrt{2}/2$	$\mp\sqrt{2}/2$	0	$\sigma_1/2$	$\sigma_1/2$
	$\pm\sqrt{2}/2$	0	$\mp\sqrt{2}/2$	$\sigma_1/2$	$\sigma_1/2$
Equi-biaxial	$\pm\sqrt{2}/2$	0	$\pm\sqrt{2}/2$	$\sigma_1/2$	$\sigma_1/2$
	0	$\pm\sqrt{2}/2$	$\mp\sqrt{2}/2$	$\sigma_1/2$	$\sigma_1/2$

the normal stress, play an important role in crack initiation and early crack growth (Stage I). It is interesting to note that on all the maximum shear stress planes identified in Table 4.3, the sum of shear stress and normal stress magnitudes are the same. Yet the crack planes and fatigue lives are quite different. Thus, the critical plane theories which advocate damage functions of the form (4.10), e.g.

$$\psi = \tau_{\text{max}} + k\,\sigma_{\text{n}} \tag{4.23}$$

are not suitable unless the material parameter k is made to change with the loading condition.

An extensive experimental and analytical investigation was carried out by Grubisic and Simbürger [31] which indicated that, irrespective of the change of direction of the principal axes (proportional or non-proportional loading), slipping on all planes (not only the critical plane) influences fatigue failure.

In order to treat in a coherent manner the history that a material experiences during a multiaxial loading, we will use the local strain energy input as a damage parameter. The severity of the input energy depends on the nature of the imposed constraint and the level of tensile mean stress.

Once cracks are initiated in the local slip bands, energy is required to drive the cracks. It is clear that both phases, i.e. initiation and propagation, require energy and we have demonstrated in Chapter 3 that this energy is related to the cyclic strain energy, ΔW. It should be noted that the strain energy expression involves both shear stress/strain and normal stress/strain components and thus could properly represent the initiation and propagation phases of a fatigue crack. An advantage of the energy formulation is that once the states of stress and strain in a material element are known, the calculated energy is a scalar, and thus independent of the orientation of reference axes. Another attractive feature of this energy quantity is its ability to unify high- and low-cycle regimes.

In this chapter we will extend the energy concept to multiaxial stress states. The first step in such an extension is to calculate the strain energy per cycle. We will use the incremental separation of the total strain into the elastic and plastic components, i.e.

$$d\varepsilon_{ij} = d\varepsilon_{ij}^{e} + d\varepsilon_{ij}^{p}, \tag{4.24}$$

and calculate the elastic and plastic parts of the strain energy separately.

4.3.1 ELASTIC STRAIN ENERGY

The elastic strain increment is related to the stress increment through the generalized Hooke's law (Chapter 2, (2.29))

$$d\varepsilon_{ij}^{e} = \frac{1+\nu}{E} d\sigma_{ij} - \frac{\nu}{E} d\sigma_{kk}\delta_{ij}, \tag{4.25}$$

where ν is Poisson's ratio, E is Young's modulus, and δ_{ij} is the Kronecker delta equal to unity when $i=j$, otherwise nil. The elastic strain energy density (energy per unit volume) is determined from

$$W^{e} = \int_{0}^{\varepsilon_{ij}} \sigma_{ij} d\varepsilon_{ij}^{e}. \tag{4.26}$$

Substituting from (4.25) into the above we can carry out the integration which results in

$$W^{e} = \frac{1+\nu}{2E}\sigma_{ij}\sigma_{ij} - \frac{\nu}{2E}(\sigma_{kk})^{2}. \tag{4.27}$$

The stress tensor, σ_{ij}, can be separated into the deviatoric and spherical parts, i.e.

$$\sigma_{ij} = s_{ij} + \tfrac{1}{3}\sigma_{kk}\delta_{ij}\frac{\delta_{ij}}{3}. \tag{4.28}$$

Introducing the above into (4.27), the elastic strain energy can be written as

$$W^{e} = \frac{1+\nu}{2E}s_{ij}s_{ij} + \frac{1-2\nu}{6E}(\sigma_{kk})^{2}. \tag{4.29}$$

It is convenient, at times, to express (4.29) in terms of the equivalent stress, σ_{eq}. From Table 4.1 and (4.4) the von Mises equivalent stress is given by

$$\sigma_{eq}^{2} = \tfrac{3}{2}s_{ij}s_{ij} = 3J_{2}, \tag{4.30a}$$

and the first stress invariant, I_1, is defined as

$$I_{1} = \sigma_{kk} = \sigma_{1} + \sigma_{2} + \sigma_{3}. \tag{4.30b}$$

Substituting from (4.30) into (4.29), we get

$$W^{\mathrm{e}} = \frac{1+v}{3E}\sigma_{\mathrm{eq}}^2 + \frac{1-2v}{6E}I_1^2 . \tag{4.31}$$

The first term on the RHS of (4.29) or (4.31) is the distortion energy and the second term is the energy associated with the volumetric change. In the case of fully plastic deformation, $v = 0.5$, and therefore the energy due to the volume change (the second term on the RHS of (4.29) or (4.31)) becomes zero.

To give an indication of the two energy parts, i.e. the strain energy and distortion energy, let us consider a plane stress condition where $\sigma_{i3} = 0$. In principal stress space, σ_1, σ_2, (4.29) reduces to

$$W^{\mathrm{e}} = \frac{1}{2E}[(\sigma_1 - \sigma_2)^2 + 2(1-v)\sigma_1\sigma_2] \tag{4.32a}$$

and the corresponding distortion energy is

$$W_{\mathrm{d}}^{\mathrm{e}} = \frac{1+v}{3E}[(\sigma_1 - \sigma_2)^2 + \sigma_1\sigma_2] . \tag{4.32b}$$

Now, let us equate each energy part to its uniaxial value ($\sigma_1 = \sigma_u$, $\sigma_2 = 0$), i.e.

$$W^{\mathrm{e}} = \frac{1}{2E}[(\sigma_1 - \sigma_2)^2 + 2(1-v)\sigma_1\sigma_2] = \frac{1}{2E}\sigma_{\mathrm{u}}^2 \tag{4.33a}$$

$$W_{\mathrm{d}}^{\mathrm{e}} = \frac{1+v}{3E}[(\sigma_1 - \sigma_2)^2 + \sigma_1\sigma_2] = \frac{1+v}{3E}\sigma_{\mathrm{u}}^2 . \tag{4.33b}$$

We can thus plot the above equations in a dimensionless stress space as shown in Fig. 4.3. The experimental data for the onset of plastic deformation in monotonic loading, is generally closer to the distortion energy plot in Fig. 4.3, which is the von Mises criterion (see also Chapter 6). The high-cycle fatigue data are generally close to the W^{e} plot.

The strain energy per load reversal, i.e. strain energy range, may be obtained from (4.29) by adding a prefix Δ in front of the stress measures, i.e.

$$\Delta W^{\mathrm{e}} = \frac{1-v}{2E}\Delta s_{ij}\Delta s_{ij} + \frac{1-2v}{6E}(\Delta\sigma_{kk})^2 , \tag{4.34}$$

where

$$\Delta s_{ij} = s_{ij}^{\max} - s_{ij}^{\min} \quad \text{and} \quad \Delta\sigma_{kk} = \sigma_{kk}^{\max} - \sigma_{kk}^{\min} .$$

Note that the above definitions apply only for proportional loading.

In the case of cyclic loading the elastic energy is recovered during a complete cycle. However, in a series of tests [32, 33] it was reported that

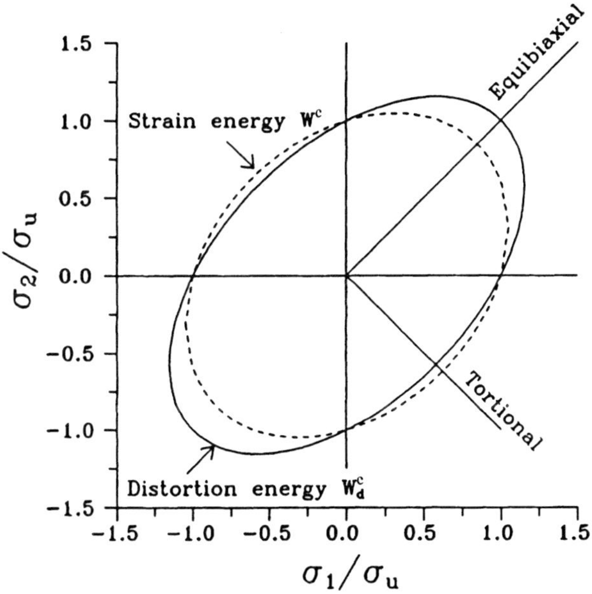

Fig. 4.3 A plot of dimensionless elastic strain energy, W^e, and elastic distortion energy, W^e_d (4.33).

superposed hydrostatic tension decreased the fatigue life, whereas that of compression increased it. Thus, in contrast to the yielding of metals, the effect of hydrostatic pressure has to be accounted for in the fatigue life prediction. When the fatigue life is related to the plastic strain energy [27–29], the effect of the hydrostatic pressure cannot be taken into account, since $d\varepsilon^p_{kk} \equiv 0$ (no change in volume during plastic deformation). Whilst the contribution of the tensile hydrostatic pressure to the total energy may be small in the low-cycle fatigue regime, this may not be true at the high-cycle region. One way to account for the hydrostatic pressure effect in energy calculations is to consider the positive elastic strain energy density during a load cycle, i.e.

$$\Delta W^{e+} = \int_{H(\sigma_i^{\min})\sigma_i^{\min}}^{H(\sigma_i^{\max})\sigma_i^{\max}} \sigma_i \, d\varepsilon_i^e \quad (i = 1, 2, 3) \tag{4.35}$$

where the σ_i's are the principal stress components, and H is the Heaviside function with the following properties:

$$\begin{aligned} H(\sigma_i) &= 1 \quad \text{for} \quad \sigma_i \geqslant 0 \\ H(\sigma_i) &= 0 \quad \text{for} \quad \sigma_i < 0 . \end{aligned} \tag{4.36}$$

Substituting from (4.25) into (4.35) and carrying out the integration, we get

$$\Delta W^{e+} = \frac{1}{2E}[(I_1^{max})^2 - (I_1^{min})^2] - \frac{1+\nu}{E}[I_2^{max} - I_2^{min}] \tag{4.37}$$

where

$$I_1^{max} = \sigma_1^{max}H(\sigma_1^{max}) + \sigma_2^{max}H(\sigma_2^{max}) + \sigma_3^{max}H(\sigma_3^{max}),$$

and

$$I_2^{max} = \sigma_1^{max}\,\sigma_2^{max}H(\sigma_1^{max})\,H(\sigma_2^{max}) + \sigma_2^{max}\,\sigma_3^{max}\,H(\sigma_2^{max})\,H(\sigma_3^{max})$$
$$+ \sigma_3^{max}\,\sigma_1^{max}\,H(\sigma_3^{max})\,H(\sigma_1^{max}),$$

with identical expressions for I_1^{min} and I_2^{min}. For example, when σ_i^{min}s are compressive then (4.37) reduces to

$$\Delta W^{e+} = \frac{1}{2E}[(I_1^{max})^2 - 2(1+\nu)I_2^{max}]. \tag{4.38}$$

In the case of the uniaxial condition, $\sigma_2 = \sigma_3 = 0$, the above reduces to

$$\Delta W^{e+} = \frac{1}{2E}(\sigma_1^{max})^2, \tag{4.39}$$

which is the area shown in Fig. 4.4 corresponding to the tensile part of the elastic strain energy.

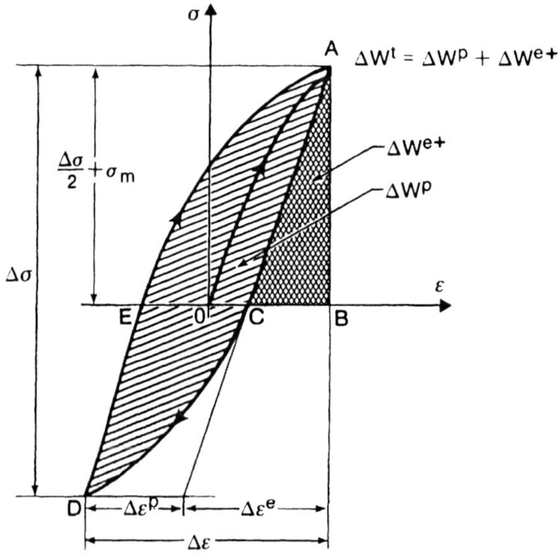

Fig. 4.4 Elastic and plastic strain energies, and the nomenclature for a uniaxial stress state.

Equations (4.37) or (4.38) may be expressed in terms of the equivalent stress, σ_{eq}, by noting that

$$I_2 = -J_2 + \tfrac{1}{3}I_1^2 \quad \text{and} \quad J_2 = \tfrac{1}{2}s_{ij}s_{ij} = \tfrac{1}{3}\sigma_{eq}^2 . \tag{4.40}$$

Substitution from (4.40) into (4.38) results in

$$\Delta W^{e+} = \frac{1+\nu}{3E}(\sigma_{eq}^{max})^2 + \frac{1-2\nu}{6E}(I_1^{max})^2 . \tag{4.41}$$

Again, the first term on the RHS of (4.41) is the distortion energy and the second term is the energy associated with the volumetric change. When there is a mean stress present, the principal stresses may be decomposed to the alternating and mean components, i.e.

$$\sigma_i^{max} = \sigma_i^a + \sigma_i^m \quad (i = 1,2,3) , \tag{4.42}$$

where the superscripts a and m refer to the amplitude and mean stress, respectively. Equation (4.35) becomes

$$\Delta W^{e+} = \frac{1+\nu}{3E}(\sigma_{eq}^{max})^2 + \frac{1-2\nu}{6E}\left[\sum_{i=1}^{3}(\sigma_i^a + \sigma_i^m)\right]^2 . \tag{4.43}$$

It can be seen that the positive part of the strain energy contains the mean stress in an explicit manner. However, the effect of the mean stress will be discussed later. Here it suffices to mention that for a small mean stress, the experimental data near the fatigue limit correlate fairly well with the strain energy density (4.41) or (4.43).

4.3.2 CYCLIC PLASTIC STRAIN ENERGY

The plastic strain energy density per cycle, ΔW^p, can be calculated from

$$\Delta W^p = \int_{cycle} \sigma_{ij}\,d\varepsilon_{ij}^p , \tag{4.44a}$$

where the integration is carried out over the closed cyclic loop (Fig. 4.4). The incompressibility of the material during the plastic deformation $(d\varepsilon_{kk}^p) = 0)$ enables us to write (4.44a) as

$$\Delta W^p = \int_{cycle} s_{ij}\,d\varepsilon_{ij}^p , \tag{4.44b}$$

where the deviatoric stress, s_{ij}, is given by (4.28).

A stable cyclic constitutive relation for the proportional loading was given in Chapter 2 (2.32) in the form of

$$\varepsilon_{ij}^p = \tfrac{3}{}(2K^*)^{-1/n^*}(\sigma_{eq}^*)^{(1-n^*)/n^*}s_{ij}^* , \tag{4.45}$$

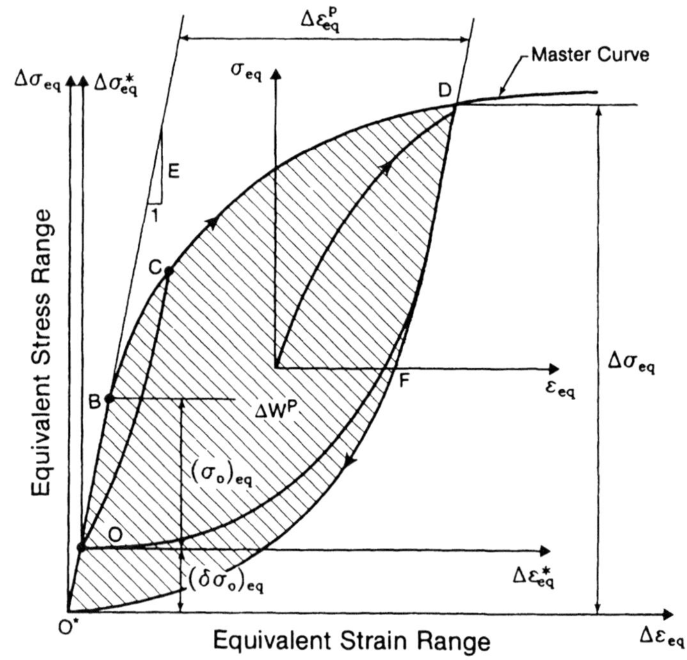

Fig. 4.5 Master curve, various coordinate systems, and notations in the equivalent stress–equivalent strain space.

where the origin of the master curve at 0* is depicted in Fig. 4.5. The use of the above relation is justified during stable cyclic loading where stress and strain components increase (or decrease) in an almost proportional manner. In an incremental form (4.45) is given by

$$d\varepsilon^p_{ij} = 3(2K^*)^{-1/n^*}\left[\frac{1-n^*}{n^*}(\sigma^*_{eq})^{(1-2n^*)/n^*}s^*_{ij}\,d\sigma^*_{eq} + \sigma^{*(1-n^*)/n^*}_{eq}\,ds^*_{ij}\right]. \quad (4.46)$$

Substituting from (4.46) into (4.44b) and noting that from (4.40)

$$\tfrac{3}{2}s^*_{ij}\,ds^*_{ij} = \sigma^*_{eq}d\sigma^*_{eq}, \quad (4.47)$$

we then get

$$\Delta W^p = \frac{2(2K^*)^{-1/n^*}}{n^*}\int_{cycle}(\sigma^*_{eq})^{1/n^*}d\sigma^*_{eq}. \quad (4.48)$$

The stress–plastic strain relation (4.45), when written in the equivalent plastic strain–equivalent stress form, is given by ((2.28) in Chapter 2. Thus

$$d\varepsilon^p_{eq} = \frac{2(2K^*)^{-1/n^*}}{n^*}(\sigma^*_{eq})^{(1-n^*)/n^*}d\sigma^*_{eq}. \quad (4.49)$$

Introducing (4.49) into (4.48) we obtain

$$\Delta W^{\mathrm{p}} = \int_{\mathrm{cycle}} \sigma^{*}_{\mathrm{eq}} \mathrm{d}\varepsilon^{\mathrm{p}}_{\mathrm{eq}} \, . \tag{4.50}$$

Referring to Fig. 4.5, it can be easily seen that the area 0*BCD0* is given by

$$\Delta W^{*\mathrm{p}}_{\mathrm{M}} = \Delta\sigma^{*}_{\mathrm{eq}} \Delta\varepsilon^{\mathrm{p}}_{\mathrm{eq}} - 2\int_{0}^{\Delta\sigma^{*}_{\mathrm{eq}}} \Delta\varepsilon^{\mathrm{p}}_{\mathrm{eq}} \mathrm{d}(\Delta\sigma^{*}_{\mathrm{eq}}) \, . \tag{4.51}$$

The above is for a Masing-type material where the upper (0*CD) and lower (D0*) branches of the hysteresis loops are geometrically identical. For a non-Masing type of behaviour, the addition area 0*DF0* in Fig. 4.5 has to be evaluated (Chapter 3). This additional area is given by [34],

$$\Delta W^{*\mathrm{p}}_{\mathrm{NM}} = (\delta\sigma_0)_{\mathrm{eq}} \Delta\varepsilon^{\mathrm{p}}_{\mathrm{eq}} \, , \tag{4.52}$$

where $(\delta\sigma_0)_{\mathrm{eq}}$, the increase in the proportional limit, is determined from (2.31)

$$(\delta\sigma_0)_{\mathrm{eq}} = \Delta\sigma_{\mathrm{eq}} - \Delta\sigma^{*}_{\mathrm{eq}} = \Delta\sigma_{\mathrm{eq}} - 2K^{*}(\Delta\varepsilon^{\mathrm{p}}_{\mathrm{eq}}/2)^{n^{*}} \, . \tag{4.53}$$

Thus, the plastic strain energy density per cycle (hysteretic energy) is obtained by adding (4.51) and (4.52) and carrying out the integration. The result is

$$\Delta W^{\mathrm{p}} = \frac{1-n^{*}}{1+n^{*}} \Delta\sigma_{\mathrm{eq}} \Delta\varepsilon^{\mathrm{p}}_{\mathrm{eq}} + \frac{2n^{*}}{1+n^{*}} (\delta\sigma_0)_{\mathrm{eq}} \Delta\varepsilon^{\mathrm{p}}_{\mathrm{eq}} \, . \tag{4.54}$$

The above expression for the cyclic plastic strain energy density can be written in terms of the equivalent stress by substituting for $\Delta\varepsilon^{\mathrm{p}}_{\mathrm{eq}}$ from the equivalent plastic strain–equivalent stress relationship (2.28) in Chapter 2

$$\Delta\varepsilon^{\mathrm{p}}_{\mathrm{eq}} = \frac{1}{K^{*}} \left(\frac{\Delta\sigma^{*}_{\mathrm{eq}}}{2K^{*}} \right)^{(1-n^{*})/n^{*}} \Delta\sigma^{*}_{\mathrm{eq}} \, . \tag{4.55}$$

Introducing (4.55) into (4.54) we obtain

$$\Delta W^{\mathrm{p}} = \frac{2(1-n^{*})}{1+n^{*}} (2K^{*})^{-1/n^{*}} (\Delta\sigma^{*}_{\mathrm{eq}})^{(1+n^{*})/n^{*}}$$

$$+ 2(2K^{*})^{-1/n^{*}} (\delta\sigma_0)_{\mathrm{eq}} (\Delta\sigma^{*}_{\mathrm{eq}})^{1/n^{*}} \, . \tag{4.56}$$

The range of the equivalent stress $\Delta\sigma^{*}_{\mathrm{eq}}$, is determined with reference to the coordinate system with the origin at 0* (Fig. 4.5), where

$$\Delta\sigma_{\mathrm{eq}} = [\tfrac{3}{2}(s^{\mathrm{a}}_{ij} - s^{\mathrm{b}}_{ij})(s^{\mathrm{a}}_{ij} - s^{\mathrm{b}}_{ij})]^{1/2} \, , \tag{4.57}$$

and s_{ij}^{a} and s_{ij}^{b} are the maximum and minimum of the deviatoric stress components in one cycle. When the stress range is known, the plastic strain energy can be determined from (4.56).

For a Masing-type material description $(\delta\sigma_0)_{eq} \equiv 0$ and n^* is replaced by the cyclic stress–strain hardening exponent, n', and similarly K^* by K'. Thus (4.56) for a Masing material reduces to

$$\Delta W^{p} = \frac{2(1-n')}{1+n'}(2K')^{-1/n'}(\Delta\sigma_{eq})^{(1+n')/n'} . \tag{4.58}$$

This plastic strain energy density was used in some of the earlier studies [27–29] to correlate fatigue life in low-cycle multiaxial fatigue.

4.3.3 CYCLIC TOTAL STRAIN ENERGY

To obtain a damage parameter which applies to the entire life spectrum, we note the following. At the low-cycle regime, $N_f < 10^5$, the damaging variable is the plastic strain energy per cycle, ΔW^{p}, and in this regime, $\Delta W^{e} \ll \Delta W^{p}$. In the high-cycle regime, especially near the fatigue limit, it is difficult to measure plastic energy, and the distortion energy is related to ΔW^{e+} (4.41). Thus, in the long life regime, the damage function can be related to the positive part of the elastic strain energy, ΔW^{e+}. Therefore, a damage function which could apply to the total spectrum could be

$$\Delta W^{t} = \Delta W^{e+} + \zeta \Delta W^{p} , \tag{4.59}$$

where the first term on the RHS of (4.59) is given by (4.41) or (4.43) and the second term by (4.56), and ζ is a coefficient dependent upon the biaxiality ratio, i.e. $\zeta = \zeta(\rho)$. For the case where $\sigma_i^{min} < 0$ and $\sigma_i^{max} > 0$, (4.59) is given by

$$\begin{aligned}\Delta W^{t} = &\left\{\frac{1+\nu}{3E}(\sigma_{eq}^{max})^2 + \frac{1-2\nu}{6E}\left[\sum_{i=1}^{3}(\sigma_i^{a}+\sigma_i^{m})\right]^2\right\}\\ &+\zeta\left\{\left[\frac{2(1-n^*)}{1+n^*}(2K^*)^{-1/n^*}(\Delta\sigma_{eq}^*)^{(1+n^*)/n^*}\right]\right.\\ &\left.+2(2K^*)^{-1/n^*}(\delta\sigma_0)_{eq}(\Delta\sigma_{eq}^*)^{1/n^*}\right\} .\end{aligned} \tag{4.60}$$

For a Masing-type material (4.60) takes the form of

$$\begin{aligned}(\Delta W^{t})_M = &\left\{\frac{1+\nu}{3E}(\sigma_{eq}^{max})^2 + \frac{1-2\nu}{6E}\left[\sum_{i=1}^{3}(\sigma_i^{a}+\sigma_i^{m})\right]^2\right\}\\ &+\zeta\left\{\left[\frac{2(1-n')}{1+n'}(2K')^{-1/n'}(\Delta\sigma_{eq})^{(1+n')/n'}\right] .\right.\end{aligned} \tag{4.61}$$

The above can be expressed in the general form of

$$\Delta W^{t} = \Delta W^{t}(I_1, J_2) , \tag{4.62}$$

thus establishing the invariant property of the damage function ΔW^t. A major part of this input energy (plastic part) is dissipated into heat, and the remaining part is spent initially in mobilizing dislocation movement along the slip bands – eventually leading to formation of nucleation sites and micro-cracks. Later on the cyclic energy is required to drive the initiated cracks. The cyclic elastic strain energy included in (4.62) is that which contributes to the above process, particularly during the crack growth phase. The damage function (4.59) with $\zeta = 1$ was used in [34] and [35] to correlate multiaxial fatigue data.

4.4 THE MULTIAXIAL FATIGUE FAILURE CRITERION

Having derived various forms of strain energy as a fatigue damage parameter, we now proceed to define a failure criterion. Referring to (4.1) the LHS is now defined, and therefore the general form of the failure criterion is given by

$$\Delta W_i^t(I_1, J_2) = q(N_f)\,. \tag{4.63}$$

The functional form of $q(N_f)$ is also given by (4.2), thus (4.63) can be expressed as

$$\Delta W^t(I_1, J_2) = \kappa(2N_f)^\alpha + C\,, \tag{4.64}$$

where κ, α and C are material parameters to be determined from appropriate tests and $2N_f$ is the number of reversals to failure. Substitution from (4.60) into the LHS of (4.64) gives a failure criterion in terms of the effective stress which could easily be reduced to that of the uniaxial stress condition discussed in Chapter 3. Let us now examine criterion (4.64) in more detail.

In the low strain, high-cycle fatigue regime, the plastic strain is small ($d\varepsilon_{ij}^p \approx 0$) and thus $\Delta W^p \to 0$, and from (4.60) $\Delta W^t \to \Delta W^{e+}$. Also, $(2N_f)^\alpha \to 0$ (because $\alpha < 0$) and thus, combining (4.60) and (4.64) we obtain

$$\frac{1+\nu}{3E}(\sigma_{eq}^{max})^2 + \frac{1-2\nu}{6E}(I_1^{max})^2 = C\,. \tag{4.65}$$

The above is a modified form of the von Mises yield condition (Table 4.1 and Fig. 4.3). In the case of the uniaxial stress condition, $\sigma_{eq}^{max} = I_1^{max} = \sigma_u^{max} = 2\tau_u^{max}$; thus, the LHS of (4.65) reduces to

$$C = \frac{1}{2E}(\sigma_u^{max})^2 = \frac{1}{2E}(2\tau_u^{max})^2\,. \tag{4.66}$$

We can calculate (4.65) for any other stress state, for example in the case of **equi-biaxial** conditions, $\sigma_1 = \sigma_2$ and $\tau_{bi}^{max} = \frac{1}{2}\sigma_1$, and (4.65) reduces to

$$C = \frac{1-\nu}{E}\sigma_1^2 = \frac{1-\nu}{E}(2\tau_{bi}^{max})^2\,. \tag{4.67a}$$

For the case of **torsion**, $\sigma_1 = -\sigma_2$ and $\tau_t^{max} = \sigma_1$, and (4.65) becomes

$$C = \frac{1+\nu}{E}\sigma_1^2 = \frac{1+\nu}{E}(\tau_t^{max})^2 . \tag{4.67b}$$

In comparing (4.66) and (4.67) we note that the shear stress amplitude required to produce the same amount of elastic strain energy will be smaller for equi-biaxial and larger for torsion than the corresponding values for the uniaxial loading. For example, for $\nu = 0.3$, $\tau_{bi}^{max} = 0.85\ \tau_u^{max}$ and $\tau_t^{max} = 1.24\tau_u^{max}$. The salient feature to remember is that for the same amount of elastic strain energy, the fatigue life will be the same for any stress condition.

In the low-cycle, high-strain regime, $\Delta W^{e+} \ll \Delta W^p$, i.e. $\Delta W^t \approx \Delta W^p$, and thus the failure criterion (4.64) may be expressed as

$$\Delta W^p = \kappa(2N_f)^{\alpha_p} . \tag{4.68a}$$

In the uniaxial stress condition (4.68a) becomes

$$\Delta W_u^p = \kappa_u(2N_f)^{\alpha_p} , \tag{4.68b}$$

therefore, α_p is the slope of the straight line in the $\log \Delta W^p$ vs. $\log(2N_f)$ plot [36].

The parameter κ cannot in general be deduced from uniaxial tests alone, it depends on the controlled triaxial strain (or stress) state. Since fatigue cracks generally initiate from a free surface, the constraint imposed on layers parallel to the free surface will influence the initiation and propagation of cracks. Various parameters have been proposed, e.g. [34]. Here we will use the definition given in [35] for the multiaxial constraint factor

$$\bar{\rho} = (1+\bar{\nu})\frac{\hat{\varepsilon}_{max}}{\hat{\gamma}_{max}} \tag{4.69a}$$

with

$$\hat{\varepsilon}_{max} = \max[\varepsilon_a, \varepsilon_t] \tag{4.69b}$$

$$\hat{\gamma}_{max} = \max[|\varepsilon_a - \varepsilon_r|, |\varepsilon_t - \varepsilon_r|] . \tag{4.69c}$$

$\bar{\nu}$ is an effective Poisson's ratio and is calculated from (see Chapter 5 for its derivation)

$$\bar{\nu} = \frac{\nu_p(1-\nu_e)(\varepsilon_a+\varepsilon_t) + (\nu_e - \nu_p)(\varepsilon_a^e + \varepsilon_t^e)}{(1-\nu_e)(\varepsilon_a+\varepsilon_t) + (\nu_e - \nu_p)(\varepsilon_a^e + \varepsilon_t^e)}; \quad \varepsilon_a \neq -\varepsilon_t . \tag{4.70}$$

In the above ε_a and ε_t are principal in-plane strain (axial and transversal) parallel to the 'free' surface, and ε_r (radial strain) is perpendicular to the 'free' surface, and subscripts (or superscripts) e and p stand for elastic and plastic values, respectively. The radial strain, ε_r (perpendicular to the 'free' surface), is given by (Chapter 5)

$$\varepsilon_r = \frac{-\bar{\nu}}{(1-\bar{\nu})}(\varepsilon_a + \varepsilon_t)\,. \tag{4.71}$$

Equations (4.70) and (4.71) were obtained by assuming that the third principal stress (radial stress), σ_r, can be neglected in comparison to the other two principal stresses, σ_a and σ_t. However, if σ_r is not negligible, then the elastic and plastic components of ε_r can be calculated from the generalized Hooke's law and from the incompressibility condition, respectively (Chapter 5).

The multiaxial constraint factor, $\bar{\rho}$, defined by (4.69a), demonstrates the importance of the orientation of the 'free' surface with respect to the imposed principal strains. The $\bar{\rho}$ also has an important physical interpretation regarding the fatigue damage process. Using $\bar{\rho}$ provides a means to differentiate between the fully-reversed torsion, uniaxial and equi-biaxial stress states shown in Fig. 4.2. For

$$\begin{aligned}
&\text{Torsion:} && \varepsilon_a = -\varepsilon_t && \text{and} \quad \bar{\rho} = 1+\bar{\nu}\,,\\
&\text{Uniaxial:} && \varepsilon_t = \varepsilon_r = -\bar{\nu}\varepsilon_a && \bar{\rho} = 1\,,\\
&\text{Equi-biaxial:} && \varepsilon_a = \varepsilon_t && \bar{\rho} = 1-\bar{\nu}\,.
\end{aligned} \tag{4.72}$$

Thus $\bar{\rho}$ distinguishes between states where cracks grow along the surface (e.g. torsional loading) and those where cracks grow into the material (equi-biaxial loading), i.e. in the terminology of Brown and Miller [23], between Case A and Case B loading. Therefore, $1/\bar{\rho}$ indicates the severity of a particular multiaxial loading. The actual 'driving force' of a fatigue damage accumulation is accounted for through an appropriate form of the multiaxial constraint function, $f(\bar{\rho})$. Since $f(\bar{\rho}) = 1$ for uniaxial loading, then a simple function form could be a power law type, i.e.

$$f(\bar{\rho}) = (\bar{\rho})^n\,. \tag{4.73}$$

Thus, κ on the RHS of (4.64) can be written as

$$\kappa = (\bar{\rho})^n \kappa_u \tag{4.74}$$

which reduces to $\kappa = \kappa_u$ in the case of uniaxial loading by virtue of (4.72). It has been shown [37] that $n = 1$ correlates fairly well with the experimental data, i.e. (4.74) can be further simplified to

$$\kappa = \bar{\rho}\,\kappa_u\,. \tag{4.75}$$

The failure criterion (4.64) is now completely defined. Given the stress or strain ranges and cyclic properties of the material, the LHS of (4.64) can be determined from (4.60) or (4.61). From the uniaxial data α and C_u can be obtained (Chapter 3).

An alternative form of representing the multiaxial fatigue criterion is through definition of a damage function for constant life curves, i.e. $N_f =$ constant. This is equivalent to defining contours of the failure surface with planes $N =$ constant in Fig. 4.1. The damage function Ψ can be obtained from (4.59), (4.65), (4.68) and (4.75) in the form of

$$\Psi = \frac{\Delta W^{p}}{\bar{\rho}} + \Delta W^{e+} = \kappa_u (2N_f)^{\alpha} + C_u\,. \tag{4.76}$$

The LHS of (4.76) is the 'dividing energy' which causes fatigue damage. A plot of Ψ versus N_f (or $2N_f$) will yield a unique curve for all multiaxial stress states. This will be discussed in the next section.

4.5 MULTIAXIAL FATIGUE LIFE PREDICTION

To predict the life under multiaxial stress conditions from (4.76), the first requirement is to calculate the energy quantities, ΔW^{p} and ΔW^{e+}. In general, this can be done through a constitutive relation and numerical calculation of the areas shown in Fig. 4.4 (Chapter 6). Alternatively, for the proportional loading we can use expressions (4.58) or (4.56) to calculate ΔW^{p} for Masing or non-Masing materials, respectively, and ΔW^{e+} is calculated from (4.37) or (4.41). In the case of the Masing material, data on the uniaxial cyclic stress–strain curve are required from which E, n' and K' can be obtained. These data are generally available in material handbooks, e.g. [38, 39]. For a material exhibiting non-Masing behaviour, the master curve equation has to be known so that n^*, K^* and $\delta\sigma_0$ can be determined. These are not readily available, and as described in Chapter 2, a minimum of two fully-reversed uniaxial tests will be required to determine n^*, K^* and $\delta\sigma_0$. These two tests can also be used to determine α (or α_p), the slope of the ΔW^{t} (or ΔW^{p}) versus $2N_f$ uniaxial fatigue life curve. It should be noted that experimental data with various imposed in-plane ratios ($\rho = \varepsilon_t/\varepsilon_a$) indicate parallel lines (the same slope) when they are plotted in terms of a damage parameter Ψ versus life (Fig. 5.11). Therefore, in (4.76) the slope $\alpha \approx \alpha_u$.

The constant κ_u (intercept energy axis) can also be determined from the two tests mentioned above. Thus, once the uniaxial stress–strain and fatigue life curves are available (or constructed through at least two tests) E, n^* (n'), ν, $\delta\sigma_0$, $K^*(K')$, κ_u and α can be determined. The value of C, (4.65), can also be found from the fatigue life curve or available data on the fatigue limit, σ_{FL}. Alternatively, a long life test with $2N_f \geqslant 2 \times 10^6$ can be performed to evaluate C_u. Note, however, that the value of C will be a function of the mean stress, and this will be discussed in the next section.

For a specified strain range, the corresponding stress range is obtained from the stress–strain relationship (section 2.11). Consequently, ΔW^{p} and ΔW^{e+} on the LHS of (4.76) are calculated from (4.56) and (4.34) or (4.37).

Knowing the imposed constraint ratio, $\bar{\rho}$, the LHS of (4.76) is now completely specified. The RHS of (4.76) is determined from the uniaxial life curve. Therefore, we can plot multiaxial data in the Ψ versus $2N_f$ coordinates.

The experimental data for the carbon steel ASTM A-516 Gr. 70 are reported in Chapter 5 for three in-plane biaxial ratios: $\rho = \varepsilon_t/\varepsilon_a = 1.0$, 0.5 and 0.0, and the uniaxial loading $\rho = -\bar{\nu}$. The number of cycles to failure range from $N_f \approx 10^3$–2×10^5 cycles, i.e. in the range of the low-cycle fatigue regime. Therefore, we will first present a comparison between the plastic energy, ΔW^p, and N_f, i.e. by assuming ΔW^{e+} in (4.76) is negligible.

A master life curve, i.e. one curve for all the stress states, is obtained by plotting $\Psi = \Delta W^p/\bar{\rho}$ versus $2N_f$. Using (4.76) the master life curve can be represented by

$$\frac{\Delta W^p}{\bar{\rho}} = \kappa_u (2N_f)^{\alpha} . \tag{4.77}$$

The fatigue failure criterion given by (4.77) states that for multiaxial loading in the low-cycle fatigue regime, all data should collapse into the uniaxial fatigue curve, if the ΔW^p for each loading case is divided by its multiaxial constraint factor, $\bar{\rho}$ (4.69a).

Figure 4.6 shows the correlation of biaxial fatigue results for in-phase loading reported in Chapter 5. In this figure the ordinate is ($\Delta W^p/\bar{\rho}$) and the abscissa is the number of cycles to failure, N_f. The uniaxial fatigue curve from Chapter 3 (Fig. 3.12) is shown by a solid line. It can be seen that most of the multiaxial data for various in-plane applied strain ratios, ρ, fall within a narrow scatter band of the uniaxial fatigue curve, thus confirming

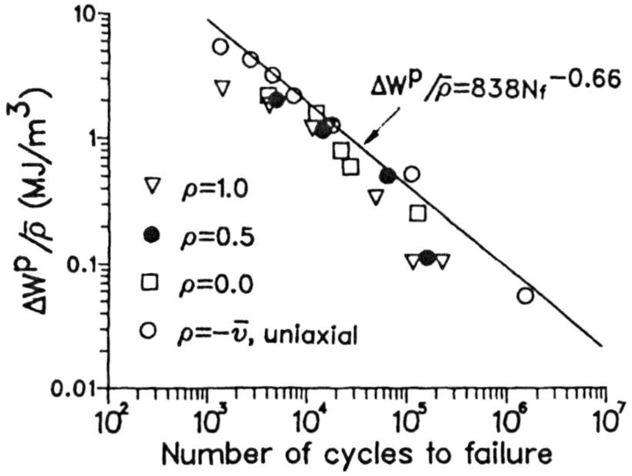

Fig. 4.6 A plot of the plastic strain energy divided by the multiaxial constraint factor versus number of cycles to failure for four different controlled in-plane strain ratios. The solid line is the uniaxial curve.

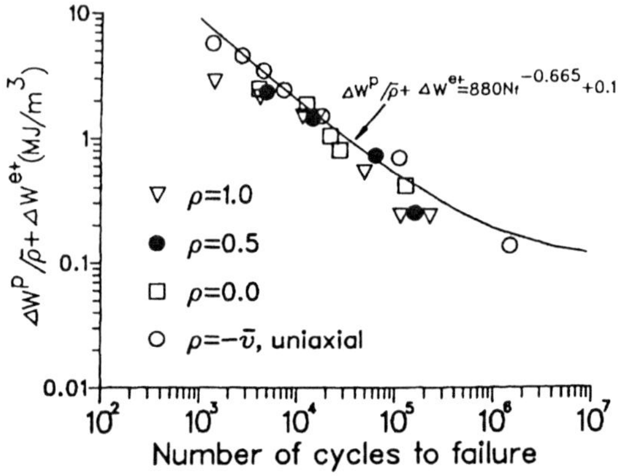

Fig. 4.7 A plot of the plastic strain energy divided by the multiaxial constraint factor plus the positive elastic strain energy versus number of cycles to failure. The solid line is the uniaxial curve. Reproduced with permission from ASME from [35].

(4.77). For fatigue lives $N_f > 10^5$, the positive elastic strain energy, ΔW^{e+}, is not negligible compared to the plastic energy. Thus (4.76) is more appropriate to use when $N_f > 10^5$. The experimental data from Fig. 4.6 are replotted in Fig. 4.7 using $(\Delta W^p/\bar{\rho}) + \Delta W^{e+}$, RHS of (4.76) as the ordinate. The solid line in this figure represents the uniaxial data with $C_u = 0.1$ MJ m^{-3} (Fig. 3.15). It can be seen that the scatter band is now well within the acceptable range for the entire life regime.

The multiaxial criterion represented by (4.76) has considerable practical significance, since all the required data on the RHS are determined from the simple uniaxial tests. The LHS for a given multiaxial stress state could be calculated from (4.56) and (4.37) or (4.41) or using a suitable constitutive relation (Chapter 6) to evaluate integrals (4.44a) and (4.35).

4.6 EFFECT OF MEAN STRESS – PROPORTIONAL LOADING

The effect of mean stress on the fatigue life in a multiaxial stress state has been studied by a number of investigators. For example, Sines [40] proposed a criterion based on the octahedral shear amplitude, $\tau_{a,oct}$ and a mean hydrostatic pressure p_m, in the form of

$$\tau_a + a\,p_m = b\,, \tag{4.78a}$$

where

$$p_m = \tfrac{1}{3}(\sigma_1^m + \sigma_2^m + \sigma_3^m)\,, \tag{4.78b}$$

and a and b are parameters. Crossland [41] has suggested a criterion similar to that of Sines' except that the maximum value of the hydrostatic pressure p_{max} was considered instead of the mean value p_m, i.e.

$$p_{max} = \tfrac{1}{3}(\sigma_1^{max} + \sigma_2^{max} + \sigma_3^{max}) . \tag{4.79}$$

Dang Van [42] has put forward a criterion which uses local variables based on the cyclic steady state material response for a constant amplitude test, i.e. the shear stress amplitude acting on the plane of maximum shear, $\tau_{a,max}$, and the maximum hydrostatic pressure, p_{max}.

Using p_{max} instead of p_m provides a means to differentiate between the fully-reversed torsion and tension (or bending) fatigue tests. Note that $p_{max} = 0$ and $p_{max} = \sigma_{max}/3$, in pure alternating torsion and tension (or bending), respectively. It can be seen from (4.78b) that the mean value of pressure, p_m, would be equal to zero in both cases. The above criteria are generally applicable to high-cycle regimes. On the other hand, the criterion proposed by Ellyin and co-workers [34, 35] and described in section 4.3, is applicable to both low- and high-cycle regimes and contains the maximum hydrostatic pressure $p_{max} = I_1^{max}/3$, see (4.60) and (4.61). A recent investigation [37] includes the mean-stress effect in a different manner, and is briefly described below.

An alternative form of expressing the damage function Ψ in (4.76) is:

$$\Psi = \Psi[\Delta W_d, \bar{\rho}, \bar{\sigma}_m] , \tag{4.80}$$

where ΔW_d is the range of the distortion energy given by (4.44b), except that it is calculated for a half cycle (Fig. 4.8), $\bar{\rho}$ is the multiaxial constraint factor (4.69a) and $\bar{\sigma}_m$ is a measure of the mean stress given by (cf. (4.43))

$$\bar{\sigma}_m = \sum_i \sigma_i^m = \sigma_1^m + \sigma_2^m + \sigma_3^m . \tag{4.81}$$

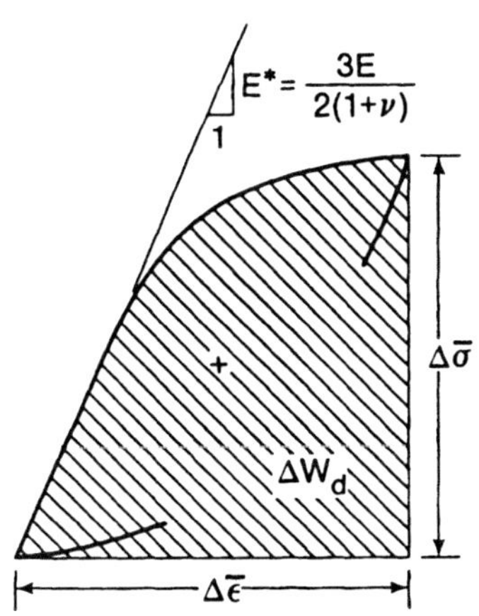

Fig. 4.8 Definition of the distortion energy range, ΔW_d, and equivalent stress, equivalent total strain ranges.

An explicit form of (4.80) was suggested by Ellyin and Kujawski [37] in the form of

$$\Psi = \frac{\Delta W_{\mathrm{d}}[1 + \eta\, \bar{\sigma}_{\mathrm{m}}/(\Delta \sigma_{\mathrm{eq}}/2)]}{\bar{\rho}}, \tag{4.82}$$

where the coefficient η $(0 \leqslant \eta \leqslant 1)$ characterizes a material's sensitivity to the mean stress.

Multiaxial experimental data with mean stress are very limited, and those available in the literature generally do not provide sufficient material data to enable a meaningful comparison. However, correlation with mean stress data in the high-cycle regime for both uniaxial and biaxial experiments is fairly good [37].

A more general approach regarding the effect of mean stress, including ratcheting strain, will be given in 4.8.

4.7 NON-PROPORTIONAL CYCLIC LOADING

In a multiaxial loading, the cycling stress or strain components may not vary in a proportional manner. For example, strains caused by mechanical loads may not be in phase with those caused by a thermal fluctuation. Experimental results reported in Chapters 5 and 6 indicate that non-proportional loading causes more damage than an equivalent proportional loading. For example, there is an extra strain hardening associated with a non-proportional path. This then implies that to achieve a certain plastic deformation, more stress has to be applied in a non-proportional loading case than a proportional one. Experimental evidence is reported in the next two chapters and we will not repeat it here. The objective of this section is to show how the theory presented in section 4.5 can be extended to non-proportional cyclic loading.

It should be noted that non-proportional loading will only affect plastic deformation, since the elastic part is path independent. In other words, a non-proportional cyclic loading will have a pronounced effect in the high-strain, low-cycle regime and a negligible effect in the region of bulk elastic response.

The multiaxial fatigue criterion (4.76) applies equally for non-proportional loading, except that some of the definitions have to be generalized [43]. For example, in the case of the multiaxial constraint factor, $\bar{\rho}$ (4.69a), the ratio $\hat{\varepsilon}/\hat{\gamma}$ is constant at any time during a cycle for a proportional loading. However, in a non-proportional loading the value of this ratio will change with time. For a non-proportional loading the generalization of (4.69a) as proposed in [43] is as follows. To obtain a unique value of $\bar{\rho}$, the ratio of $\hat{\varepsilon}/\hat{\gamma}$ is taken at the instant when the shear strain, $\hat{\gamma}$, in the direction

45° to the surface reaches its maximum value, i.e.

$$\bar{\rho} = (1 + \bar{\nu}) \left[\frac{\hat{\varepsilon}(t)}{\hat{\gamma}(t)} \right]_{\hat{\gamma} = \gamma_{\max}} . \tag{4.83}$$

This definition reduces to (4.69a) in the case of proportional loading. To illustrate this, let us consider a general biaxial out-of-phase straining of the form,

$$\begin{aligned} \varepsilon_a &= A \cos \omega t \\ \varepsilon_t &= \rho A \cos (\omega t - \phi) \end{aligned} \tag{4.84}$$

where A is the amplitude, ω is the frequency of oscillation, ϕ is the phase angle between ε_a and ε_t and $\rho = \Delta\varepsilon_t / \Delta\varepsilon_a$ is the ratio of the tangential strain range to the axial strain range. The relation (4.83) is then given by

$$\bar{\rho} = \frac{(1 - \bar{\nu}^2)(1 + \bar{\nu}\rho \cos \phi)}{(1 + \bar{\nu}\rho \cos \phi)^2 + (\bar{\nu}\rho \sin \phi)^2} . \tag{4.85}$$

It can easily be verified that (4.85) reduces to (4.69a) for the in-phase loading when $\phi = 0$.

In the LHS of (4.76), ΔW^{p} is calculated from (4.44) by using an appropriate constitutive model (Chapter 6). The positive elastic strain energy, $\Delta W^{\mathrm{e}+}$, is calculated from

$$\Delta W^{\mathrm{e}+} = \int_t^{t+T} H(\sigma_I) H(\mathrm{d}\varepsilon_I^{\mathrm{e}}) \sigma_I \mathrm{d}\varepsilon_I^{\mathrm{e}} , \tag{4.86}$$

where T is the period of the cyclic loading, and $H(x)$ is the Heaviside function defined in (4.36).

Figure 4.9 shows a plot of damage parameter $\Psi = (\Delta W^{\mathrm{p}} / \bar{\rho}) + \Delta W^{\mathrm{e}+}$ versus number of cycles to failure, N_{f}, of the data included in Fig. 4.7 and additional out-of-phase results with $\phi = 60°$ and $90°$. As before, the solid line is the uniaxial curve. From Fig. 4.9 it can be seen that the plot of the energy-based fatigue damage parameter Ψ versus N_{f} leads to a master life curve applicable to uniaxial and multiaxial loading conditions with proportional and non-proportional loading paths.

4.8 EFFECTS OF MEAN STRESS AND RATCHETING DEFORMATION

In an elastoplastic regime, a material which experiences a stress-controlled cyclic loading with a non-zero mean stress, will exhibit an accumulated ratcheting (or cyclic creep) deformation (sections 2.3, 3.9 and 6.6.5). This ratcheting deformation causes further damage in the material. Thus, a complete theory of fatigue should be capable of evaluating the effect of

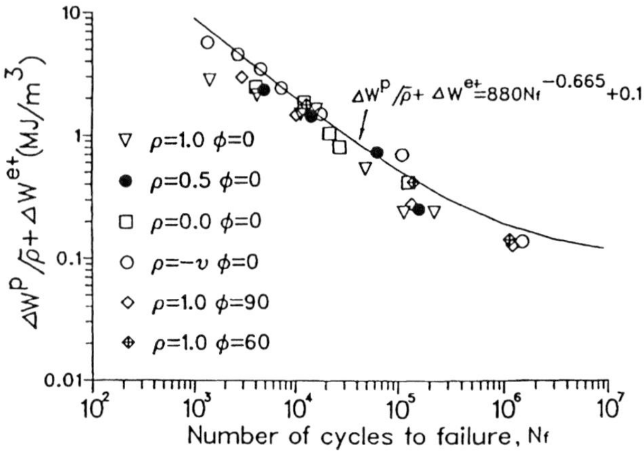

Fig. 4.9 Correlation of multiaxial in-phase and out-of-phase experimental data for ASTM A-516 Gr. 70 steel using $(\Delta W^p/\bar{\rho}) + \Delta W^{e+}$ versus number of cycles to failure, from [44]. Reprinted with permission from Kluwer Academic Publishers.

the mean stress as well as the ratcheting strains. Such a theory was presented in section 3.9. Here we will extend the theory to non-proportional multiaxial loading conditions.

Consider first the effect of mean stress by itself. Relation (4.82) indicates that it is not the magnitude of the mean stress but the ratio of the mean stress to the alternating stress which affects the fatigue life. Due to the scalar nature of the energy, relation (3.73) derived for uniaxial loading can be extended to multiaxial loading. In essence we modify the positive elastic part of (4.76) to reflect a normalized mean stress part, i.e.

$$\Psi_m = \Delta W^p/\bar{\rho} + f(\sigma_m/\sigma_a)\Delta W_a^e = \kappa_u(2N_{fm})^\alpha + C_u \tag{4.87}$$

where $f(\sigma_m/\sigma_a)$ is a mean stress function defined in a similar way to (3.74), i.e.

$$f(\sigma_m/\sigma_a) = 1 + \eta(\sigma_m/\sigma_a) \tag{4.88a}$$

and

$$\Delta W_a^e = \frac{\sigma_a^2}{2E}\,. \tag{4.88b}$$

In the above η is a material constant and N_{fm} is the fatigue life influenced by the mean stress. The other symbols on the RHS of (4.87) have all been defined earlier on, and the $\bar{\rho}$ in the LHS is determined from (4.83). In (4.88) σ_m and σ_a are equivalent mean stress and equivalent stress amplitude values. There is no ambiguity in defining these quantities for the uniaxial and proportional multiaxial loadings (sections 3.9 and 4.6). For a general non-proportional loading, they have to be defined appropriately.

In a multiaxial stress space, the geometric centre of a cyclic stress path will be defined as the mean stress point, $\sigma_{ij,\mathrm{m}}$. It can be determined from [45]:

$$\sigma_{ij,\mathrm{m}} = \int_t^{t+T} \sigma_{ij} \mathrm{d}s \Big/ \int_t^{t+T} \mathrm{d}s \tag{4.89}$$

where T is the period of the cyclic loading and

$$\mathrm{d}s = (\mathrm{d}\sigma_{ij}\mathrm{d}\sigma_{ij})^{1/2}. \tag{4.90}$$

The stress amplitude, σ_a, can be defined as the maximum distance between a point on the cyclic stress path, σ_{ij}, and the mean stress point, $\sigma_{ij,\mathrm{m}}$, based on the equivalent total elastic energy:

$$\sigma_\mathrm{a} = \max \{[(1+v)(\sigma_{ij} - \sigma_{ij,\mathrm{m}})(\sigma_{ij} - \sigma_{ij,\mathrm{m}}) - v(\sigma_{kk} - \sigma_{kk,\mathrm{m}})^2]^{1/2}\} \tag{4.91}$$

where v is the Poisson's ratio. Furthermore, noting that the mean-stress effect depends on the volumetric part of the stress tensor, the equivalent mean stress value is defined as

$$\sigma_\mathrm{m} = \frac{[(1+v)\sigma_{ij,\mathrm{m}}\sigma_{ij,\mathrm{m}} - v\sigma_{kk,\mathrm{m}}^2]^{1/2}.(\sigma_{1,\mathrm{m}} + \sigma_{2,\mathrm{m}} + \sigma_{3,\mathrm{m}})}{(\sigma_{1,\mathrm{m}}^2 + \sigma_{2,\mathrm{m}}^2 + \sigma_{3,\mathrm{m}}^2)^{1/2}} \tag{4.92}$$

where $\sigma_{1,\mathrm{m}}$, $\sigma_{2,\mathrm{m}}$, and $\sigma_{3,\mathrm{m}}$ are the principal values of the mean stress $\sigma_{ij,\mathrm{m}}$.

It should be noted that for a uniaxial cyclic loading, the definitions (4.91) and (4.92) reduce to the conventional ones, and for a fully-reversed cyclic loading, the damage function (4.87) reduces to (4.76).

Up to this point the ratcheting effect has not been explicitly included in the formulation of the damage parameter. The ratcheting behaviour of metals is a relatively complex phenomenon. It depends on a number of factors, namely mean stress, stress amplitude, frequency and loading history, etc. [46]. For example, it was shown in section 3.9 that for the same mean stress and stress amplitude, the ratcheting strain is quite different for different preload histories (see also examples in section 6.6.5). Following the developments in section 3.9, we introduce an energy-type parameter for the ratcheting damage in the form of:

$$\Psi_\mathrm{r} = \sum \sigma_{ij,\max} \dot{\varepsilon}_{ij,\mathrm{aver}}^\mathrm{r} \quad \text{(no summation over } i, j) \tag{4.93}$$

where $\dot{\varepsilon}_{ij,\mathrm{aver}}^\mathrm{r}$ is the average ratcheting strain-rate during the entire life of each strain component, and $\sigma_{ij,\max}$ is the extreme value of the corresponding cyclic stress component in the direction of $\dot{\varepsilon}_{ij,\mathrm{aver}}^\mathrm{r}$. Similar to the development in section 3.9, we assume

$$\Psi_\mathrm{r} = \kappa_\mathrm{r}(2N_\mathrm{fr})^\beta + C_\mathrm{r} \tag{4.94}$$

where N_fr can be envisaged as the 'pure ratcheting' life, and κ_r, β and C_r are material constants related to the ratcheting effect. The constant C_r is a threshold value of Ψ_r, i.e. the ratcheting effect could be neglected if $\Psi_r \leqslant C_r$. Then the total life of the material, N_f, can be predicted by

$$\frac{1}{N_f} = \frac{1}{N_{fm}} + \frac{1}{N_{fr}} \tag{4.95}$$

where N_{fm} and N_{fr} are calculated from (4.87) and (4.94), respectively. We have already discussed how the material constants on the RHS of (4.90), κ_u, α and C_u can be determined. The material constant η on the RHS of (4.88a) is evaluated from the uniaxial cyclic tests with the mean stress but without appreciable ratcheting deformation. The constants κ_r, β and C_r in (4.94) are determined from the cyclic tests with mean stress and ratcheting deformation as explained in section 3.9.

Multiaxial fatigue experiments with mean stress and ratcheting deformation are very limited and a fatigue life comparison at this stage is not possible. However, a comparison with the uniaxial test shows a fairly good agreement (Fig. 3.40).

REFERENCES

1. Lanza, G. (1866) Strength of shafting subjected to both twisting and bending. *Trans. ASME*, **8**, 130–44.
2. Irwin, G.K. (1957) Analysis of stresses and strains near the end of a crack traversing a plate. *J. Applied Mech., Trans. ASME*, **24**, 361–75.
3. Ellyin, F. (1988) Recent developments in predicting multiaxial fatigue failure. *Res. Mechanica*, **25**, 1–23.
4. Frost, N.E., Marsch, K.J. and Pook, L.P. (1974) *Metal Fatigue*, Clarendon Press, Oxford, UK.
5. Ellyin, F. and Valaire, B. (1985) Development of fatigue failure theories for multiaxial high strain conditions. *SM Archives*, **10**, 45–85.
6. Guest, J.J. (1943) The problem of combined stress. *Engineering*, **155**, 21–3, 101–2, 281–2, 303–4.
7. Gough, H.J. and Pollard, H.V. (1935) Strength of metals under combined alternating stresses. *Proc. Inst. Mech. Engrs.*, **131**, 3–54.
8. Stanfield, G. (1935) Discussion of Gough and Pollard (1935), *Proc. Inst. Mech. Engrs.*, **131**, 59.
9. Findley, W.N. (1957) Fatigue of metals under combinations of stress. *Trans. ASME*, **79**, 1337–8.
10. McDiarmid, D.L. (1974) A new analysis of fatigue under combined bending and twisting. *Aero. J.*, **78**, 325–9.
11. Fuchs, H.O. (1974) Fatigue research with discriminating specimens. *Fatigue Engng. Mater. Struct.*, **2**, 207–15.
12. Sines, G. and Ohgi, G. (1981) Fatigue criteria under combined stresses or strains. *J. Engng. Mater. Technol., Trans. ASME*, **103**, 82–90.
13. Tavernelli, J.F. and Coffin Jr., L.F. (1962) Experimental support for generalized equation predicting low cycle fatigue. *J. Basic Engng., Trans. ASME*, **84**, 533–7.
14. Manson S.S. (1962) Discussion of Tavernelli and Coffin (1962), *J. Basic Engng., Trans. ASME*, **84**, 537–41.

15. Zamrik, S.Y. and Goto, T. (1968) The use of octahedral shear strain in biaxial low cycle fatigue, in *Materials Technology on Inter-American Approach*, American Society of Mechanical Engineers, New York, pp. 551–62.
16. Andrews, J.M.H. and Ellison, E.G. (1973) A testing rig for cycling at high biaxial strains. *J. Strain Anal.*, **8**, 168–75.
17. Libertiny, G.Z. (1967) Short-life fatigue under combined stresses. *J. Strain Anal.*, **2**, 91–5.
18. Havard, D.G., Williams, D.P. and Topper, T.H., Biaxial fatigue of mild steel: Data synthesis and interpretation. *Ontario Hydro Res. Quart.*, **27**, 11–18.
19. Lefebvre, D.F. (1989) Hydrostatic pressure effect on life prediction in biaxial low-cycle fatigue, in *Biaxial and Multiaxial Fatigue*, EGF3 (eds M.W. Brown and K.J. Miller), Mechanical Engineering Publications, London, pp. 511–33.
20. Zamrik, S.Y. and Frismuth, R.E. (1973) The effect of out-of-phase biaxial strain cycling on low-cycle fatigue. *Exp. Mech. SESA*, **13**, 204–8.
21. Pascoe, K.J. and DeVilliers, J.W.R. (1967) Low cycle fatigue of steels under biaxial straining. *J. Strain Anal.*, **2**, 117–26.
22. Mowbray, D.F. (1980) A hydrostatic stress-sensitive relationship for fatigue under biaxial stress conditions. *J. Testing Eval.*, **8**, 3–8.
23. Brown, M.W. and Miller, K.J. (1973) A theory for fatigue failure under multiaxial stress-strain conditions. *Proc. Inst. Mech. Engrs.*, **187**, 745–55.
24. Ellyin, F. and Valaire, B. (1982) High strain multiaxial fatigue. *J. Engng. Mater. Technol., Trans. ASME*, **104**, 165–73.
25. Makinde, A. and Neale, K.W. (1989) A general criterion for low-cycle multiaxial fatigue failure. *J. Engng. Mater. Technol., Trans. ASME*, **111**, 263–9.
26. Socie, D. (1993) Critical plane approaches for multiaxial fatigue damage assessment, in *Advances in Multiaxial Fatigue*, ASTM STP 1191 (eds D.L. McDowell and R. Ellis), American Society for Testing and Materials, Philadelphia, PA, pp. 7–36.
27. Ellyin, F. (1974) A criterion for fatigue under multiaxial states of stress. *Mech. Res. Commun.*, **1**, 219–24.
28. Lefebvre, D., Neale, K.W. and Ellyin, F. (1981) A criterion for low-cycle fatigue under biaxial states of stress. *J. Engng. Mater. Technol., Trans. ASME*, **103**, 1–6.
29. Garud, Y.S. (1981) A new approach to the evaluation of fatigue under multiaxial loading. *J. Engng. Mater. Technol., Trans. ASME*, **103**, 118–25.
30. Ellyin, F. (1989) Cyclic strain energy density as a criterion for multiaxial fatigue failure, in *Biaxial and Multiaxial Fatigue, EGF3* (eds M.W. Brown and K.J. Miller), Mechanical Engineering Publications, London, pp. 571–83.
31. Grubisic, V. and Simbürger, A. (1976) *Fatigue under Combined Out-of-Phase Multiaxial Stresses.* Proc. Int. Conf. on Fatigue Testing and Design, Society of Environmental Engineers, London, pp. 27.1–27.8.
32. White, D.J., Crossland, B. and Parry, J.S.C. (1959) Effect of hydrostatic pressure on direct-stress fatigue strength of an alloy steel. *J. Mech. Engng. Sci.*, **1**, 29–49.
33. Morrison, J.L.M., Crossland, B. and Parry, J.S.C. (1960) The strength of thick cylinders subjected to repeated internal pressure. *J. Engng. Ind., Trans. ASME*, **82**, 143–53.
34. Ellyin, F. and Golos, K. (1988) Multiaxial fatigue damage criterion. *J. Engng. Mater. Technol., Trans. ASME*, **110**, 63–18.
35. Ellyin. F., Golos, K. and Xia, Z. (1991) In-phase and out-of-phase multiaxial fatigue. *J. Engng. Mater. Technol., Trans. ASME*, **113**, 112–18.

36. Ellyin, F. and Kujawski, D. (1984) Plastic strain energy in fatigue failure. *J. Pressure Vessel Technol., Trans ASME*, **106**, 342–7.
37. Ellyin, F. and Kujawski, D. (1993) A multiaxial fatigue criterion including mean-stress effect, in *Advances in Multiaxial Fatigue*, ASTM STP 1191 (eds D.L. McDowell and R. Ellis), American Society for Testing and Materials, Philadelphia, PA, pp. 55–66.
38. SAE Handbook (1983) Vol. 1, Materials, Society of Automotive Engineers, Warrendale, PA, pp. 3.57–3.59.
39. Boller, C.H.R. and Seeger, T. (1987) *Material Data for Cyclic Loading*, Elsevier, Amsterdam.
40. Sines, G. (1955) Failure of Materials under Combined Repeated Stresses with Superimposed Static Stresses. NACA TN3495, National Advisory Committee for Aeronautics.
41. Crossland, B. (1956) *Effect of Large Hydrostatic Pressure on the Torsional Fatigue Strength of an Alloy Steel.* Proc. Int. Conf. on Fatigue of Metals, Institute of Mechanical Engineers, London, pp. 138–49.
42. Dang Van, K. (1973) Sur la résistance à la fatigue des métaux. *Sciences Techniques Armament*, **47**, 641–722.
43. Ellyin, F. and Xia, Z. (1993) A general theory of fatigue with application to out-of-phase cyclic loading. *J. Engng. Mater. Technol., Trans. ASME*, **115**, 411–16.
44. Ellyin, F., Kujawski, D. and Xia, Z. (1993) *Modelling of Multiaxial Fatigue. Fatigue 93*, Proc. 5th Int. Conf. on Fatigue and Fatigue Thresholds, Vol. III, May 1993, Montreal, Canada (eds J.P. Bailon and J.I. Dickinson). EMAS, Warley, UK, pp. 1505–14.
45. Ellyin, F., Xia, Z. and Kujawski, D. (1995) *A Multiaxial Fatigue Theory Including Mean Stress and Ratcheting Effects.* Proc. Int. Conf. on Computational Engineering Science (ICES '95, Hawaii) (eds S.N. Atluri, G. Yagawa and T.A. Cruse). Springer, Berlin, pp. 1146–51.
46. Xia, Z. and Ellyin, F. (1995) *A Constitutive Model with Capability to Simulate Complex Multiaxial Ratcheting Behaviour of Materials*, in Dynamic Plasticity and Structural Behaviours (Proc. Plasticity '95, Sakai, Japan), (eds S. Tanimura and A.S. Khan). Gordon and Breach Publishers, Amsterdam, pp. 681–4.

Multiaxial experimental facilities 5

5.1 INTRODUCTION

In this chapter we will describe the development of a multiaxial testing facility. The equipment was designed to supply basic fatigue data and to permit a comparison with the proposed theories. However, before embarking on the detailed description of the test facility, it is appropriate to provide a background to various test methods and their limitations.

Uniaxial fatigue tests in the low-strain region where the specimen response is in the elastic domain, are relatively straightforward to perform on simple load reversal equipment. This type of test facility is available in many laboratories. In the high-strain region where the response of the specimen is elasto-plastic and could vary due to cyclic hardening or softening, it may be necessary to perform the tests under deformation (strain) controlled mode. This form of control is not an artificial one. In many applications the highly strained material is loaded under strain control by the elastic matrix surrounding the highly strained region, e.g. at the root of a notch. Also, the risk of structural instability is greatly reduced under deformation controlled mode. Various test results have been presented in Chapters 2 and 3 for uniaxial loading conditions and we will not pursue this topic any further.

When multiaxial fatigue tests are performed at stress levels above the yield point of the material, the complexity of the test is greatly increased. As in the case of uniaxial tests, load reversal mechanisms are not suitable and the more elaborate specimen and loading systems may introduce undesirable side effects. The required equipment depends on the chosen specimen shape, and this will be discussed in the next section.

5.2 SPECIMEN GEOMETRY

A large variety of specimen geometries have been used in fatigue testing. In earlier investigations, the tests were aimed at obtaining design curves for typical structural geometries subject to actual load environments. Tests of this type involved solid cylinders [1–4], thick-walled cylinders [5], and beams and plates [6–9] among other shapes. The resulting stress and strain distributions were generally non-uniform, and thus the data were not well suited for basic fatigue investigations.

There are two specimen types which meet the requirements of uniform strain distribution, and thus are suitable for basic fatigue studies, namely the thin-walled cylindrical specimen subject to combined loading, and the so-called cruciform specimen subject to biaxial in-plane loading. Each specimen type has advantages and disadvantages, and the selection depends on the fatigue phenomenon under investigation. As pointed out earlier, the irreversible plastic strain component is the dominant parameter in the fatigue damage process, and to separate elastic and plastic components of the total strain, it is necessary to have a reasonably accurate estimate of the stresses.

The cruciform specimen permits a free choice of surface environment and continuous observation of the surface, and is often the choice for crack growth studies. The major disadvantages with this type of specimen are the inability to determine the stresses with any degree of accuracy and the difficulty of monitoring the strains throughout the test. It has also been found that at high levels of strain, the control region becomes subject to buckling. To avoid this, the gauge length is often heavily reinforced with concave spherical surfaces, rather than flat surfaces, leading to stress gradients. Results of investigations on cruciform specimens are reported in [10–14], among others. A review of cruciform testing machines can be found in recent papers by Makinde *et al.* [15] and Boëhler *et al.* [16].

There are two approaches to the use of thin-walled tubes. They can be subjected to the combined effects of torsion and axial load. With this type of loading continuous observation of the surface is possible but the range of strain ratios is limited and the direction of the principal strains varies with the strain ratio. For example, the ratio of the in-plane principal strains, $\rho = \varepsilon_t/\varepsilon_a$, is limited to a small negative range $-1 \leqslant \rho \leqslant -\nu$ (Fig. 5.1). Investigations on this type of loading have been reported in [17–22] to mention but a few. A review of axial–torsional fatigue life investigations is given by Kalluri and Bonacuse [23].

Alternatively, the thin-walled cylindrical specimens can be subjected to axial load and a pressure difference across the wall thickness. This type of loading overcomes the limitations of torsional and axial combination, but requires a pressurized environment and prohibits surface observation during loading. However, for thin-walled cylindrical specimens under any type of loading, the stresses can be accurately determined.

It should be noted that application of an internal pressure in conjunction with the axial load will result in a mean tensile stress in the circumferential direction, e.g. McDiarmid's experiments [24].

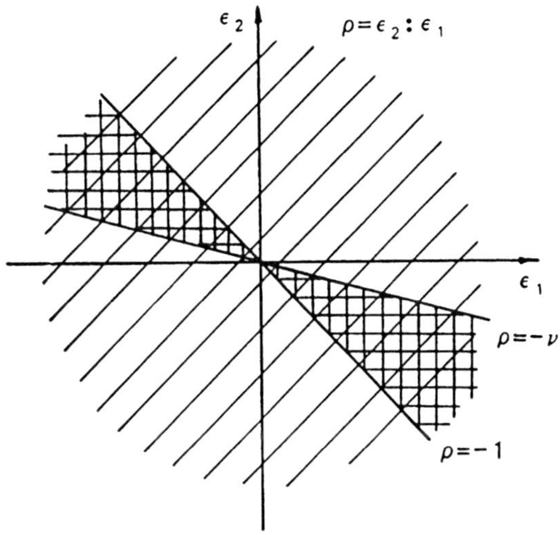

Fig. 5.1 Biaxial strain plane, and the domain of the axial ($\rho = -\nu$) and torsional ($\rho = -1$) loading. No positive strain ratios can be tested by an axial–torsional loading combination.

From the foregoing discussion it is clear that compromises are involved in the selection of specimen type and loading system in a multiaxial fatigue investigation. In the following analysis we will concentrate on the thin-walled cylindrical specimen subject to the axial load and differential pressure. The main advantages motivating such a choice are the availability of the full range of strain ratios (Fig. 5.1) and the relatively large, uniformly strained gauge length which permits the installation of strain transducers to control the experiment. In addition, the ability to measure the loads and pressures and hence determine the state of stress and the multiaxial cyclic stress–strain characteristic of the material are of major importance.

The first test apparatus to apply differential pressure (constant external pressure and pulsating internal pressure) and axial load (an initial load to compensate pressure loading and pulsating axial load) was reported by Rotvel [25]. Similar facilities were reported by Havard and Topper [26], Andrews and Ellison [27], and Ellyin and Valaire [28]. An indication of the available test facilities may be found in [29].

5.3 ANALYSIS OF THIN-WALLED CYLINDRICAL SPECIMENS

When testing under multiaxial conditions there are additional unavoidable variables to be taken into account. Apart from the introduction of the second controlled strain, there are the complications resulting from a more complex specimen geometry. The two conflicting requirements of uniform strain field and resistance to buckling have to be balanced. Another important consideration is the validity of the specimen particularly at high strains. A specimen which is heavily reinforced outside the gauge length would develop a 'barrel shape' on loading so that complex bending strains would be superimposed on the specimen.

To eliminate the effects of surface roughness, high quality finishes are generally used, although these effects at high strains appear to be much less critical than at low strains. The highly polished surfaces lose their finish after a few high strain reversals due to the slip band formation in the surface grains and the production of intrusions and extrusions (Chapter 1).

A variety of specimen designs were investigated by Ellyin and Valaire [28] and three profiles with varying wall thicknesses are shown in Fig. 5.2. The investigation started with profile 'A' but the most satisfactory designs were found to be profiles 'B' and 'C'. Tests were conducted at progressively higher strain levels until it became apparent that the specimen was acting in an unstable manner.

An analysis of the deformed gauge length and the strains resulting from the 'barrel distortion' (BD) are reported in [28]. Figure 5.3 shows the variation of the tangential strain ε_{tBD} and axial strain ε_{aBD} due to the barrel distortion. Note that at high applied strains the strains due to barrelling could be of the same magnitude as the applied strains, cf. profile A in Fig. 5.3 [28].

5.3.1 COMPUTED STRESSES AND STRAINS IN THE GAUGE LENGTH

Figure 5.4 shows the gauge length of a specimen subjected to internal and external pressures (p_i and p_e) and an axial force, F. Assuming that the stresses are distributed almost uniformly throughout the thickness, we can write

$$\sigma_a = \frac{F}{A} \tag{5.1a}$$

$$\sigma_t = \left(\frac{d_e + d_i}{d_e - d_i}\right)(p_i - p_e) \tag{5.1b}$$

$$\sigma_r \approx -\left(\frac{p_i + p_e}{2}\right) \tag{5.1c}$$

where d_e and d_i are external and internal diameters of the cylindrical gauge length and $A = \pi(d_e^2 - d_i^2)/4$ is the cross-sectional area. However, since $\sigma_r \ll \sigma_a, \sigma_t$ we may assume that the stress state is approximately biaxial.

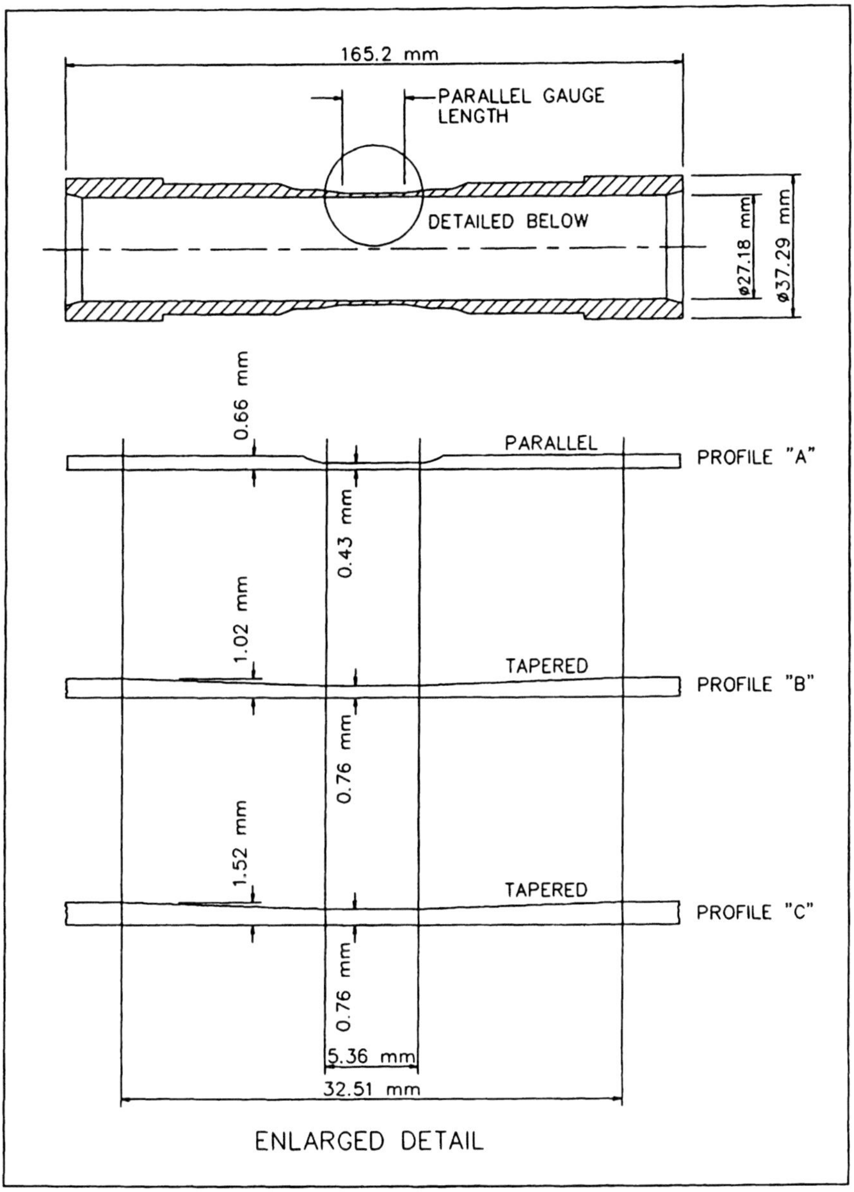

Fig. 5.2 Three profiles of test sections of thin-walled circular cylindrical specimens analysed in [28].

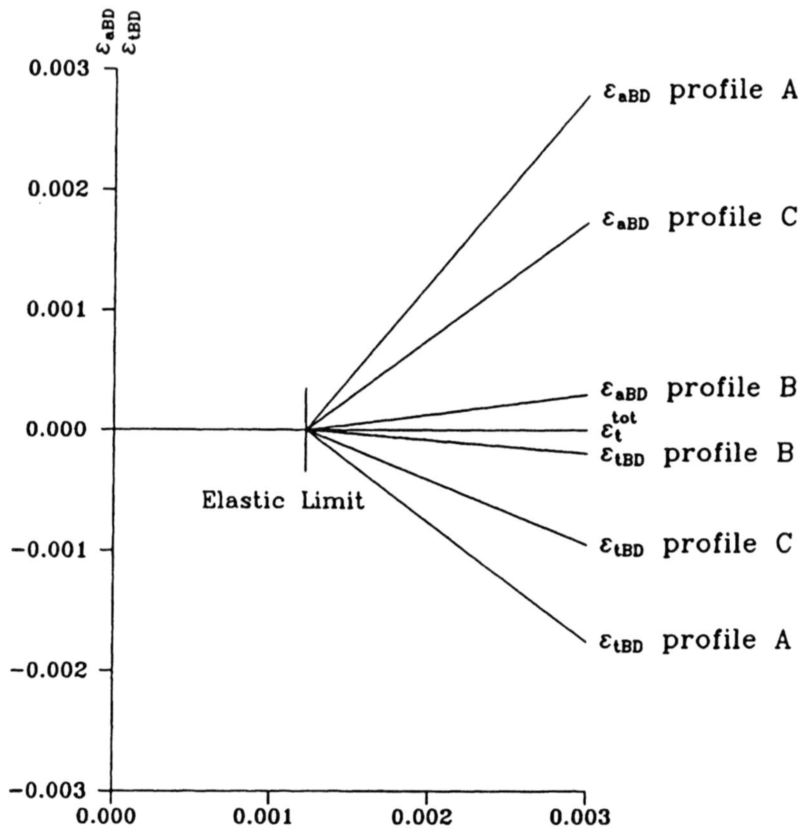

Fig. 5.3 Tangential and longitudinal strains due to the barrel distortion versus the applied strain for the three profiles in Fig. 5.2. The applied tangential strain is indicated on the abscissa while strains due to barrel deformation are shown on the ordinate, after [28].

The strains in the gauge length are determined from

$$\varepsilon_a = \frac{\delta \ell}{\ell} \tag{5.2a}$$

$$\varepsilon_t = \frac{\delta d_e}{d_e} \tag{5.2b}$$

$$\varepsilon_r = \frac{\delta (d_e - d_i)}{d_e - d_i} \tag{5.2c}$$

where the prefix δ before a symbol indicates its variation. It is relatively easy to measure the change in length, $\delta \ell$, and the change in diameter, δd_e,

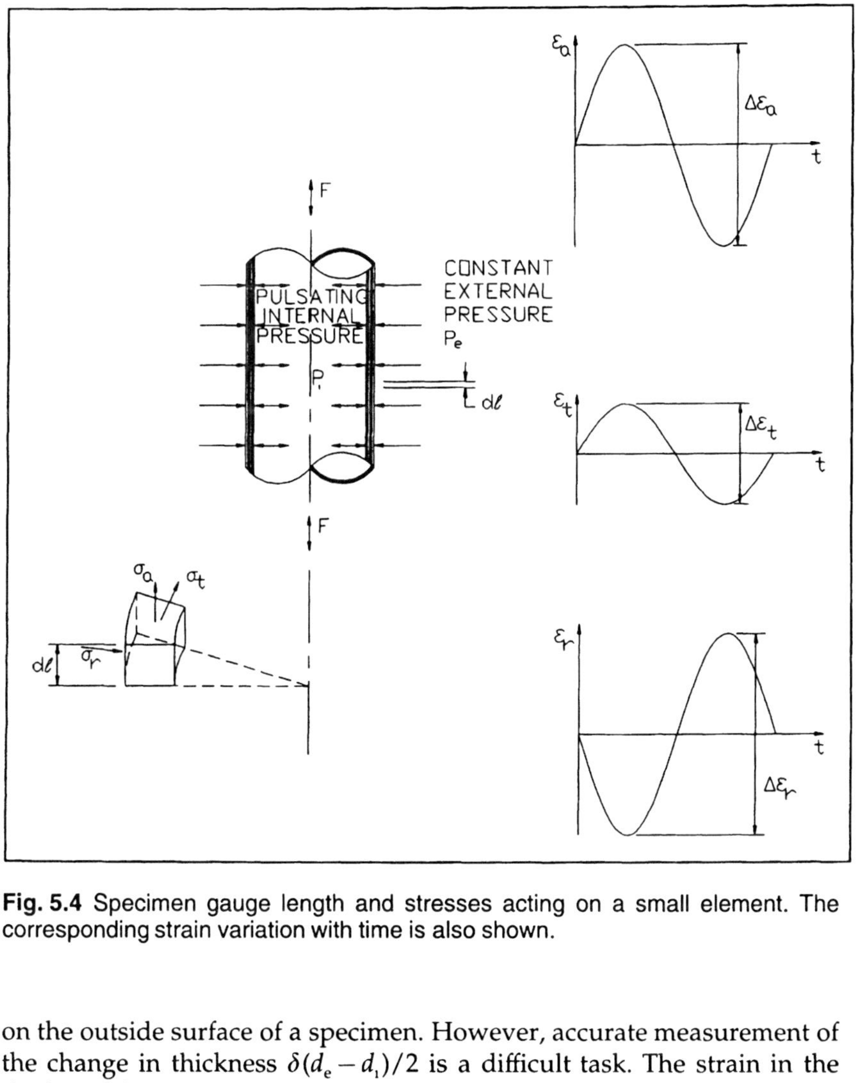

Fig. 5.4 Specimen gauge length and stresses acting on a small element. The corresponding strain variation with time is also shown.

on the outside surface of a specimen. However, accurate measurement of the change in thickness $\delta(d_e - d_i)/2$ is a difficult task. The strain in the thickness direction, ε_r, in the elastic range, can be calculated in terms of the other two principal strains as follows. From Hooke's law,

$$\varepsilon_r^e = \frac{1}{E}[\sigma_r - \nu(\sigma_a + \sigma_t)], \tag{5.3}$$

and

$$\varepsilon_r^e + \varepsilon_a^e + \varepsilon_t^e = \frac{1}{3K}(\sigma_r + \sigma_a + \sigma_t), \tag{5.4}$$

where the bulk modulus $K = E/3(1 - 2\nu)$. From (5.3) and (5.4) we get

$$\varepsilon_r^e = \frac{(1-2\nu)(1+\nu)}{(1-\nu)}\left(\frac{\sigma_r}{E}\right) - \frac{\nu}{1-\nu}(\varepsilon_a^e + \varepsilon_t^e). \tag{5.5}$$

At the outer surface of the cylinder, $\sigma_r = -p_e$, thus (5.5) can be written as

$$\varepsilon_r^e = \frac{(1-2\nu)(1+\nu)}{(1-\nu)}\left(-\frac{p_e}{E}\right) - \frac{\nu}{1-\nu}(\varepsilon_a^e + \varepsilon_t^e). \tag{5.6}$$

Therefore, if ε_a^e and ε_t^e and p_e are specified, then ε_r^e can be calculated from (5.6). For most steels, $\nu \approx 0.3$, $E \approx 200\,\text{GPa}$, in which case $(1-2\nu)(1+\nu)p_e/(1-\nu)E = 0.372 \times 10^{-5} p_e$ (MPa). For example, when $p_e = 35\,\text{MPa}$ ($\approx 5000\,\text{psi}$), then the contribution of the first term on the RHS of (5.6) is 0.01×10^{-2}. For tests near the elastic limit of the material, under equibiaxial loading, the second term on the RHS of (5.6) is about 0.1×10^{-2}, i.e. an order of magnitude higher than the first term. It should be noted, however, that ε_r^e is of the same order of magnitude as the other strain components, ε_a^e and ε_t^e.

Similarly, during plastic deformation the volume change is negligible, i.e.

$$\frac{\delta V}{V} = \varepsilon_r^p + \varepsilon_a^p + \varepsilon_t^p = 0, \tag{5.7a}$$

from which

$$\varepsilon_r^p = -(\varepsilon_a^p + \varepsilon_t^p). \tag{5.7b}$$

Relation (5.7b) could have been obtained from (5.6) by setting $\nu = 0.5$.

From (5.6) and (5.7b) it is clear that the strain component ε_r is of the same order as the other principal strains ε_a and ε_t, and thus not negligible. During a high-strain test, Poisson's ratio ν changes from its elastic value ν^e to that of plastic ν^p. If the first term in (5.6) is neglected, the strain in the thickness direction can be expressed in the form of

$$\varepsilon_r = -\frac{\bar{\nu}}{1-\bar{\nu}}(\varepsilon_a + \varepsilon_t), \tag{5.8}$$

where $\bar{\nu}$ is an 'effective' Poisson's ratio. The value of $\bar{\nu}$ can be obtained from (5.6) and (5.7) and noting that $\varepsilon_r = \varepsilon_r^e + \varepsilon_r^p$, and recalling that the first term on the RHS of (5.6) is negligible in comparison with the second one. The result is

$$\bar{\nu} = \frac{\nu^p(1-\nu^e)(\varepsilon_a + \varepsilon_t) + (\nu^e - \nu^p)(\varepsilon_a^e + \varepsilon_t^e)}{(1-\nu^e)(\varepsilon_a + \varepsilon_t) + (\nu^e - \nu^p)(\varepsilon_a^e + \varepsilon_t^e)}, \quad (\varepsilon_a \neq -\varepsilon_t). \tag{5.9}$$

Knowing the elastic components of the imposed strains ε_a and ε_t, we can calculate the effective Poisson's ratio $\bar{\nu}$ from (5.9) and consequently ε_r from (5.8). Hysteresis loops can thus be generated for σ_r versus ε_r through variation of (5.1c) and (5.8). In the case of $\varepsilon_a = -\varepsilon_t$, the effective Poisson's ratio, $\bar{\nu}$, is obtained from (5.9) through a limiting process, e.g. L'Hopital's rule.

5.4 THE TEST SYSTEM

The test system is based on an electrohydraulic testing facility using three control circuits schematically shown in Fig. 5.5(a). Control system A is used to maintain a constant hydraulic pressure in the outer pressure chamber which houses the specimen, transducers and axial load cell. This pressure acts on the outside surface of the specimen (Fig. 5.5(b)). Control

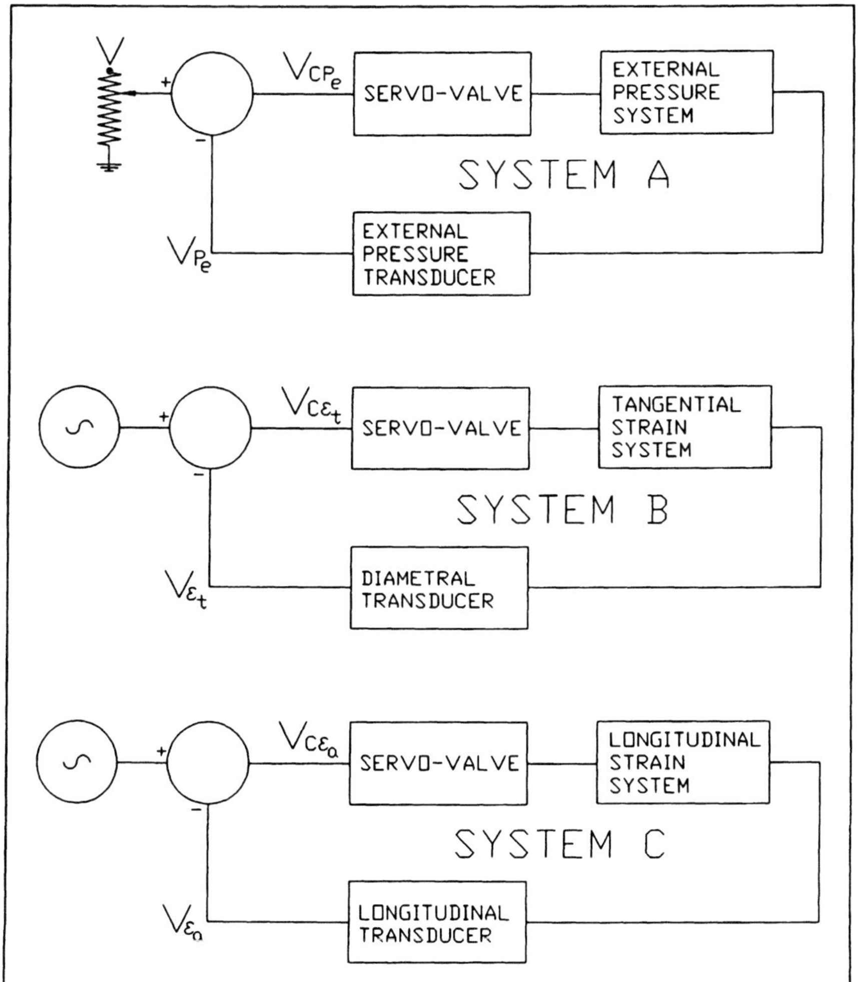

(a)

Fig. 5.5(a) Schematic of control circuits for the testing system in strain control: external pressure system (A), internal pressure system (B) and axial load system (C).

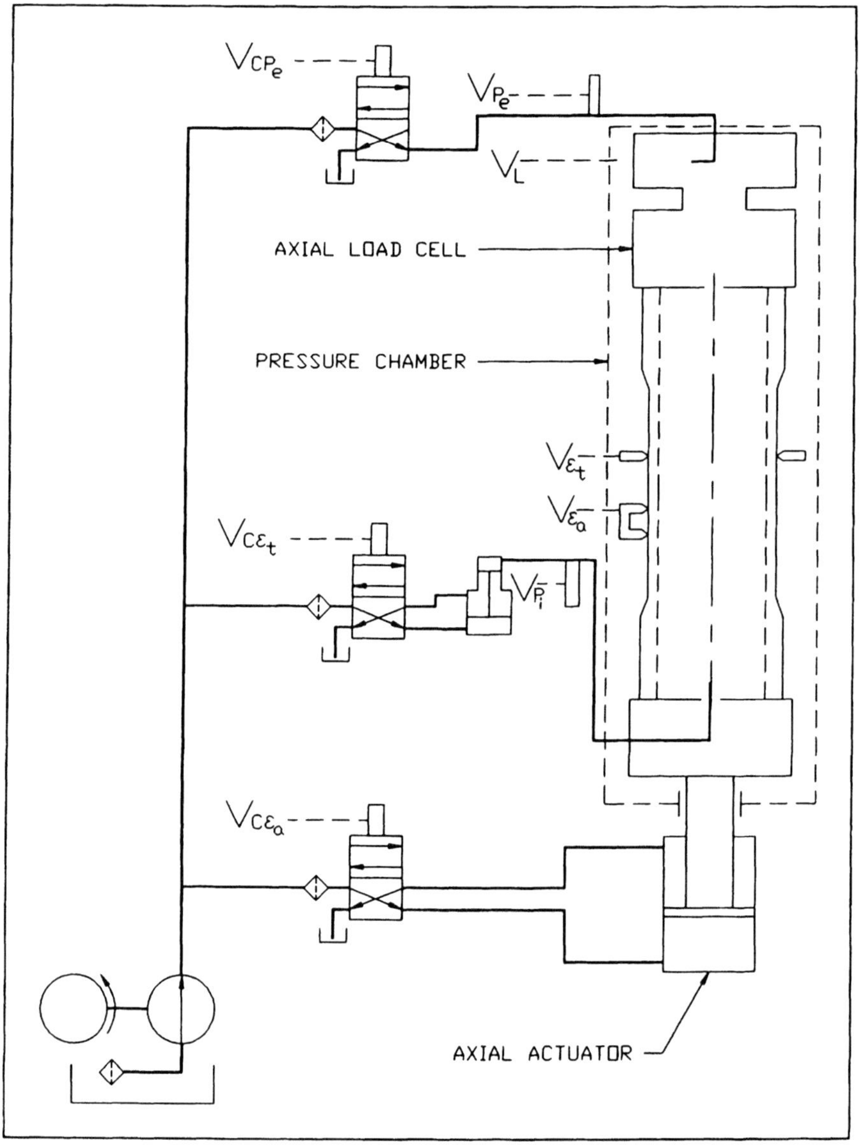

Fig. 5.5(b) Schematic of the hydraulic curcuit for the testing system including specimen and transducers.

system B, using the diametral extensometer as a feedback element, drives the internal pressure above and below the external pressure via a pressure intensifier so that the measured tangential (hoop) strain, ε_t, follows the command signal. Control system C, connecting the longitudinal strain extensometer and the axial actuator, forces the measured longitudinal (axial) strain, ε_a, to follow the command signal.

Figure 5.6 is a photograph of the test facility showing the specimen holder, axial load cell, longitudinal and diametral extensometers, and part of the pressure chamber. This assembly was designed to overcome some of the inconveniences of an earlier design [28]. The problems encountered in the former facility were the length of time involved in changing the specimen, the care necessary to ensure that the extensometers did not move, and the difficulty of ensuring an accurate alignment. All these problems are alleviated by the present design as shown in Fig. 5.6. Here, the specimen and extensometers are quickly assembled in the final position, and the pressure chamber is then lowered over the complete assembly. A load frame is not necessary for this design as the rods in the specimen assembly system act as a closed frame. The machine frame in this instance (Fig. 5.6), is only used to lift and lower the pressure chamber. The axial load cell is located within the pressure chamber so that no low friction seals are required, resulting in a more accurate load reading [28]. Another advantage is that the length of the assembly in the longitudinal direction is kept short, thus helping to minimize misalignment in the axial direction which causes unwanted bending stresses. In fact, the upper grip assembly is of a floating type and is connected to the load cell when the mounted specimen and upper grip touches it.

To demonstrate the improved alignment quality, let us compare the results of the uniaxial fatigue tests of thin-walled and solid specimens. Figure 5.7 shows the results of tests performed in our previous facility at the Université de Sherbrooke [30] and Fig. 5.8 shows the result of the present test facility at the University of Alberta. A comparison of these two sets of results clearly shows the improved performance. Firstly, the intercept with the ordinate is increased by a factor of more than 3, ($\Delta W^t|_{N_f=1}$ = 880 versus 261) and secondly, the slope of the curve is steeper ($\alpha = -0.665$ versus -0.54). Most of the recently reported multiaxial test facilities, e.g. [29], are similar to that of [28] and thus suffer from the same type of problem, i.e. inaccurate reading of the axial load to which the specimen is subjected, and undesired bending stresses due to the unavoidable misalignment of the long rod connecting the actuator to the load cell.

5.5 MEASURING DEVICES

Special extensometers using strain gauges were developed and are shown in Fig. 5.9. The longitudinal extensometer is fitted with razor blade edges and is held against the specimen with springs (Fig. 5.9(b)). The diametral extensometer is of a closed box shape with needlepoint screws. It is held firm against the specimen by a spring, and is weight balanced with respect

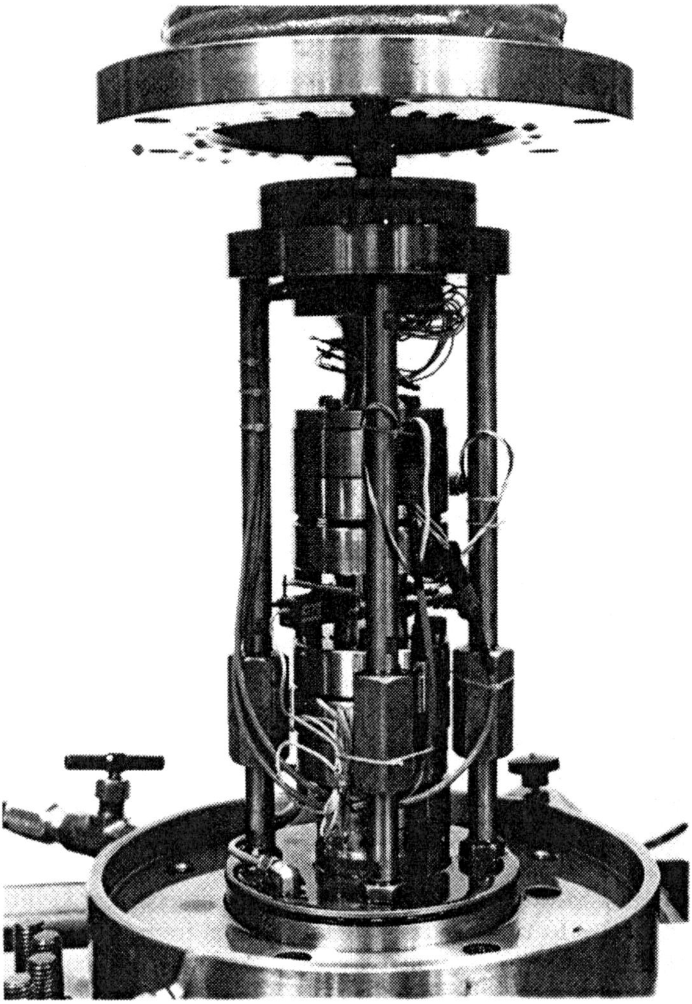

Fig. 5.6 A photograph of the specimen ready for testing, before lowering of the pressure chamber.

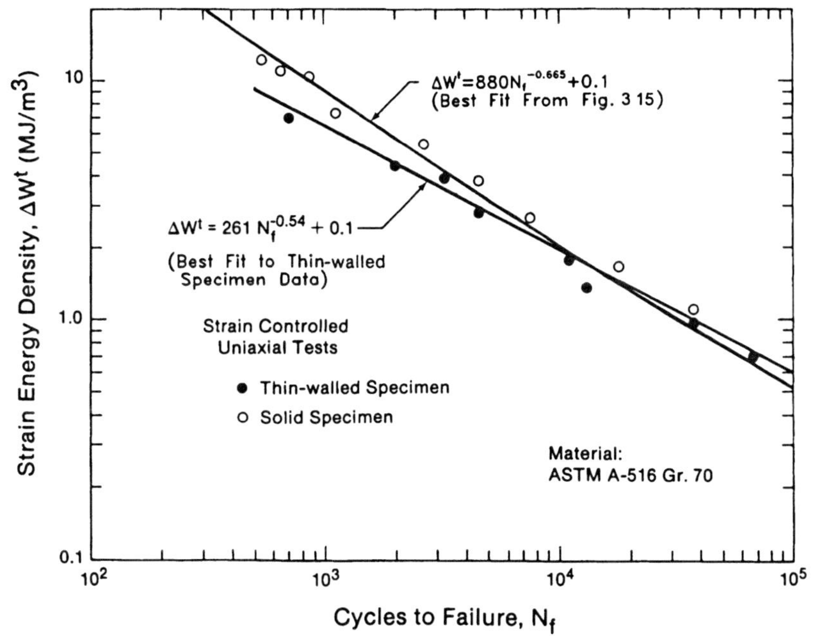

Fig. 5.7 Comparison of uniaxial fatigue results of solid and thin-walled specimens [30]. Best fit to the thin-walled specimens is shown. Also shown is the best fit curve of solid specimens from Fig. 3.15 of Chapter 3.

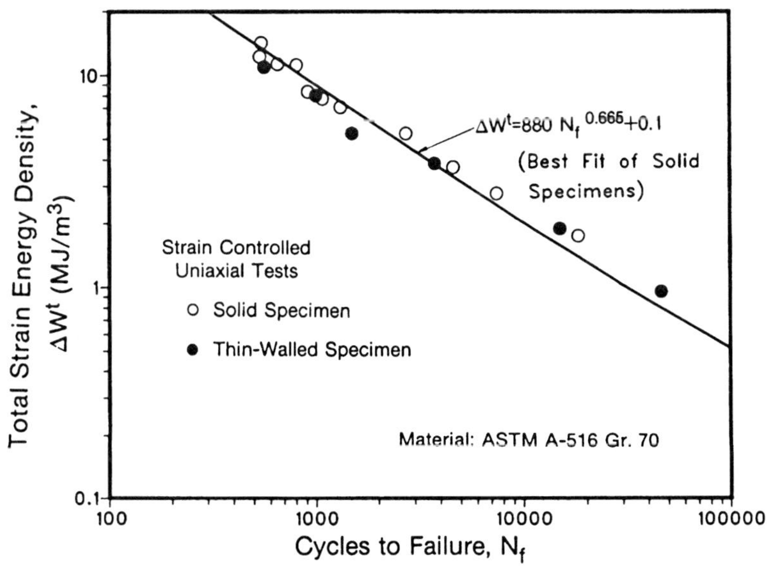

Fig. 5.8 Comparison of uniaxial fatigue results of solid and thin-walled specimens. Reproduced with permission from ASME from [43]. Best fit curve of solid specimens from Fig. 3.15 of Chapter 3 is also shown.

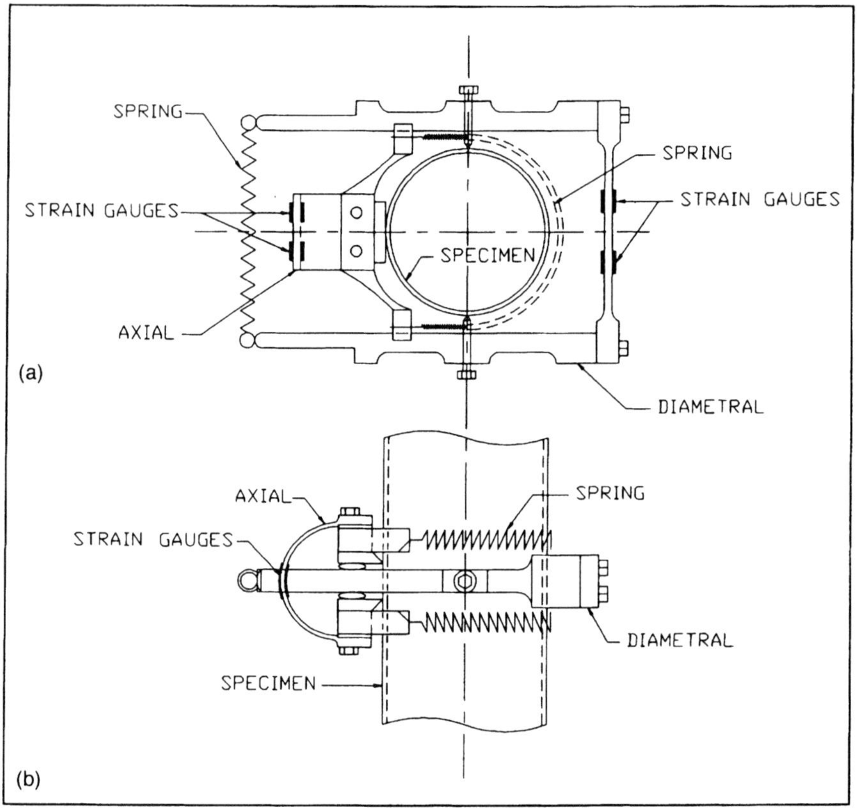

Fig. 5.9 Longitudinal and diametral extensometers in mounted position: (a) plane view; (b) side view.

to the specimen axis (Fig. 5.9(a)). This type of diametral extensometer is more accurate in comparison to that described in [28].

The extensometers are placed in a constant pressure environment (Fig. 5.5(b)) and are calibrated with precise measuring equipment and give a linear output.

5.6 TEST PROCEDURE

The assembly and test procedure is as follows. A prepared specimen is clamped in the grips by tightening the load assembly in place and then bolting the bottom grip to the axial actuator and the top grip to the load cell. The extensometers are then mounted and the specimen is filled with oil. The oil is injected into the bottom port while the air is bled through the top port. The pressure chamber is then lowered, secured and filled with oil in a similar manner.

Before commencing a fatigue test, the pressure conditions have to be raised to those of a point in the load cycle. The procedure adopted was to raise the external pressure with the command signals (systems B and C in Fig. 5.5(a)) set at zero. To maintain the axial and tangential strains (ε_a and ε_t) at zero, the feedback system will automatically adjust the internal pressure and axial load. The signal generators (Fig. 5.5(b)) are then started and the amplitude is increased gradually from zero to the desired level.

A pair of X–Y recorders, a pair of peak detecting digital voltmeters and an oscilloscope are used to measure and record the specimen response to the applied strains. A computer data acquisition system is also used to record data and to display specific hysteresis loops at intervals specified by the user. Axial load F, axial strain ε_a, tangential (hoop) strain ε_t, and internal pressure p_i, are collected by the data acquisition system.

When a fatigue crack penetrates the specimen wall thickness, hydraulic oil flows from the interior to the exterior system. Due to this loss of oil, the cycling hydraulic intensifier (Fig. 5.5(a), System B) reaches the limit of its stroke and becomes unable to supply the pressure necessary to maintain the required tangential strain amplitude. This results in a large error signal between the command and output signals and the machine is set to shut down under these conditions.

As mentioned earlier, during each test four channels are monitored. They record axial load, axial strain, internal pressure and hoop strain. A counter, connected to the synchronic pulse on the function generator is also monitored. The number of load cycles that a specimen has undergone is monitored by using the counter. Very large amounts of data can be collected by the computer, but it was decided that not every loop was necessary and so the program collects a complete hysteresis loop according to a user predefined sequence. The sequence must have gaps between loops to allow for the computer to write to the disk. For example, the sampling sequence used in the tests reported here was designed to obtain loops fairly close together for the first 100 cycles and then take 10 loops every log 10 cycles, i.e. to take 10 loops between 100 and 1000 cycles, and 10 loops between 1000 and 10 000 cycles, etc. It was decided to write the data to the disk drive every time a loop was collected to ensure that not all of the data is lost in the event of a power failure. The number of points per hysteresis loop to be collected is set by the user at the start of the test. These data are collected at a high speed to avoid time lag effects between the four channels. When the program is not collecting data or writing to the disk, the data for the previous loop can be plotted on the screen. After the test is over, the data is analyzed. The software developed in our laboratory allows for the plotting of hysteresis loops, peak stresses versus number of cycles and calculation of the hysteresis loop area or the plastic strain energy per cycle, ΔW^p.

5.7 TYPICAL MULTIAXIAL TEST RESULTS

Under a deformation (strain) controlled condition, we set the limits on the strains and then record the material response. For a thin-walled circular cylindrical specimen subject to differential pressure and axial load, the principal strains are

$$\varepsilon_1 = \varepsilon_a, \quad \varepsilon_2 = \varepsilon_t \quad \text{and} \quad \varepsilon_3 = \varepsilon_r . \tag{5.10}$$

However, as explained earlier, we could only accurately control ε_a and ε_t. Furthermore, when the strains alternate about a datum line, their variation in time and the position of the datum line have to be specified (Fig. 5.4). For example, in a sinusoidally varying mode, we may express

$$\varepsilon_a = A \sin 2\pi ft, \quad \varepsilon_t = \rho A \sin (2\pi ft - \phi), \tag{5.11}$$

where A is the amplitude of the axial strain, $\rho = \Delta\varepsilon_t / \Delta\varepsilon_a$ is the ratio of the tangential to axial strain range (Fig. 5.1), f is the frequency of oscillation, and ϕ is the phase angle. The position of the datum line about which the oscillation takes place is denoted as the mean strain, $\varepsilon_m = (\varepsilon_{max} + \varepsilon_{min})/2$. As discussed in Chapter 3, the effect of the mean strain is not significant in the short life (low-cycle) region. Therefore, to perform a multiaxial test we have to fix four parameters, namely amplitudes of strains (A and ρ), frequency of load application (f) and phase difference between the strain components (ϕ).

5.7.1 IN-PHASE LOADING CONDITIONS ($\phi = 0$)

In this type of test the strains oscillate in-phase, and the three parameters to be specified are A, ρ and f (5.11). The frequency at which the tests can be conducted is predetermined. The dynamic characteristics of transducers, the maximum flow rate of the hydraulic pressure supply, and the speed of the recording devices impose limits on the frequency. For the results reported here, the test frequency was $f = 0.15$ Hz. Four strain ratios were tested ($\rho = -\bar{\nu}$, 0.0, 0.5 and 1.0). The strain amplitude, A, varied from 0.1% to 0.25%.

Similar to the uniaxial case, the hysteresis energy does not vary appreciably during the cyclic life. Figure 5.10 shows the variation of the hysteresis energy (absorbed energy) in the axial and tangential directions for a typical test. This result confirms the assumption made in Chapter 4 that the cyclic strain energy is an appropriate correlating parameter in the multiaxial loading condition.

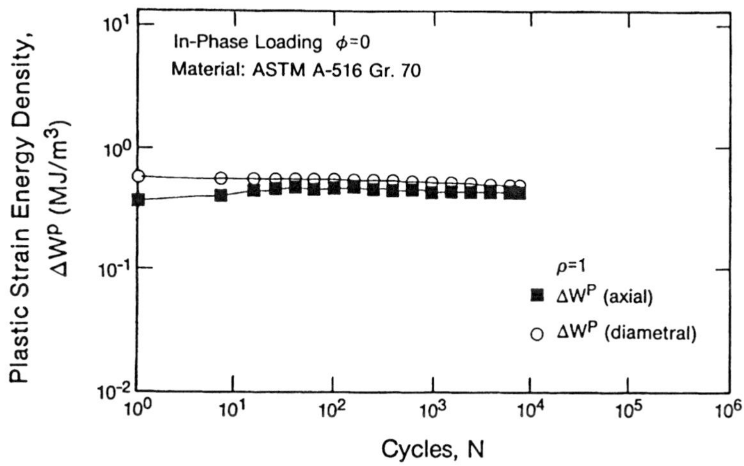

Fig. 5.10 Variation of hysteresis loop area, ΔW^p, with number of cycles, N, for in-phase loading, $\rho = 1$, $\Delta\varepsilon_a = \Delta\varepsilon_t = 0.3\%$. Reproduced with permission from ASME [43].

The life versus plastic strain energy (ΔW^p) curves for four strain ratios are depicted in Fig. 5.11. It can be seen that the curves are indeed parallel, and that ΔW^p is a good correlating factor. It is also observed that the $\rho = 1$ (equi-biaxial) condition is the most severe case. This was also predicted by the theory presented in Chapter 4.

The parallel nature of the life curves on the log–log scale suggests that through an appropriate definition of the multiaxial constraint factor

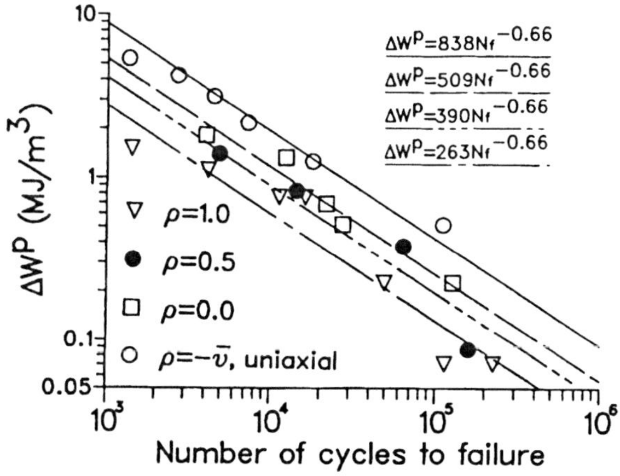

Fig. 5.11 Variation of plastic strain energy density, ΔW^p, with number of cycles to failure, N_f, for four strain ratios.

(MCF), $\bar{\rho}$, they could be collapsed into a unique curve. Indeed, this is the case when we use the definition:

$$\bar{\rho} = (1 + \bar{v}) \frac{\hat{\varepsilon}}{\hat{\gamma}} \tag{5.12}$$

where

$$\hat{\varepsilon} = \max\,[\varepsilon_a \text{ or } \varepsilon_t]$$

and

$$\hat{\gamma} = \max\,[(\varepsilon_a - \varepsilon_r) \quad \text{or} \quad (\varepsilon_t - \varepsilon_r)]$$

and the effective Poisson's ratio, $\bar{v}$, is given by (5.9). This has been demonstrated in Chapter 4 (Fig. 4.6), where the data for the four strain ratios were plotted as $\Delta W^p/\bar{\rho}$ versus N_f. This type of curve is useful in design situations where the life under multiaxial loading can be predicted from the uniaxial test data.

A relationship between the MCF, $\bar{\rho}$, and the applied in-plane strain ratio, ρ, can be found from substituting (5.11) and (5.8) into (5.12), and taking $\phi = 0$. The result is

$$\bar{\rho} = \frac{1 - \bar{v}^2}{1 + \rho \bar{v}}, \tag{5.13}$$

which indicates that $\bar{\rho}$ depends not only on the applied in-plane strain ratio, but also on their magnitude through (5.9).

5.7.2 OUT-OF-PHASE LOADING CONDITIONS ($\phi \neq 0$)

When the imposed strains ε_a and ε_t are not in-phase, i.e. $\phi \neq 0$ in (5.11), then the direction of the maximum principal strain/stress changes by 180° during every cycle. In this process various crystalline slip planes are activated during out-of-phase loading which results in additional strain hardening. Thus, the cyclic stress–strain curve of the material may not be unique, and it could depend on the straining path and accumulated plastic strain. The odelling of the deformation behaviour under non-proportional loading will be discussed in the next chapter, and therefore we will not pursue this topic further. However, during an out-of-phase loading, the energy required to produce a given plastic deformation is increased in comparison to that of the proportional loading. This then results in a decreased fatigue life. Depending upon the phase angle and strain ratio, fatigue life reduction of up to an order of magnitude is observed in comparison to that of the equivalent proportional straining.

The following is a brief summary of the experimental investigations on the effect of out-of-phase loading on fatigue life.

Earlier investigations into out-of-phase cyclic loading were performed on shafts subject to torsion and bending [31–33]. Later on, Lee [34] supplemented data on the high cycle out-of-phase torsion and bending by extending it to the intermediate life cycle ($3.5 \times 10^5 \geqslant N_f \geqslant 1.2 \times 10^4$). He concluded that both the Tresca and von Mises type criteria could seriously underestimate or overestimate the fatigue damage under 90° out-of-phase torsion and bending, depending upon dominance of the applied bending or torsional stress, respectively. A design criterion similar to Gough's ellipse quadrant was proposed which contained three material parameters. It correlated fairly well with the fatigue data for medium and long lives [34].

Experimental data on torsion bending, while of interest to designers, are not appropriate for fundamental studies, since the stress and strain distributions have gradients through the thickness. To overcome this problem, tests on thin-walled tubes have to be performed. Results of such an investigation under a range of out-of-phase angles and frequency ratios were reported by McDiarmid [35]. He concluded that the maximum shear stress criterion can be adapted for use in high-cycle fatigue, independent of both phase angle and frquency ratio.

Test results in the low-cycle fatigue regime have been reported by a number of investigators [36–42]. In most of these investigations, additional strain hardening is observed over and above the stable cyclic state in proportional loading. Life estimates based on the amplitude of strain do not yield satisfactory results for non-proportional straining. Various strain measures have been proposed ranging from integrating the octahedral shear strains [36], to maximum value of total strain [37], to some function of the amplitude of shear strain and the amplitude of normal strain acting on a critical plane [38, 41, 42]. It is worth emphasizing that the fatigue damage is not only a function of the strain amplitude, but also of the the strain path. Therefore, any attempt at defining an 'effective strain' measure without due regard to the path and the resulting stress level will be of limited use. Furthermore, all the published experimental data in the low-cycle fatigue regime have been confined to the relatively small region of tension–torsion strain ratio $-1 \leqslant \rho \leqslant -\bar{\nu}$ (Fig. 5.1). The positive strain ratio has not been investigated except for McDiarmid's results in the high-cycle region where the effect of the out-of-phase loading condition is not very pronounced due to the small magnitude of the plastic strain. In the following we will present some results of the 90° out-of-phase straining for the strain ratio $\rho = \Delta\varepsilon_t/\Delta\varepsilon_a = 1$, the most critical fatigue loading.

The hysteresis loops for the axial and circumferential directions are plotted in Figs 5.12 and 5.13, respectively, for $\rho = 1$ and $\Delta\varepsilon_a = \Delta\varepsilon_t = 0.30\%$ and $\phi = 0$ and 90°. From these figures we note a pronounced strain hardening due to the out-of-phase loading. In addition, the shape and size of the hysteresis loops are quite different, indicating the influence of

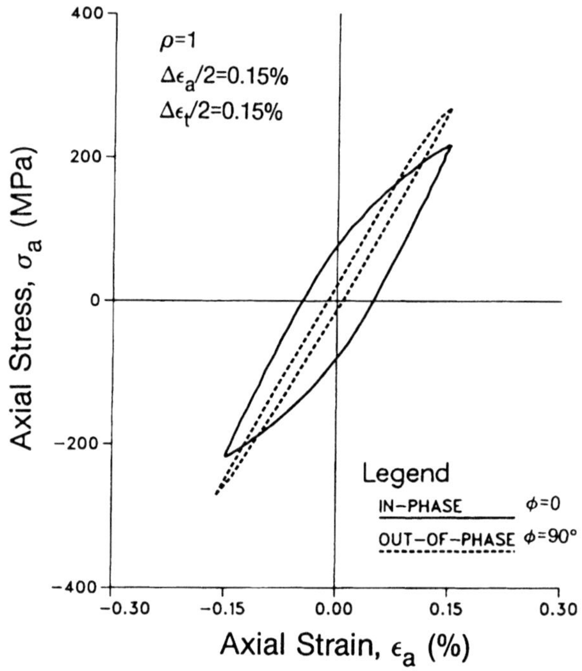

Fig. 5.12 Stable hysteresis loops of the axial stress–axial strain (σ_a vs. ε_a) for the in-phase ($\phi = 0$) and 90° out-of-phase loading of ASTM A-516 Gr. 70 low alloy steel. Reproduced with permission from ASME from [43].

out-of-phase (non-proportional) loading which induces strong anisotropy in the material. Figure 5.14 shows the variation of the hysteresis loop for 90° out-of-phase loading with the number of cycles. When this figure is compared with Fig. 5.10, we note the strong influence of induced anisotropy due to the non-proportional loading. Note that in the case of in-phase loading the strain path is linear, while in the case of 90° out-of-phase strain path is a circular one. Further examples of the effect of out-of-phase loading on the hysteresis loops are given in section 6.6.4, Figs 6.24 and 6.25.

The results of the out-of-phase loading for $\rho = 1$, $\phi = 90°$, are compared with the in-phase data for the same in-plane strain ratio in Fig. 5.15 [43]. It can be seen that the out-of-phase loading is much more damaging and results in lower lives. This can be attributed to the circular strain path which attempts to force dislocation movements along all possible crystal slip planes. The resistance to such a motion causes extra cyclic hardening which eventually leads to more damage accumulation compared to the proportional (in-phase) loading. However, as the plastic strain decreases, the effect of out-of-phase loading also decreases. Data in Fig. 5.15 indicate

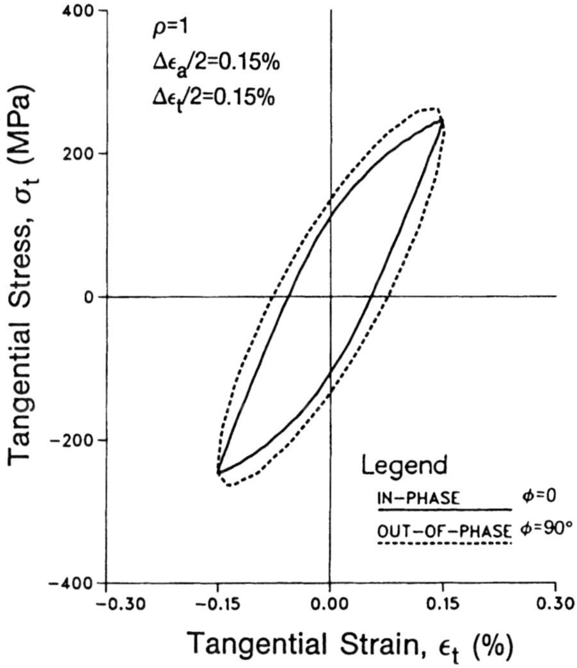

Fig. 5.13 Stable hysteresis loops of the tangential stress–tangential strain (σ_t vs. ε_t) for the in-phase and 90° out-of-phase loading of ASTM A-516 Gr. 70 low alloy steel. Reproduced with permission from ASME from [43].

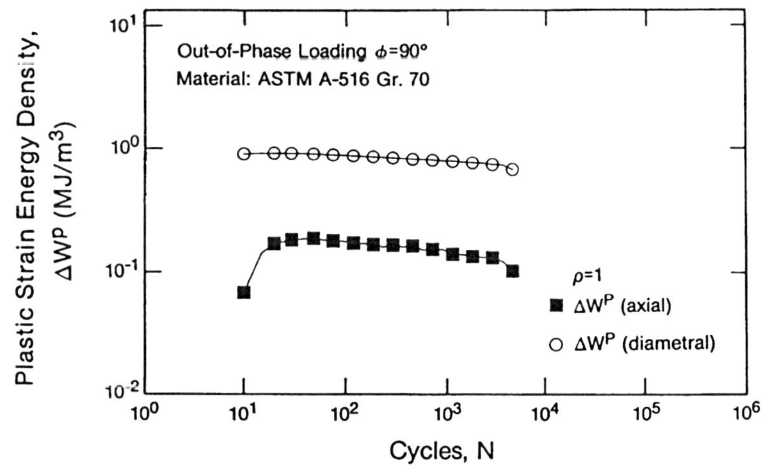

Fig. 5.14 Variation of hysteresis loop area, ΔW^p, with number of cycles, N, for 90° out-of-phase loading, $\rho = 1$, and $\Delta\varepsilon_a = \Delta\varepsilon_t = 0.3\%$. Reproduced with permission from ASME from [43].

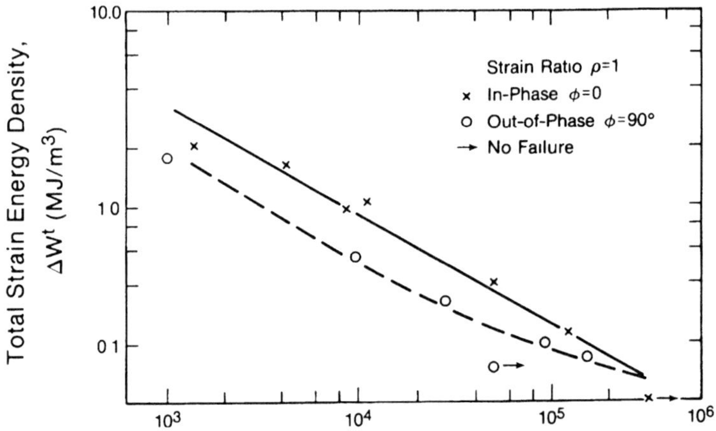

Fig. 5.15 A comparison between cycles to failure of the in-phase and out-of-phase tests in terms of the absorbed total strain energy density, ΔW^t, for ASTM A-516 Gr. 70 low alloy pressure vessel steel. Reproduced with permission from ASME from [43].

that at lower strains, the effect of out-of-phase loading decreases and the best fit curves for in-phase and out-of-phase may asymptotically approach the fatigue limit of the material. This type of behaviour is to be expected when considering that the dislocation movement is generated at stress levels above the material yield limit, and that the dislocation structure is a function of the plastic deformation. Test results reported by McDiarmid in the high-cycle regime [35] support the above conclusion.

Table 5.1 shows a comparison between in-phase and out-of-phase tests for the same strain amplitudes and ratios. It should be noted that lives to failure could change by an order of magnitude as a result of non-proportional loading.

Table 5.1 Comparison between lives of in-phase and out-of-phase tests for $\rho = 1$

$\Delta\varepsilon_a/2$ (%)	$\Delta\varepsilon_t/2$ (%)	ϕ (deg.)	N_f (cycle)
0.150	0.150	0	11 300
0.153	0.150	0	8 802
0.152	0.151	90	1 009
0.105	0.100	0	49 980
0.101	0.103	90	10 800
0.075	0.074	0	115 000
0.077	0.075	90	34 500

It is worth emphasizing once again the importance of the axial load alignment in the design of multiaxial test facilities. A comparison of test results reported in [30] and those presented here shows a considerable increased life for the tests performed in our improved test facility ([43], Table 3).

To predict the plastic energy per cycle and correlate the results with the experimental data, an appropriate constitutive relation is required. This will be described in the next chapter. Here it will suffice to mention that such a comparison is reported in [44].

5.8 OTHER TEST FACILITIES

In section 5.2 it was pointed out that the tubular specimens are either subjected to the combined axial and torsional loads, or axial and differential pressure. In the case of proportional (in-phase) straining, the principal axes are fixed in the latter type of loading set-up, whereas the direction of the principal stresses changes with the change of the strain ratio in an axial–torsional loading set-up. Of course, the direction of principal stresses changes during each cycle when the straining is out-of-phase (non-proportional loading). If one wishes to combine the advantages of viewing the specimen surface for crack propagation with the ability to apply positive strain ratios, then a torsional actuator has to be added to the test facility described in section 5.4. Such a system has been developed in our laboratory. It is capable of applying axial load, differential pressure and torsion to a tubular specimen under independent servo-hydraulic control. Details of the experimental apparatus are given in [45].

A recent ASTM symposium (details of proceedings are given in [45]) on the multiaxial testing contains some interesting recent developments in the field. An interested reader is encouraged to consult this volume.

REFERENCES

1. Wöhler's Experiments on the Fatigue of Metals, *Engineering*, **11**, 1871, p. 199.
2. Lanza, G. (1886) Strength of shafting subjected to both twisting and bending. *Trans. ASME*, **8**, 130–44.
3. Gough, H.J. (1935) The strength of metals under combined alternating stresses. *Proc. Inst. Mech. Engrs.*, **131**, 3–54.
4. Gough, H.J. (1950) Engineering steels under combined cyclic and static stresses. *J. Appl. Mech., Trans. ASME*, **72**, 113–25.
5. Gough, H.J. and Pollard, H.V. (1935) The effect of specimen form on the resistance of metals to combined alternating stresses, *Proc. Inst. Mech. Engrs.*, **132**, 549–73.
6. Das, P.K., Chandler, D.C. and Foster, B.K. (1973) The plastic bending of beams and their failure by low-cycle fatigue. *J. Engng. Mater. Technol., Trans. ASME*, **95**, 161–9.

7. Ives, K.S., Kooistra, L.F. and Tucker, J.T. (1966) Equi-biaxial low-cycle fatigue properties of typical pressure-vessel steels. *J. Basic Engng., Trans. ASME*, **88**, 745–54.
8. Matake, T. (1977) An explanation on fatigue limit under combined stress. *Bulletin JSME*, **20**, 257–63.
9. Zamrik, S.Y. and Shabara, M.A. (1977) The effect of stress ratio on fatigue crack growth in a biaxial stress field. *J. Pressure Vessel Technol., Trans. ASME*, **99**, 137–43.
10. Pascoe, K.J. and DeVilliers, J.W.R. (1967) Low-cycle fatigue of steels under biaxial straining. *J. Strain Anal.*, **2**, 117–26.
11. Wilson, J.H. and White, D.J. (1971) Cruciform specimens for biaxial fatigue tests. *J. Strain Anal.*, **6**, 27–37.
12. Parsons, M.W. and Pascoe, K.J. (1975) Development of a biaxial fatigue testing rig. *J. Strain Anal.*, **10**, 1–9.
13. Tanaka, K., Hoshide, T., Yamada, A. and Taira, S. (1974) Fatigue crack propagation in biaxial stress fields. *Fatigue Engng. Mater. Struct.*, **2**, 181–94.
14. Kitagawa, H., Yuuki, R. and Tohgo, K. (1974) A fracture mechanics approach to high-cycle fatigue crack growth under in-plane biaxial loads. *Fatigue Engng. Mater. Struct.*, **2**, 195–206.
15. Makinde, A., Thibodeau, L. and Neale, K.W. (1992) Development of an apparatus for biaxial testing using cruciform specimens. *Exp. Mech. SESA*, **32**, 138–44.
16. Boëhler, J.P., Demmerle, S. and Koss, S. (1994) A new direct biaxial testing machine for anisotropic materials. *Exp. Mech. SESA*, **34**, 1–9.
17. Findley, W.N. (1954) Experiments in fatigue under ranges of stress in torsion and axial load from tension to extreme compression. *Proc. ASTM*, **54**, 836–52.
18. Taira, S., Inoue, T. and Yoshida, T. (1969) *Low-Cycle Fatigue Under Multiaxial Stresses (in the Case of Combined Cyclic Tension-Compression and Cyclic Torsion at Room Temperature)*. Proc. 12th Japan Congress on Material Research. Soc. Mater. Sci., Kyoto, Japan, pp. 50–5.
19. Brown, M.W. and Miller, K.J. (1979) Biaxial cyclic deformation behaviour of steels. *Fatigue Engng. Mater. Struct.*, **1**, 93–106.
20. Ellyin, F., Lefebvre, D. and Neale, K.W. (1980) High-strain biaxial fatigue of 2024-T351 aluminium under combined axial stress and torsion, in *Advances in Materials Technology in the Americas — 1980*, Vol. 2, American Society of Mechanical Engineers, New York, 17–21.
21. Hua, C.T. and Socie, D.F. (1984) Fatigue damage in 1045 steel under constant amplitude biaxial loading. *Fatigue Engng. Mater. Struct.*, **7**, 165–79.
22. Tanaka, K., Matsuoka, S. and Kimura, M. (1984) Fatigue strength of 7075-T6 aluminum alloy under combined axial loading and torsion. *Fatigue Engng. Mater. and Struct.*, **7**, 195–211.
23. Kalluri, S. and Bonacuse, P.J. (1993) In-plane and out-of-phase axial-torsional fatigue behavior of Haynes 188 Superalloy at 760°C, in *Advances in Multiaxial Fatigue*, ASTM STP 1191 (eds D.L. McDowell and R. Ellis), American Society for Testing and Materials, Philadelphia, PA, 133–50.
24. McDiarmid, D.L. (1972) Failure criteria and cumulative damage in fatigue under multiaxial stress conditions. Ph.D. Thesis, The City University, London.
25. Rotvel, J. (1970) Biaxial fatigue tests with zero mean stresses using tubular specimens. *Int. J. Mech. Sci.*, **12**, 597–613.

26. Havard, D.G. and Topper, T.H. (1969) *Biaxial Fatigue of 1018 Mild Steel at Low Endurance.* Proc 1st Int. Conf. on Pressure Vessel Technology, Delft, The Netherlands, Vol. II. American Society of Mechanical Engineers, New York, pp. 1267–77.
27. Andrews, J.M.H. and Ellison, E.G. (1973) A testing rig for cycling at high biaxial strains. *J. Strain Anal.*, **8**, 168–75.
28. Ellyin, F. and Valaire, B. (1982) *High Strain Biaxial Fatigue Test Facility.* Proc. 1982 Joint Conf. on Experimental Mechanics, Hawaii, May 1982, SESA-JSME, Society for Experimental Stress Analysis, Brookfield Center, CT, pp. 136–43.
29. Brown, M.W. and Miller, K.J. (eds) (1988) *Biaxial and Multiaxial Fatigue*, EGF 3, Mechanical Engineering Publication, London.
30. Lefebvre, D., Chebl, C. and Khazzari, E. (1984) *Multiaxial High-Strain Fatigue of A-516 Gr. 70 Steel.* Proc. 2nd Conf. on Fatigue Threshold, Birmingham, UK.
31. Manson, W. and Delaney, W.J. (1921) Alternating Combined Stress Experiments, Reports of the British Association for the Advancement of Science, pp. 329–41.
32. Nishihara, T. and Kawamoto, M. (1995) The Strength of Metals Under Combined Alternating Bending and Torsion with Phase Difference, Memoirs, College of Engineering, Kyoto University, **11**, 145–9.
33. Dietmann, H. and Lempp, W. (1974) Untersuchengen zum festigkeitsverhalten von stahlen bei mechrachsiger phasenverschobener daverschwingbeansprunchung. *Konstruction*, **31**, H5, 191–200.
34. Lee, D.N. (1985) A criterion for fully-reversed out-of-phase torsion and bending, in *Multiaxial Fatigue*, ASTM STP 853 (eds K.J. Miller and M.W. Brown), American Society for Testing and Materials, Philadelphia, PA, pp. 553–68.
35. McDiarmid, D.L. (1985) Fatigue under out-of-phase biaxial stresses of different frequencies, in *Multiaxial Fatigue*, ASTM STP 853 (eds K.J. Miller and M.W. Brown), American Society for Testing and Materials, Philadelphia, PA, pp. 553–68.
36. Taira, S., Inoue, T. and Yoshida, T. (1968) *Low Cycle Fatigue under Multiaxial Stresses (in the Case of Combined Cyclic Tension-Compression and Cyclic Torsion Out-of-Phase at Elevated Temperature).* Proc. 11th Japan Congress on Material Research. Soc. Mater. Sci., Kyoto, Japan, pp. 60–5.
37. Zamrik, S.Y. and Frismuth, R.E. (1973) The effect of out-of-phase biaxial strain cycling on low cycle fatigue. *Exp. Mech. SESA*, **13**, 204–8.
38. Kanazawa, K., Miller, K.J. and Brown, M.W. (1977) Low cycle fatigue under out-of-phase loading conditions. *J. Engng. Mater. Technol., Trans. ASME*, **99**, 222–8.
39. Jordan, E.H., Brown, M.W. and Miller, K.J. (1985) Fatigue under severe non-proportional loading, in *Multiaxial Fatigue*, ASTM STP 83 (eds K.J. Miller and M.W. Brown), American Society for Testing and Materials, Philadelphia, PA, pp. 569–85.
40. Sonsino, C.M. and Grubisic, V. (1985) Fatigue behaviour of cyclically softening and hardening steels under multiaxial elastic-plastic deformation, in *Multiaxial Fatigue*, ASTM STP 83 (eds K.J. Miller and M.W. Brown), American Society for Testing and Materials, Philadelphia, PA, pp. 586–604.
41. Andrews, R.M. and Brown, M.W. (1988) Elevated temperature out-of-phase fatigue behaviour of a stainless steel, in *Biaxial and Multiaxial Fatigue* (eds M.W. Brown and K.J. Miller), Mechanical Engineering Publications, London, pp. 641–58.

42. Fatemi, A. and Socie, D. (1988) A critical plane approach to multiaxial fatigue damage including out-of-phase loading. *Fatigue Fract. Engng. Mater. Struct.*, **11**, 149–65.
43. Ellyin, F., Golos, K. and Xia, Z. (1991) In-phase and out-of-phase multiaxial fatigue. *J. Engng. Mater. Technol., Trans. ASME*, **113**, 112–18.
44. Xia, Z. and Ellyin, F. (1991) Nonproportional multiaxial cyclic loading: Experiments and constitutive modelling. *J. Appl. Mech., Trans. ASME*, **58**, 317–25.
45. Ellyin, F. and Wolodko, J. (1996) Testing facilities for multiaxial loading of tubular specimens, in *Multiaxial Fatigue and Deformation Testing Techniques*, ASTM STP 1269 (eds S. Kalluri and P.J. Bonacuse), American Society for Testing and Materials, Philadelphia, PA.

Constitutive laws for transient and stable behaviour of inelastic solids

6

6.1 INTRODUCTION

The stress–strain relationship for simple deformation paths (proportional paths) was outlined in Chapter 2. In this chapter we will develop a more general class of elastic–plastic constitutive equations for arbitrary loading paths. The term constitutive equations (or law) is used to indicate the set of equations which predict the material response to a prescribed range of loading conditions. Constitutive equations can generally be derived in three ways, namely: (a) through examination of microscopic (or crystal) structure, (b) by thermodynamic consideration of the irreversible plastic deformation, and (c) using macroscopic continuum mechanics theory.

In the first approach, one examines the process of deformation in micro-scale, e.g. using dislocation or crystallography theories, and an 'averaging' process to derive macroscopic stress–strain relationships. Examples of this type of approach are given in [1–8] where references to other works are cited.

The second approach, based on the thermodynamics of irreversible processes, attempts to describe the plastic deformation in terms of energy exchanges as prescribed in the first law, and the thermal energy (entropy production) of the second law of thermodynamics (see [9–12] where further references to similar investigations are given).

The third approach is based on hypotheses and assumptions of a phenomenological character. This type of constitutive relation uses experimentally observed phenomena to formulate a stress (stress-rate) and strain (strain-rate) relationship. This approach is relatively simple compared to the other two, and has been developed to an extent that most experimentally observed responses can be modelled with relative ease. Once the general characteristics of a class of materials are identified, it then becomes possible to construct a framework for the constitutive relation of inelastic solids.

In the following we will confine ourselves to describing essential features of the phenomenological theories of elastic–plastic solids applicable to complex loading paths including cyclic loading, and rate effects.

6.2 REQUIREMENTS FOR A CONSTITUTIVE MODEL

Three fundamental constituents are necessary to describe the behaviour of an inelastic solid under general states of stress: an initial yield criterion to specify the state of stress for which plastic flow first begins; a hardening rule to specify the behaviour of the yield or flow surface as work hardening takes place; and a flow rule to connect the inelastic strain-rate with the stress and stress-rate. In the following, we will briefly discuss each of the above requirements.

6.2.1 INITIAL YIELD SURFACE

Based on his experiments on extrusion of metals, Tresca proposed in 1864 a maximum shear stress criterion in the form of:

$$\varphi_0 = \max\left[|\sigma_1 - \sigma_2|, |\sigma_2 - \sigma_3|, |\sigma_3 - \sigma_1|\right] - 2\tau_0 = 0, \tag{6.1}$$

where $\sigma_1, \sigma_2, \sigma_3$ are principal stresses and τ_0 is the yield stress in shear. Huber in 1904 and von Mises in 1913 proposed a modified form of Tresca's yield condition in the form of

$$\varphi_0 = \tfrac{1}{2}\left[(\sigma_1 - \sigma_2)^2 + (\sigma_2 - \sigma_3)^2 + (\sigma_3 - \sigma_1)^2\right] - \sigma_0^2 = 0, \tag{6.2}$$

where σ_0 is now the yield stress in tension. Equation (6.2) can also be written in terms of the deviatoric stress components, $s_{ij} = \sigma_{ij} - \sigma_{kk}\delta_{ij}/3$, i.e.

$$\varphi_0 = 3J_2 - \sigma_0^2 = \tfrac{3}{2}s_{ij}s_{ij} - \sigma_0^2 = 0, \tag{6.3}$$

where J_2 is the second invariant of the deviatoric stress.

From (6.1) we obtain the relation between the yield stress in tension ($\sigma_1 = \sigma_0, \sigma_2 = \sigma_3 = 0$) and pure shear, i.e.

$$\sigma_0 = 2\tau_0, \tag{6.4}$$

whereas from (6.2), we get ($\sigma_1 = \tau_0, \sigma_2 = 0, \sigma_3 = -\tau_0$, i.e. $\tau_{max} = \tau_0$) the relation

$$\sigma_0 = \sqrt{3}\,\tau_0. \tag{6.5}$$

The geometric representation of (6.1) is a right hexahedral cylinder with the axis $\sigma_1 = \sigma_2 = \sigma_3$, perpendicular to the deviatoric plane ($\sigma_1 + \sigma_2 + \sigma_3 = 0$). The projection of the cylinder on the stress deviatoric plane (or Π-plane) is a hexagon as shown in Fig. 6.1. The intersection of the cylinder given by (6.2) with the Π-plane is a circle which circumscribes the hexagon.

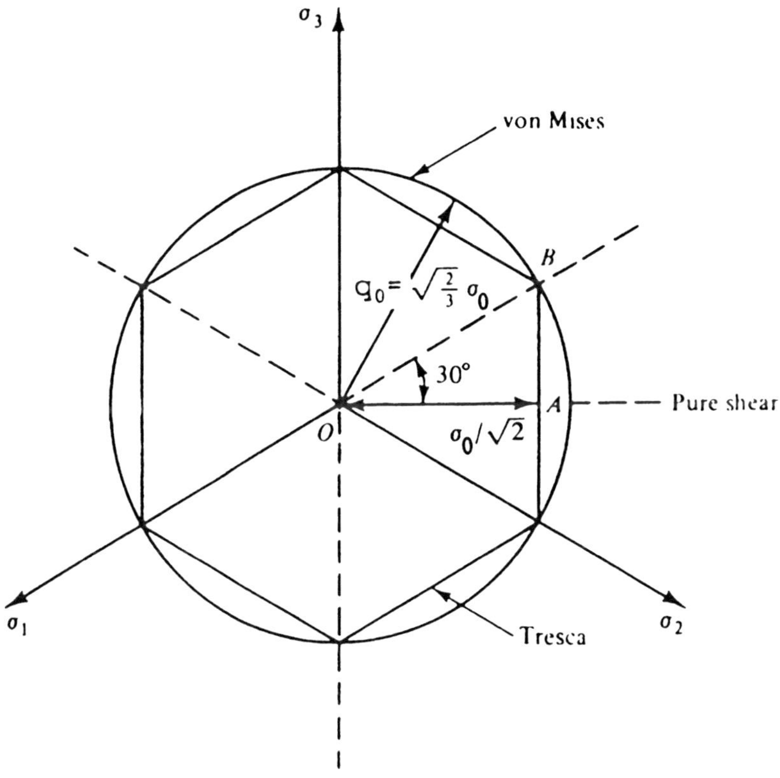

Fig. 6.1 Tresca and von Mises yield surfaces projected on the Π-plane.

Initial yield surfaces ((6.1) and (6.2)) are for homogeneous, isotropic materials. For an initially anisotropic material, Hill [13] introduced a general quadratic function

$$\varphi_0 = A_{ijkl}\sigma_{ij}\sigma_{kl} - 2K^2 = 0\,, \tag{6.6}$$

where the symmetry condition $A_{ijkl} = A_{jikl} = A_{ijlk}$ reduces the material constants to 15 in the general case. For the specific case of orthotropy (i.e. three mutually orthogonal planes of symmetry) the criterion (6.6) reduces to

$$\begin{aligned}\varphi_0 = F(\sigma_y - \sigma_z)^2 + G(\sigma_z - \sigma_x)^2 + H(\sigma_x - \sigma_y)^2 + 2L\,\tau_{yz}^2 \\ + 2M\tau_{zx}^2 + 2N\tau_{xy}^2 - 2K^2 = 0\,,\end{aligned} \tag{6.7}$$

where $F, G, H, 2L, 2M$ and $2N$ are parameters specifying the current state of anisotropy. If the x, y, z are all axes of material symmetry, then (6.7) reduces to (6.2).

6.2.2 HARDENING RULE – SUBSEQUENT YIELD SURFACES

The existing incremental theories of plasticity differ from one another in the choice of the hardening rule which specifies the change of the yield surface during plastic deformation. The simplest hardening rule is the so-called 'single-curve' hypothesis. Experiments have shown that for the proportional or nearly proportional loading paths, there exists a one-to-one relationship between the equivalent (effective) stress, σ_{eq}, and the corresponding strain, ε_{eq}, i.e.

$$\sigma_{eq} = F(\varepsilon_{eq}) , \tag{6.8}$$

and the condition of the material hardening is

$$\frac{d\sigma_{eq}}{d\varepsilon_{eq}} = E_t > 0 . \tag{6.9}$$

This type of hardening for cyclic loading was discussed in Chapter 2 and in this section we will focus our attention on more general types of hardening rules.

(a) Isotropic hardening rule

The initial yield surface is assumed to expand uniformly with the increasing plastic strain while its centre remains fixed. This type of hardening rule may be expressed as

$$\phi_y = f_y(\sigma_{ij}) - q^2(H_\alpha) = 0 , \tag{6.10}$$

where q^2 is a monotonous and positive definite scalar function of the history dependent parameter H_α. Depending upon the choice of H_α, various hardening functions are obtained.

The concept of isotropic hardening is not supported by the experimental evidence, e.g. it is in direct conflict with the observed Bauschinger effect. In addition, the process of plastic flow induces anisotropy due to rotation and alignment of crystals of an initially isotropic material, and cannot be represented by a symmetric expansion of the yield surface. It should be noted, however, that for a monotonically increasing load, the isotropic hardening rule is a useful analytical tool describing the loading surface.

(b) Kinematic hardening rule

In this model the size and shape of the yield surface remain the same, but it translates in the stress space as plastic deformation progresses. This type of hardening rule may be expressed as

$$\phi_y = f_y(\sigma_{ij} - \alpha_{ij}) - q_0^2 = 0 , \tag{6.11}$$

where α_{ij} are coordinates of the centre of the current yield surface, and q_0 is a constant denoting the 'radius' of the initial yield surface. For an increment of $d\alpha_{ij}$, we must specify a differential relationship. For example, we may assume

$$d\alpha_{ij} = c\, d\varepsilon^{p}_{ij}, \tag{6.12}$$

where c could be either a constant or dependent on the stress or strain. This type of translational rule is generally known as Prager's hardening rule [14], although it has also been credited to others [15]. Ziegler's modification of (6.12) is given in [16] as

$$d\alpha_{ij} = d\mu(\sigma_{ij} - \alpha_{ij}), \tag{6.13}$$

where the scalar $d\mu$ is determined from the consistency condition, i.e. imposing that the loading point remains on the yield surface.

(c) A combined hardening rule

A combined form of the isotropic and kinematic rules may lead to a better description of the subsequent yield surface. This can be written as

$$\phi_y = f_y(\sigma_{ij} - \alpha_{ij}) - q^2(H_\alpha) = 0, \tag{6.14}$$

where the scalar q^2 is some measure of the isotropic hardening. The yield surface described by (6.14) in general could expand (or contract) and translate by retaining its original shape.

(d) A field of constant hardening moduli

In the preceding hardening rules it was assumed that a single yield function can describe the plastic behaviour of a solid. Experiments indicate that the material hardening process is more complex than that described by (6.10) to (6.14), and depends on the history of deformation.

Mróz [17] assumed that the state of hardening is defined not only by the initial yield surface, but also by a set of surfaces of constant hardening moduli, i.e.

$$\phi_0 = f_0(\sigma_{ij} - \alpha^{(0)}_{ij}) - q_0^2 = 0, \tag{6.15a}$$

and

$$\phi_k = f_k(\sigma_{ij} - \alpha^{(k)}_{ij}) - q_k^2 = 0, \quad k = 1, 2, \ldots n, \tag{6.15b}$$

where f_0 and f_k are homogeneous functions of order two, each having a constant hardening modulus. The motion of the constant moduli surfaces is along a path which avoids their intersection (Fig. 6.2). Now suppose that the stress point $P(\sigma^{(k)}_{ij})$ on f_k is subjected to a stress increment $d\sigma_{ij}$. Then f_k will translate along PR, where R is a point on f_{k+1} whose outward

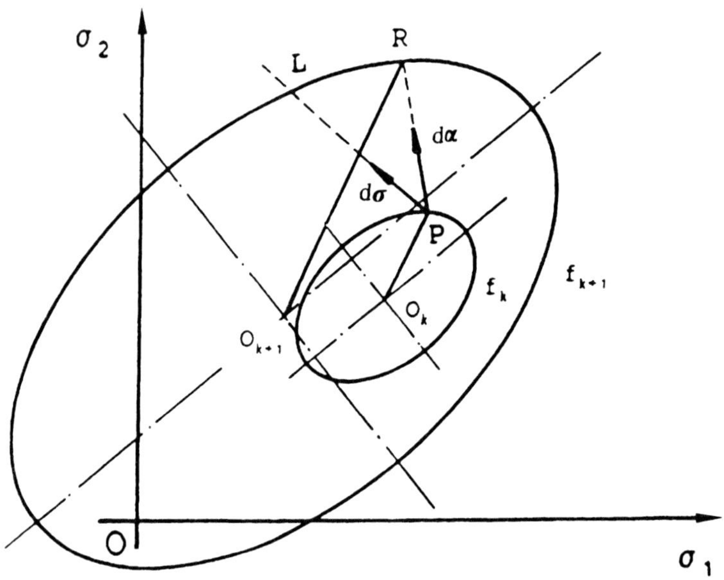

Fig. 6.2 Translation rule of Mróz, surface f_k moves along PR [17].

normal is parallel to that of P. From Fig. 6.2 and self-similarity of two surfaces, we get

$$\sigma_{ij}^{(k+1)} - \alpha_{ij}^{(k+1)} = \frac{q_{k+1}}{q_k}(\sigma_{ij}^{(k)} - \alpha_{ij}^{(k)}) \tag{6.16}$$

and

$$\mathrm{d}\alpha_{ij}^{k} = \frac{\mathrm{d}\mu}{q_k}[(q_{k+1} - q_k)\sigma_{ij}^{(k)} - (\alpha_{ij}^{(k)} q_{k+1} - \alpha_{ij}^{(k+1)} q_k)] \tag{6.17}$$

where $\mathrm{d}\mu$ is a scalar.

(e) Bounding surface models

The model of Mróz was subsequently simplified by Krieg [18], and Dafalias and Popov [19]. In these models the family of constant moduli surfaces was replaced by two surfaces: an outer bounding surface and an inner yield surface. The yield surface, representing the elastic domain, translates inside the bounding surface according to a specified rule. Similarly, the bounding surface translates mainly as a function of the uniaxial tangent modulus.

Referring to Fig. 6.3(b), a stress increment $\mathbf{d\sigma}$ is applied at point A of the current yield surface, and the vector $\mathbf{d\alpha}$ represents the motion of the yield

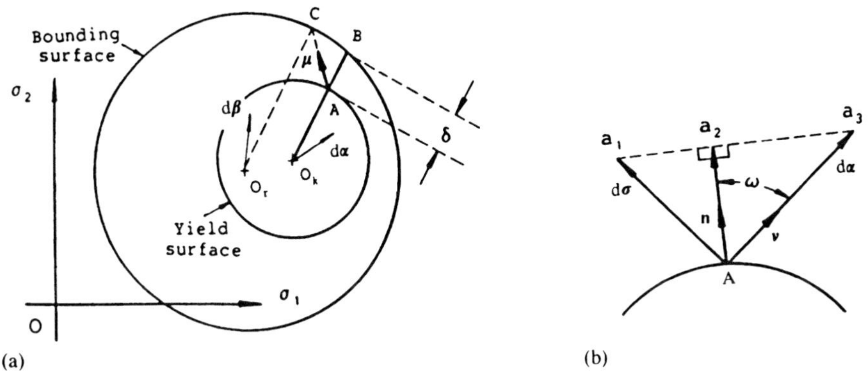

Fig. 6.3 Translation rule of yield and bounding surface of Dafalias and Popov [19].

surface. Within a second order of approximation, we get

$$\mathrm{d}\alpha_{ij} = \frac{\mathrm{d}\sigma}{\cos\omega}\mathbf{v}_{ij} \tag{6.18}$$

where **v** is a unit vector in the direction of $\mathrm{d}\boldsymbol{\alpha}$, ω is the angle between v and $\boldsymbol{n}$, the unit normal vector to the yield surface and $\mathrm{d}\sigma$ is the projection of $\mathrm{d}\boldsymbol{\sigma}$ on the $\boldsymbol{n}$.

The translation of the bounding surface is similar to Mróz's rule, i.e. the two surfaces must approach one another along the unit vector $\boldsymbol{\mu}$ (Fig. 6.3(a)). In this figure the outward normal at point C on the bounding surface has the same direction as that at point A on the yield surface. If the two surfaces are not to intersect when they come into contact ($\delta = 0$), then the motion of the bounding surface can be expressed as

$$\mathrm{d}\beta_{ij} = \mathrm{d}\sigma\left[\frac{1}{\cos\omega}v_{ij} - \left(1 - \frac{k_0}{k}\right)\mu_{ij}\right], \tag{6.19}$$

where the value of k is obtained from a functional relation

$$k = \hat{k}(\delta, W^{\mathrm{p}}) \tag{6.20a}$$

$$k_0 = \hat{k}(0, W^{\mathrm{p}}) \tag{6.20b}$$

and

$$W^{\mathrm{p}} = \int \sigma_{ij}\,\mathrm{d}\varepsilon_{ij}^{\mathrm{p}}. \tag{6.20c}$$

For a discussion of other types of hardening rules, the reader could consult the book by Desai and Siriwardane [20], among others.

6.2.3 FLOW RULE

In the theory of elasticity, the strain energy density, $W = \int \sigma_{ij}\, d\varepsilon_{ij}$, can be expressed either in terms of the stress or of the strain components, i.e. $W = W(\sigma_{ij})$ or $W = W(\varepsilon_{ij})$. This quadratic function is also referred to as the elastic potential function. It then follows that

$$\frac{\partial W}{\partial \varepsilon_{ij}} = \sigma_{ij} \tag{6.21a}$$

and

$$\frac{\partial W}{\partial \sigma_{ij}} = \varepsilon_{ij}\,. \tag{6.21b}$$

In an analogous fashion, von Mises in 1928 introduced the concept of plastic potential, $\Phi(\sigma_{ij})$. The plastic strain increment can be derived from the plastic potential by taking the partial derivative of the plastic potential function, i.e.

$$d\varepsilon_{ij}^{p} = d\lambda \frac{\partial \Phi(\sigma_{ij})}{\partial \sigma_{ij}}\,. \tag{6.22}$$

The equation $\Phi(\sigma_{ij}) = \text{constant}$ defines a hypersurface in the stress space, and the direction cosines of the normal to this surface are proportional to $\partial \Phi / \partial \sigma_{ij}$. Therefore, (6.22) implies that the plastic strain increment, $d\varepsilon_{ij}^{p}$, is directed along the normal to the plastic potential surface. For metals, we generally assume that

$$\Phi = \phi_y\,, \tag{6.23}$$

i.e. the plastic potential surface coincides with the yield surface. This condition is termed the **associated flow rule** of plasticity. The consistency condition will impose that the plastic strain increment also be proportional to the quantity (Kachanov [21])

$$df_y = \frac{\partial f_y}{\partial \sigma_{kl}} d\sigma_{kl}\,, \tag{6.24}$$

where the increment df_y is carried out while the plastic strain is held constant. Combining (6.22) to (6.24) and noting that $\partial \phi_y / \partial \sigma_{ij} = \partial f_y / \partial \sigma_{ij}$, we get

$$\begin{aligned} d\varepsilon_{ij}^{p} &= g \frac{\partial f_y}{\partial \sigma_{ij}} \frac{\partial f_y}{\partial \sigma_{kl}} d\sigma_{kl} \quad &&\text{for} \quad df_y \geqslant 0 \quad \text{and} \quad \phi_y = 0 \\ d\varepsilon_{ij}^{p} &= 0 &&\text{for} \quad df_y < 0 \quad \text{or} \quad \phi_y < 0 \end{aligned} \tag{6.25}$$

where the proportionality factor 'g' is a function of the material hardening.

6.3 EXPERIMENTAL DEFINITION OF YIELD POINT AND YIELD LOCI

Several definitions have been proposed to define the yield point when the material does not exhibit a sharp transition from the elastic to inelastic response, e.g. Haythornthwaite [22]. In order to detect the elastic limit, one has to load beyond it. In doing so, a small amount of plastic strain is introduced. For the uniaxial case one may write

$$\varepsilon^{\mathrm{p}} = \varepsilon^{\mathrm{t}} - \frac{\sigma}{E} = f(\sigma)\,. \tag{6.26}$$

The imposed plastic strain ε^{p} or $f(\sigma)$ may be determined by inspecting Fig. 6.4. There exists a band of measuring accuracy of $\Delta\varepsilon$ at the linear (or reversible) zone. Therefore, the lower limit of the imposed plastic strain has to be greater than $2\Delta\varepsilon$. An upper limit may be imposed such that it does not cause a change in the orientation of the crystalline structure. In experiments performed by Ellyin and Grass [23] $\Delta\varepsilon$ was about 2×10^{-6} (0.0002%) and the upper limit was set at 20×10^{-6} (0.002%) whereupon if exceeded, an automatic unloading occurred. Therefore, the bounds of the permanent deformation may be

$$2.5\,\Delta\varepsilon \leqslant \varepsilon^{\mathrm{p}} < 10\,\Delta\varepsilon\,, \tag{6.27}$$

where $\Delta\varepsilon$ is the band of uncertainty associated with the strain measurement.

For each value of ε^{p} or $f(\sigma)$ there corresponds a point on the σ–ε curve. The intersection of the tangent at the point with the linear response (backward extrapolation) defines a point Q^{T} on the stress axis (Fig. 6.4). As shown in the figure, the value of Q^{T} is relatively insensitive to the variation of ε^{p} between the two bounds defined in (6.27). This definition is similar to that proposed by Phillips and co-workers [24].

In a biaxial test (axial force and torsion) the vector Q^{T} $(Q_x, Q_{x\theta})$ is obtained by its mutually orthogonal components defined by

$$Q_x = \frac{F}{2\pi R_{\mathrm{m}} t}, \quad Q_{x\theta} = T/2\pi R_{\mathrm{m}}^2 t\,, \tag{6.28}$$

where F and T are axial force and torsional moment applied on a tubular specimen, R_{m} is the mean radius of the tubular specimen and t is its thickness. Note that when $t \ll R_{\mathrm{m}}$, then $Q_x \approx \sigma_x$ and $Q_{x\theta} \approx \tau_{x\theta}$. The loading path and the sequence of constructing a yield surface is shown in Fig. 6.5. Starting from an interior point with coordinates (F_1, T_1), the first loading path is a positive torsional stress superimposed on the initial stress state. The next point is the neighbouring state 2, and so on. The corresponding

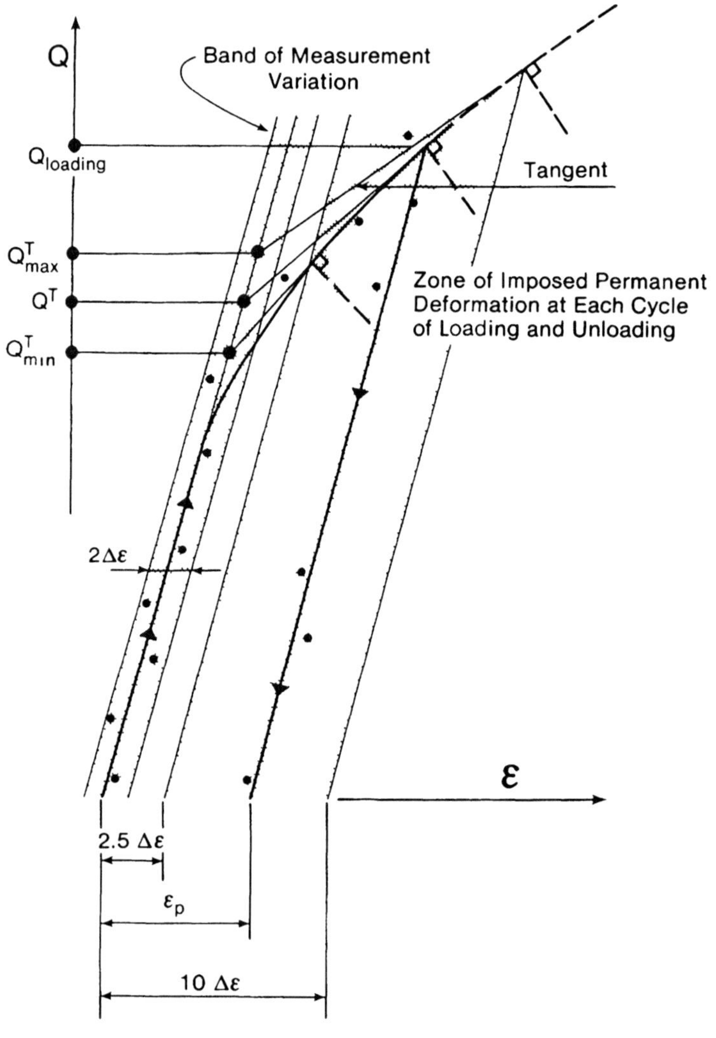

Representation of the Definition Q^T

Fig. 6.4 Experimental definition of yield point, from [23].

axial and shear strains are determined from

$$\varepsilon_x = \frac{\Delta L}{L_0} \quad \text{and} \quad \gamma_{x\theta} = \frac{\pi R_{ex} \Delta\theta}{(180\, L_0)}, \tag{6.29}$$

where ΔL is the measured change in the gauge length L_0 and $\Delta\theta$ is the measured relative end rotation of the specimen in degrees [23].

The required precision of the instrumentation must be such that the starting and closing points 1 and 17 are sufficiently close. This procedure of determining yield surface is preferable to that advocated by Phillips and co-workers, where the next point on the yield surface is obtained by a reversed loading, i.e. point 9 instead of 2 in Fig. 6.5 (see [23] for a discussion of the two methods). The definition of the yield point and the accurate determination of the yield locus is a very important step in any experimental investigation of yield surfaces. Some of the discrepancies noted in the technical literature arise from the confusion regarding the proper definition and determination of the yield point. It is worth emphasizing that a yield surface is the locus of the elastic response **not** an arbitrarily determined plastic strain magnitude, and as such care should be taken to arrive at a very small off-set strain which is of the order of 10^{-6} (Fig. 6.4).

It is not within the scope of this chapter to review the contributions made by numerous investigators on the yield surfaces. Review articles by Paul [25], Ikegami [26], Hecker [27] and Phillips [28] should be consulted for a list of such contributions. In the following we will present results of an experimental study by the writer which is the basis of constitutive model to be described later on.

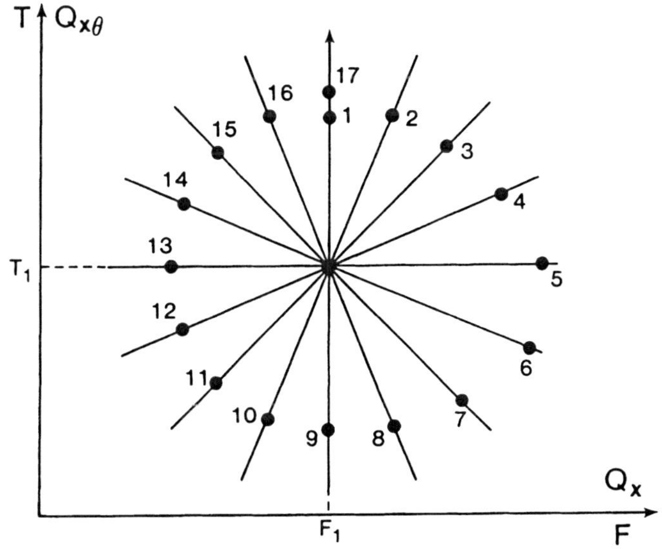

Fig. 6.5 Loading paths to construct a yield surface, from [23].

6.4 EXPERIMENTAL OBSERVATIONS

Thin-walled circular cylindrical specimens made of commercially pure titanium were subjected to the loading program shown in Fig. 6.6. Each path contains a proportional loading path, unloading and non-proportional loading. The chemical composition and mechanical properties of the material are given in [23].

Figure 6.7 shows an example of the initial yield surface. The von Mises yield surface is also traced by a dashed line for an equivalent stress,

$$\sigma_{eq} = (Q_x^2 + 3Q_{x\theta}^2)^{1/2}, \tag{6.30}$$

of 241 MPa. It can be seen that the agreement is fair, despite the initial anisotropy of the material. Also, the plastic strain vector appears to deviate slightly from the normal to the surface. However, it could be said that the yield condition (6.3) and the flow rule (6.22, 6.23) are good approximations of the observed experimental response.

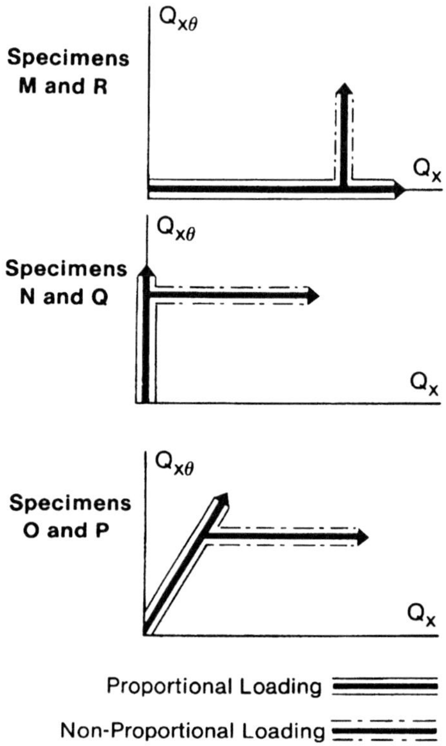

Fig. 6.6 Loading path for various specimens, from [23].

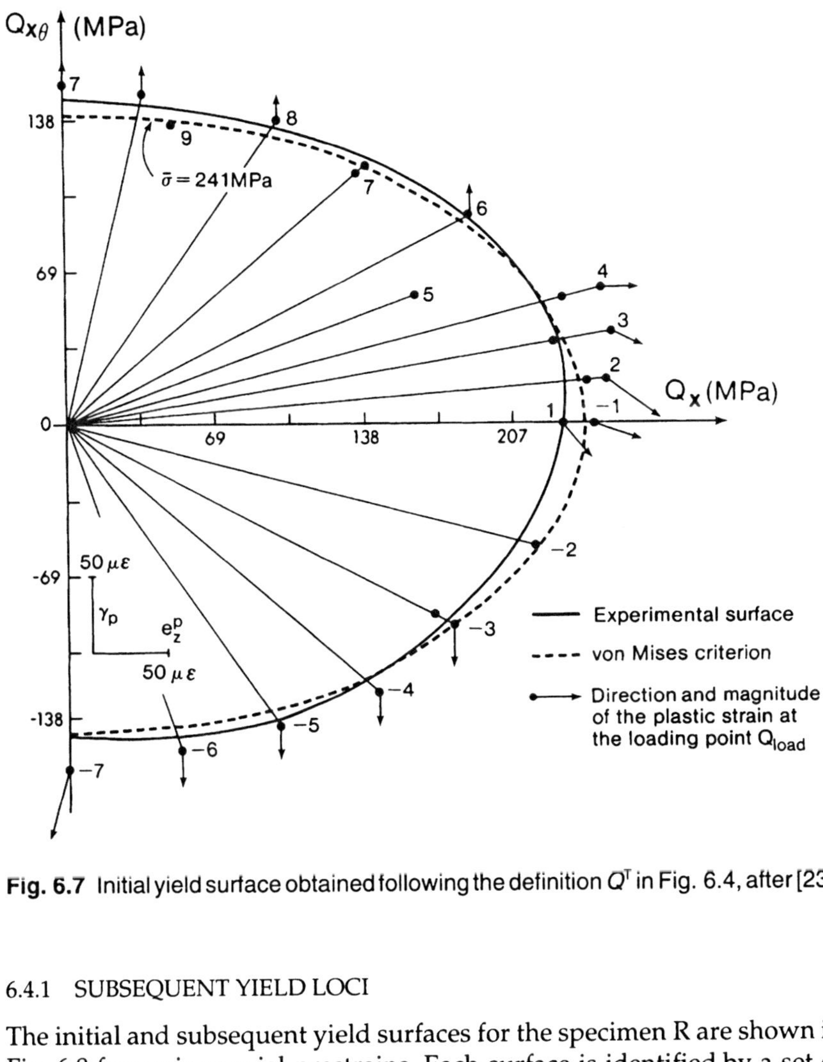

Fig. 6.7 Initial yield surface obtained following the definition Q^T in Fig. 6.4, after [23].

6.4.1 SUBSEQUENT YIELD LOCI

The initial and subsequent yield surfaces for the specimen R are shown in Fig. 6.8 for various axial prestrains. Each surface is identified by a set of letters and numbers, Γ (R-*i*a) where R identifies the specimen and $i = 2$–5 denotes the subsequent yield surfaces; Γ (R-1a) being the initial yield surface. The sequence of loading for this test was ABCDEFGH. Two types of surfaces can be identified in this figure. A set of self-similar surfaces passing through the maximum loading point are termed **memory surfaces.** (The term memory surface was used by Ellyin [29, 30] and independently by Tseng and Lee [31], as opposed to Phillips' loading surface [32].) The memory surface at first coincides with the initial surface and expands uniformly and remains self-similar with the increased deformation; thus it obeys the isotropic hardening model (6.10). A memory surface is the locus

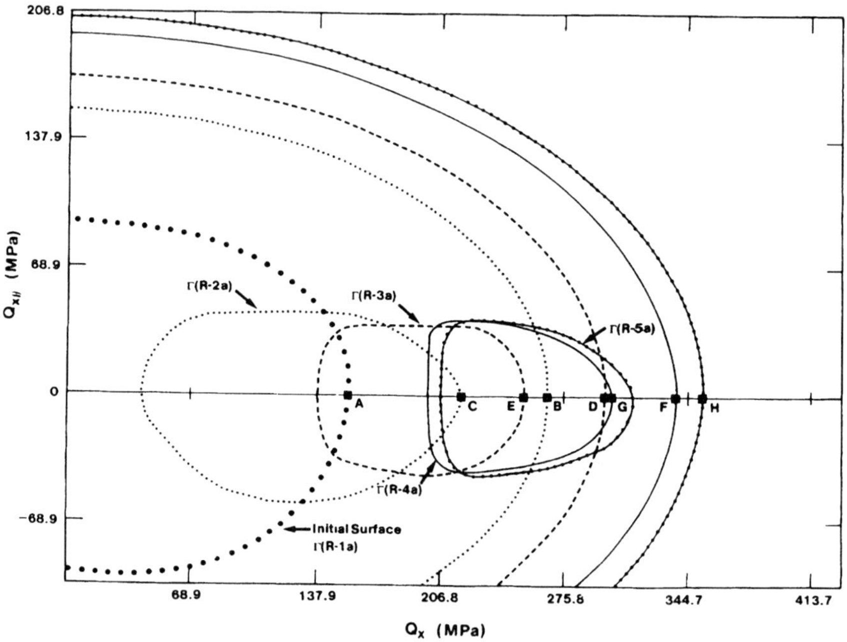

Fig. 6.8 Evolution of yield and memory surfaces with axial prestress. Reproduced with permission from [30].

of the maximum equivalent stress experienced by the material, and has a constant hardening modulus.

The subsequent yield surfaces (locus of the elastic domain) change in size and shape, and move in the stress space, but they always remain inside the current memory surface (Fig. 6.8). In the case of quasi-static loading, the subsequent yield loci gradually contract and tend towards a limiting size. Therefore, they do not, in general, follow the kinematic hardening rule.

To illustrate further the evolution of the yield surface, let us examine specimen 'N' which is pretorqued initially, path ABCDEFGHIJK' (Fig. 6.9). The behaviour for the torsional prestrain is similar to that of the axial prestrain (Fig. 6.8). The last yield surface corresponding to the torsional path J is denoted by Γ(N-6a). Upon application of a non-proportional loading path, K'KMM'LNPQRSS', the corresponding yield surfaces are denoted by Γ(N-7a) to Γ(N-10a). It can be seen that the subsequent yield surfaces change in shape and move in such a manner that they always remain inside the memory surface. Note that once the loading exceeds the previous maximum effective stress, point J, the subsequent yield loci Γ(N-9a) and Γ(N-10a) move almost parallel to the prestress path, QS.

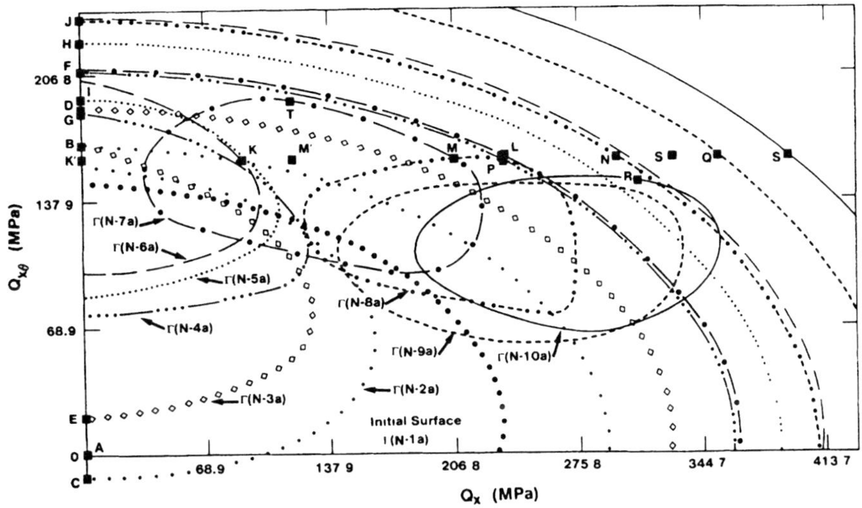

Fig. 6.9 Evolution of yield and memory surfaces with shear prestrain followed by an axial prestrain. Reproduced with permission from [30].

It should be noted that in Figs 6.8 and 6.9 the imposed plastic prestrain was in the moderate deformation range, the final equivalent plastic strain for the proportional parts of specimens R and N were: $(e^{\mathrm{p}}_{\mathrm{eq}})_{\mathrm{R}} \approx 4\%$ and $(e^{\mathrm{p}}_{\mathrm{eq}})_{\mathrm{N}} \approx 4.9\%$. Further discussion of the experimental results is given in [23, 30].

From the results presented in Figs 6.8 and 6.9, and those in [23, 30], we observe that for quasi-static loading, there is a contraction of the initial yield surface as a result of increased plastic prestrain. This type of behaviour is observed in the uniaxial tests (Fig. 2.1). The size of a yield locus is highly dependent on the definition of the yield point and accuracy of measuring devices. This change in size of subsequent yield surfaces appears to be confined to the transition (bend) region of the elastic to predominantly plastic deformation where there is a significant change in the tangent modulus.

6.4.2 HARDENING MODULUS CURVE

An investigation of the experimental results discussed above indicates that there exists a relationship between the tangent modulus, E_t, of the material and the ratio of the yield surface area (or memory surface) to that of the initial yield surface [33]. This relationship is termed the **hardening modulus curve** and will be discussed below. Figure 6.10 is a plot of the area ratio of the subsequent yield surface (or memory surface) in the Π-plane

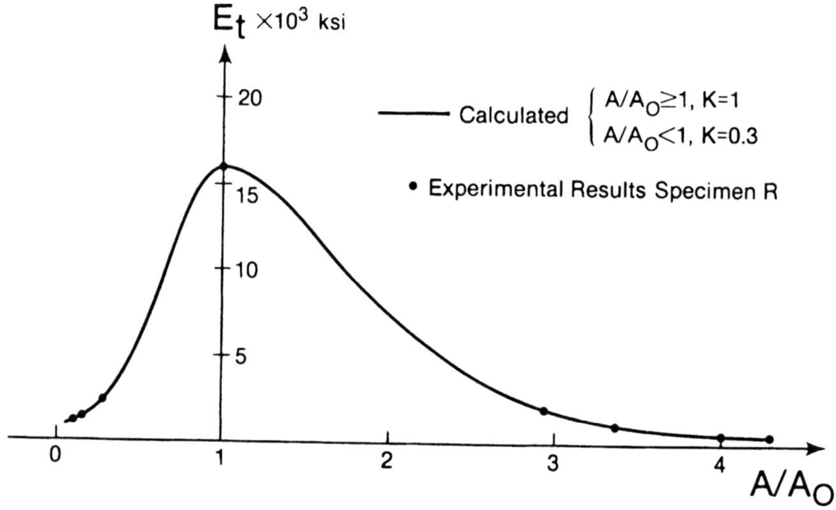

Fig. 6.10 Hardening modulus curve. Tangent modulus vs. ratio of the area of yield or memory surface to the initial yield surface. Reproduced with permission from ASME from [33].

(A) to the initial yield surface (A_0) versus tangent modulus, E_t. When $A/A_0 > 1$, the curve represents the relationship between the memory surface and the tangent modulus at the maximum load, E_t. For $A/A_0 < 1$, the curve depicts the relationship of the subsequent yield surface with the same tangent modulus. We observe that for this material the slope of the curve at the $A/A_0 < 1$ region is steeper than that at $A/A_0 > 1$. It should be noted that the tangent modulus is constant when moving on a given memory surface.

The following relationship applies for the hardening modulus curve [33]:

$$E_t(\omega, k) = E_0 \exp[-(\omega - 1)^2/2k^2], \quad \omega \geqslant 0, \tag{6.31}$$

where $\omega = A/A_0$ is the area ratio; k is a parameter indicating the extent of hardening, and E_0 is the tangent modulus at the point of the initial yield. For a von Mises material, the memory surface to the initial yield surface ratio is

$$\omega = \frac{A}{A_0} = \frac{\sigma_{eq}^2}{(\sigma_0)_{eq}^2}, \tag{6.32}$$

where $(\sigma_0)_{eq}$ is the equivalent stress at the yielding point of the material.

Two different values for k are obtained to represent the asymmetry of the curve about the $A/A_0 = 1$ axis (Fig. 6.10). For a given material, k is constant and can be determined through a suitable uniaxial test involving loading and unloading.

In the limit when $k \to 0$, the hardening modulus curve collapses to line $A/A_0 = 1$, and we obtain the kinematic hardening rule. On the other hand, when the hardening modulus curve is symmetric with respect to line $\omega = 1$, and is mirrored to one side, we then recover the isotropic hardening rule. It is, therefore, clear that the hardening modulus curve (Fig. 6.10, (6.31)) can be viewed as a general description of the material hardening process, and the isotropic and kinematic hardening laws are but two particular cases of it. Whether (6.31) applies to more complicated strain paths than those shown in Fig. 6.6 remains to be verified. One limiting case could be a class of materials whose stable cyclic curve is path dependent (see section 6.5.2 (e)). It is also worth mentioning that other type of relations than (6.31) could be found to fit the trend shown in Fig. 6.10. Relation (6.31) is merely used to demonstrate the observed experimental trend. (In subsection 6.5.2 (d) we will describe a more general relation for the variation of yield surface radius.)

6.4.3 EVOLUTION OF YIELD AND MEMORY SURFACES

The hardening modulus curve described in the previous section indicates the relationship between the size of the memory or yield surfaces and the tangent modulus at the maximum load, E_t, of the stress–strain curve. It does not, however, give the change of position of the yield surface during the loading process.

It was indicated earlier that subsequent yield surfaces move along the prestressing path for proportional loading (Fig. 6.8). For non-proportional loading, the motion of the subsequent yield surface is more complex, e.g. Fig. 6.9. The direction of its motion deviates from that of the loading path. From the analysis of the experimental results, it was deduced that the motion of the subsequent yield surface is as follows [30]. The centre of the yield surface moves parallel to the direction of the vector connecting the current loading point to a point on the corresponding memory surface whose exterior normal is parallel to the outward normal of the instantaneous yield surface at the loading point. The above rule applies as long as the yield surface remains inside the memory surface. Figure 6.11 is a schematic representation of the motion of the subsequent yield surfaces with various prestressing paths (Q_1 and Q_2 are averaged principal stresses, $Q_1 \approx \sigma_1$ and $Q_2 \approx \sigma_2$ in a thin-walled section). For example, the motion of the subsequent yield surface Γ_{AB} when prestressed along DF is obtained by finding point E on the memory surface whose exterior normal $\mathbf{n}_E$, is parallel to the outward normal at D, $\mathbf{n}_D$. The centre of the subsequent yield surface

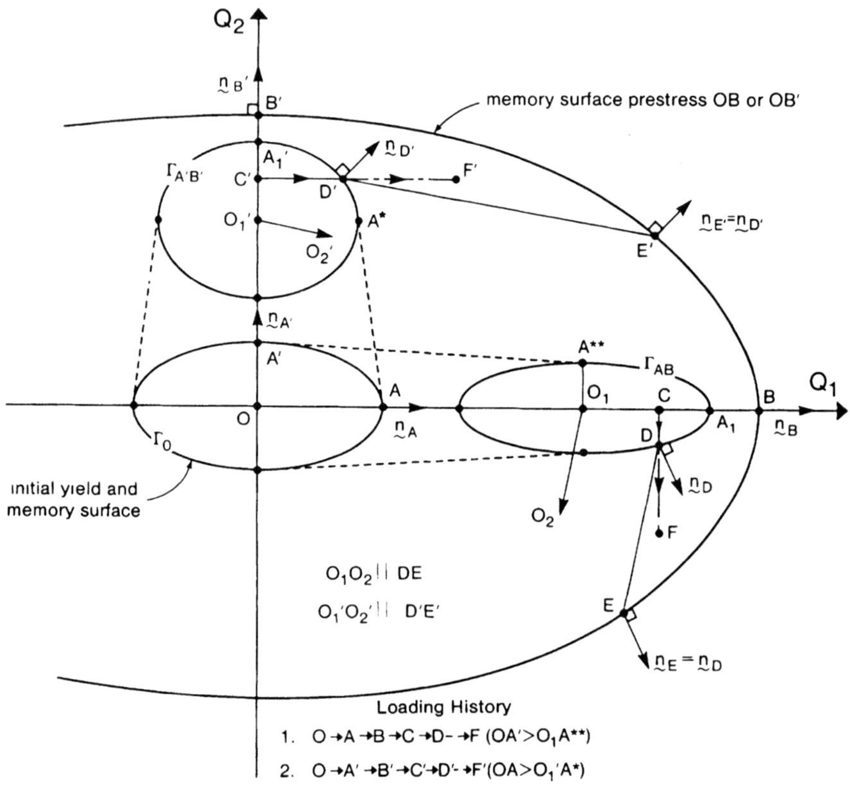

Fig. 6.11 Schematic representation of the motion of yield loci inside the memory surface. Reproduced with permission from [30].

will then move in the direction parallel to DE and will remain inside the memory surface.

The evolution law thus described is, in a sense, a generalization of that of Phillips and Lee [32], i.e. in the case of monotonic loading, the yield surface moves along the stress path and remains tangent to the memory surface. However, once there is a reversed loading, the motion is according to the rule described by Fig. 6.11, which is of the Mróz type. The necessity of expressing two hardening rules is well described in the concluding part of Phillips' review of his experimental work [28]. This, of course, could not be done by plasticity theories with either a single surface or those with bounding surfaces. The mathematical expressions to describe the above-mentioned evolution rules will be given later.

6.5 A CONSTITUTIVE MODEL FOR TRANSIENT NON-PROPORTIONAL PLASTICITY – RATE-INDEPENDENT BEHAVIOUR

In the previous section we presented some experimental results for quasi-static proportional and non-proportional loadings. A constitutive model simulating the observed behaviour has been developed by the writer and his collaborators. The case of quasi-static and stable cyclic loading has been presented in [34–36] and will not be discussed here. In this section we will be concerned with the more general case of transient non-proportional loading.

The rate-independent constitutive models for inelastic solids under cyclic loading may be grouped into two categories. The first group are those that use a single surface to delineate elastic and plastic responses, e.g. Eisenberg [37], Chaboche *et al.* [38], Drucker and Palgen [39]. These models are relatively simple and are more suitable for uncomplicated loading paths. The second group of models use two or multi-surfaces, e.g. Mróz [40], Dafalias [41], Ohno [42], McDowell [43], Tanaka *et al.* [44], Ellyin and Xia [45], among others. These models generally predict better results than those of a single surface for non-proportional paths.

6.5.1 SOME EXPERIMENTAL OBSERVATIONS REGARDING TRANSIENT HARDENING

In Chapter 2, we discussed material responses to some well-defined uniaxial loading paths. The case of the transient response was also discussed, albeit briefly, and in this section we will further elaborate on it. One of the well documented transient responses is the so called 'strain range effect'. After stabilizing under a cyclic strain level, more transient hardening may be observed for cycling under a higher strain level. Decreasing the strain level to the lower (initial) value, one observes an additional hardening, e.g. [38]. Thus it would appear that each higher strain level activates a new hardening phase. We pointed out in Chapter 2 that the dislocation cell sizes were inversely proportional to the strain range. Therefore, once a material is stabilized at a higher strain range, the size of the stable cells will be dictated by the highest strain range, and the hardening characteristic at the lower strains will be a function of the maximum strain experienced by the material.

The above-mentioned phenomenon of 'strain-range effect' indicates a loading history dependency. The effect of the loading path cannot be investigated in a uniaxial test, it requires a biaxial stress field. In such a test one notes that for a given equivalent strain range, a non-proportional path will cause an increased level of hardening as compared to a proportional path. The maximum hardening for an out-of-phase axial–torsional loading occurs when the ratio of the shear strain amplitude to the axial strain amplitude is about 1.5 [46, 47]. In this case, the strain path in the strain Π-plane is almost a circle, which implies a constant equivalent strain at all times. A circular strain path with constant equivalent strain amplitude would activate slip planes in various directions causing additional barriers to dislocation movements. As a result, the force required for the dislocation movement will be increased, manifesting a further strain hardening.

In addition to the extra hardening during out-of-phase cycling, some materials exhibit a 'cross-effect' when the direction of straining is altered while keeping the equivalent strain constant [47–49]. However, this cross-effect is not observed if the material is cyclically saturated in an out-of-phase loading. That is, after an out-of-phase cyclic saturation, there is no further strain hardening for subsequent proportional or non-proportional cyclic paths at the same or lower strain levels. This observation was termed 'erasure of memory property' by Lamba and Sidebottom [46].

The constitutive model to be described below will reproduce the experimental observations discussed above under the prescribed conditions, as well as other discriminating experiments.

6.5.2 DESCRIPTION OF THE CONSTITUTIVE MODEL

In this model we define three hypersurfaces: a yield surface, a stress memory surface and a strain memory surface. As noted in section 6.4, the yield surface which is the locus of the elastic regime depends on the loading history. In the stress space, it is expressed as (cf. (6.11)):

$$\varphi_{\mathrm{y}} = f_{\mathrm{y}}(\sigma_{ij} - \alpha_{ij}) - q^2 = 0\,, \tag{6.33}$$

where as before α_{ij} and q specify the location of the centre and the 'radius' of the yield surface, respectively. In the case of the von Mises material

$$f_{\mathrm{y}} = \tfrac{3}{2}\bar{s}_{ij}\bar{s}_{ij}\,, \tag{6.34}$$

where $\bar{s}_{ij} = \bar{\sigma}_{ij} - \delta_{ij}\bar{\sigma}_{kk}/3$ and $\bar{\sigma}_{ij} = \sigma_{ij} - \alpha_{ij}$.

The stress memory surface is expressed as:

$$\phi^{\sigma}_{\mathrm{m}} = f^{\sigma}_{\mathrm{m}}(\sigma_{ij} - \beta_{ij}) - R^2_{\mathrm{mem}}(\sigma_{\mathrm{eq,\,max}}) = 0\,. \tag{6.35}$$

For the von Mises material

$$f^{\sigma}_{\mathrm{m}} = \tfrac{3}{2}\hat{s}_{ij}\hat{s}_{ij}\,, \tag{6.36}$$

where $\hat{s}_{ij} = \hat{\sigma}_{ij} - \delta_{ij}\hat{\sigma}_{kk}/3$ and $\hat{\sigma}_{ij} = \sigma_{ij} - \beta_{ij}$. The size of the stress memory surface, R_{mem}, is determined by the maximum equivalent stress level experienced by the material during its previous loading history. The β_{ij} is the centre of the stress memory surface. Its evolution rule will be described later.

The strain memory surface specifies the maximum equivalent (effective) strain experienced by the material during its previous loading history, i.e.

$$\phi_{\mathrm{m}}^{\varepsilon} = f_{\mathrm{m}}^{\varepsilon}(\varepsilon_{ij}) - C^2(\varepsilon_{\mathrm{eq,max}}) = 0 . \tag{6.37}$$

For the von Mises material (6.37) becomes

$$f_{\mathrm{m}}^{\varepsilon} = \tfrac{2}{3} e_{ij} e_{ij} , \tag{6.38}$$

where $e_{ij} = \varepsilon_{ij} - \delta_{ij}\varepsilon_{kk}/3$, is the deviatoric strain tensor. On the strain Π-plane, the strain memory surface is represented by a circle. This circle expands isotropically with the increase of the maximum equivalent strain.

(a) Stress–strain relation

It is assumed that the total strain increment can be decomposed into elastic and plastic parts, i.e.

$$\mathrm{d}\varepsilon_{ij} = \mathrm{d}\varepsilon_{ij}^{\mathrm{e}} + \mathrm{d}\varepsilon_{ij}^{\mathrm{p}} . \tag{6.39}$$

The increment of the elastic strain is related to the stress increment through the generalized Hooke's law:

$$\mathrm{d}\varepsilon_{ij}^{\mathrm{e}} = \frac{1}{E}[(1+\nu)\delta_{ik}\delta_{jl} - \nu\delta_{ij}\delta_{kl}]\mathrm{d}\sigma_{kl} . \tag{6.40}$$

From (6.25) the plastic strain increment is given by

$$\mathrm{d}\varepsilon_{ij}^{\mathrm{p}} = cg \frac{\partial f_{\mathrm{y}}}{\partial \sigma_{ij}} \frac{\partial f_{\mathrm{y}}}{\partial \sigma_{kl}} \mathrm{d}\sigma_{kl} , \tag{6.41}$$

where

$$c = 1 \quad \text{for} \quad \frac{\partial f_{\mathrm{y}}}{\partial \sigma_{kl}} \mathrm{d}\sigma_{kl} \geqslant 0 \quad \text{and} \quad f_{\mathrm{y}}(\sigma_{ij} - \alpha_{ij}) - q^2 = 0$$

and

$$c = 0 \quad \text{for} \quad \frac{\partial f_{\mathrm{y}}}{\partial \sigma_{kl}} \mathrm{d}\sigma_{kl} < 0 \quad \text{or} \quad f_{\mathrm{y}}(\sigma_{ij} - \alpha_{ij}) - q^2 < 0 .$$

In the case of von Mises material the stress–strain relation (6.39)–(6.41) can be written as [45]

$$\mathrm{d}\sigma_{ij} = L_{ijkl}\mathrm{d}\varepsilon_{kl} , \tag{6.42a}$$

where

$$L_{ijkl} = \frac{E}{1+\nu}\left(\delta_{ik}\delta_{jl} + \frac{\nu}{1-2\nu}\delta_{ij}\delta_{kl}\right) - ch(3\bar{s}_{ij})(3\bar{s}_{kl}) \tag{6.42b}$$

and

$$h = \left(\frac{E}{1+\nu}\right) g \bigg/ \left[\left(\frac{1+\nu}{E}\right) + 6q^2\right]. \tag{6.42c}$$

The constant $c = 1$ if the stress point is on the yield surface and $\bar{s}_{ij}\mathrm{d}\varepsilon_{ij} \geqslant 0$, and $c = 0$ if the stress point is inside the yield surface or $\bar{s}_{ij}\mathrm{d}\varepsilon_{ij} < 0$.

The hardening function g in (6.42c) can be determined by specializing (6.42a) to the uniaxial stress condition. In this way we get

$$g = \frac{1}{4q^2}\left(\frac{1}{E_t} - \frac{1}{E}\right), \tag{6.43}$$

where E is the Young's modulus and E_t is the instantaneous tangent modulus.

In this model, two types of plastic loading are distinguished by the introduction of the stress memory surface (Fig. 6.12(a)). The first type is called monotonic loading (ML), in which the stress memory surface expands with the movement of the yield surface, and the two surfaces remain tangent to each other at the current stress point (path AB, Fig. 6.12(a)). The second type is termed reloading (RL). After an elastic unloading (path BC, Fig. 6.12(a)), plastic deformation takes place again. But in this case, the current stress point is inside the stress memory surface and thus the maximum equivalent stress value experienced remains the same, i.e. the size of the stress memory surface does not change during this loading stage (path CD, Fig. 6.12(a)). When the current stress point touches the memory surface again and moves it through the expansion, we incorporate the latter stage into the ML case.

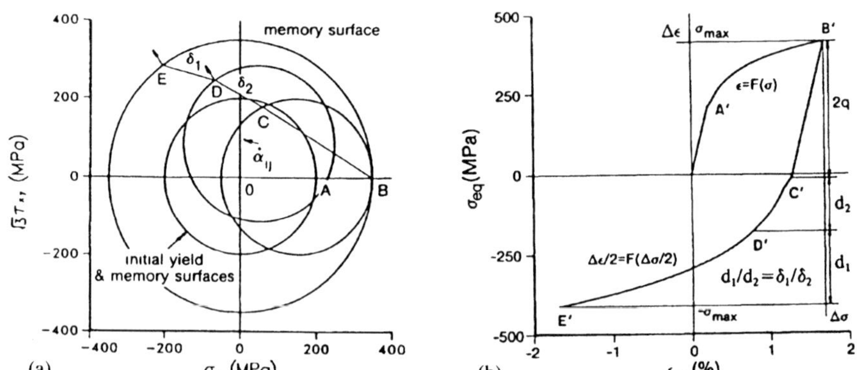

Fig. 6.12 Determination of tangent modulus, E_t, for loading and reversed loading cases. (a) biaxial stress state; (b) uniaxial stress–strain curve. Reproduced with kind permission from [45].

The evolution rule of the yield surface and the calculation of the tangent modulus, E_t, are also different for the two plastic loading cases.

(b) Evolution of the yield surface

As mentioned in section 6.4, based on a review of experimental work, Phillips pointed out the necessity of using different hardening rules for monotonic loading and reversed loading cases [28]. The lack of correlation obtained with the use of Prager's (6.12) or Ziegler's (6.13) rule has been demonstrated especially for the reversed loading case. In the present model, for the RL case, a Mróz type rule is used, i.e.

$$d\alpha_{ij} = d\mu(\sigma_{ij}^{m} - \sigma_{ij}^{y}), \tag{6.44}$$

where σ_{ij}^{y} is the current stress point D (Fig. 6.12(a)), and σ_{ij}^{m} is a point on the stress memory surface, E, whose exterior normal is parallel to the outward normal at point σ_{ij}^{y}. Geometrically, it means that the centre of the yield surface moves along the direction of line DE. By using this rule, the yield surface and the stress memory surface will be tangent at the current stress point when the former approaches the latter. Experimental observations by Ellyin [30] and McDowell [43] have confirmed that the above rule is the appropriate one for the reversed loading case.

For the ML case, the difference in using various proposed rules becomes significant only in the case of non-proportional loading. The question of which rule is to be used may depend on a material and its loading history. For example, Phillips *et al.* [24, 28, 32] reported that for a commercially pure aluminium, the motion of the yield surface was in the direction of $d\boldsymbol{\sigma}$ when the angle between it and the normal $\boldsymbol{n}$ to the yield surface at the current stress point was small. If the angle was large, the motion of the yield surface was generally between $d\boldsymbol{\sigma}$ and $d\boldsymbol{n}$, but with $d\boldsymbol{\sigma}$ predominating the motion.

Here, we will first introduce an evolution equation, which has the same form as (6.44) i.e.

$$d\alpha_{ij} = d\mu(\sigma_{ij}^{\mathrm{lim}} - \sigma_{ij}^{y}) \tag{6.45a}$$

$$\sigma_{ij}^{\mathrm{lim}} = \frac{R_{\mathrm{lim}}}{q}(\sigma_{ij}^{y} - \alpha_{ij}) \tag{6.45b}$$

where R_{lim} is a constant. The above equation is equivalent to introducing another von Mises surface with its centre at the origin and with a radius of R_{lim},

$$\phi_{\mathrm{lim}}^{\sigma} = f_{\mathrm{lim}}^{\sigma} - R_{\mathrm{lim}}^{2} = \tfrac{3}{2}s_{ij}s_{ij} - R_{\mathrm{lim}}^{2} = 0\,. \tag{6.46}$$

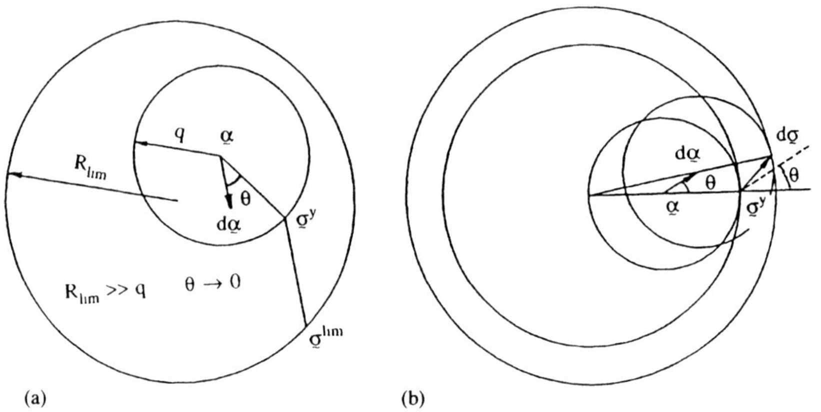

Fig. 6.13 Evolution of yield surface during monotonic loading within a limit surface. (a) when $R_{lim} \gg q$, Ziegler's rule; (b) when R_{lim} is not large compared to q then yield surface may touch the limit surface; and the direction of motion is between $d\boldsymbol{\sigma}$ and normal $\boldsymbol{n}$. Reproduced with kind permission from [50].

The σ_{ij}^{lim} is a point on the above surface whose normal is parallel to the normal at point σ_{ij}^{y}. When R_{lim} is taken to be a much larger constant than q, then Ziegler's rule (6.13) is recovered (Fig. 6.13(a), $R_{lim} \gg q$, $\theta \to 0$). When R_{lim} is not large, then at a certain moment, the yield surface may eventually become tangent with the limit surface at the current stress point because of (6.45b). With further loading, the limit surface will expand, but the two surfaces will remain tangent at the loading point, i.e.

$$\begin{cases} \mathrm{d}\left(\dfrac{\partial f_y}{\partial \sigma_{ij}} - \kappa \dfrac{\partial f_{lim}}{\partial \sigma_{ij}}\right) = 0 \\ \kappa = \dfrac{q}{R_{lim}} \end{cases} \tag{6.47}$$

It can be seen from Fig. 6.13(b) that by using (6.47) the direction of $\mathrm{d}\alpha_{ij}$ will be between $\mathrm{d}\sigma_{ij}$ and $(\sigma_{ij} - \alpha_{ij})$, but generally closer to $\mathrm{d}\sigma_{ij}$, which is in agreement with the Phillips' experimental observations.

In summary, for the ML case, we use (6.45) first. When the yield surface and the limit surface become tangent to each other, we switch to (6.47). The R_{lim} is a material-dependent constant, e.g. it may be viewed as the material's ultimate stress in uniaxial tension. For the RL case, (6.44) is used. With the additional consistency condition

$$\mathrm{d}\phi_y = \frac{\partial f_y}{\partial \sigma_{ij}}(\mathrm{d}\sigma_{ij} - \mathrm{d}\alpha_{ij}) - 2q\,\mathrm{d}q = 0 \tag{6.48}$$

the $\mathrm{d}\alpha_{ij}$'s are completely determined for each loading case.

(c) Determination of the tangent modulus, E_t

The hardening function g, given by (6.43), requires calculation of the instantaneous tangent modulus. The tangent modulus, E_t, is determined from the equivalent stress–equivalent strain curve. Similar to the evolution of the yield surface, two cases can be distinguished. In the ML case, E_t is directly defined as a function of the equivalent stress, σ_{eq}. The functional relation is obtained from the uniaxial stress–strain curve (Fig. 6.12(b))

$$\varepsilon = F(\sigma)\,. \tag{6.49}$$

In the RL case it is obvious that the equivalent stress value is no longer a suitable variable to calculate the E_t. For any given point in the multiaxial stress space, we can find a ratio

$$r = \frac{\delta_1}{\delta_2}\,, \tag{6.50}$$

where δ_1 is the distance measured from the loading point D (Fig. 6.12(a)) to a corresponding point on the memory surface E, whose exterior normal is parallel to that at point D, and δ_2 is the distance from the point of the onset of plastic flow C to point D. The distance between two points in the stress space is defined as

$$\mathbf{d} = [(s_{ij}^{(2)} - s_{ij}^{(1)})(s_{ij}^{(2)} - s_{ij}^{(1)})]^{1/2}\,, \tag{6.51}$$

where $s_{ij}^{(1)}$, $s_{ij}^{(2)}$ are the deviatoric stress values of the two stress points, respectively.

Once r is determined from (6.50), a corresponding point D′ on the reversed uniaxial curve B′C′E′, is determined (Fig. 6.12(b)). Masing's assumption (discussed in Chapter 2) states that the uniaxial unloading stress–strain branch can be described by the equation

$$\frac{\Delta\varepsilon}{2} = F\left(\frac{\Delta\sigma}{2}\right), \tag{6.52}$$

where the prefix Δ refers to the range of the variable. Knowing the coordinates of point D′ and function form (6.49), the current tangent modulus can be determined from the uniaxial stress–strain curve. For example, when the stress–strain curve of the material is fitted with the Ramberg–Osgood type relation (Chapter 2)

$$\varepsilon_{eq} = \left(\frac{\sigma_{eq}}{E}\right) + \left(\frac{\sigma_{eq}}{K}\right)^{1/n}, \tag{6.53}$$

then E_t is given by [34, 45]

$$E_t = \begin{cases} \dfrac{E}{\{1 + (E/nK)(\sigma_{eq}/K)^{(1-n)/n}\}} & \text{for ML case} \\ \dfrac{E}{\{1 + (E/nK)[(\sigma_{eq,max} + rq)/(1+r)K]^{(1-n)/n}\}} & \text{for RL case} \end{cases} \tag{6.54}$$

The above description of the tangent modulus during reversed loading has several advantages in comparison to previously proposed methods [18, 19, 40, 43, 46]. A discussion of this is given in [34].

(d) Evolution of the stress memory surface

As mentioned above, in the ML case, the stress memory surface should be tangent to the yield surface at the current stress point. To satisfy this condition, β_{ij} is determined from the following equations:

$$\begin{cases} \dfrac{\partial f_y(\sigma_{ij} - \alpha_{ij})}{\partial \sigma_{ij}} = \lambda \dfrac{\partial f_m^\sigma(\sigma_{ij} - \beta_{ij})}{\partial \sigma_{ij}} \\ f_m^\sigma(\sigma_{ij} - \beta_{ij}) - R^2 = 0 \end{cases} \tag{6.55}$$

It should be noted that for a proportional monotonic loading, $\beta_{ij} = 0$, i.e. the stress memory surface is an isotropic one. However, for a non-proportional loading, the stress memory surface may become anisotropic, i.e. $\beta_{ij} \neq 0$.

For the RL case, the current stress point is inside the stress memory surface. To simulate the ratcheting behaviour of materials under asymmetric stress cycling, we specify the following rule for the evolution of the stress memory surface:

$$\frac{\partial \beta_{ij}}{\partial \ell_p} = \zeta(\sigma_{ij}^{mean} - \beta_{ij}) \tag{6.56}$$

where

$$\ell_p = \int \left(\frac{2}{3} d\varepsilon_{ij}^p d\varepsilon_{ij}^p\right)^{1/2} \tag{6.57}$$

is the accumulated plastic length. The coefficient ζ is a material constant.

The mean stress σ_{ij}^{mean} is defined as the geometrical centre of the cyclic stress path in the stress space. The geometrical centre of the cyclic stress path can be expressed as

$$\sigma_{ij}^{\text{mean}} = \frac{\int_t^{t+T} \sigma_{ij}\, d\ell}{\int_t^{t+T} d\ell}, \tag{6.58}$$

where T is the period of the cyclic loading and $d\ell$ is the differential length of the curve. Note that (6.56) is applied only in the RL loading case, and the radius of the stress memory surface, R_{mem}, does not change. It was indicated earlier that the value of E_t is related to a ratio of two distances in the stress space, which is dependent on the relative position of the yield and memory surfaces. If the stress memory surface remains isotropic ($\beta_{ij} = 0$), then the ratcheting rate of a non-symmetric cyclic loading ($\sigma_{ij}^{\text{mean}} \neq 0$) will converge to a constant value. By using the evolution rule (6.56), one gets a decreasing ratcheting rate with increasing cycles which is observed in experiments (section 6.6.5).

(e) Transient hardening

Experimental observations show that transient hardening is manifested in two ways. First, there is a change in the elastic domain, an indication of the expansion or contraction of the yield surface. Second, the stress–strain curve of the material varies depending upon the history of loading, thus causing a variation in the tangent modulus. It would be difficult to simulate the experimental observation accurately by considering only the change in the yield surface. We use the accumulated plastic length, ℓ_p, as an internal variable to measure the change of the radius q of the yield surface and the change of the tangent modulus, E_t. At the virgin state, the radius of the yield surface is denoted by q_0 and the tangent modulus by $E_{t,0}$, and the corresponding stable values by q_{st} and $E_{t,st}$. The evolution rules for the elastic domain and tangent modulus are given by [45]

$$\frac{\partial q}{\partial \ell_p} = \xi (q_{st} - q) \tag{6.59}$$

$$\frac{\partial E_t}{\partial \ell_p} = \xi (E_{t,st} - E_t) \,. \tag{6.60}$$

The transient hardening rule specified by (6.59) and (6.60) indicates the existence of a family of uniaxial curves corresponding to specific values of the accumulated plastic length, ℓ_p. The aforementioned curves are bounded between the monotonic and the stable cyclic stress–strain curves (Fig. 6.14). The stable state is gradually attained as ℓ_p increases. Depending upon the material description (Masing or non-Masing), the stable curve

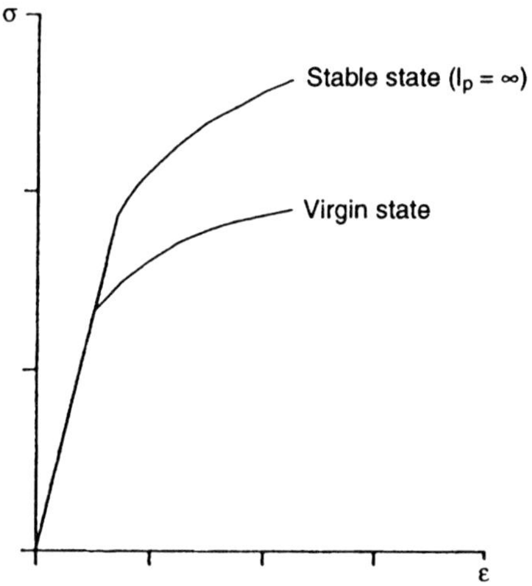

Fig. 6.14 Two limits of a family of uniaxial curves indicating transient response as a function of the accumulated plastic length, ℓ_p.

could be either the cyclic stress–strain curve or the master curve (Chapter 2, Figs 2.18 and 2.19).

To describe the cross-effect discussed in section 6.5.1, we introduce another internal variable F, which is defined as

$$F = \frac{A^{\varepsilon}_{sw}}{A^{\varepsilon}_{m}}, \tag{6.61}$$

where A^{ε}_{m} is the area of the strain memory surface, ϕ^{ε}_{m}, (6.37), projected on the octahedral (Π)-plane, and A^{ε}_{sw} is the area of a polygon connecting all vertices of the projection of the strain vector (Fig. 6.15). The value of F is bounded, $0 \leqslant F \leqslant 1$, and $F = 0$ when the strain path is proportional.

The stable values of q_{st} and $E_{t,st}$ in (6.59) and (6.60) should be a function of $\varepsilon_{eq,max}$ and F, in order to reflect the strain range and cross-effects, i.e.

$$q_{st} = q_{st}(\varepsilon_{eq,max}, F) \tag{6.62}$$

$$E_{t,st} = E_{t,st}\,[\varepsilon_{eq,max}, F, \sigma_{eq}\ (\text{or } \sigma_{eq,max} \text{ and } r)]\,. \tag{6.63}$$

A simple linear form in F was proposed in [45] leading to

$$q_{st} = (1 + \eta_1 F)\, q^{in}_{st}\,(\varepsilon_{eq,max}) \tag{6.64}$$

$$E_{t,st} = (1 + \eta_2 F)\, E^{in}_{t,st}\,[\varepsilon_{eq,max}, \sigma_{eq}\ (\text{or } \sigma_{eq,max} \text{ and } r)]\,, \tag{6.65}$$

where the superscript 'in' means in-phase (proportional) cyclic loading, η_1 and η_2 are material constants reflecting the degree of cross-effect of the

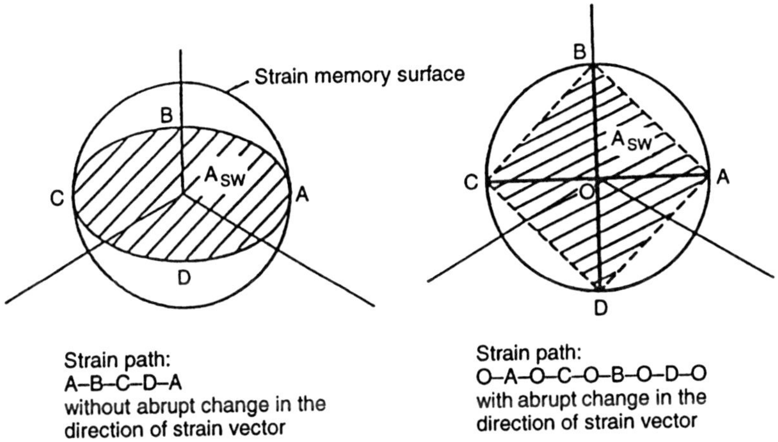

Fig. 6.15 Projection of the strain memory surface and the area swept by the strain vector on the Π-plane. (a) no abrupt change in the direction of strain vector; (b) abrupt change in the strain vector direction. Reproduced with permission from ASME from [51].

material. They can be evaluated from an out-of-phase biaxial test. A further linearization of (6.64) is suggested by the test results of stable cyclic yield surfaces [45, 52], i.e.

$$q_{st} = (1 + \eta_1 F)(A\varepsilon_{eq,max} + B), \tag{6.66}$$

where A and B are material constants which could be determined from a uniaxial test.

6.6 CORRELATION WITH SOME EXPERIMENTAL RESULTS

The constitutive model described in section 6.5 is applicable to monotonic, transient and stable non-proportional cyclic loading. It is relatively simple compared to the existing models in the literature. In the most general case, in addition to the uniaxial stress–strain curves of the material at the virgin and stable cyclic states, the material constants ν (Poisson's ratio), $\xi, \zeta, \eta_1, \eta_2$, A and B have to be specified. By performing uniaxial cyclic tests with increasing strain amplitude, the virgin and saturated uniaxial curves and the constants ξ, A and B can be obtained. A biaxial out-of-phase test is needed to determine the constants η_1 and η_2. Furthermore, a ratcheting test (cycling with non-zero mean stress) is needed to determine ζ.

The predictions of the model have been compared with the available experimental results from uniaxial tests with varying histories to biaxial and multiaxial tests with non-proportional and out-of-phase cyclic strain histories [34, 35, 36, 45, 50, 53]. The comparison shows that the model is capable of simulating the observed transient and stable behaviour of various metals. In this section we will only present a few representative examples to give an indication of the versatility of the model.

6.6.1 EFFECT OF STRAIN RANGE

Results of experiments conducted by Chaboche *et al.* [38] are shown in Fig. 6.16(c) and (d). They show that after stabilizing at a lower strain level,

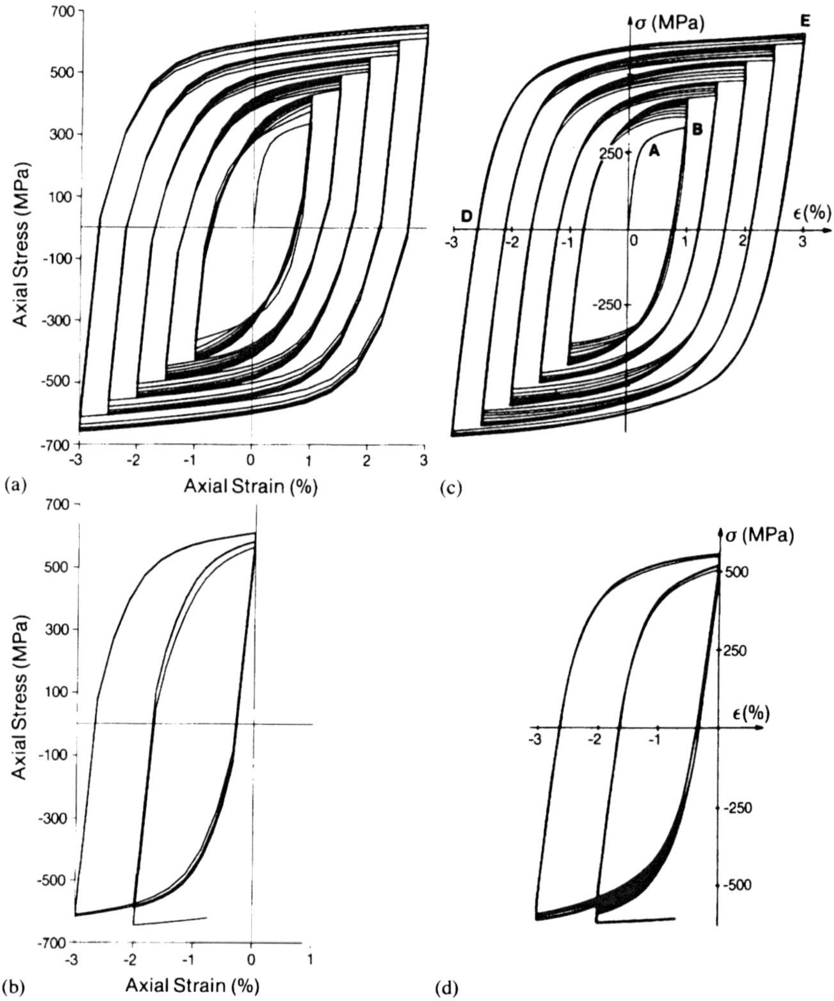

Fig. 6.16 Cyclic straining of 316 L stainless steel at room temperature. Right column, (c) and (d) experimental results [38]; left column, predicted results, reprinted with kind permission from [45]. (a) increasing strain level tests, (b) cycling after $\pm$ 3% prior strain.

more transient hardening is observed under a subsequent higher cyclic strain level. Conversely, after saturation at a large strain level, cycling at a smaller strain level shows more hardening than that without the previous prestrain (Fig. 6.16(d)). The predicted results are depicted in Fig. 6.16(a) and (b) showing very good agreement. Details of material properties and the evaluation of material constants are given in [45]. These results confirm the functional form of the evolution rule (6.59) at least for the uniaxial case.

6.6.2 EFFECT OF STRAIN HISTORY

Methods for determining a stable cyclic stress–strain curve were reviewed in Chapter 2 (Table 2.1). Electrolyte polycrystal copper shows a very strong transient hardening (Fig. 6.17(b)). When this material is cycled under a constant strain amplitude (single step test, SST) the saturated hysteresis loops indicate a non-Masing behaviour (Fig. 6.18(b)). The predicted results using the constitutive model developed in section 6.5 are shown in Figs 6.17(a) and 6.18(a). The material parameter deduced from the monotonic and stable cyclic curves are: $E = 122\,600$, $q_0 = 20.83$, $A = 6361$, $B = 49.14$ (all in MPa), and $\nu = 0.5$ and $\xi = 10.0$.

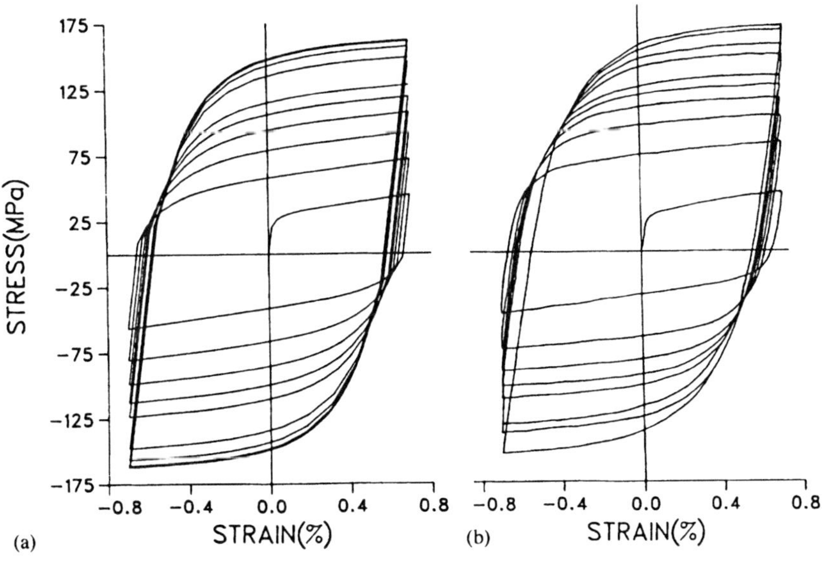

Fig. 6.17 Transient response of electrolyte polycrystal copper under constant amplitude ($\varepsilon_{max} = \pm 0.7\%$) strain control. (a) predicted results; (b) experimental data courtesy of Dr. Schubert [54].

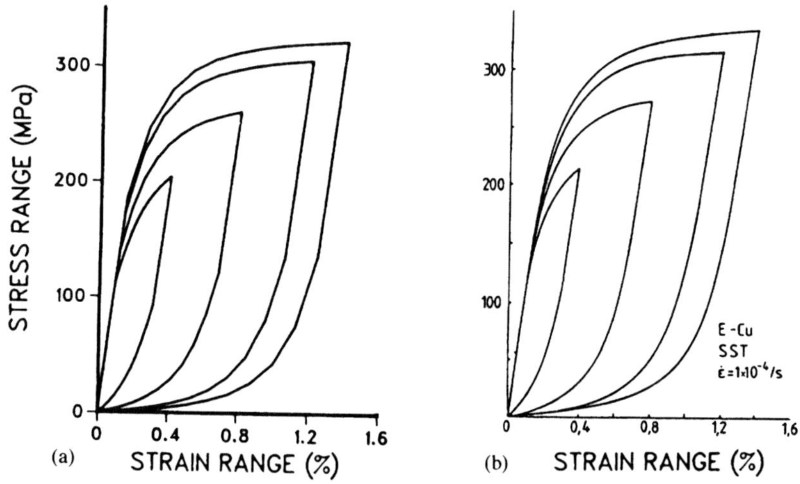

Fig. 6.18 Stable hysteresis loops of electrolyte polycrystal copper under constant amplitude strain control. (a) predicted results; (b) experimental data courtesy of Dr. Schubert [54].

A comparison between calculated and experimental results for the transient hardening (Fig. 6.17) and stable hysteresis loops (Fig. 6.18) indicates a very good agreement.

When the saturated (stable) cyclic stress–strain curve is obtained by the cyclically varying amplitude test (IST), Fig. 6.19, we observe a Masing type of behaviour (Fig. 6.20(b)). The apparent difference between these two sets of results is clearly due to the strain history – constant versus varying strain amplitudes. In fact, this behaviour reflects the strain-range effect discussed earlier. The question which arises is whether the polycrystal copper is a Masing or non-Masing material. In Chapter 2 we briefly discussed the issue of experimental determination of a 'stable cyclic curve'. From the results presented in Figs 6.18 and 6.20, we note the non-unique nature of the saturated curve. In essence, all the so-called saturated curves are transient because, as pointed out earlier in this chapter, the plastic response is a path and history dependent phenomenon. However, from the practical point of view, it becomes important to investigate whether specifying the properties for a given set of conditions, one could then predict the response under other circumstances. This is indeed the case with the constitutive model developed in the previous section. That is, given the material response for a constant amplitude condition (Fig. 6.17(a)) we are able to predict the response for any other strain history, e.g. Figs 6.19(a) and 6.20(a).

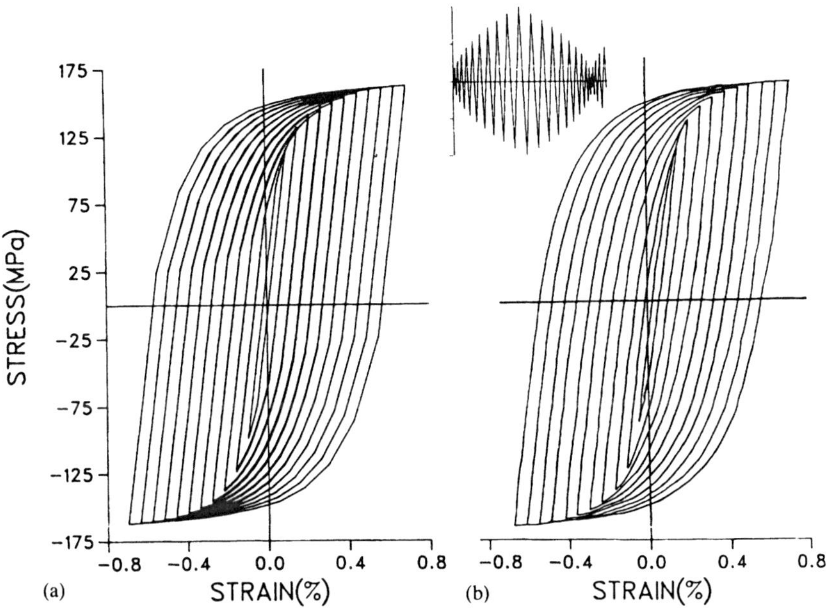

Fig. 6.19 Stable cyclic response of electrolyte polycrystal copper under incremental strain history. (a) predicted results; (b) experimental data courtesy of Dr. Schubert [54].

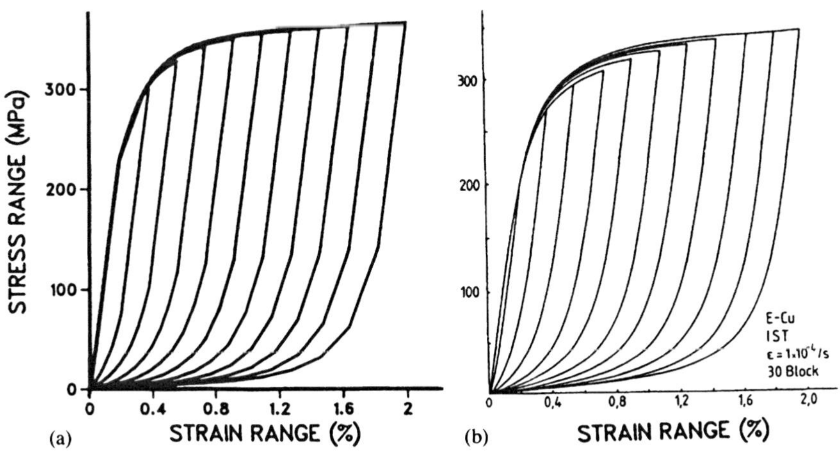

Fig. 6.20 Stable hysteresis loops of electrolyte polycrystal copper under incremental strain history. (a) predicted results; (b) experimental data courtesy of Dr. Schubert [54].

6.6.3 EFFECT OF NON-PROPORTIONAL STRAIN PATH

An annealed copper was stabilized under a 90 degree out-of-phase axial strain–shear strain test conducted by Lamba and Sidebottom [46]. Following this a complex non-proportional strain path as shown in the inset of Fig. 6.21 was applied. No transient hardening was observed during the second loading path. It would appear, therefore, that a true stabilized (saturated) state is only arrived at when the strain path coincides with the strain memory surface (Fig. 6.15).

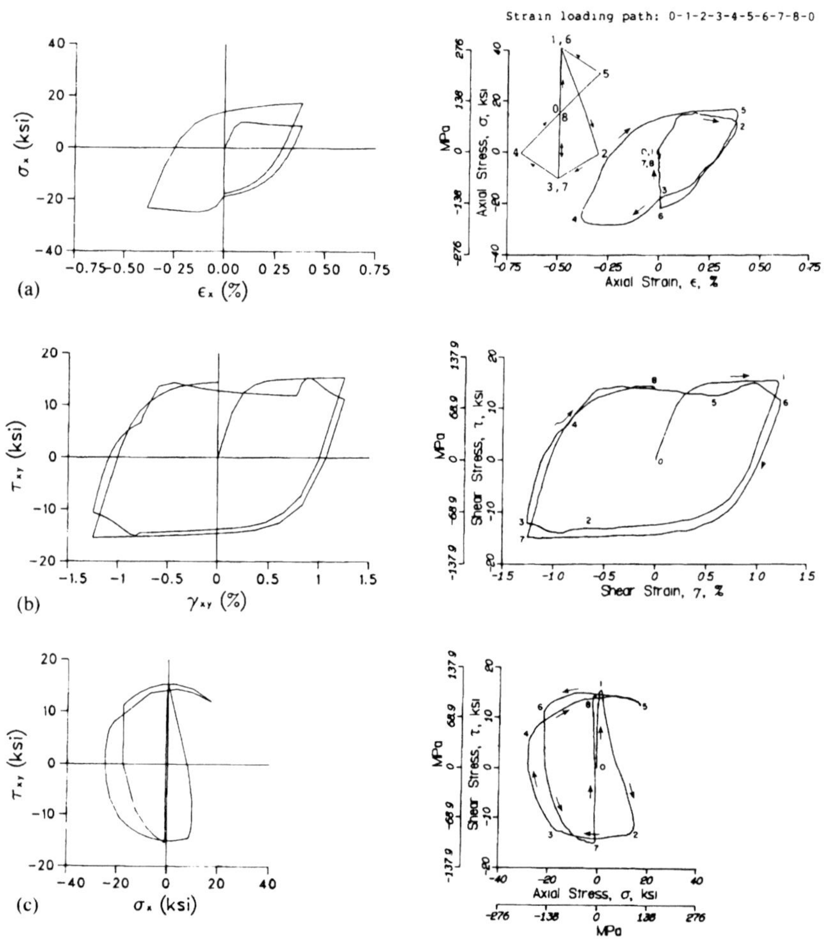

Fig. 6.21 Non-proportional strain control tests of copper. (a) axial behaviour; (b) torsional behaviour; (c) stress response. Right column: experimental results after [46]; left column: predicted results from [45], reprinted with kind permission.

Results of a non-proportional strain cycling test on a type 304 stainless steel specimen has been presented by McDowell [43]. The strain path with the maximum normal strain $\varepsilon_x = 0.0041$ and shear strain $\gamma_{xy} = 0.0060$ is shown in the inset of Fig. 6.22. The transient hardening phenomenon is obvious in this test.

Figures 6.21 and 6.22 show the predicted (left column) and experimental (right column) stress–strain responses of the above-mentioned tests, taken from [45]. It can be seen that despite the very complex loading path, the predicted results are in very good agreement with the experimental data.

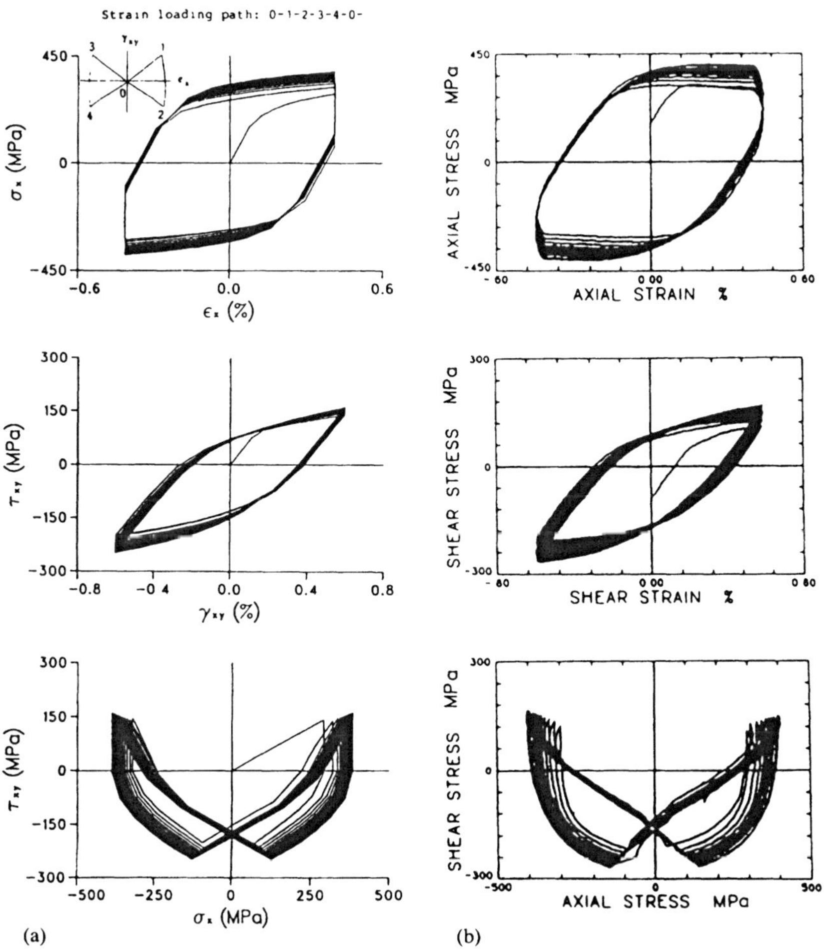

Fig. 6.22 Non-proportional strain-controlled cycling. Right column: experimental results after [43]; left column: predicted results from [45], reprinted with kind permission.

6.6.4 EFFECT OF OUT-OF-PHASE LOADING

We have discussed the effect of out-of-phase loading on the fatigue life in Chapters 4 and 5. Here we will be concerned with predicting the stress–strain response in a biaxial out-of-phase loading.

Most data on biaxial loading have been obtained by subjecting thin-walled circular cylindrical specimens to cyclic axial and torsional loading. Results of such experiments are presented for example in [43, 55, 56] where reference to other works may be found. A test result for an out-of-phase cyclic strain path with $\varepsilon_x = 0.005 \cos \omega t$, $\gamma_{xy} = 0.0075 \cos(\omega t - \pi/6)$, taken from [43], is shown in Fig. 6.23, right column. The predicted result by the

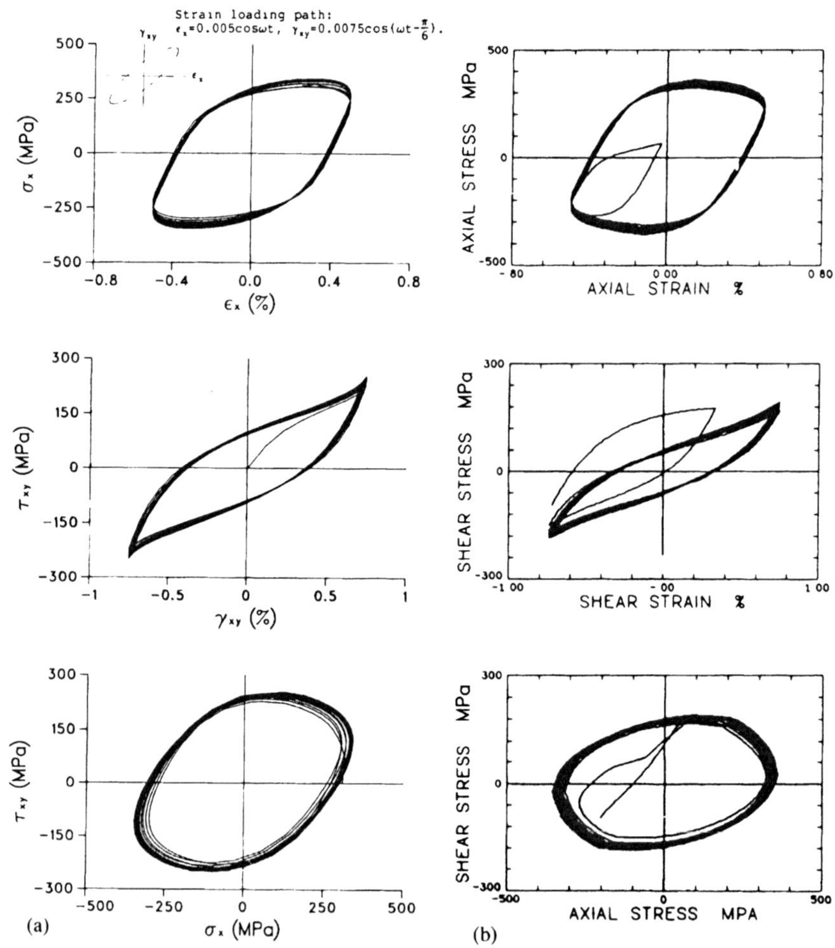

Fig. 6.23 Out-of-phase strain-controlled cyclic loading. Right column: experimental rlesults after [43]; left column: predicted results from [45], reprinted with kind permission.

present constitutive model is also shown in the left column. The evaluation of material constants is given in [45].

A more elaborate test facility would be required in order to apply cyclic tension–compression loading in both the longitudinal and tangential direction of a thin-walled tubular specimen. A series of biaxial tension–compression out-of-phase cyclic tests on A-516 Gr. 70 steel specimens have been carried out in our laboratory. In the following we will present some interesting experimental results.

Figure 6.24 shows the stable cyclic hysteresis loops in the axial and tangential directions of a 90 degree out-of-phase loading ($\varepsilon_a = a \cos \omega t$, $\varepsilon_t = a \cos(\omega t - \pi/2)$) for three different strain amplitudes. For the smallest strain amplitude ($a = 0.0005$, Fig. 6.24(a)) the material is within the elastic range and the axial and tangential loops are almost the same in shape and size, but rotate in the opposite directions. The total work done in one cycle is zero since the algebraic sum of the areas of the two loops is nil. When the strain amplitude is increased to 0.1%, a certain amount of irreversible plastic deformation takes place. The axial loop becomes narrower than the tangential one (Fig. 6.24(b)) although they still rotate in opposite directions. The plastic work done in one cycle is equal to the difference in area of the two loops. If the strain amplitude is further increased to 0.2%, we can observe from Fig. 6.24(c) that the direction of rotation of the axial loop has now changed to the same direction as that of the tangential loop, and more plastic work, which is the sum of the area of the two loops, is done per cycle.

Figure 6.25 shows three experimental results under different phase angles ($\phi = 0°, 30°, 90°$) but with the same strain amplitude, $a = 0.15\%$. For $\phi = 0°$, it is a proportional strain loading path. The axial and tangential loops are similar to each other (Fig. 6.25(a)) duc to the symmetry of the strain path with respect to the two directions. With increasing phase angle the loops begin to differ from each other, the highest difference occurring for $\phi = 90°$ (Fig. 6.25(b) and (c)). It should be noted that while the maximum equivalent strain decreases with the increasing phase angle, the stress range increases at the same time. This then indicates an increased level of hardening caused by the non-proportional strain path.

The predicted results for the above-mentioned tests are also presented in Figs 6.24 and 6.25 by dashed lines. An interested reader can find the details in [53].

6.6.5 RATCHETING UNDER CYCLIC LOADING WITH MEAN STRESS

Mechanical ratcheting behaviour of materials is one of the problems which has recently attracted considerable interest. A comparative study of 10 types of constitutive models was made by Inoue *et al.* [57]. The study revealed that most constitutive models have a poor capability of predicting

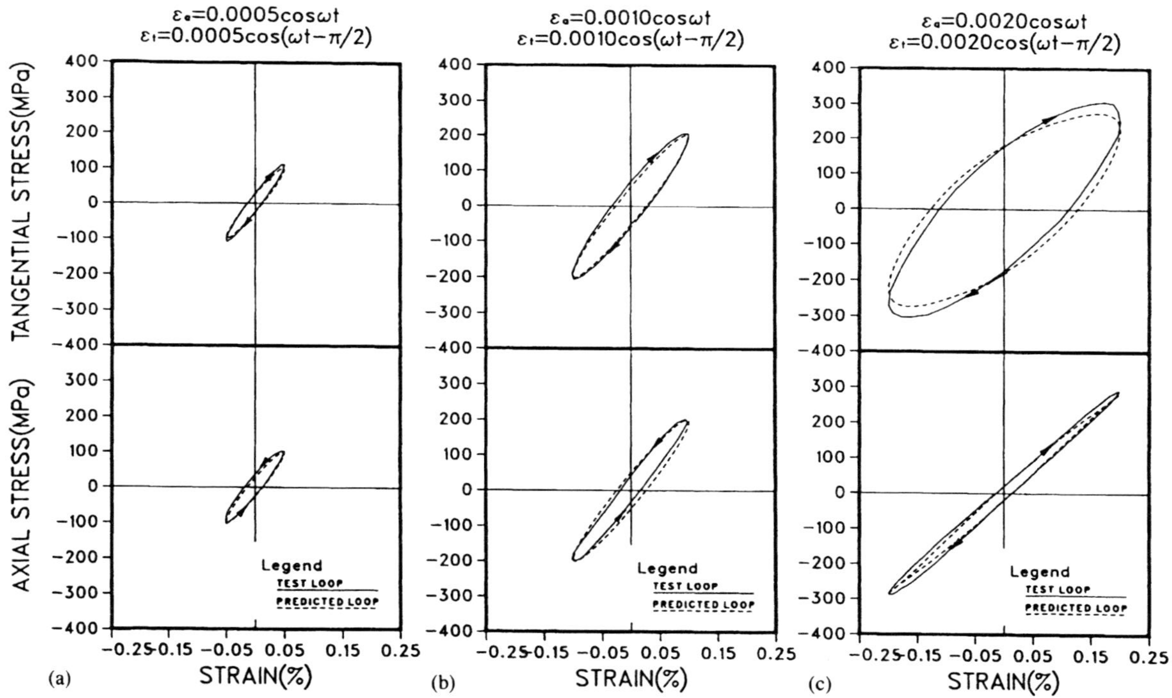

Fig. 6.24 Hysteresis loops of a 90 degree out-of-phase loading with different strain amplitudes. (a) $a = 0.0005$; (b) $a = 0.001$; (c) $a = 0.002$. Reproduced with permission from ASME from [53].

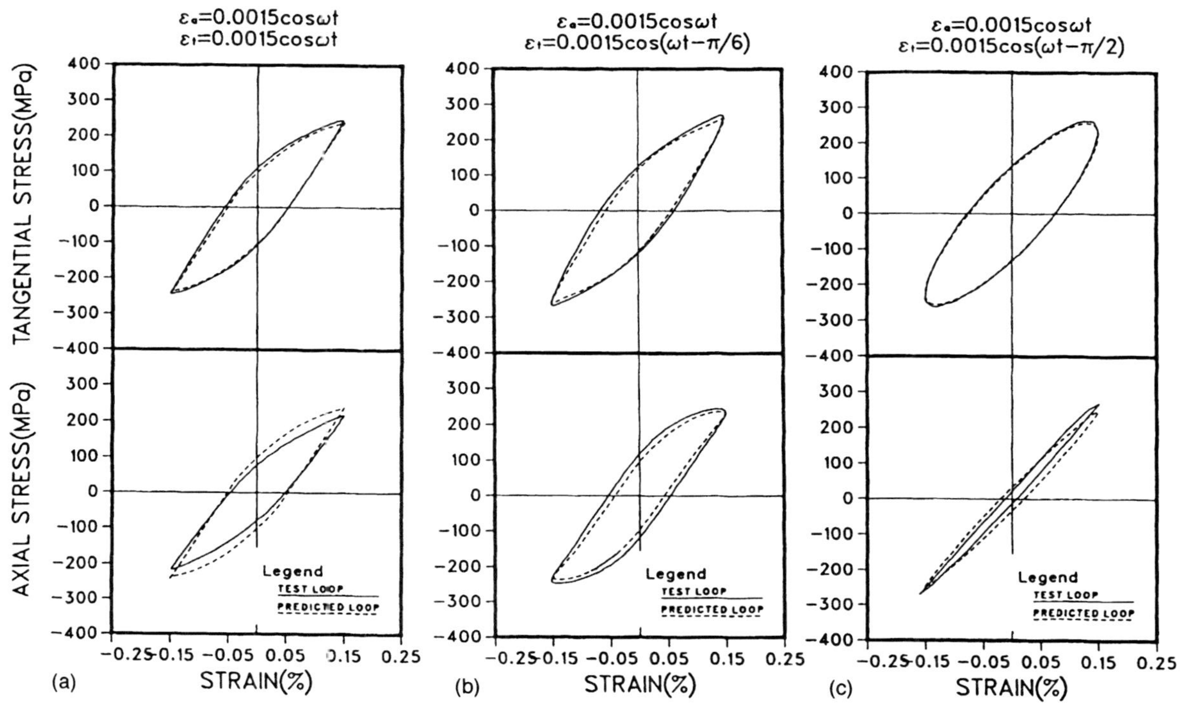

Fig. 6.25 Hysteresis loops of out-of-phase loading with different phase angles, but the same amplitude. (a) $\phi = 0°$; (b) $\phi = 30°$; (c) $\phi = 90°$. Reproduced with permission from ASME from [53].

ratcheting strain in the uniaxial and multiaxial cases. To simulate practical loading conditions of pressure vessel structures, biaxial ratcheting tests under strain- or stress-controlled axial cycling with constant hoop stress were conducted on thin-walled specimens [58]. Figure 6.26(a) shows the test results of type 304 stainless steel tubular specimens under strain-controlled axial cycling ($\Delta\varepsilon_a/2 = 0.2\%$) with a constant hoop stress ($\sigma_t =$ 80 MPa). It can be seen that the hoop strain increases with each cycle, but the rate of ratcheting decreases with the increasing number of cycles. Figure 6.27(a) shows the results of stress-controlled axial cycling ($\Delta\sigma/2 =$ 180 MPa) with a constant hoop stress ($\sigma_t = 80$ MPa). It is observed that a positive hoop ratcheting strain and a negative axial ratcheting strain are developed at the same time. The ratcheting rate decreases with increasing cycles.

The results predicted by the present model for the above-mentioned tests are presented in Fig. 6.26(b) and Fig. 6.27(b). More details are given in [58].

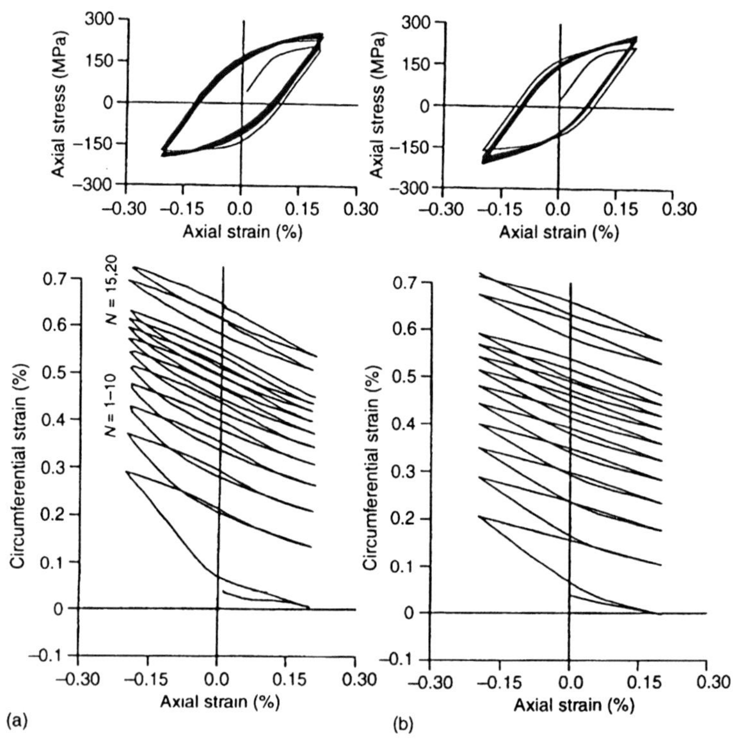

Fig. 6.26 Circumferential strain versus axial strain for ratcheting test with constant hoop stress under cyclic axial strain. (a) test result; (b) predicted result from [58], reprinted with kind permission from ASME.

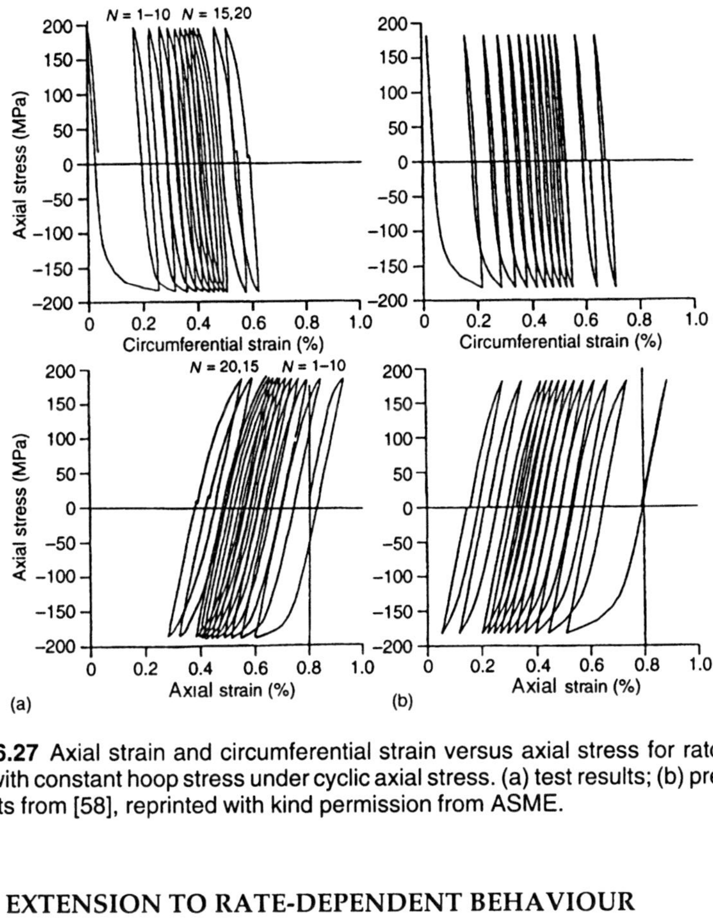

Fig. 6.27 Axial strain and circumferential strain versus axial stress for ratcheting test with constant hoop stress under cyclic axial stress. (a) test results; (b) predicted results from [58], reprinted with kind permission from ASME.

6.7 EXTENSION TO RATE-DEPENDENT BEHAVIOUR

In this section we extend the rate-independent constitutive model described earlier to include rate effects. The effect of prior creep on the subsequent plastic deformation is also incorporated into the model. The creep process itself will be treated in another section. The constitutive model to be described below can thus be used for inelastic processes with variable loading rates. The framework to be established is that any loading sequence can be predicted by two separate but coupled models. This is to be interpreted as the total strain rate, being the sum of the elastic and plastic parts or the sum of the elastic and creep portions, i.e.

$$\dot{\varepsilon}^{t} = \dot{\varepsilon}^{e} + (\dot{\varepsilon}^{p} \text{ or } \dot{\varepsilon}^{c})\,. \tag{6.67}$$

For a given boundary-value problem, if the body forces and assigned boundary conditions are kept constant, the creep equation will be used, otherwise the plastic equation will be applied.

6.7.1 BACKGROUND

Experiments have shown that engineering alloys such as type 304 and 316 stainless steels, high strength titanium alloys, etc. exhibit rate-dependent behaviour even at room temperature, e.g. Krempl [59], Kujawski *et al.* [60], Kujawski and Krempl [61], Ikegami and Ni-Itsu [62], Xia and Ellyin [63], among others. The main manifestations of rate-dependent deformation are: loading rate sensitivity, creep, relaxation, and creep–plasticity interaction. A constitutive relation for rate-dependent materials must be able to simulate the above-mentioned behaviour.

The rate-independent theory of plasticity has been shown to give good results for those applications where the loading rate is constant or its variation is small. The applicability of this type of theory to any non-constant strain rate history is doubtful. The classical creep equations are generally formulated quite separately from the plastic deformation and the effect of prior plasticity is not usually included in the constitutive equations. For a loading sequence which includes alternating plastic and creep deformations, both theories, therefore, cease to be effective.

In recent years a number of viscoplastic constitutive models have appeared in the technical literature [64–68]. This type of formulation is sometimes referred to as a 'unified theory', since all aspects of inelastic deformation such as plasticity, creep and recovery are treated by one set of equations. Temperature dependence of material properties is included in these models. A description of the general characteristics of the unified theories, and a comparison of different proposed models can be found in literature surveys by Chan *et al.* [69], and Bammann and Kreig [70]. An attractive and desirable feature of the unified theories is their broad range of application. By introducing appropriate interval variables and their evolution rules, most experimental observations may **qualitatively** be simulated. However, as rightly pointed out by Miller [65], there is an inevitable trade-off between the breadth and accuracy. The **quantitative** results of various simulated tests may not be as good. For example, in simulating a simple test (plastic deformation or creep) the predicted results will not generally be as accurate as those given by explicit equations for that kind of deformation. Another shortcoming of the unified theories is that there is no standard or recommended test procedure for the evaluation of material constants or function relations introduced in the theoretical expressions. A 'trial and error' method is used in most cases, and this becomes more difficult with the increased number of material constants required for a broad-based constitutive model.

An important basis for a unified model is the belief that plastic and creep deformations undergo common microstructural changes, mainly caused by dislocaion movements. However, experimental results indicate that, at least in the macro-scale, the effect of prior deformation on the subsequent creep or plastic deformation is quite different. For example, a plastic prestrain has a greater contribution to creep hardening than the same amount of creep prestrain [71]. Such a distinction cannot be made in a unified creep–plasticity model since, as stated earlier, creep and plastic strains are not separable from each other in these models. The underlying concept of the unified creep–plasticity models without a yield surface precludes specification of an elastic region, and as such the governing set of differential equations are 'stiff' in the numerical sense. For example, at small strains or large strain-rates, the inelastic strain must be essentially zero to approximate a linear response, and it must increase rapidly with increasing stress. In comparing a unified creep–plasticity (UCP) with a 'yield surface' type model, Bammann and Kreig [70] found that the UCP solution required over two orders of magnitude more time than the yield surface plasticity model. It was reported that in the case of the UCP model, 95% of the solution time was spent in the constitutive subroutine.

Results of experimental investigations into creep–plasticity interaction are reported by Xia and Ellyin [63], Pugh and Robinson [72], Pugh [73], and Kawai and Ohashi [71], among others. From the results of these experiments it would appear that the effect of creep (or prior plastic deformation) can be incorporated into plastic (or creep) constitutive equations. We may thus treat a complex loading history by a set of two coupled constitutive equations. (Further discussion of creep–plasticity models will be given in section 6.9.)

6.7.2 FORMULATION OF A RATE-DEPENDENT ELASTIC–PLASTIC CONSTITUTIVE MODEL

Typical uniaxial stress–strain diagrams at different strain-rates are shown in Fig. 6.28 [59]. A characteristic feature of this type of response is an increase in the initial yield stress with the increasing strain-rate. This may be explained through the required increase in dislocation density and higher velocity of movement at higher strain-rates. Another observed feature relates to the change of slope of various curves at the elastic–plastic transition region with the strain (or stress) rate change.

The rate-dependent yield surface and rate-dependent tangent modulus will be assumed to be a generalization of the rate-independent theory, i.e.

$$q = q(\ell_p, \dot{\varepsilon}_{ep}) \tag{6.68a}$$

$$E_t = E_t[\ell_p, \sigma_{eq} \text{ (or } \sigma_{eq,max} \text{ and } r), \dot{\varepsilon}_{eq}], \tag{6.68b}$$

where $\dot{\varepsilon}_{eq} = (2\dot{e}_{ij}\dot{e}_{ij}/3)^{1/2}$ is the equivalent strain rate and e_{ij} has been defined earlier (6.38). Note that $\dot{\varepsilon}_{eq}$ thus defined will always be positive.

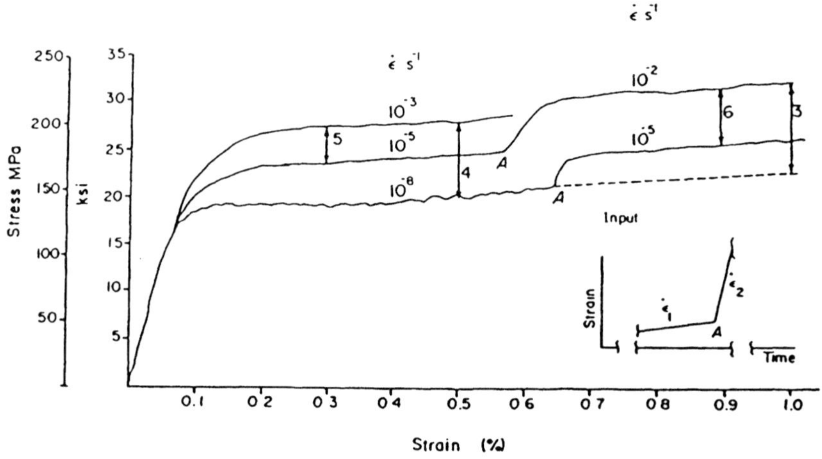

Fig. 6.28 Uniaxial stress–strain curves at different constant strain-rates for annealed AISI type 304 stainless steel. Reprinted with kind permission from Krempl [59].

The expression (6.68a) indicates a family of strain-rate-dependent yield surfaces. Each surface in the family corresponds to a certain constant strain-rate and all these surfaces have a common centre (Fig. 6.29). This family of yield surfaces moves and changes its size with the development of the accumulated plastic length, ℓ_p. For a reference strain-rate, $\dot{\varepsilon}_{0,\mathrm{eq}}$, the evolution of the yield surface radius, q, is expressed by

$$\frac{\partial q^0}{\partial \ell_p} = \xi (q^0_{\mathrm{st}} - q^0) , \tag{6.69}$$

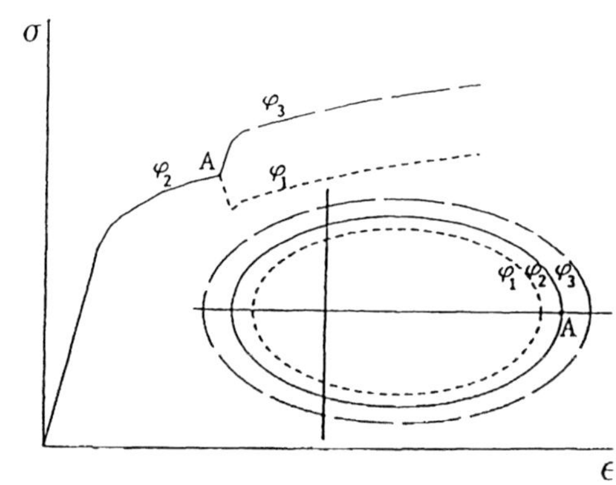

Fig. 6.29 The effect of instantaneous change of strain-rate. Reproduced with permission from ASME from [51].

where ξ is a material constant and the subscript 'st' refers to the saturated (stable) value of q^0. Incorporating a logarithmic strain-rate dependence [74] for all other surfaces in the family, we get

$$q^1 = q^0\{1 + C\log(\dot{\varepsilon}^1_{eq}/\dot{\varepsilon}^0_{eq})\, H[(\dot{\varepsilon}^1_{eq}/\dot{\varepsilon}^0_{eq}) - 1]\} \tag{6.70a}$$

$$q^1_{st} = q^0_{st}\{1 + C\log(\dot{\varepsilon}^1_{eq}/\dot{\varepsilon}^0_{eq})\, H[(\dot{\varepsilon}^1_{eq}/\dot{\varepsilon}^0_{eq}) - 1]\} \tag{6.70b}$$

where C is a material constant, q^1, q^1_{st} are the instantaneous and saturated values of the yield surface radius with the constant strain rate $\dot{\varepsilon}^1_{eq}$ and $H(x)$ is the Heaviside function equal to one when $x \geqslant 0$ and zero when $x < 0$. From (6.69) and (6.70) we get

$$\frac{\partial q^1}{\partial \ell_p} = \xi(q^1_{st} - q^1)\,. \tag{6.71a}$$

Likewise, the evolution of the rate-dependent tangent modulus E_t, is given by

$$\frac{\partial E^1_t}{\partial \ell_p} = \xi\,(E^1_{t,st} - E^1_t)\,. \tag{6.71b}$$

Note that the above two equations (6.71a) and (6.71b) are similar to those of (6.59) and (6.60) for a constant strain-rate. In a similar manner, (6.64) to (6.66) could be used to describe the cross-effect for a rate-dependent material.

By constructing a family of uniaxial curves in terms of accumulated plastic strain (ℓ_p) and equivalent strain-rate ($\dot{\varepsilon}^1_{eq}$), i.e. $\sigma_{eq} = \sigma_{eq}(\ell_p, \dot{\varepsilon}^1_{eq}, \varepsilon_{eq})$ we can evaluate q^1 and E^1_t in (6.71). The curves can be constructed through a series of uniaxial cyclic tests with increasing strain range for various constant strain-rates. In this manner, a family of initial and stable cyclic curves is obtained at various strain-rates. The evolution from the initial to the stable state of the above family of uniaxial curves is controlled by (6.71).

Often it is difficult to accurately fit an experimental stress–strain curve with a single analytical expression for the whole range. In such a case, we may construct a function table for certain points on the curve (e.g. [45, 51]) and the value of E_t is then obtained by using a two-dimensional interpolation.

All other feature of the rate-independent model described in section 6.5.2, i.e. stress memory surface, strain memory surface, distinction of two types of plastic loading cases, etc. are used in the rate-dependent model. When using the consistency condition (6.48), for a loading with continuously varying total strain-rate, the increment in the size of the yield surface, dq, should include two parts, i.e. from (6.68a)

$$\mathrm{d}q = \frac{\partial q}{\partial \ell_p}\,\mathrm{d}\ell_p + \frac{\partial q}{\partial \dot{\varepsilon}_{eq}}\,\mathrm{d}\dot{\varepsilon}_{eq}\,. \tag{6.72}$$

Loading sequences with abrupt changes in the strain-rate or with interaction of creep strain are discussed below.

(a) Instantaneous change of strain-rate

Suppose the strain-rate is increased instantaneously (a positive jump) when the stress state is at point A in Fig. 6.29. From (6.70) the yield surface will change abruptly from ϕ_2 to ϕ_3, in which case the point A becomes an interior point with respect to the instantaneous yield surface ϕ_3. Thus, the subsequent loading will produce elastic deformation, i.e. $E_t = E$, until the stress point touches the yield surface again.

If a negative jump of the strain-rate is imposed at point A (Fig. 6.29), then the yield surface will change abruptly from ϕ_2 to ϕ_1. In such an event the stress point A becomes exterior to the instantaneous yield surface ϕ_1. We define such a state as an unstable one since the subsequent stress–strain curve will have a negative slope, and the slope is assumed to be $E_t = -E$, until the stress point reaches the yield surface again. Experimental results appear to support the above description (e.g. Fig. 6.33(b)).

(b) Effect of creep on subsequent plastic behaviour

The effect of creep prestrain on the subsequent plastic deformation was investigation by Ikegami and Ni-Itsu [62] by performing step-up creep tests. The material was type 304 stainless steel and the specimens were subjected to pure tension or torsion at room or elevated temperature (Fig. 6.30).

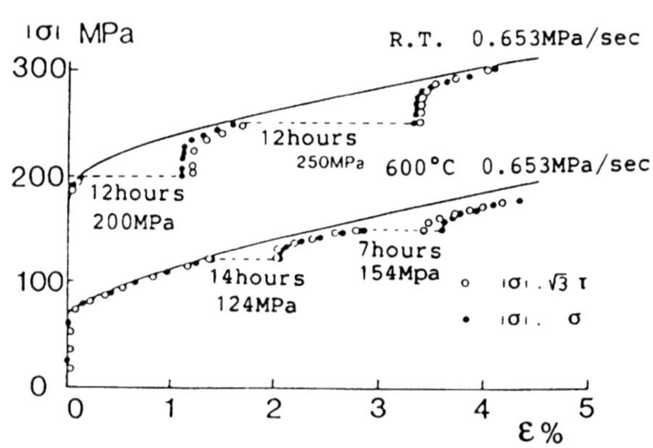

Fig. 6.30 The influence of creep on subsequent plastic behaviour for AISI type 304 stainless steel at room temperature or at elevated temperature (after Ikegami and Ni-Itsu [62], reproduced with permission from ASME).

Kujawski *et al.* [60, 61] performed similar step-up creep tests whereby the specimens of type 304 stainless steel and a titanium alloy were subjected to tension and cyclic loading at room temperature (Fig. 6.36(b) and 6.37(b)). The experimental data indicate that subsequent to the creep or relaxation, the initial response is almost elastic until the increasing stress approaches the stress–strain curve corresponding to the current strain-rate. Thereafter, the flow stress coincides with that of the corresponding rate curve, and the effect of prior creep strain fades away. At room temperature, the hardening due to creep strain is almost equal to that produced by an equivalent plastic strain.

The above experimental observations could be described by the shift of the yield surface during the creep process. The shift value, $\Delta\alpha_{ij}$, is set equal to the stress increment value $\Delta\sigma_{ij}$, resulting from the same quantity of plastic strain increment as that of the creep strain. For example, in a stress-controlled loading process, shown in Fig. 6.31, the creep process starts at the loading point A and a creep strain increment $\Delta\varepsilon^c$ is accumulated during the process, and the loading is subsequently resumed. During the creep process, AB, the loading point A remains stationary while the yield surface shifts with the creep strain (Fig. 6.31). The shift value, $\Delta\alpha_{ij}$, is then calculated from the elastic–plastic constitutive equation (6.42a)

$$\Delta\alpha_{ij} = \Delta\sigma_{ij} = \int L_{ijkl}\,\mathrm{d}\varepsilon^{c}_{kl}\,. \tag{6.73}$$

In a relaxation process, the total strain increment is zero, thus the creep strain increment $\Delta\varepsilon^c_{ij} = -\Delta\varepsilon^e_{ij}$. Apart from a small shift of the yield surface determined from (6.73), the loading point initially on the yield surface

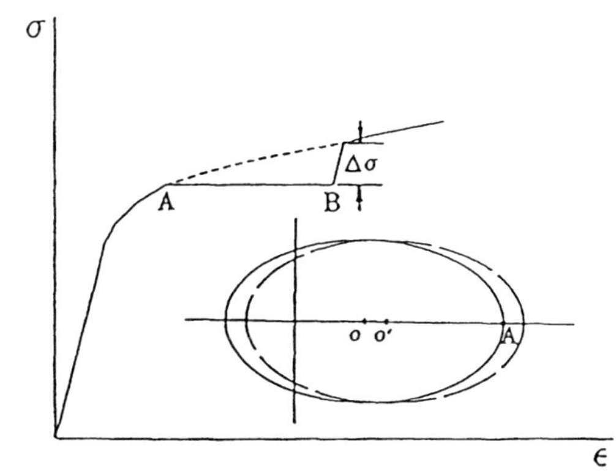

Fig. 6.31 Shift of the yield surface during the creep process and subsequent elasto-plastic deformation. Reproduced with permission from ASME from [51].

moves inside it during the relaxation process. Thus, upon reloading after a creep or relaxation process, an elastic deformation will ensue until the loading point reaches the current yield surface.

6.8 CORRELATION WITH SOME RATE-DEPENDENT EXPERIMENTAL OBSERVATIONS

The experimental data used in this section are taken from papers by Krempl [59], Kujawski *et al.* [60, 61], and Inoue *et al.* [68]. Materials tested include type 304 stainless steel (at room or elevated temperature) and titanium alloys. For the type 304 stainless steel at room temperature the experimental curves in Fig. 6.28 are considered to be the virgin state of the uniaxial stress–strain curves. The continuous curve (dashed line) in Fig. 6.33(b) is used to define the stable master curve. The other stable uniaxial stress–strain curves are obtained from the master curve by a transition along the linear response portion. The material constants and function tables on the above family of uniaxial stress–strain curves are given in [51].

6.8.1 STRESS–STRAIN RESPONSE AT LOAD OR DEFORMATION CONTROLLED MODES

Six uniaxial stress–strain curves at constant strain-rates ranging from 10^{-8} to $10^{-3}\,s^{-1}$ and one at a constant stress-rate predicted by the model are shown in Fig. 6.32. It should be noted that the transition from the elastic to inelastic response is more gradual in the stress (load) controlled case. Experimental data indicate the same trend and this can be explained as

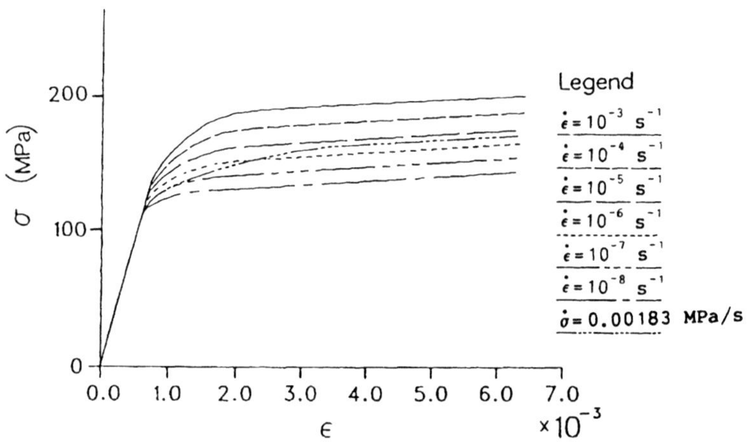

Fig. 6.32 Prediction of uniaxial stress–strain curves for different constant strain-rates or constant stress-rate for annealed AISI type 304 stainless steel. Reproduced with permission from ASME from [51].

follows. For the unaixial stress state, the constitutive equation reduces to

$$\dot{\varepsilon} = \frac{1}{E_t(\sigma, \dot{\varepsilon})} \dot{\sigma} . \tag{6.74}$$

In a load (stress) controlled test $\dot{\sigma}$ = constant. Upon exceeding the elastic limit, E_t decreases with increasing ε and consequently $\dot{\varepsilon}$ increases gradually. Within the transition zone, E_t is an increasing function of the variable $\dot{\varepsilon}$, therefore as a result of the two opposing trends, the stress–strain curve will tend to have a more gradual and large transition zone at the constant stress-rate. The stress-rate for the curve in Fig. 6.32 is $\dot{\sigma} = 1.8272 \times 10^{-3}$ MPa s^{-1} (2.65×10^{-4} ksi s^{-1}) which is equivalent to a strain-rate $\dot{\varepsilon} = 10^{-8} s^{-1}$ in the elastic regime. It is thus seen that the model is capable of reproducing the overall rate sensitivity features of metals as observed experimentally.

6.8.2 RESPONSE TO REPEATED STRAIN-RATE CHANGES

The response of a cyclically saturated specimen to repeated strain-rate changes was reported by Krempl [59] and is shown in Fig. 6.33(b). The specimen was saturated at fully-reversed strain-controlled mode with $\varepsilon_{max} = 0.004$, therefore from (6.62) and (6.70) the radius of the stable cyclic yield surfaces, with $\dot{\varepsilon} = 10^{-3} s^{-1}$ and $\dot{\varepsilon} = 10^{-5} s^{-1}$ are

$$q_{st}^{1} = q_{st}^{o}[1 + C \log(\dot{\varepsilon}_{eq}^{1}/\dot{\varepsilon}_{eq}^{o})] = \begin{cases} 188.2 \text{ MPa} & \text{for} \quad \dot{\varepsilon}_{eq} = 10^{-3} s^{-1} \\ 170.9 \text{ MPa} & \text{for} \quad \dot{\varepsilon}_{eq} = 10^{-5} s^{-1} \end{cases}$$

The predicted response to the repeated strain-rate changes is shown in Fig. 6.33(a). The proposed model simulates fairly well the characteristic behaviour of the material to the abrupt changes of strain-rate.

6.8.3 EFFECT OF STRAIN-RATE HISTORY

It was reported by Ellyin *et al.* [75] that not only the strain-rate, but also the strain-rate history, may influence the cyclic hardening behaviour. Uniaxial and non-proportional biaxial cyclic tests with a constant strain amplitude but with different sequences of strain-rates were conducted on tubular specimens of 304 stainless steel at room temperature. Each test comprised three levels of strain-rates applied in a predetermined order. In each step, 50 cycles were applied before switching to another rate to ensure stable material response at each strain-rate level. Figure 6.34(a) shows the relationship between the stress amplitude and number of cycles of three uniaxial cyclic tests with different strain-rate histories, F–M–S, S–M–F and F–S–F, where the letters F, M, S, refer to a Fast, a Medium, and a Slow strain rate, respectively. The stable stress–strain loops of these tests at each step

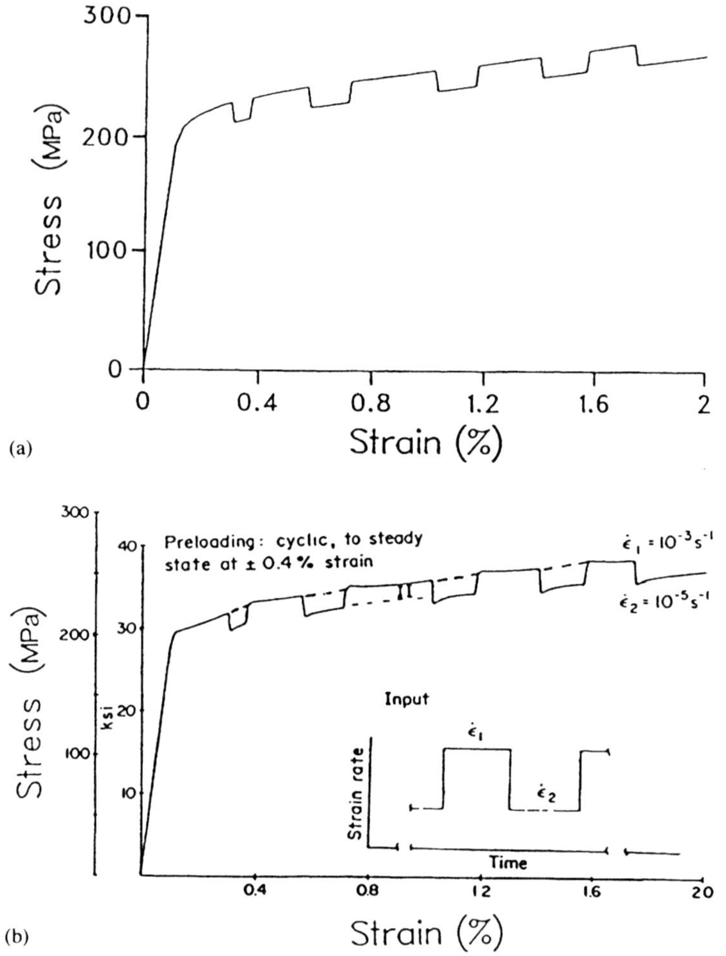

Fig. 6.33 Response to repeated strain-rate changes in a tensile test for AISI type 304 stainless steel. Cyclically prestrained specimen. (a) predicted results from [51], reproduced with permission from ASME; (b) experimental results from Krempl [59], reprinted with kind permission.

of the strain-rate are depicted in Fig. 6.35(a). For the test F–M–S, a decrease in the stress amplitude with decreasing strain-rate was observed. However, no appreciable change is observed in the test S–M–F.

In the biaxial cyclic tests a similar behaviour was observed with respect to the strain-rate history (Fig. 6.34(b)), i.e. once a material is stabilized under a lower strain-rate, the stress–strain response is not appreciably affected by a jump to a faster strain-rate. A possible explanation is that at a given stress level, the plastic strain is larger at a lower strain-rate than at a higher one. The dislocation substructures of a larger plastic strain are

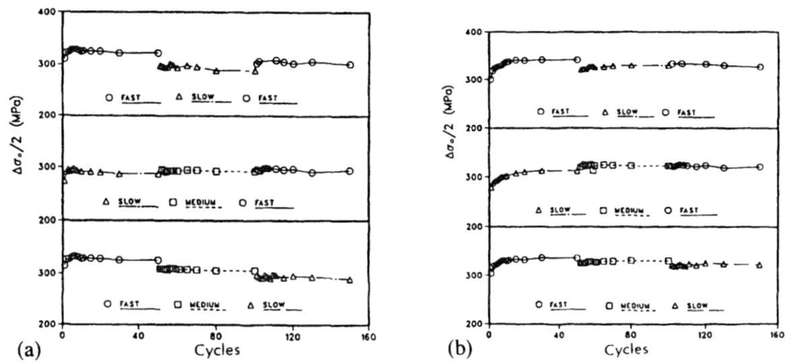

Fig. 6.34 Stress amplitude versus number of cycles at three strain-rates for uniaxial and biaxial cyclic tests. Reprinted with kind permission from [75]. (a) uniaxial cyclic test; (b) biaxial cyclic test.

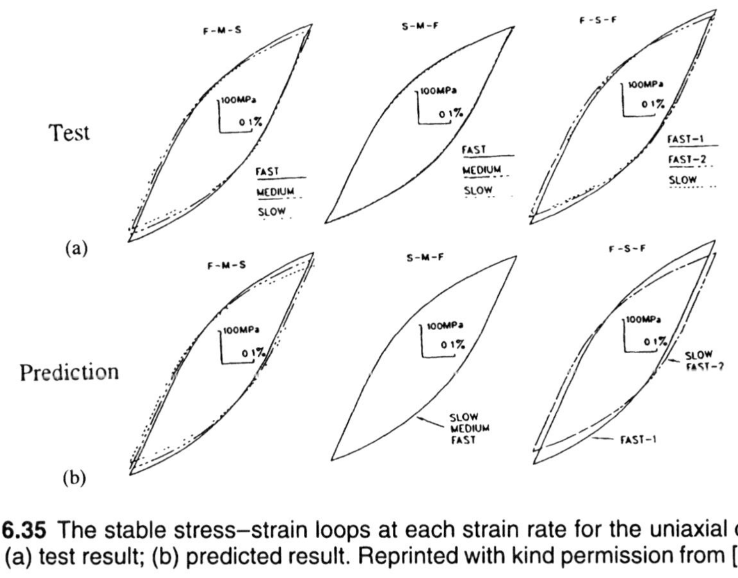

Fig. 6.35 The stable stress–strain loops at each strain rate for the uniaxial cyclic test. (a) test result; (b) predicted result. Reprinted with kind permission from [75].

most stable, and subsequent change to a higher strain-rate would not affect the stabilized dislocation structure and, consequently, the stress–strain response. To simulate the strain-rate history effect on the stable stress–strain responses as observed in the above tests, we consider that the material can memorize the lowest cyclically saturated equivalent strain-rate, $\dot{\varepsilon}_{eq,min}$, ever experienced.

Thus the equation (6.70) is replaced by

$$q^{\mathrm{i}}_{\mathrm{st}} = q^{\mathrm{o}}_{\mathrm{st}}\left\{1 + C\log\left(\frac{\dot{\varepsilon}_{\mathrm{eq,min}}}{\dot{\varepsilon}^{\mathrm{o}}_{\mathrm{eq}}}\right)H\left[\left(\frac{\dot{\varepsilon}^{\mathrm{i}}_{\mathrm{eq}}}{\dot{\varepsilon}^{\mathrm{o}}_{\mathrm{eq}}}\right) - 1\right]\right\}. \tag{6.75}$$

Therefore, for the test with sequence F–M–S, the stress amplitude will decrease due to the decrease of the q^1_{st} after a drop in strain-rate, whereas for the sequence S–M–F the stress amplitude will not change because the $\dot{\varepsilon}_{eq,min}$ and consequently the q^1_{st} remain unchanged after a jump in strain-rate. The predicted results of this model for the uniaxial cyclic tests are shown in Fig. 6.35(b). For more details regarding the tests and the material constants used in the modelling, the reader is encouraged to consult [75].

6.8.4 STRESS–STRAIN RESPONSE DURING STEP-UP CREEP TESTS

Two stress–strain curves derived from step-up creep tests were reported by Kujawski *et al.* [60] and are shown in Fig. 6.36(b). The predicted response is shown in Fig. 6.36(a). At this stage, our purpose is to demonstrate

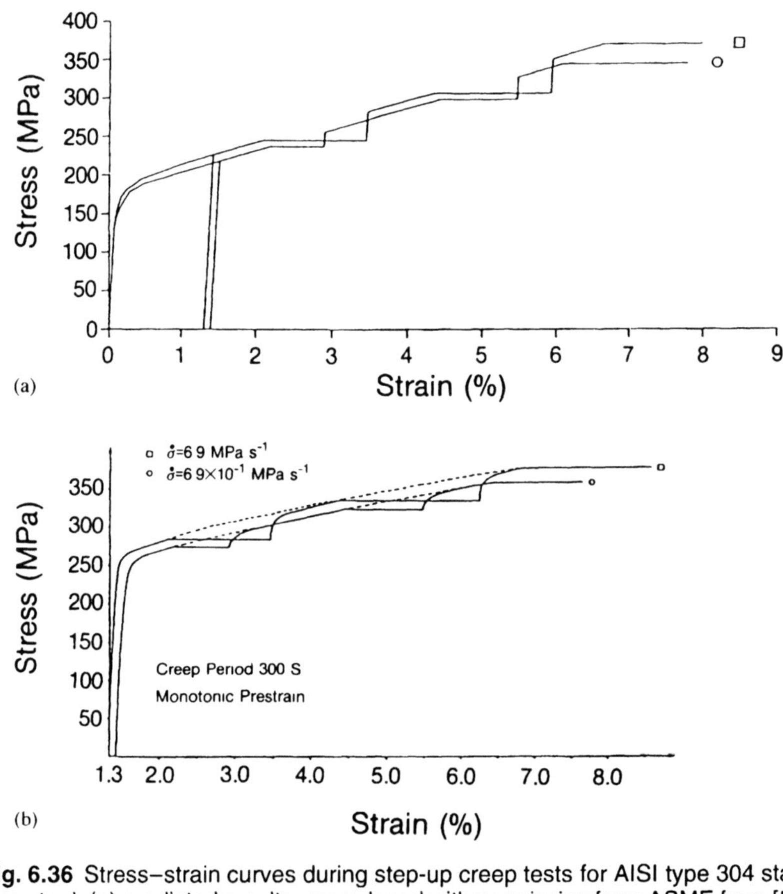

Fig. 6.36 Stress–strain curves during step-up creep tests for AISI type 304 stainless steel. (a) predicted results, reproduced with permission from ASME from [51]; (b) experimental results after Kujawski *et al.*, reproduced with kind permission [60].

the predictive capabilities of the rate-dependent elastic–plastic constitutive relation, therefore to avoid unwarranted complication, the creep strain increments in each step are measured directly from the experimental curves. (The model for the creep deformation will be discussed in section 6.9.) It can be seen from comparing Figs 6.36(a) and (b) that not only the effect of creep strain on the subsequent loading process, but also the monotonic prestrain history is well predicted by the constitutive model.

6.8.5 STRAIN-CONTROLLED CYCLIC TESTS WITH HOLD-TIME (STEP-UP RELAXATION TESTS)

A specimen of titanium alloy was subjected to 29 cycles at a strain amplitude $\varepsilon_a = 1.2\%$ with different strain-rates. Subsequently, 2 minute relaxation periods commencing at each arrow (Fig. 6.37(b)) followed by a strain-controlled loading at $|\dot{\varepsilon}| = 10^{-3}\,s^{-1}$, were applied (Kujawski and Krempl [61]. The experimental result is shown in Fig. 6.37(b). The uniaxial stress–strain curve of the material is taken from another paper by Krempl and Lu [76], Fig. 6.37(c), curve 0A. Figure 6.37(a) shows the model predictions for the step-up relaxation test program of Fig. 6.37(b). The predicted results are in very good agreement with the experimental data.

6.8.6 TRANSIENT HARDENING BEHAVIOUR FOR ALTERNATING AXIAL–TORSION CYCLING

In most previous examples of this section, the state of stress was uniaxial, and they were used to demonstrate the capability of the constitutive model to simulate various rate-dependent features of the material behaviour. The model, however, is applicable to arbitrary complex non-proportional multiaxial loading paths, including cyclic loading. Most transient hardening behaviour of materials, such as the strain amplitude hardening effect, cross-hardening effect, erasure of memory, etc. can be simulated by the model. In essence, the rate-dependent model extends greatly the range of applicability of the rate-independent model presented in section 6.5.

Multiaxial rate-dependent test data are very limited, and in the following we will describe a cross-hardening behaviour. Krempl and Lu [76] conducted experiments at room temperature on type 304 stainless steel by subjecting virgin specimens to one of the following loadings: (i) continuous axial cycling, (ii) continuous torsional cycling, and (iii) alternating axial–torsional cycling for one cycle each for a total of twenty cycles. After each cycle, the stress and strain were returned to zero. The tests were conducted under the strain-controlled mode with the equivalent strain amplitude $(\varepsilon_a)_{eq} = 0.004$ and constant equivalent strain-rate, $\dot{\varepsilon}_{eq} = 3 \times 10^{-4}\,s^{-1}$. The experimental results are shown in Fig. 6.38(b), where the

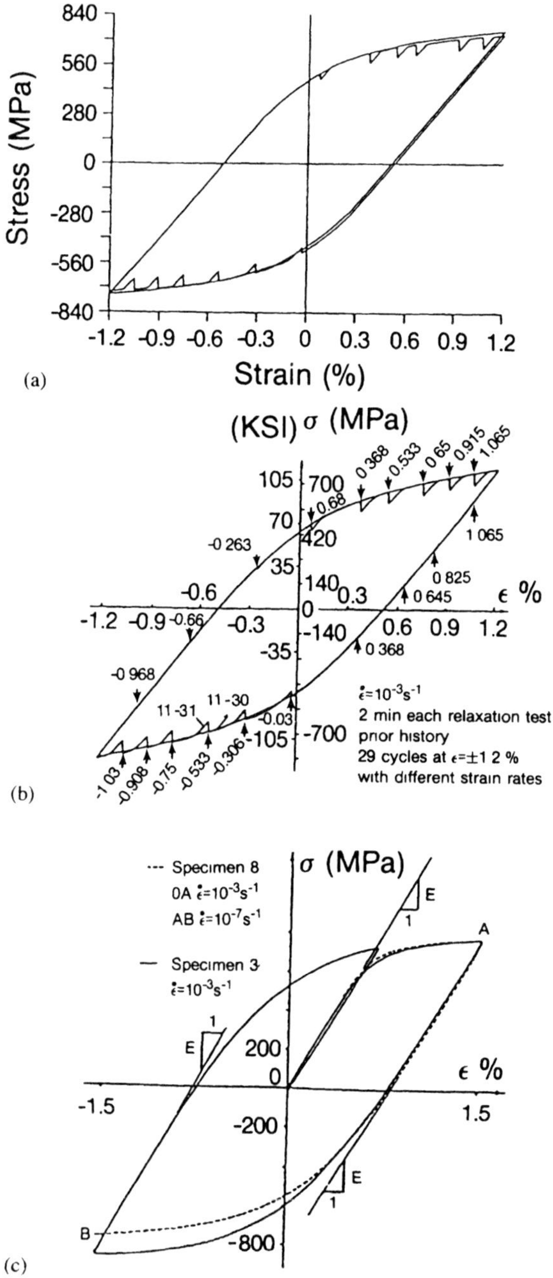

Fig. 6.37 Strain-controlled cyclic test with hold-time (step-up relaxation test) for Ti-7Al-2Cb-1Ta titanium alloy. Reproduced with permission of ASME from [51]. (a) predicted results; (b) experimental results from Kujawski and Krempl [61]; (c) uniaxial loading and unloading branches of annealed specimens, experimental results from Krempl and Lu [76].

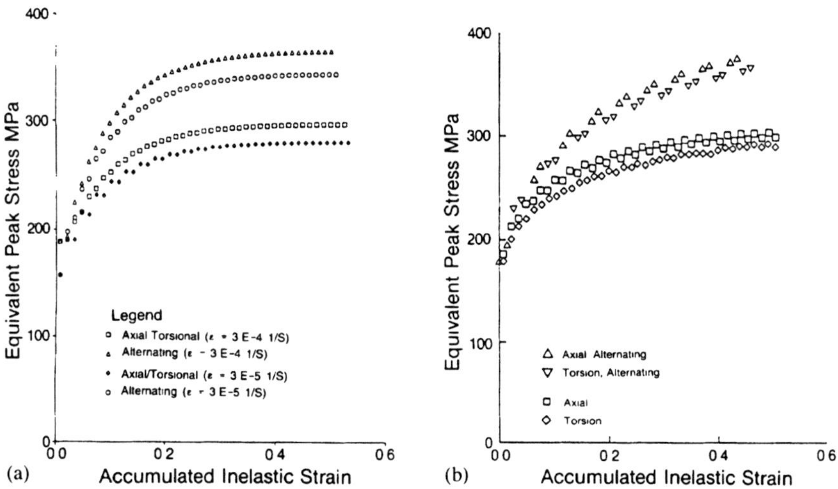

Fig. 6.38 The equivalent peak stress as a function of the accumulated inelastic strain for AISI type 304 stainless steel. Reproduced with permision of ASME from [51]. (a) predicted results; (b) experimental results from Krempl and Lu [76].

equivalent peak stress values increase with the increasing accumulated inelastic equivalent strain and tend towards an asymptotic value. The alternating axial–torsional cycling shows additional hardening in comparison to the other two modes of loading (Fig. 6.38(b)).

To predict the alternating axial–torsional cyclic loading, we note from (6.61) that $F = 0.67$ for this case, and $F = 0$ for the other two cases. The predicted results are depicted in Fig. 6.38(a). In the same figure results for a lower strain-rate, $\dot{\varepsilon}_{eq} = 3 \times 10^{-5}\,\mathrm{s}^{-1}$, are also shown, thus indicating the rate effect in a non-proportional cross-hardening case.

6.8.7 MATERIAL RESPONSE DUE TO CHANGE IN STRAIN TRAJECTORY AND STRAIN-RATE

Experimental data in this section are taken from Inoue *et al.* [68]. The material is SUS-304 stainless steel and tests were conducted at an elevated temperature of 650°C. In their paper a superposition model and three types of unified models were examined by comparing predicted results with the test data. A biaxial test result is shown in Fig. 6.39 where a proportional loading in a 45° direction was imposed in stage I, followed by stage II where the ratio of the strain-rate was $\dot{\varepsilon}/(\dot{\gamma}/\sqrt{3}) = 5$. Subsequently, the direction in stage III was changed to the same as in stage I. To investigate the effect of strain-rate, two kinds of equivalent strain-rates were selected as shown in the inset table in Fig. 6.39. The method of evaluating material constants is similar to that described at the beginning of this section. From

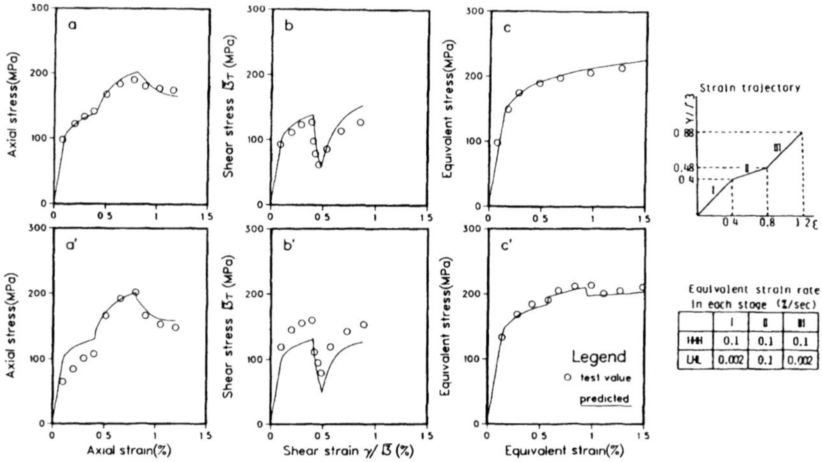

Fig. 6.39 Stress–strain response due to change in strain trajectory and strain rate (RHS inset). Reproduced with permission from ASME from [51]. Figures (a), (b) and (c) are for program HHH; and (a′), (b′) and (c′) are for program LHL. Experimental data from [68].

the uniaxial curves at different strain-rates and a uniaxial cyclic test we obtain the family of initial and stable uniaxial stress–strain curves. From these curves we obtain the elastic modulus, tangent modulus, E_t, and initial and stable radii of the yield surfaces. Other material constants are taken as: $\nu = 0.5$, $\xi = 40$, and $\eta_1 = \eta_2 = 0.65$. The predicted results of the model are shown in Fig. 6.39 by a full line. Comparing these results with the predictions of four other models presented in Fig. 6 of [68] it can be seen that the present results are in better agreement with the experimental data.

6.9 A CONSTITUTIVE MODEL FOR CREEP DEFORMATION INCLUDING PRIOR PLASTIC STRAIN EFFECTS

In section 6.7 we described a rate-dependent constitutive model, and the framework for the elastic–plastic–creep deformation process. In the two examples dealing with step-up creep and relaxation tests (sections 6.8.4 and 6.8.5) the creep strain was taken from the experimental curves. It is the objective of this section to develop a model whereby the creep deformation can be predicted, taking into account the influence of prior plastic or creep strain history.

Referring to an arbitrary hysteresis loop shown in Fig. 6.40, the creep strain, $\varepsilon^c = B'C'$, will be influenced by the prior plastic prestrain, $\varepsilon^p = A'B'$. Similarly, the reversed plastic flow C′D′ would be affected by the prior creep strain history, and so on. The strain-rate partition, (6.67), applies to

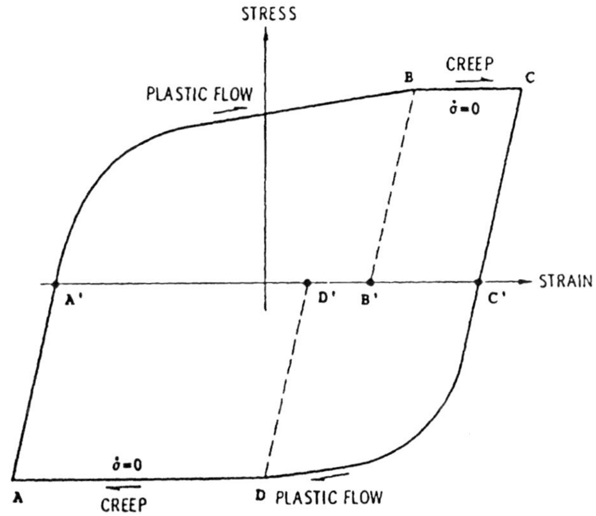

Fig. 6.40 A hysteresis loop depicting plastic and creep strains. Reproduced with permission from ASME from [77].

a situation shown in Fig. 6.40, and the constitutive relation

$$\dot{\sigma}_{ij} = L_{ijkl}\dot{\varepsilon}_{kl} \tag{6.76}$$

applies to loading paths AB and CD. In (6.76) L_{ijkl} is given by (6.42b), q by (6.71a) and E_t by (6.71b).

In this section we will concern ourselves with developing a constitutive model for creep deformation. The approach based on the viscoplastic theory, nowadays referred to as 'unified creep–plasticity' constitutive equations, has been discussed in section 6.7.1 and will not be dealt with further. It suffices to mention that a cross-section of this type of constitutive relation has been assembled in a monograph by Miller [78] which also contains a summary and critique by Bammann and Kreig [70]. There have also been attempts to formulate coupled constitutive equations [71–73]. In the following a creep constitutive model incorporating the effects of prior plastic strains and strain-rate will be presented. Thus, for loading sequences including alternating plastic and creep deformations, such as in Fig. 6.40, the two coupled models could be used to predict the overall material response.

6.9.1 DESCRIPTION OF THE CREEP MODEL

A typical uniaxial strain response with time is shown in Fig 6.41. There is an initial elastic–plastic strain due to prior loading, thereafter, the load level is kept constant and the strain–time response is comprised of three

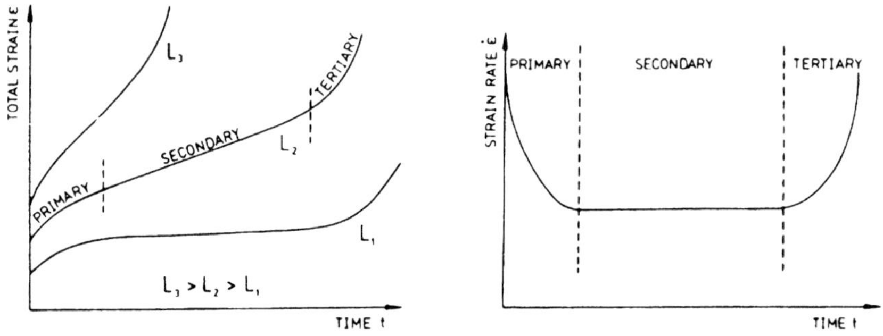

Fig. 6.41 Creep strain, ε^c, and strain-rate, $\dot{\varepsilon}^c$, curves as a function of time. Reproduced with permission from ASME from [77].

parts: a decreasing strain-rate portion called primary creep, a part of almost constant strain-rate termed secondary creep, and a final portion of increasing strain-rate named tertiary creep. Here we will be concerned with the primary and secondary creep stages which are of more interest in the design and operation of components in high-temperature environments.

For the general case of multiaxial stress states, the creep flow law is assumed to be of the form

$$\dot{\varepsilon}^c_{ij} = \lambda(s_{ij} - \alpha_{ij}), \tag{6.77}$$

where $\dot{\varepsilon}^c_{ij}$ is the creep strain-rate tensor, λ is a proportionality factor, and s_{ij} is the deviatoric stress tensor defined earlier (6.2), and α_{ij} is introduced as an internal variable which has been called 'internal stress', 'equilibrium stress', 'back stress' or 'rest stress', in the literature. During the creep process, s_{ij} is kept constant and α_{ij} evolves from an initial value to a saturated level, while creep develops from the primary to the secondary stage.

Let us now multiply (6.77) by itself to obtain

$$\lambda = \frac{3}{2}\frac{\dot{\varepsilon}^c_{eq}}{\bar{\sigma}_{eq}}, \tag{6.78}$$

where

$$\dot{\varepsilon}^c_{eq} = (\tfrac{2}{3}\dot{\varepsilon}^c_{ij}\dot{\varepsilon}^c_{ij})^{1/2} \quad \text{and} \quad \bar{\sigma}_{eq} = [\tfrac{3}{2}(s_{ij} - \alpha_{ij})(s_{ij} - \alpha_{ij})]^{1/2}. \tag{6.79}$$

If we assume a creep power law for the uniaxial stress state (a Norton type)* i.e.

$$\dot{\varepsilon}^c_{eq} = A(\bar{\sigma})^{1/n}_{eq}, \tag{6.80}$$

* Here we have used exponent n to conform with that used in the fatigue literature. It is the inverse of that used in the creep nomenclature.

then from (6.77), (6.78) and (6.80) we get

$$\dot{\varepsilon}^{c}_{ij} = \tfrac{3}{2}A(\bar{\sigma}_{eq})^{(1-n)/n}(s_{ij} - \alpha_{ij})\,. \tag{6.81}$$

The internal variable, α_{ij}, plays a key role in the present model and through its evolution rule, various phenomena observed during creep experiments can be simulated.

First let us use an exponential function to describe the evolution process of the internal stress, α_{ij}. The evolution of α_{ij} is given by

$$\dot{\alpha}_{ij} = \psi\left(\alpha_s \frac{s_{ij}}{\sigma_{eq}} - \alpha_{ij}\right)\dot{\bar{\varepsilon}}^{c}_{1}\,, \tag{6.82}$$

where $\sigma_{eq} = (3s_{ij}s_{ij}/2)^{1/2}$ is the equivalent stress and $\bar{\varepsilon}^{c}_{1}$ is a hardening parameter which we will define later. With the increase in $\bar{\varepsilon}^{c}_{1}$, α_{ij} evolves from an initial value to a saturated value, and the creep development from the primary to the secondary stage can be simulated. In the above equation, α_s is related to the saturated values of α_{ij}, and ψ influences the rate of the creep transition process from the primary to the secondary stage.

Figure 6.42 (taken from Kawai and Ohashi [71]) shows the results of creep tests at a uniaxial tensile stress of $\sigma_B = 140\,\text{MPa}$, subjected to prior plastic prestrain of $\varepsilon^p = 0.0028, 0.01, 0.02$ and 0.03, respectively. It can be clearly seen that the subsequent creep deformation decreases significantly with the increase in the plastic prestrain, and both the initial and saturated slopes of the creep curves have been influenced by the prestrain. On the other hand, the resistance to subsequent creep deformation is less for

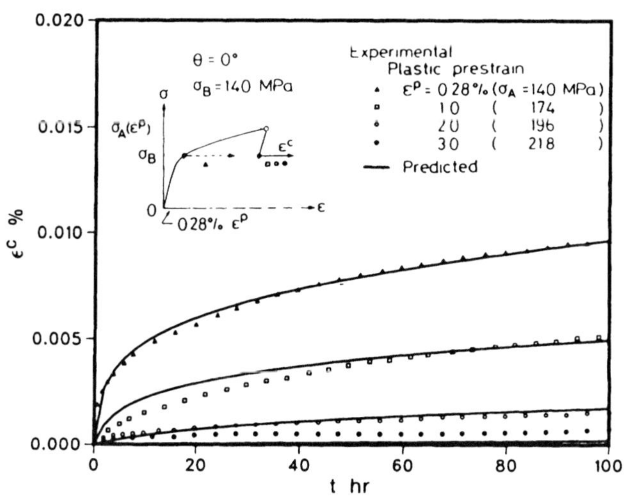

Fig. 6.42 Tensile creep following prior tensile plastic deformation, experimental data from [71] and predicted results from [77], reproduced with permission from ASME.

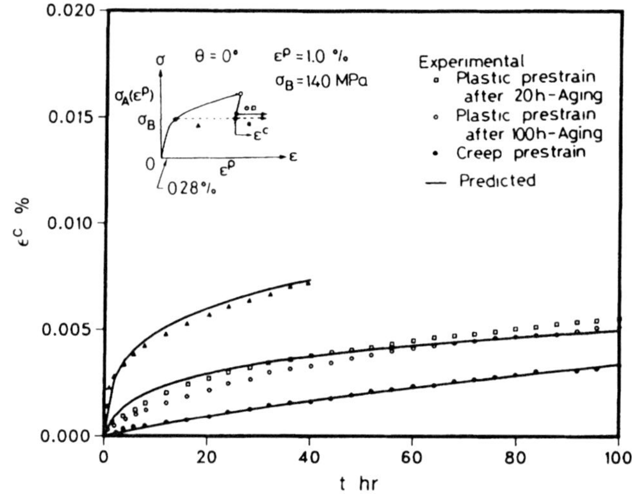

Fig. 6.43 Effect of prior creep and tensile plastic strain on the subsequent creep deformation, experimental data from [71] and predicted results from [77], reprinted with permission from ASME.

a plastic prestrain compared to the same amount of creep prestrain (Fig. 6.43).

Figure 6.44 (taken from Xia and Ellyin [63]) shows the results of another set of four carefully designed creep tests. The test sequences of the four tests are summarized in Table 6.1. Before switching to creep tests, the loading, unloading and reloading procedures were under strain-controlled mode. The stress level, pre-plastic strain and the total strain-rate just prior to the creep stage are almost the same (Table 6.1), but the creep curves (Fig. 6.44(b)) for the four cases are quite different. The difference is mainly caused by the different plastic strain-rates just prior to the creep stage. This can be shown from the different tangent modulus, E_t, at the instant before switching to the creep test, since we have the relation

$$\frac{\dot{\varepsilon}^{\mathrm{p}}}{\dot{\varepsilon}^{\mathrm{t}}} = 1 - \left(\frac{E_t}{E}\right). \tag{6.83}$$

The measured rate of E_t/E for the four cases ranges from 0.017 to 0.34 to 0.60 to 1.0 (Fig. 6.44(a)). From 6.43(b) one also can find that the four creep curves tend to become parallel when the creep time is greater than 500 minutes.

From the above observation, it follows that the saturated values of α_{ij} are related to the prior plastic strain, whereas the initial value of α_{ij} should be related to both the prior plastic strain and strain-rate. Integrating (6.82) we

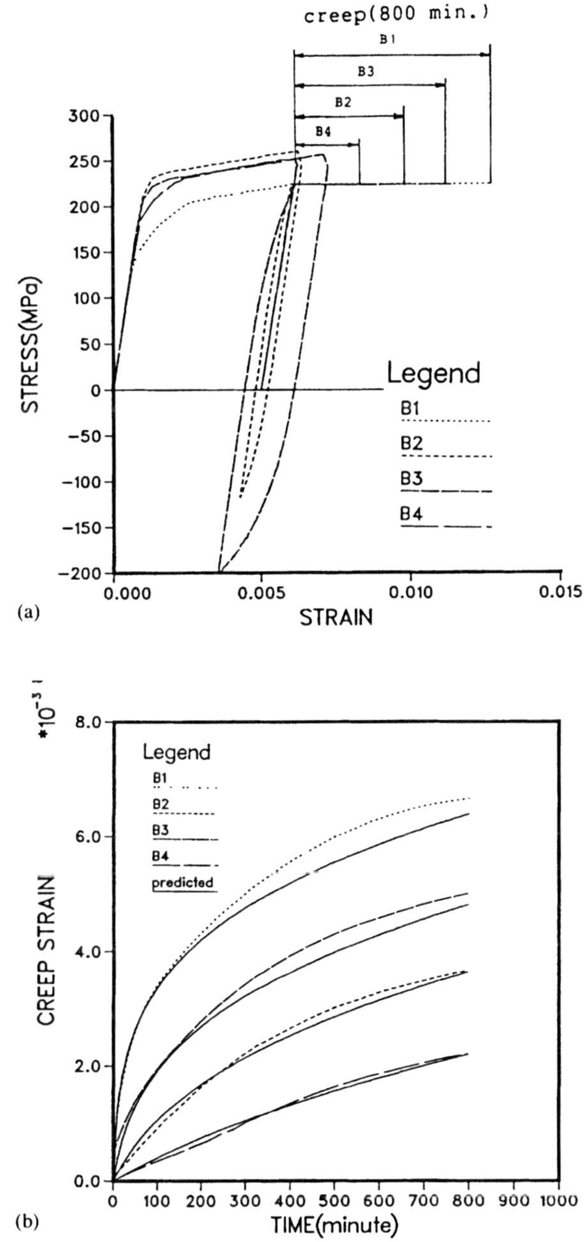

Fig. 6.44 Experiments and predicted results for tests B1–B4 from [63]. (a) stress–strain curves; (b) creep strain versus time curves.

Table 6.1 Test sequences and instantaneous conditions before switching to a creep test

Specimen	Loading rate (1/s)	Unloading rate (1/s)	Reloading rate (1/s)	Stress level (MPa)	Plastic prestrain	Tangent modulus E_t (MPa)[a]	E_t/E[a]
B1	4.0×10^{-6}	–	–	224	0.005	3.26×10^{3}	0.017
B2	3.0×10^{-4}	-4.0×10^{-6}	4.0×10^{-6}	224	0.005	1.95×10^{5}	1.0
B3	3.0×10^{-4}	-4.0×10^{-6}	4.0×10^{-6}	224	0.0049	1.17×10^{5}	0.60
B4	3.0×10^{-4}	-4.0×10^{-6}	4.0×10^{-6}	224	0.0051	6.63×10^{4}	0.34

[a] The values of E_t are obtained by measuring the slope of the stress–strain curves at the instant before switching to creep tests. The value of E is taken to be 1.95×10^{5} MPa.

get

$$\alpha_{ij} = \alpha_s \frac{S_{ij}}{\sigma_{eq}} \left[1 - C \exp(-\psi \bar{\varepsilon}_1^c) \right]. \tag{6.84}$$

We decompose α_s into two components, namely

$$\alpha_s = \alpha_s^p + \alpha_s^c , \tag{6.85}$$

and assume that

$$C = \exp(-\psi f_o^p) . \tag{6.86}$$

Thus α_s^p in (6.85) should be a function of the plastic prestrain, and f_o^p in (6.86) should be a function of the plastic prestrain and plastic prestrain-rate.

In the case of a biaxial stress condition (tension–torsion), Fig. 6.45 shows a comparison between subsequent creep–time responses in $\theta = 0$, 45, 90, 135 and 180 degree directions with respect to a common tensile plastic prestrain of $\varepsilon^p = 2\%$ at $\theta = 0°$(the experimental data are taken from Kawai and Ohashi [71]). It can be seen that similar to that of the uniaxial case, the plastic prestrain increases the resistance to the subsequent creep deformation in **all** directions. However, the intensity of resistance decreases with the increase in the directional angle from the prestrain path. It therefore follows that α_s^p and f_o^p depend on both the magnitude and direction of the plastic prestrain tensor, ε_{ij}^p.

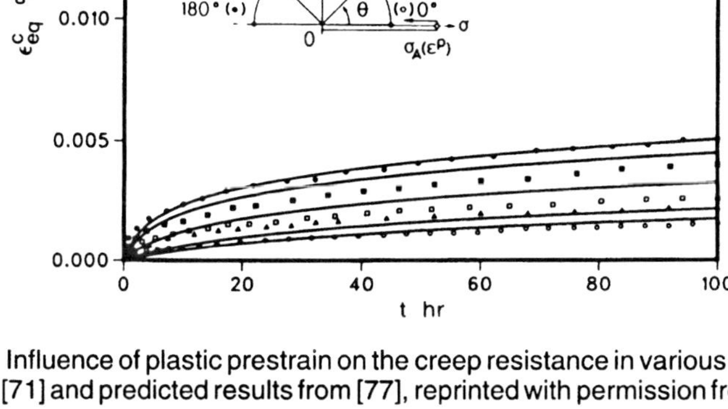

Fig. 6.45 Influence of plastic prestrain on the creep resistance in various directions, data from [71] and predicted results from [77], reprinted with permission from ASME.

Let us define an angle between the deviatoric stress tensor, s_{ij}, and the plastic strain tensor, $\varepsilon^{\mathrm{p}}_{ij}$, in the nine-dimensional space, namely

$$\cos\theta = \frac{s_{ij}\varepsilon^{\mathrm{p}}_{ij}}{(3s_{kl}s_{kl}/2)^{1/2}(2\varepsilon^{\mathrm{p}}_{mn}\varepsilon^{\mathrm{p}}_{mn}/3)^{1/2}} . \tag{6.87}$$

We now postulate, based on the experimental results discussed earlier, that

$$\alpha^{\mathrm{p}}_{\mathrm{s}} = \mu_1(1 + \nu_1\cos\theta)(\varepsilon^{\mathrm{p}}_{\mathrm{eq}})^{\gamma_1} , \tag{6.88}$$

and

$$f^{\mathrm{p}}_{\mathrm{o}} = \mu_2(1 + \nu_2\cos\theta)(\varepsilon^{\mathrm{p}}_{\mathrm{eq}})^{\gamma_2} F(\dot{\varepsilon}^{\mathrm{p}}_{ij}) . \tag{6.89}$$

In (6.89), F is a scalar function of $\dot{\varepsilon}^{\mathrm{p}}_{ij}$, i.e. it depends on both the magnitude and direction of the prior plastic strain-rate. From (6.88) and (6.89) we note that $\alpha^{\mathrm{p}}_{\mathrm{s}}$ and $f^{\mathrm{p}}_{\mathrm{o}}$ have their maximum values when the current stress tensor, s_{ij}, is coaxial with the plastic prestrain tensor, $\varepsilon^{\mathrm{p}}_{ij}$ ($\cos\theta = 1$) and the minimum values are obtained when the two tensors are in the opposite direction, i.e. $\cos\theta = -1$.

We now return to the definition of $\bar{\varepsilon}^{\mathrm{c}}_{\mathrm{I}}$ introduced in (6.82). Similar to most strain-hardening models in the literature, we may choose the instantaneous equivalent creep strain, $\varepsilon^{\mathrm{c}}_{\mathrm{eq}}$, or the accumulated equivalent creep strain, $l^{\mathrm{c}}_{\mathrm{eq}}$, as the hardening parameter, i.e.

$$\bar{\varepsilon}^{\mathrm{c}}_{\mathrm{I}} = \varepsilon^{\mathrm{c}}_{\mathrm{eq}} = \left(\frac{2\varepsilon^{\mathrm{c}}_{ij}\varepsilon^{\mathrm{c}}_{ij}}{3}\right)^{1/2} \tag{6.90}$$

or

$$\bar{\varepsilon}^{\mathrm{c}}_{\mathrm{I}} = l^{\mathrm{c}}_{\mathrm{eq}} = \int\left(\frac{2\dot{\varepsilon}^{\mathrm{c}}_{ij}\dot{\varepsilon}^{\mathrm{c}}_{ij}}{3}\right)^{1/2} \mathrm{d}t . \tag{6.91}$$

However, a strain-hardening theory cannot express the transient increase of the creep strain-rate observed just after the stress reversals or an abrupt rotation in the direction of the stress vector. To describe this transient feature of creep, we note the work carried out in the Oak Ridge National Laboratory (ORNL) in which the classical strain-hardening model was modified by an auxiliary procedure (Pugh and Robinson [72]). Later a creep hardening surface (CHS) model was proposed (Ohno *et al.* [79] where further references to this investigation are listed). We can incorporate these results directly into our model by letting

$$\bar{\varepsilon}^{\mathrm{c}}_{\mathrm{I}} = \bar{\varepsilon}^{\mathrm{H}} \qquad \text{(ORNL model)} \tag{6.92}$$

or

$$\bar{\varepsilon}^{\mathrm{c}}_{\mathrm{I}} = q \qquad \text{(CHS model)} . \tag{6.93}$$

The procedure to calculate $\bar{\varepsilon}^{\mathrm{H}}$ was described in the appendix of Pugh and Robinson's paper [72], whereas equations (2)–(6) in Ohno *et al.*'s paper [79] could be used to evaluate the parameter q. Our present model could also be used in such a way for complex creep sequences with large stress rotation or stress reversals (see example in section 6.9.3). It should be noted that for a simple creep case without rotation of the stress vector, all of the above four definitions of $\bar{\varepsilon}_{\mathrm{I}}^{\mathrm{c}}$ reduce to the equivalent creep strain, $\varepsilon_{\mathrm{eq}}^{\mathrm{c}}$.

6.9.2 BIAXIAL STRESS CONDITION

As a special case, we will develop (6.81)–(6.88) for a two-dimensional stress state arising from subjecting a thin-walled tubular specimen to an axial–torsional loading. For such a case, we have

$$\dot{\varepsilon}_x^{\mathrm{c}} = A(\sigma_x - \alpha_x)\,[(\sigma_x - \alpha_x)^2 + 3(\tau_{xy} - \alpha_{xy})^2]^{(1-n)/2n} \tag{6.94a}$$

$$\dot{\gamma}_{xy}^{\mathrm{c}} = 3A(\tau_{xy} - \alpha_{xy})\,[(\sigma_x - \alpha_x)^2 + 3(\tau_{xy} - \alpha_{xy})^2]^{(1-n)/2n} \tag{6.94b}$$

$$\alpha_x = (\alpha_{\mathrm{s}}^{\mathrm{p}} + \alpha_{\mathrm{s}}^{\mathrm{c}})\,\{1 - \exp[-\psi(f_{\mathrm{o}}^{\mathrm{p}} + \bar{\varepsilon}_{\mathrm{I}}^{\mathrm{c}})]\}\frac{\sigma_x}{\sigma_{\mathrm{eq}}} \tag{6.94c}$$

$$\alpha_{xy} = (\alpha_{\mathrm{s}}^{\mathrm{p}} + \alpha_{\mathrm{s}}^{\mathrm{c}})\,\{1 - \exp[-\psi(f_{\mathrm{o}}^{\mathrm{p}} + \varepsilon_{\mathrm{I}}^{\mathrm{c}})]\}\frac{\tau_{xy}}{\sigma_{\mathrm{eq}}} \tag{6.94d}$$

$$\alpha_{\mathrm{s}}^{\mathrm{p}} = \mu_1(1 + \nu_1\cos\theta)(\varepsilon_{\mathrm{eq}}^{\mathrm{p}})^{\gamma_1} \tag{6.94e}$$

$$f_{\mathrm{o}}^{\mathrm{p}} = \mu_2(1 + \nu_2\cos\theta)(\varepsilon_{\mathrm{eq}}^{\mathrm{p}})^{\gamma_2}F(\dot{\varepsilon}_x^{\mathrm{p}}, \dot{\gamma}_{xy}^{\mathrm{p}}) \tag{6.94f}$$

$$\cos\theta = \frac{\sigma_x\varepsilon_x^{\mathrm{p}} + \tau_{xy}\gamma_{xy}^{\mathrm{p}}}{\sigma_{\mathrm{eq}}\varepsilon_{\mathrm{eq}}^{\mathrm{p}}} \tag{6.94g}$$

$$\sigma_{\mathrm{eq}} = (\sigma_x^2 + 3\tau_{xy}^2)^{1/2} \tag{6.94h}$$

$$\varepsilon_{\mathrm{eq}}^{\mathrm{p}} = [(\varepsilon_x^{\mathrm{p}})^2 + \tfrac{1}{3}(\gamma_{xy}^{\mathrm{p}})^2]^{1/2}. \tag{6.94i}$$

6.9.3 CORRELATION WITH EXPERIMENTAL RESULTS

For the purpose of experimental correlation, we first choose the results reported by Kawai and Ohashi [71]. In their investigation, thin-walled circular cylindrical tubes made of type 316 stainless steel were tested at 650°C. Each specimen was subjected to a predetermined tensile plastic strain following which a constant creep test was carried out. For all tests (uniaxial or biaxial) the equivalent stress level was constant during the creep test. In the case of biaxial tests, the loading paths were proportional, i.e. after prestraining, the specimens were unloaded to $\sigma_{\mathrm{eq}} = 0$ and subsequently reloaded in constant proportions. Each test procedure is illustrated in the inset of the relevant figures. The proportional stressing from complete unloading state to the creep stress was performed at a constant stress rate $|\dot{\sigma}_{\mathrm{eq}}| = 80\,\mathrm{MPa\,min^{-1}}$. However, no detailed data were available to determine the value of the prior plastic strain-rate for all the tests. For simplicity, we assume that $F = 1$ in (6.94f). The constitutive model formulated here, (6.94), contains eight material constants A, n, ψ, α_s^c, μ_1, ν_1, γ_1, μ_2, ν_2, γ_2. However, they could be evaluated through several axial or torsional creep tests. First, the A, n, ψ, α_s^c could be determined from a uniaxial creep test at a stress level near the elastic limit ($\varepsilon_x^p \approx 0$). We can then perform one or two uniaxial creep tests with a specified plastic prestrain, $\varepsilon_x^p =$ constant to evaluate the values $\mu_1(1+\nu_1)$, γ_1, $\mu_2(1+\nu_2)$ and γ_2. Finally, through a torsional creep test with the same value of axial plastic prestrain, ν_1 and ν_2 could be separated from μ_1 and μ_2. For the stainless steel type 316, the material constants were found to be

$$A = 3.0 \times 10^{-15}\,\text{MPa}^{-60}\text{h}^{-1},\; n = 0.167,\; \psi = 250,\; \alpha_s^c = 90\,\text{MPa}\,,$$

$$\mu_1 = 498\,\text{MPa},\; \mu_2 = 0.039,\; \nu_1 = \nu_2 = 0.23,\; \gamma_1 = 0.70,\; \gamma_2 = 0.60\,.$$

Figure 6.42 compares the predicted results with the experimental data for a uniaxial constant stress ($\sigma_B = 140$ MPa) following four different plastic prestrains. It can be seen that the model predicts very closely the trend shown by the experiment regarding increased resistance to creep deformation with the increased plastic prestrain level.

The effect of the prior plastic and creep prestrain on the subsequent creep strain is shown in Fig. 6.43. The data depicted with open circles refers to the plastic prestrain $\varepsilon^p = 1\%$, while those with a filled circle are for 0.28% plastic prestrain plus 0.72% creep prestrain. The latter curve is, in fact, a continuation of the tensile creep curve shown with triangular symbols. We note again that the predicted results are in very good agreement with the experimental observations.

Figure 6.45 shows a comparison between the predicted and observed trend regarding decreased resistance to creep deformation as the increased angle between the loading and prestrain paths. The agreement between the experimental data and predicted results is indeed good.

Finally, the effect of the tensile plastic prestrain on the torsional creep is shown in Fig. 6.46. The experimental trend is well predicted by the model presented here.

Kawai and Ohashi [71] compared their results with the predictions of Miller's unified model [65] and their own coupled model. A comparison of Figs 1–4 in [65] with Figs 6.42, 6.43, 6.45 and 6.46 herein clearly shows that the model developed here predicts a much better agreement with the experimental results in both a quantitative and a quantitative sense [78].

Figure 6.47 shows the experimental results of creep strain under cyclic reversed torsion which is taken from Ohno *et al.* [79]. As described in section 6.9.1, we use the hardening parameter $\bar{\varepsilon}^H$ (ORNL model) or q (CHS model) for $\bar{\varepsilon}_i^c$ in (6.93c,d). The transient increase in the creep strain rate after an abrupt rotation of stress vector is well simulated by the model (Fig. 6.47).

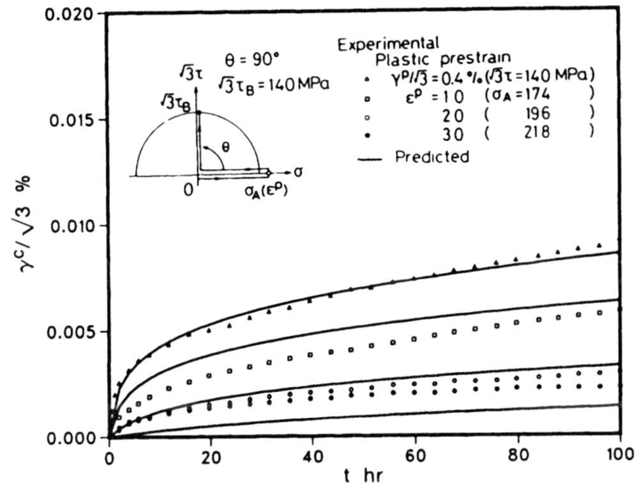

Fig. 6.46 Effect of tensile plastic prestrain on the torsional creep, test data from [68] and predicted results from [77], reprinted with permission from ASME.

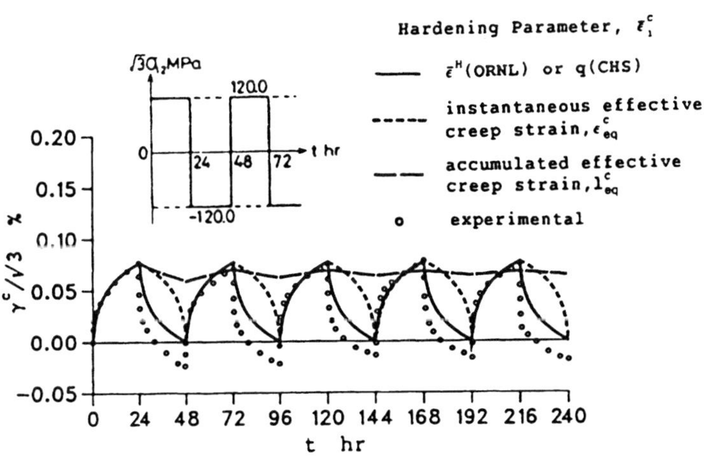

Fig. 6.47 Creep strain under cyclic reversed torsion, test data from [79] and predicted results from [77], reprinted with permission from ASME.

The material constants used in this calculation are: $A = 8.5 \times 10^{-17}\,\mathrm{MPa}^{-60}\,\mathrm{h}^{-1}$, $n = 0.167$, $\psi = 1050$, $\alpha_s^c = 90\,\mathrm{MPa}$ and the plastic prestrain ε^P is assumed to be zero. It is worth recalling that the q in the CHS model reduces to the $\bar{\varepsilon}^H$ of the ORNL model in the special case of uniaxial creep. For the purposes of comparison, the predicted results of (6.91) and (6.92), when using the instantaneous equivalent creep strain ε^c_{eq} or the accumulated equivalent creep strain l^{c}_{eq}, are also shown in Fig. 6.47. Obviously, neither of the latter two can properly predict the experimental observations.

We now turn to the experimental results reported by Xia and Ellyin [63] (Fig. 6.44) in which the data were provided to evaluate the effect of prior plastic strain-rate. For the uniaxial case, and especially for the tests B1–B4, which have the same stress level and plastic prestrain value, (6.94) can be reduced to

$$\dot{\varepsilon}^{c} = A(\sigma - \alpha)^{1/n} \tag{6.95a}$$

$$\alpha = \alpha_{s}\{1 - \exp[-\psi(f^{p}_{o} + \varepsilon^{c})]\} \tag{6.95b}$$

$$f^{p}_{o} = \mu_{2}(\varepsilon^{p})^{\gamma_2} F(\dot{\varepsilon}^{p})\,. \tag{6.95c}$$

It was indicated that the ratio of E_{t}/E is related to the plastic strain-rate through (6.83). Therefore, it may be more convenient to introduce this ratio in F instead of $\dot{\varepsilon}_{p}$. We choose a very simple functional relation for f^{p}_{o} of the form

$$f^{p}_{o} = \varepsilon^{p}_{eq}\frac{E_{t}}{E}\,. \tag{6.96}$$

Other constants are taken to have the following values:

$$n = 0.185,\ A = 7.0 \times 10^{-17}\,\mathrm{MPa}^{-5.4}\,\mathrm{Min}^{-1},\ \alpha_{s} = 150\,\mathrm{MPa},\ \psi = 350$$

The predicted creep curves are also given in Fig. 6.44(b) as solid lines. It can be seen that the agreement is quite good. For the multiaxial creep cases, it can be assumed that F is a function of ε^{p}_{eq}, E_{t}/E and θ_{1}, where θ_{1} is the angle between the vector of plastic strain-rate and stress,

$$F = F\left(\varepsilon^{p}_{eq}, \frac{E_{t}}{E}, \theta_{1}\right). \tag{6.97}$$

More test data under multiaxial creep conditions are required to obtain an explicit functional relation of (6.97).

6.10 CONCLUDING REMARKS

In this chapter we have shown that metals have a rather complex response to non-proportional loading in general, and cyclic loading in particular. For example, an extra hardening effect is observed which varies with the extent of non-proportionality. This extra hardening causes considerably more material damage in comparison to the same equivalent strain magnitude in a proportional path cyclic loading. In attempting to predict damage, a primary prerequisite is a constitutive model which is based on the underlying deformation mechanism.

Here we have presented a hierarchy of constitutive models each applicable to a class of problems and building on one another to achieve a broader class of applications. The experimental basis of the incremental elastic–plastic nonlinear hardening model is described in sections 6.3 and 6.4. A rate-independent constitutive model for transient response is then described in section 6.5. The effects of strain history, strain range and non-proportional path have been discussed and the predictions of the constitutive model are compared to a variety of experimental data. The constitutive model is subsequently extended to include rate effects which are generally observed in high temperature environments (section 6.7). The rate-dependent constitutive model simulates most of the rate-dependent features of a material response, among them rate sensitivity, abrupt changes in strain-rate, step-up creep or relaxation (hold-time) effects, mechanical ratcheting behaviour, rate history effect, etc. The effect of the precreep on the subsequent plastic deformation is also considered. The constitutive model for creep deformation is treated in section 6.9, which includes the effect of prior plastic straining and prior plastic strain-rate. The constitutive models presented in sections 6.7 and 6.9 form a coupled system capable of predicting the material response to complex loading sequences, including alternative plastic and creep or relaxation processes.

Finally, the constitutive model described here has been implemented into a commercial finite element code in a user-defined subroutine. This is summarized in [50] and has been used to predict stress/strain fields in front of a crack subject to overloads and crack closure development, etc. (Chapters 7 and 8). Furthermore, the constitutive model has been used to determine the damage parameter for the multiaxial fatigue described in Chapter 4.

REFERENCES

1. Taylor, G.I. (1938) Plastic strain in metals. *J. Inst. Metals*, **62**, 307–24.
2. Bishop, J.F.W. and Hill, R. (1951) A theoretical derivation of the plastic properties of a polycrystalline face-centered metal. *Phil. Mag.*, **42**, 414–27.
3. Lin, T.H. (1971) Physical theory of plasticiry, in *Advances in Applied Mechanics*, Vol. 11, pp. 255–311.
4. Hill, R. and Rice, J.R. (1972) Constitutive analysis of elastic-plastic crystals at arbitrary strain. *J. Mech. Phys. Solids*, **20**, 401–13.
5. Havner, K.S. and Shalaby, A.H. (1977) A single mathematical theory of finite distortional latent hardening in single crystals. *Proc. R. Soc. London* A **358**, 47–70.
6. Asaro, R.J. (1983) Crystals plasticity. *J. Appl. Mech., Trans. ASME*, **50**, 921–34.
7. Kocks, U.F. (1987) Constitutive behaviour based on crystal plasticity, in *Unified Constitutive Equations for Creep and Plasticity* (ed. A.K. Miller), Elsevier, London, pp. 1–88.

8. Toth, L.S., Jonas, J.J. and Neale, K.W. (1990) Comparison of the minimum plastic spin and rate sensitive slip theories for loading of symmetrical crystal orientation. *Proc. R. Soc. London* A **427**, 201–19.
9. Kestin, J. and Rice, J.R. (1970) Paradoxes in the application of thermodynamics to strained solids, in *A Critical Review of Thermodynamics* (eds E.B. Stuart *et al.*), Nomo Book Corp., Baltimore, pp. 275–98.
10. Rice, J.R. (1975) Continuum mechanics and thermodynamics of plasticity in relation to microscale deformation mechanisms, in *Constitutive Equations in Plasticity* (ed. A.S. Argon), MIT Press, Cambridge, MA, pp. 23–79.
11. Lehmann, Th. (ed.) (1984) *The Constitutive Law in Thermoplasticity* (CISM Course and Lecture No. 281), Springer, New York.
12. Germain, P., Nguyen, Q.S. and Suquet, P. (1983) Continuum thermodynamics. *J. Appl. Mech., Trans. ASME,* **50**, pp. 1010–20.
13. Hill, R. (1950) *The Mathematical Theory of Plasticity,* Oxford University Press, Oxford.
14. Prager, W. (1956) A new method of analysing stress and strain in work hardening plastic solids. *J. Appl. Mech., Trans. ASME,* **23**, 493–6.
15. Mróz, Z. (1973) *Mathematical Models of Inelastic Material Behaviour,* Solid Mechanics Division, University of Waterloo, Waterloo, Ontario.
16. Ziegler, H. (1959) A modification of Prager's Hardening Rule. *Quart. Appl. Math.,* **17**, 55–65.
17. Mróz, Z. (1967) On the description of anisotropic work hardening. *J. Mech. Phys. Solids,* **15**, 163–75.
18. Krieg, R.D. (1975) A practical two surface plasticity theory. *J. Appl. Mech., Trans. ASME,* **42**, 641–6.
19. Dafalias, Y. and Popov, E.P. (1975) A model for nonlinear hardening materials for complex loading. *Acta Mech.,* **21**, 173–92.
20. Desai, C.S. and Siriwardane, H.J. (1984) *Constitutive Laws for Engineering Materials with Emphasis on Geologic Materials,* Prentice-Hall, Englewood Cliffs, NJ.
21. Kachanov, L.M. (1971) *Foundation of the Theory of Plasticity,* North-Holland, Amsterdam.
22. Haythornthwaite, R.M. (1968) A more rational approach to strain hardening data, in *Engineering Plasticity* (eds J. Heyman and F.A. Leckie), Cambridge University Press, Cambridge, pp. 201–18.
23. Ellyin, F. and Grass, J.P. (1975) Détermination éxperimentale de surface d'écoulement plastique du titane 50A, Parts I and II, *Trans. CSME,* **3**, 156–69.
24. Phillips, A., Liu, C.S. and Justusson, J.W. (1972) An experimental investigation of yield surfaces at elevated temperatures. *Acta Mech.,* **14**, 119–46.
25. Paul, B. (1968) Macroscopic criteria for plastic flow and brittle fracture, in *Fracture,* vol. 2 (ed. H. Liebowitz), Mathematical Foundation, Academic Press, New York, pp. 313–496.
26. Ikegami, K. (1975) A historical perspective of the experimental study of subsequent yield surfaces for metal, Parts 1 and 2. *Japan Soc. Mater. Sci.,* **24**, 491–505 and 709–19 (BISI Trans. 14420, Sept. 1976).
27. Hecker, S.S. (1976) Experimental studies of yield phenomena in biaxially loaded metals, in *Constitutive Equations in Viscoplasticity: Computational and Engineering Aspects,* AMD, Vol. 20, American Society of Mechanical Engineers, New York, pp. 1–32.
28. Phillips, A. (1986) A review of quasistatic experimental plasticity and viscoplasticity. *Int. J. Plasticity,* **2**, 315–28.

29. Ellyin, F. (1982) *Inelastic Behaviour and Subsequent Yield Surfaces for Titanium Alloy under Multiaxial Stress States*. Presented at the 9th US National Congress of Applied Mechanics, June 21–25, 1982, Ithaca, New York.
30. Ellyin, F. (1993) On the concept of initial and subsequent yield loci, in *Failure Criteria of Structured Media* (ed. J.P. Boehler), A.A. Balkema, Rotterdam, pp. 293–304.
31. Tseng, N.T. and Lee, G.C. (1983) Simple plasticity model of the two surface type. *J. Engng. Mech., Trans. ASME*, **109**, 795–810.
32. Phillips, A. and Lee, C.W. (1979) Yield surfaces and loading surfaces, experiments and recommendations. *Int. J. Solids Struct.*, **15**, 715–29.
33. Ellyin, F. and Wu, J. (1987) A description of elastoplastic behaviour based on memory and yield surface in plane stress state, in *Constitutive Laws for Engineering Materials – Theory and Application* (eds C.S. Desai *et al.*), Elsevier, New York, pp. 531–8.
34. Ellyin, F. (1989) An anisotropic hardening rule for elastoplastic solids based on experimental observations. *J. Appl. Mech., Trans. ASME*, **56**, 499–507.
35. Ellyin, F. and Xia, Z. (1990) Elastoplastic stress–strain relation based on a new anisotropic hardening model, in *Yielding, Damage and Failure of Anisotropic Solids* (ed. J.P. Boehler), Mechanical Engineering Publications Ltd, London, pp. 155–169.
36. Ellyin, F. (1988) *An Elastoplastic Constitutive Model with Nonlinear Anisotropic Hardening*. Proc. Int. Seminar, The Inelastic Behaviour of Solids, Models and Utilizations, Besançon, France, August 1988, pp. 115–126.
37. Eisenberg, M.A. (1976) A generalization of plastic flow theory with application to cyclic hardening and softening phenomena. *J. Engng. Mater. Technol., Trans. ASME*, **98**, 221–8.
38. Chaboche, J.L., Dang-Van, K. and Gordier, G. (1979) *Modelization of the Strain Memory Effect on the Cyclic Hardening of 316 Stainless Steel*. Proc. 5th SMiRT, Paper L11/3, 1979.
39. Drucker, D.C. and Palgen, L. (1981) On stress–strain relations suitable for cyclic and other loading. *J. Appl. Mech., Trans. ASME*, **48**, 479–85.
40. Mróz, Z. (1967) An attempt to describe the behaviour of metals under cyclic loads using a more general work hardening model. *Acta Mech.*, **7**, 199–212.
41. Dafalias, Y.F. (1984) Modelling cyclic plasticity: Simplicity versus sophistication, in *Mechanics of Engineering Materials* (eds C.S. Desai and R.H. Gallagher), Wiley & Sons, New York, pp. 153–78.
42. Ohno, N. (1982) A constitutive model of cyclic plasticity with a non-hardening strain region. *J. Appl. Mech., Trans. ASME*, **49**, 721–7.
43. McDowell, D.L. (1985) A two surface model for transient nonproportional cyclic plasticity, Parts 1 and 2, *J. Appl. Mech., Trans. ASME*, **52**, 298–308.
44. Tanaka, E., Murakami, S. and Ooka, M. (1985) A Constitutive Model of Cyclic Plasticity in Multiaxial Non-Proportional Loading. Trans. 8th SMiRT, Berlin, Germany, Paper L2/4 (ed. J. Stalpaert), North-Holland, pp. 71–7.
45. Ellyin, F. and Xia, Z. (1989) A rate-independent constitutive model for transient non-proportional loading. *J. Mech. Phys. Solids*, **37**, Elsevier Science Ltd, Oxford, 71–91.
46. Lamba, H.S. and Sidebottom, O.M. (1978) Cyclic plasticity for non-proportional paths, Parts 1 and 2, *J. Engng. Mater. Technol., Trans. ASME*, **100**, 96–111.
47. Krempl, E. and Lu, H. (1984) The hardening and rate-dependent behaviour of fully annealed AISI type 304 stainless steel under biaxial in-phase and out-of-

phase strain cycling at room temperature. *J. Engng. Mater. Technol., Trans. ASME*, **106**, 376–82.
48. Nouaihas, D., Policella, H. and Kaczmarek, H. (1983) On the description of cyclic hardening under complex loading histories, in *Constitutive Laws for Engineering Materials* (eds C.S. Desai and R.H. Gallagher), Elsevier, New York, pp. 45–9.
49. Rousset, M. and Marquis, D. (1985) Sur la déformation des surface seuils en plasticité cyclic. *Comptes Rendus de l'Academie des Sciences*, Serie II, **301**, 751–4.
50. Ellyin, F., Xia, Z. and Wu, J. (1995) A new elastoplastic constitutive model inserted into user-supplied material model of ADINA. *J. Computers Struct.*, **10**, Elsevier Science Ltd, Oxford 283–94.
51. Ellyin, F. and Xia, Z. (1991) A rate-dependent inelastic constitutive model, Part I: Elastic-plastic flow. *J. Engng. Mater. Technol., Trans ASME*, **113**, 314–23.
52. Ellyin, F. and Neale, K.W. (1979) Effect of cyclic loading on the yield surface. *J. Pressure Vessel Technol., Trans. ASME*, **101**, 59–63.
53. Xia, Z. and Ellyin, F. (1991) Nonproportional multiaxial cyclic loading – Experiments and constitutive modelling. *J. Appl. Mech., Trans. ASME*, **58**, 317–25.
54. Schubert, R. (1989) *Verformungsverhalten und Rißwachstum bie Low Cycle Fatigue*, Fortchr.-Ber. VDI 18 Nr.73, VDI-Verlag, Dusseldorf.
55. Brown, M.W. and Miller, K.J. (1979) Biaxial cyclic deformation behaviour of steels. *Fatigue Fract. Engng. Mater. Struct.*, **1**, 93–106.
56. Ellyin, F., Lefebvre, D. and Neale, K.W. (1980) High-strain biaxial fatigue of 2024-T351 aluminum under combined axial stress and torsion, in *Advances in Materials Technology in the Americas–1980, Vol. 2, Material Processing and Performance*, American Society of Mechanical Engineers, New York, pp. 17–21.
57. Inoue, T., Ohno, N. *et al.* (1989) Evaluation of inelastic constitutive models under plasticity–creep interaction for $2\frac{1}{4}$Cr-1 Mo steel at 600°C. *J. Nucl. Engng. Design*, **114**, 295–309.
58. Xia, Z. and Ellyin, F. (1994) Biaxial ratcheting under strain- or stress-controlled axial cycling with constant hoop stress. *J. Appl. Mech., Trans. ASME*, **61**, 422–8.
59. Krempl, E. (1979) An experimental study of room-temperature rate-sensitivity, creep and relaxation of AISI type 304 stainless steel. *J. Mech. Phys. Solids*, **27**, Elsevier Science Ltd, Oxford, 363–75.
60. Kujawski, D., Kallianpur, V. and Krempl, E. (1980) Uniaxial creep, cyclic creep and relaxation of AISI type 304 stainless steel at room temperature. *J. Mech. Phys. Solids*, **28**, Elsevier Science Ltd, Oxford, 129–48.
61. Kujawski, D. and Krempl, E. (1981) The rate (time)-dependent behaviour of Ti-7Al-2Cb-1Ta titanium alloy at room temperature under quasi-static monotonic and cyclic loading. *J. Appl. Mech., Trans. ASME*, **48**, 55–63.
62. Ikegami, K. and Ni-Itsu, Y. (1983) *Experimental Evaluation of the Interaction Effect Between Plastic and Creep Deformation*. Presented at Plasticity Today Symposium, Udine, Italy, 27–30 June, 1983.
63. Xia, Z. and Ellyin, F. (1993) An experimental study on the effect of prior plastic straining on creep behaviour of 304 stainless steel. *J. Engng. Mater. Technol., Trans. ASME*, **115**, 200–3.
64. Hart, E.W. (1976) Constitutive relations for non-elastic deformations of metals. *J. Engng. Mater. Technol., Trans. ASME*, **98**, 193–202.
65. Miller, A.K. (1976) An inelastic constitutive model for monotonic, cyclic and creep deformation: Parts I & II – Equations development and analytical procedures. *J. Engng. Mater. Technol., Trans. ASME*, **99**, 97–105.

66. Kreig, R.D., Swearengen, J.C. and Rode, R.W. (1978) A physically-based internal variable model for rate-dependent plasticity, in *Inelastic Behaviour of Pressure Vessel and Piping Components*, PVP-PB-028 (eds T.Y. Chang and E. Krempl), American Society of Mechanical Engineers, New York, pp. 15–28.
67. Chaboche, J.L. and Rousselier, G. (1983) On the plastic and viscoplastic constitutive equations, Parts I and II, *J. Pressure Vessel Technol., Trans. ASME*, **105,** 153–64.
68. Inoue, T. *et al.* (1985) Inelastic behaviour of $2\frac{1}{4}$ Cr-1 Mo steel under plasticity–creep interaction condition. *J. Nucl. Engng. Design*, **90**, 287–97.
69. Chan, K.S., Lindholm, U.S., Bodner, S.R. and Walker, K.P. (1984) *A Survey of Unified Constitutive Theories.* Proc. 2nd Symp. on Nonlinear Constitutive Relations for High Temperature Applications, Cleveland, OH, pp. 1–23.
70. Bammann, D.J. and Kreig, R.D. (1987) Summary and critique, in *Unified Constitutive Equations for Creep and Plasticity* (ed. A.K. Miller), Elsevier, London, pp. 303–36.
71. Kawai, M. and Ohashi, Y. (1986) *Creep-Plasticity Interaction of Austenitic Stainless Steel at Elevated Temperature.* Proc. Int. Conf. on Creep, April 14–18, Tokyo, 1986, pp. 459–64.
72. Pugh, C.E. and Robinson, D.N. (1978) Some trends in constitutive equation model development for high-temperature behaviour of fast-reactor structural alloys. *J. Nucl. Engng. Design*, **48,** 269–76.
73. Pugh, C.E. (1975) Constitutive equations for creep analysis of LMFBR components, in *Advances in Design for Elevated Temperature Environment*, ASME Publ. No. G00092, American Society of Mechanical Engineers, New York, pp. 1–15.
74. Drysdale, W.H. (1984) *An Extension of the Theory of Plasticity for Rate Effects.* Presented at 3rd Conf. on Mechanical Properties at High Rate of Strain, Oxford, England.
75. Ellyin, F., Xia, Z. and Sasaki, K. (1993) Effect of rate and rate history on plastic deformation: Experiments and constitutive modelling. *Int. J. Plasticity*, **9**, Elsevier Science Ltd, Oxford, 951–9.
76. Krempl, E. and Lu, H. (1989) The path and amplitude dependence of cyclic hardening of type 304 stainless steel at room temperature, in *Biaxial and Multiaxial Fatigue*, EGF3 (eds M.W. Brown and K.J. Miller), Mechanical Engineering Publications, London, pp. 89–106.
77. Xia, Z. and Ellyin, F. (1991) A rate-dependent inelastic constitutive model, Part II: Creep deformation including prior plastic strain effects. *J. Engng Mater. Technol., Trans. ASME*, **113**, 324–8.
78. Miller, A.K. (ed.) (1987) *Unified Constitutive Equations for Creep and Plasticity*, Elsevier, London.
79. Ohno, N., Murakami, S. and Ueno, T. (1985) A constitutive model of creep describing creep recovery and material softening caused by stress reversals. *J. Engng. Mater. Technol., Trans. ASME*, **107**, 1–6.

Fatigue crack growth 7

7.1 INTRODUCTION

In this chapter we will first present a synopsis of the theoretical foundation of linear and nonlinear fracture mechanics. In recent years a number of monographs, textbooks and conference proceedings have been published on the subject of fracture mechanics (see for example [1–10] where further references are listed). Our objective is not to present a review of these works, but to provide sufficient background so that the treatment of crack growth – the main objective of this chapter – can be followed with relative ease. In pursuing this objective we have been heavily influenced by general articles and book chapters by Rice [11, 12] and Hutchinson [13, 14]. In particular, Hutchinson's booklet [15] provides a remarkably concise but full treatment of nonlinear fracture mechanics. Since some aspects of nonlinear fracture mechanics are a straightforward generalization of linear elastic fracture mechanics (LEFM) we will begin this chapter with a brief summary of the foundation of LEFM. No attempt is made to give either a historical account of earlier contributions, or to provide a guide to the existing comprehensive literature. (An interested reader could consult some of the major works such as Liebowitz's Advanced Treatise on Fracture in seven volumes [1], or more recent monographs and review articles [4, 8–10].)

In the following we begin with a brief exposé of the procedures used to determine crack tip stress, strain and displacement fields for the linear and nonlinear material descriptions. To keep the preliminary part to a manageable size, we will concern ourselves with the mathematical foundation of fracture mechanics to derive the crack tip fields. References will be given for the experimental and micromechanical justifications, but without an in-depth discussion.

7.2 LINEAR ELASTIC FRACTURE MECHANICS

7.2.1 DESCRIPTION OF STRESS, STRAIN AND DEFORMATION IN CRACKED BODIES

In the case of two-dimensional problems (plane strain or plane stress), Muskhelishvili [16] has shown that for in-plane loading (Fig. 7.1) the stresses and displacements are given by

$$\begin{aligned} \sigma_x + \sigma_y &= \sigma_r + \sigma_\theta = 4\,\mathrm{Re}\,[\phi(z)] \\ \sigma_y - \sigma_x + 2\mathrm{i}\tau_{xy} &= (\sigma_\theta - \sigma_r + 2\mathrm{i}\tau_{r\theta})\exp(-2\mathrm{i}\theta) = 2[\bar{z}\phi'(z) + \psi(z)] \\ u_x + \mathrm{i}u_y &= (u_r + \mathrm{i}u_\theta)\exp(\mathrm{i}\theta) = \frac{1}{2G}\left[k\int\phi(z)\,\mathrm{d}z - z\overline{\phi(z)} - \int\overline{\psi(z)}\,\mathrm{d}\bar{z}\right. \end{aligned} \tag{7.1}$$

where x and y refer to the cartesian coordinate system and r, θ to the polar coordinate system. The functions $\phi(z)$ and $\psi(z)$ are analytical functions of the complex variable $z = x + \mathrm{i}y$, a bar over a quantity denotes its complex conjugate, and 'Re' means the real part of. (Note on the notation, it is also customary to use the derivative function $\psi'(z) = \phi(z)$ and $\chi'(z) = \psi(z)$ in

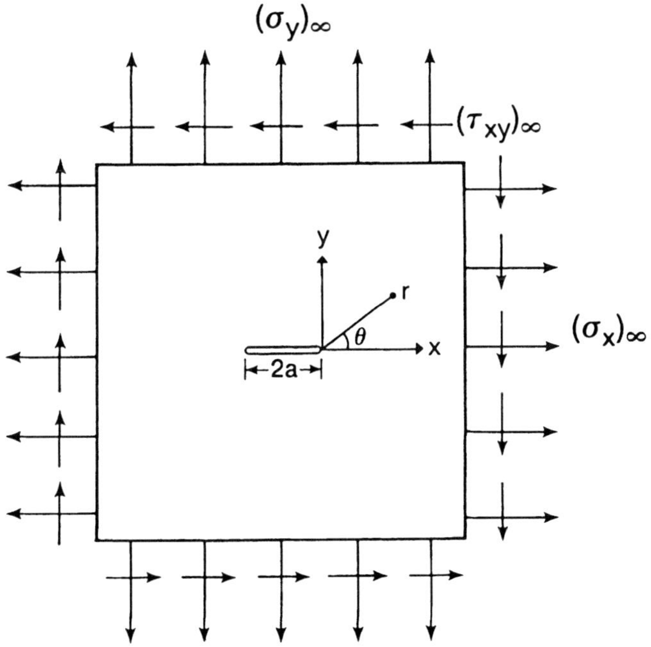

Fig. 7.1 A cracked infinite body subject to in-plane stresses and the coordinate systems.

(7.1).) The shear modulus $G = E/[2(1+\nu)]$, where E is Young's modulus and ν is Poisson's ratio, $k = (3-4\nu)$ in plane strain and $k = (3-\nu)/(1+\nu)$ for the generalized plane stress. Two plane conditions are defined by

$$\begin{aligned} &\sigma_z = \tau_{zx} = \tau_{zy} = 0\,, && \text{for plane stress,} \\ &\sigma_z = \nu(\sigma_x + \sigma_y)\,,\quad \tau_{zx} = \tau_{zy} = 0\,, && \text{for plane strain.} \end{aligned} \tag{7.2}$$

In the case of anti-plane loading (Fig. 7.2)

$$\begin{aligned} &\tau_{yz} + \mathrm{i}\tau_{xz} = (\tau_{\theta z} + \mathrm{i}\tau_{rz})\exp(-\mathrm{i}\theta) = \Omega(z), \\ &u_z = \frac{1}{G}\,\mathrm{Im}\left[\int \Omega(z)\,\mathrm{d}z\right], \end{aligned} \tag{7.3}$$

where 'Im' means imaginary part of.

For Figs 7.1 and 7.2, when the applied remote stresses are uniformly distributed, the complex stress potentials are given by (Rice [11])

$$\begin{aligned} &\phi(z) = \tfrac{1}{2}[(\sigma_y)_\infty - \mathrm{i}(\tau_{xy})_\infty][z^{-1/2}(z+2a)^{-1/2}(z+a-1)] + \tfrac{1}{4}[(\sigma_x)_\infty + (\sigma_y)_\infty] \\ &\psi(z) = \bar{\phi}(z) - \phi(z) - z\phi'(z) + \tfrac{1}{2}[(\sigma_y)_\infty - (\sigma_x)_\infty + \mathrm{i}(\tau_{xy})_\infty] \\ &\Omega(z) = (\tau_{yz})_\infty z^{-1/2}(z+2a)^{-1/2}(z+a) + \mathrm{i}(\tau_{xy})_\infty \end{aligned} \tag{7.4}$$

where $\bar{\phi}(z) = \overline{\phi(\bar{z})}$ and the branch cut of $z^{-1/2}(z+2a)^{-1/2}$ is chosen along the crack line so that the combination behaves as z^{-1} for large z. (For a crack

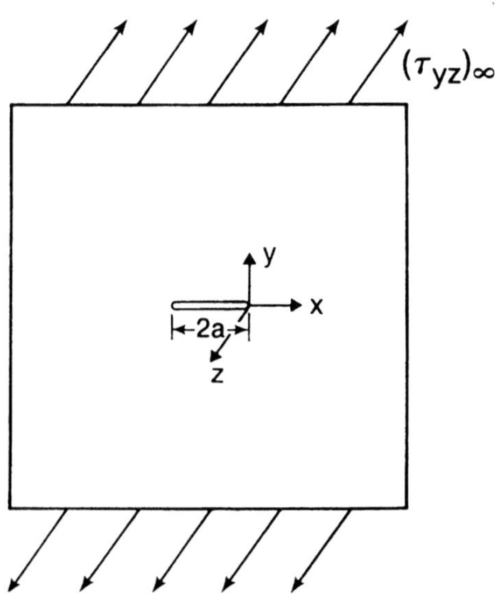

Fig. 7.2 A cracked infinite body subject to out-of-plane shear stresses.

oriented along the x-axis ($y = 0$), stress continuity across the x-axis permits the elimination of the complex potential $\psi(z)$ leading to the Westergaard type of single stress function as evident from (7.4), for details see [12].)

The crack tip stresses and displacements are computed from (7.1) and (7.2) by appropriate substitution from (7.4). For example, when the remotely applied stress is $(\sigma_y)_\infty$, we get the singular terms

$$\begin{Bmatrix} \sigma_x \\ \tau_{xy} \\ \sigma_y \end{Bmatrix} = (\sigma_y)_\infty \left(\frac{a}{2}\right)^{1/2} r^{-1/2} \cos\left(\frac{\theta}{2}\right) \begin{Bmatrix} 1 - \sin\left(\frac{\theta}{2}\right)\sin\left(\frac{3\theta}{2}\right) \\ \sin\left(\frac{\theta}{2}\right)\cos\left(\frac{3\theta}{2}\right) \\ 1 + \sin\left(\frac{\theta}{2}\right)\sin\left(\frac{3\theta}{2}\right) \end{Bmatrix}, \tag{7.5}$$

and

$$\tau_{xz} = \tau_{yz} = 0,\ \sigma_z = 0 \text{ (generalized plane stress)*}$$

$$\sigma_z = \nu(\sigma_x + \sigma_y) \text{ (plane strain)}.$$

We note in the above the inverse square root character of the crack tip stress singularity ($r^{-1/2}$) which is common to all linear elastic solutions of cracked bodies. In addition to the singular terms (7.5) there are other higher order terms, $O_{ij}(r^0) + O_{ij}(r^{1/2}) + \ldots$, which are non-singular, e.g. [2–4, 8–10, 12]. The zero order term, $O_{ij}(r^0)$, which does not vanish as $r \to 0$, is sometimes referred to as the T-term and is thought to be of significance in short and inclined cracks (see Chapter 9 for a discussion of small cracks).

In general, the components of the stress tensor σ_{ij} $(i, j = x, y, z)$, with the usual notation of $\sigma_{xx} = \sigma_x$, $\sigma_{xy} = \tau_{xy}$, $\sigma_{yy} = \sigma_y$, etc. can be represented as

$$\sigma_{ij} = r^{-1/2}[K_{\mathrm{I}} f^{\mathrm{I}}_{ij}(\theta) + K_{\mathrm{II}} f^{\mathrm{II}}_{ij}(\theta) + K_{\mathrm{III}} f^{\mathrm{III}}_{ij}(\theta)] \tag{7.6}$$
$$+ \text{ other non-singular terms at the crack tip,}$$

where the symbols I, II and III refer to the displacement discontinuities induced along the crack surface by tensile (Mode I), in-plane shear (Mode II), and anti-plane shear (Mode III) stresses, (Fig. 7.3). The dimensionless functions f^{I}_{ij}, f^{II}_{ij} and f^{III}_{ij} are functions of the orientation angle θ, and K_{I}, K_{II} and K_{III} are termed *stress intensity factors* for the appropriate crack tip deformation mode. The stress intensity factors K_i (i = I, II, III) were originally introduced by Irwin [17] and have dimensions of $FL^{-3/2}$ (F denoting force and L denoting length) in contrast to stress which has dimensions of FL^{-2}.

Of particular interest are stresses acting near the crack tip along the crack line, i.e. $(x, 0)$ for $x > 0$ (Fig. 7.1). These are summarized below.

*Strictly speaking only generalized plane stress is a two-dimensional problem, and this is implied when the words plane stress are used hereafter.

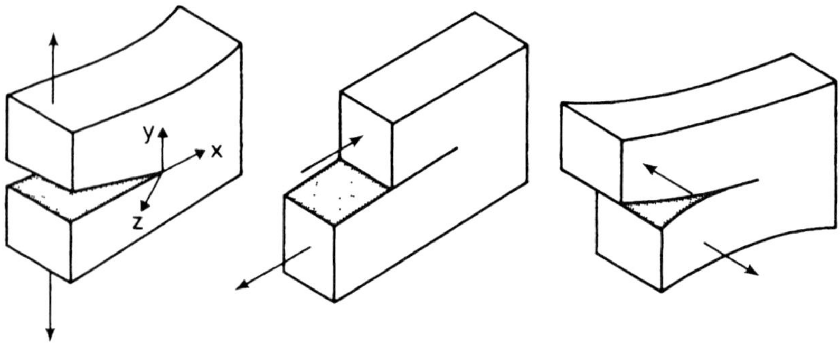

Fig. 7.3 Three basic loading (deformation) modes for a cracked body. (a) tensile (opening) Mode I; (b) in-plane shear (sliding) Mode II; (c) anti-plane shear (tearing) Mode III.

Mode I (tensile)

$$\sigma_y(x,0) = K_{\text{I}}(2\pi x)^{-1/2}, \quad \tau_{xy}(x,0) = 0, \quad \tau_{zy}(x,0) = 0. \tag{7.7}$$

Mode II (in-plane shear)

$$\sigma_y(x,0) = 0, \quad \tau_{xy}(x,0) = K_{\text{II}}(2\pi x)^{-1/2}, \quad \tau_{zy}(x,0) = 0. \tag{7.8}$$

Mode III (anti-plane shear)

$$\sigma_y(x,0) = 0, \quad \tau_{xy}(x,0) = 0, \quad \tau_{zy} = K_{\text{III}}(2\pi x)^{-1/2}. \tag{7.9}$$

In general, in any plane problem the singular stress field at the crack tip is a linear superposition of Mode I and Mode II whereas in a three-dimensional problem the crack tip singular stress field at any point along the crack edge is a superposition of Modes I and II (plane strain) and Mode III, e.g. Hutchinson [15].

The stress intensity factors K_{I}, K_{II} and K_{III} are available in the literature for a wide variety of crack configurations, e.g. Tada, Paris and Irwin [2] where a comprehensive catalogue exists with appropriate references. For example, in the case of a finite crack in an infinite plane (Figs 7.1 and 7.2)

$$K_{\text{I}} = (\sigma_y)_\infty(\pi a)^{1/2}, \quad K_{\text{II}} = (\tau_{xy})_\infty(\pi a)^{1/2}, \quad K_{\text{III}} = (\tau_{yz})_\infty(\pi a)^{1/2}. \tag{7.10}$$

For an edge crack in a semi-infinite plane in tension (Fig. 7.4)

$$K_{\text{I}} = 1.122\,(\sigma_y)_\infty(\pi a)^{1/2}. \tag{7.11}$$

For line cracks in tension (also Mode III) (Fig. 7.5)

$$K_{\text{I}} = (\sigma_y)_\infty(\pi a)^{1/2}\left[\frac{2b}{\pi a}\tan\left(\frac{\pi a}{2b}\right)\right]. \tag{7.12}$$

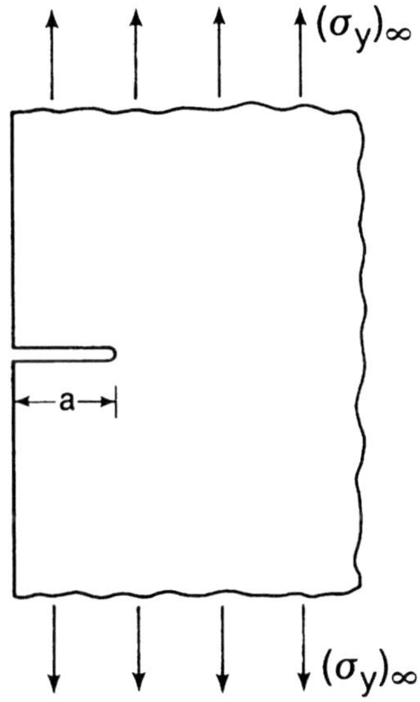

Fig. 7.4 Edge crack in a semi-infinite plane subject to tensile stresses.

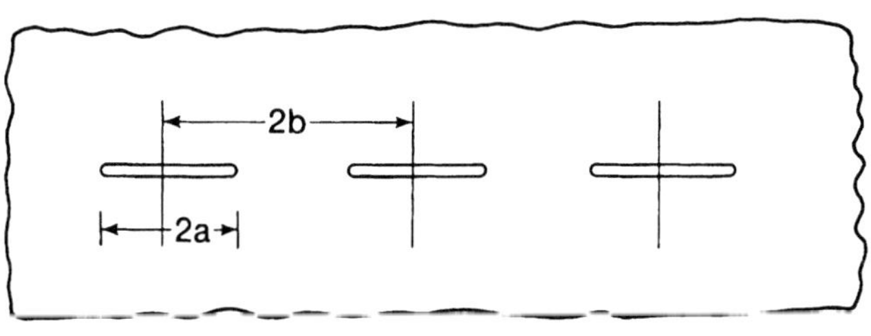

Fig. 7.5 Line cracks in an infinite plane subject to tensile stresses.

For cracks in finite width strips (Fig. 7.6)

$$K_I = \sigma(\pi a)^{1/2} F\left(\frac{a}{b}, \frac{h}{b}\right), \tag{7.13}$$

see [2] for the plot of $F(a/b, h/b)$ versus a/b.

The relationship between Irwin's stress intensity factors and the stress concentration factors (used in the analysis of notches, Chapter 8) was derived by Paris and Erdogan [18]. For an ellipse with length $2a$ and end radius of curvature ρ subject to an in-plane tension $(\sigma_y)_\infty$, they found the maximum tensile stress concentration to be

$$K_\sigma = \frac{\sigma_{\max}}{(\sigma_y)_\infty} = \left[1 + 2\left(\frac{a}{\rho}\right)^{1/2}\right]. \tag{7.14}$$

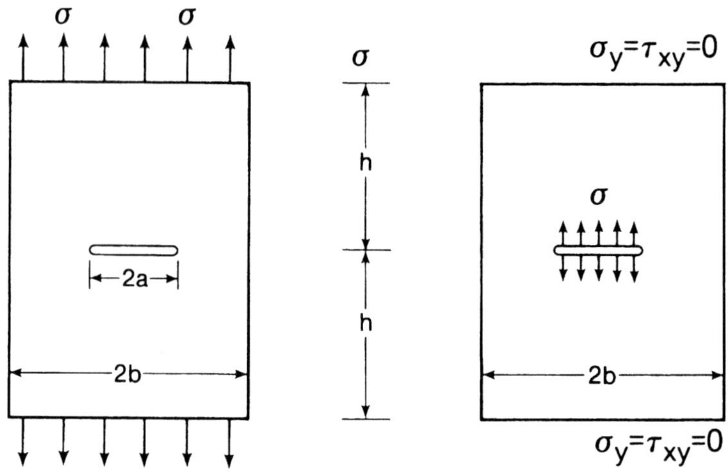

Fig. 7.6 Cracks in finite width strips. (a) applied load at the far-field; (b) applied load along the crack faces.

Comparing the above with (7.10), for $\rho \ll a$ (narrow ellipse) we obtain the following relationship:

$$K_I = \tfrac{1}{2}(\pi\rho)^{1/2}\sigma_{max} \tag{7.15a}$$

or

$$\frac{K_I}{K_\sigma} = \frac{1}{2}(\sigma_y)_\infty(\pi\rho)^{1/2}. \tag{7.15b}$$

7.2.2 ENERGY RELEASE RATE

Consider a cracked body in Mode I as shown in Fig. 7.7, where a is the crack length and P is the applied load per unit thickness. Let the compliance of the cracked body be denoted by

$$C = \frac{\Delta}{P}, \tag{7.16}$$

and that of the spring (loading device) by

$$C_M = \frac{\Delta_M}{P}, \tag{7.17}$$

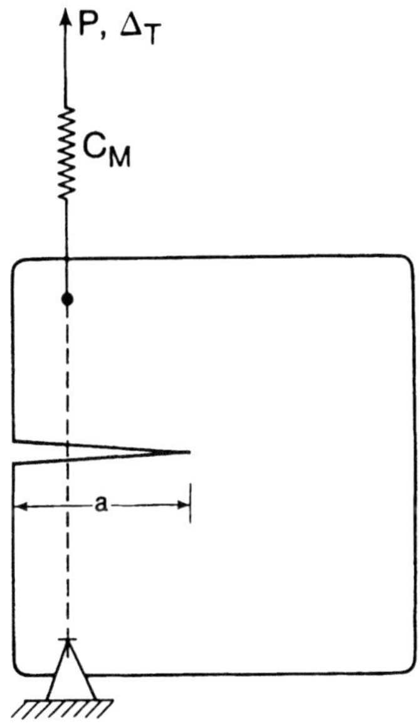

Fig. 7.7 A cracked body under Mode I compliance loading.

If the total prescribed displacement is Δ_T, then

$$\Delta_T = \Delta + \Delta_M = \Delta + \left(\frac{C_M}{C}\right)^* \Delta\,, \tag{7.18}$$

where the last step is obtained from combining (7.16) and (7.17).

Under the prescribed displacement (i.e. compliance loading), the potential energy (PE)* of the system in Fig. 7.7 is the sum of the strain energy in the body and spring, i.e.

$$\mathrm{PE} = \tfrac{1}{2}P\Delta + \tfrac{1}{2}P\Delta_M = \tfrac{1}{2}C^{-1}\Delta^2 + \tfrac{1}{2}C_M^{-1}(\Delta_T - \Delta)^2\,, \tag{7.19}$$

where the last step is obtained from using (7.16), (7.17) and (7.18).

The energy release rate under compliance conditions is

$$\mathscr{G} = -\left[\frac{\partial \mathrm{PE}}{\partial a}\right]_{\Delta_T} = -\,[C^{-1}\Delta - C_M^{-1}(\Delta_T - \Delta)]\left[\frac{\mathrm{d}\Delta}{\mathrm{d}a}\right]_{\Delta_T} + \frac{1}{2}C^{-2}\Delta^2\frac{\mathrm{d}C}{\mathrm{d}a}\,. \tag{7.20}$$

* In general, the potential energy of an elastic body is defined as

$$\mathrm{PE} = \int_V [W(\varepsilon) - F_i u_i]\,\mathrm{d}v - \int_{S_T} T_i u_i\,\mathrm{d}s \quad i = 1, 2, 3$$

where S_T is the portion of the boundary over which tractions are prescribed. The body forces, F_i, and tractions, T_i, are regarded as given. Thus PE is a function of the displacement, u_i.

Substituting from (7.18) into the above, the first term on the RHS vanishes, therefore the energy release rate for the Fig. 7.7 configuration is given by

$$\mathcal{G} = \frac{1}{2} C^{-2} \Delta^2 \frac{dC}{da} = \frac{1}{2} P^2 \frac{dC}{da}. \tag{7.21}$$

The above relation indicates that the energy release rate is independent of the compliance C_M of the loading system. In other words, the energy release rate of a specimen at a given load level is the same whether the load ($C_M \to \infty$) or the displacement ($C_M \to 0$) is held constant during the increment of crack advance. When the load is kept constant, then from (7.16) and (7.21) we get

$$\mathcal{G} = \frac{1}{2} P \left[\frac{d\Delta}{da} \right]_P. \tag{7.22}$$

Physically, $\mathcal{G}$ can be seen as the energy made available for crack extension resulting from the release of the strain energy and/or the work from displacement of loading forces.

A graphical interpretation of the energy release rate, (7.21, 7.22), is given in Figs 7.8(a) and 7.8(b). To determine $\mathcal{G}$ experimentally, one uses (7.21), Fig. 7.8(a). The compliance of two specimens' identical except for slightly different crack lengths, is measured. The change in compliance dC/da is determined, and the energy release rate is then estimated from (7.21). Rice [12] gives a more general derivation of the potential energy difference in cracked elastic bodies.

A relationship can be derived between the stress intensity factor and the energy release rate $\mathcal{G}$ per unit length of the crack, e.g. Tada *et al.* [2]. For a three-dimensional (plane strain) body with a crack edge subject to K_I, K_{II}

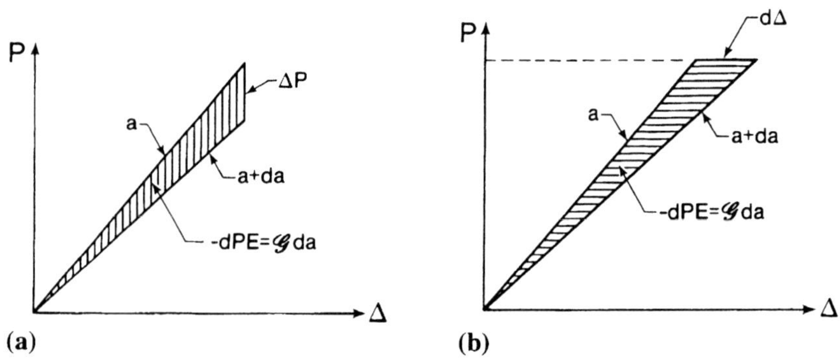

Fig. 7.8 Load deflection curves for (a) deformation control and (b) load control tests.

and K_{III} loading modes, we get

$$\mathcal{G} = \frac{1-\nu^2}{E}(K_I^2 + K_{II}^2) + \frac{1}{2\mathcal{G}}K_{III}^2 \,. \tag{7.23}$$

We could also find a relationship between $\mathcal{G}$ and strain energy per unit thickness, for an infinite strip of width $2b$ for $x \to \infty$, in the form of [15]

$$\mathcal{G} = 2b\,W_{\infty} \,. \tag{7.24}$$

We conclude from linear elastic fracture mechanics that the state of stress in the vicinity of an elastic crack body is controlled by the stress intensity factor K_I which is a function of the applied remote stress and crack size. Thus, given the crack configuration and the applied stress (load), we need only a material parameter related to stress intensity (fracture toughness) to design against a brittle fracture.

7.3 NONLINEAR FRACTURE MECHANICS

Inelastic deformation is observed in front of a sharp notch even at small applied loads, especially in the case of ductile materials. Nonlinear fracture mechanics is concerned with inelastic effects and specification of the demarcation region, where LEFM predictions are reasonable approximations even when there is a small zone of confined plastic region in front of the crack (section 7.4). At the microscopic level, the mechanics of ductile and even brittle separation of material involves plastic deformation. Therefore, elastic–plastic solutions are required if a connection is to be made between the micro- and macro-mechanisms of fracture.

In the following we will focus our attention on two types of elastic–plastic crack tip fields which have been successfully used to solve many important practical problems.

7.3.1 YIELDING ON DISCRETE SURFACES – PLANE STRESS CONDITION

Approximate solutions of crack tip plasticity have been proposed by several authors (see Rice [12] for a list of earlier works, and [8–10, 14] for more recent ones). In some of these solutions, the plastic flow in front of the crack is modelled as either slip or tensile yielding on discrete surfaces. In this section we will only deal with a model known as the Dugdale–Barenblatt model. Barenblatt's original work [19] was concerned with the influence of molecular cohesive forces on the form of deformation near a crack tip in brittle materials. Dugdale [20] observed that in thin sheets of mild steel, the plastic zone in front of the crack is essentially composed of two 45° slip bands as shown in Fig. 7.9(a). He then effectively replaced Barenblatt's cohesive forces by the known yield stress, σ_0, thus restraining

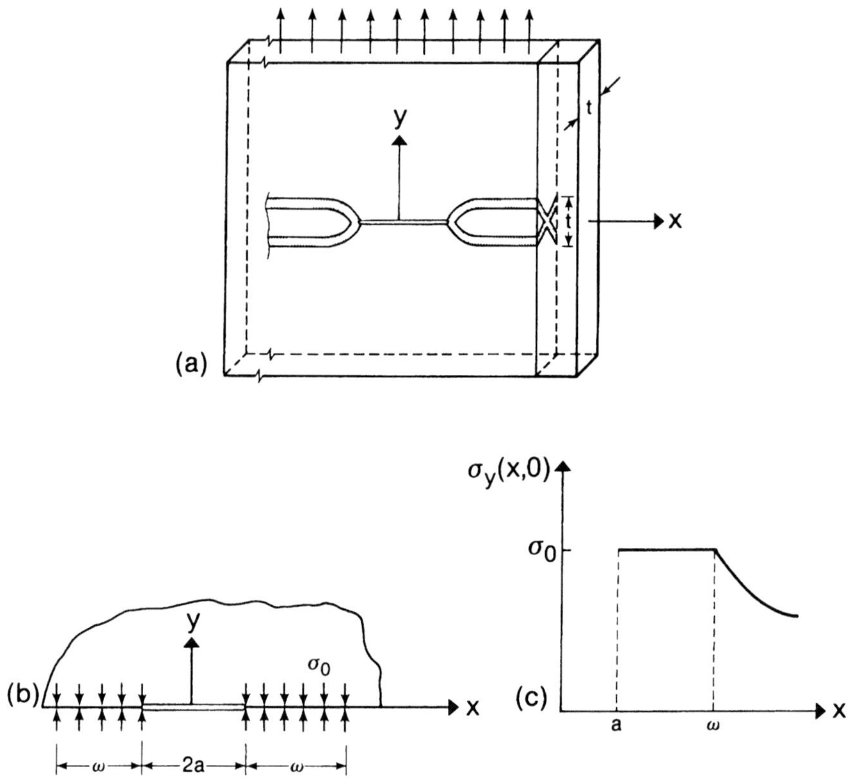

Fig. 7.9 A thin centre cracked plate under Mode I tensile loading. (a) plastic zone strip; (b) Dugdale's constraining forces; (c) distribution of the normal stress along the crack plane.

the crack opening over the plastic zone $\omega \gg t$ (Fig. 7.9(b)) (a clear exposition of this theory is given by Goodier [21]).

The boundary conditions along $y = 0$ (Fig. 7.9(b)) are: $\tau_{xy} = 0$ along x and $\sigma_y = 0$ for $|x| < a$, and $\sigma_y = \sigma_0$ for $a \leqslant |x| \leqslant \omega + a$, and $u_y = 0$ for $|x| \geqslant \omega + a$. The analysis is entirely elastic and the plastic zone size, ω, is determined from the condition which makes the stresses bounded at the outer tip of the plastic zone (Fig. 7.9(c)).

The complex potential $\phi(z)$ and $\psi(z)$ to determine stresses and displacements (7.1) for the above problem are given in Muskhelishvili [16]:

$$\phi(z) = \frac{\sigma_0}{\pi} \operatorname{arc\,tan}\left[\left(\frac{\omega}{z-\omega}\right)^{1/2}\right], \tag{7.25}$$

$$\psi(z) = z\phi'(z).$$

The distribution of the normal stress $\sigma_y(x,0)$ along the crack plane is shown in Fig. 7.9(c). The plastic zone size is then given by

$$\omega = a\left[\sec\left(\pi\frac{\sigma_\infty}{2\sigma_0}\right) - 1\right], \tag{7.26}$$

and the crack opening displacement, δ_t, is found from

$$\delta_t = 2u_y(a,0) = a\left(8\frac{\sigma_0}{\pi E}\right)\ln\left[\sec\left(\frac{\pi\sigma_\infty}{2\sigma_0}\right)\right]. \tag{7.27a}$$

For Modes II and III, we get identical expressions with σ_∞/σ_0 replaced by τ_∞/τ_0. When $\sigma_\infty/\sigma_0 \ll 1$, then (7.27a) asymptotically reaches

$$\delta_t = \frac{K_I^2}{E\sigma_0}. \tag{7.27b}$$

Following Rice [22] and taking the contour of the path-independent integral J (see (7.34) in next section) to surround the crack tip (Fig. 7.10) it can be shown that

$$J = \int_0^{\delta_t} \sigma(\delta)\,d\delta. \tag{7.28}$$

From Fig. 7.10, the average plastic strain is approximately

$$\bar{\varepsilon}^p \approx \frac{\delta}{t}, \tag{7.29}$$

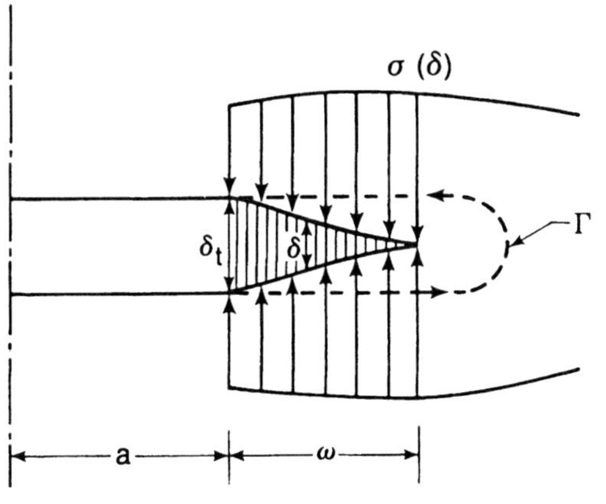

Fig. 7.10 The contour enclosing the crack in the Dugdale–Barenblatt model (Rice [22]).

where t denotes thickness. Substituting from (7.29) into (7.28) we get

$$J = t\int_0^{\delta_t/t} \sigma(\bar{\varepsilon}^{\mathrm{p}})\,\mathrm{d}\bar{\varepsilon}^{\mathrm{p}}\,. \tag{7.30}$$

Knowing the stress–strain relation σ versus $\bar{\varepsilon}^{\mathrm{p}}$ the above can easily be integrated. However, the integrand in (7.30) is the plastic strain energy density in front of the crack, thus (7.30) could be written as

$$\frac{J}{t} = \int_0^{\delta_t/t} \mathrm{d}W^{\mathrm{p}} = (W^{\mathrm{p}})_{\mathrm{pz}} \tag{7.31}$$

where pz here refers to the plastic zone ahead of the crack (Fig. 7.9). For an elastic–perfectly plastic material, (7.30) reduces to $J = \sigma_0\delta_t = t(W^{\mathrm{p}})_{\mathrm{pz}}$.

This model has been extensively used in the literature to describe various aspects of crack growth, including fatigue crack propagation and crack closure to be discussed later. Its success stems from the fact that the gradient of the plastic strain component perpendicular to the crack direction, $\varepsilon_y(x, 0)$, is so steep near the crack tip that assuming all the cracked body is elastic except for a small strip along the crack (weakest cross-section) gives a good representation of the local deformation state.

7.3.2 CRACK FIELDS FOR PLASTICALLY DEFORMED SOLIDS – HRR SINGULARITY FIELDS

The solution to a cracked body in a homogeneous material undergoing plastic deformation was given by Hutchinson [23] and Rice and Rosengren [24], within the context of the deformation theory of plasticity (i.e. small strain, nonlinear elastic solid). The stress, strain and deformation fields thus obtained are now commonly referred to as HRR crack tip singularity fields. The material description was obtained by generalizing the uniaxial Ramberg–Osgood stress–strain relation (Chapter 2, (2.7)) to the multiaxial state by the J_2 deformation theory (Chapter 2, (2.34)), i.e.

$$\frac{\varepsilon_{ij}}{\varepsilon_0} = \frac{1+\nu}{\sigma_0}s_{ij} + \frac{1-2\nu}{3\sigma_0}\sigma_{kk}\delta_{ij} + \frac{3}{2}\alpha\left(\frac{\sigma_{\mathrm{eq}}}{\sigma_0}\right)^{(1-n)/n}\frac{s_{ij}}{\sigma_0}, \tag{7.32}$$

where the expressions for the deviatoric stress tensor s_{ij} and equivalent (effective) stress σ_{eq} are given in Chapter 2 ((2.27) and (2.30)) as well as in Chapter 6. In the above relation the first two terms on the RHS give the elastic part of the strain and the last one is the plastic part.

Let us denote the strain energy density of the material by $W(\varepsilon)$. For the above nonlinear material (7.32), the stresses can be calculated from

$$\sigma_{ij} = \frac{\partial W(\varepsilon)}{\partial \varepsilon_{ij}}\,. \tag{7.33}$$

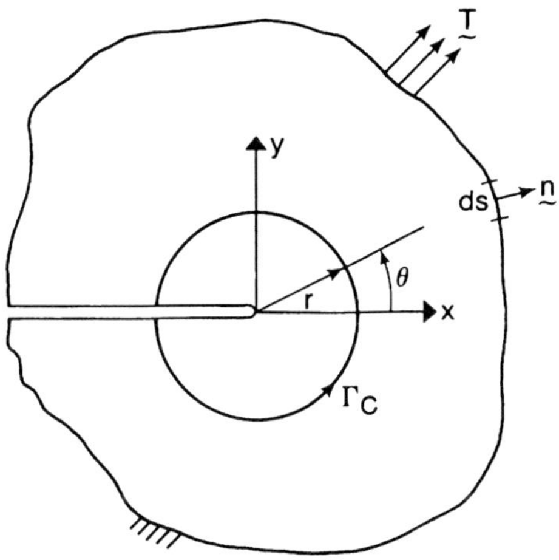

Fig. 7.11 A cracked body with applied traction, and a contour enclosing the crack.

Consider a cracked body in plane strain or plane stress condition (Fig. 7.11) and let Γ be any contour around the crack tip in the counter-clockwise direction. The *J*-integral defined by Rice [22],

$$J = \int_{\Gamma_C} \left(W n_x - \sigma_{ij} n_j \frac{\partial u_i}{\partial x} \right) \mathrm{d}s, \quad (i, j = x, y, z), \tag{7.34}$$

has the property that it is independent of the path Γ taken around the crack. Take the line integral path Γ to be a circle with radius *r* centred at the crack tip. Now let $r \to 0$, then for a non-zero value of *J*, the integrand in (7.34) should have an r^{-1} singularity [15] so that

$$r\left(W n_x - \sigma_{ij} n_j \frac{\partial u_i}{\partial x} \right) = f(\theta) \quad \text{as} \quad r \to 0. \tag{7.35}$$

Thus

$$J = \int_{-\pi}^{\pi} f(\theta) \mathrm{d}\theta. \tag{7.36}$$

In particular for the power law part of the stress–strain relation (7.32), an r^{-1} singularity in *W* (7.33) would imply an $r^{-n/(1-n)}$ singularity in stress and an $r^{-(1-n)/n}$ singularity in strain and an $r^{n/(1+n)}$ variation in the displacement. (In the nonlinear fracture mechanics literature, the power law exponent '*n*' is the inverse of that used here and in the fatigue literature.) Thus, the stress, strain and displacement components near the crack tip can be expressed as ([13–15])

$$\sigma_{ij} = \sigma_0 \left(\frac{J}{\alpha \sigma_0 \varepsilon_0 I_n r} \right)^{n/(1+n)} \tilde{\sigma}_{ij}(\theta; n) \tag{7.37a}$$

$$\varepsilon_{ij} = \alpha \varepsilon_0 \left(\frac{J}{\alpha \sigma_0 \varepsilon_0 I_n r} \right)^{1/(1+n)} \tilde{\varepsilon}_{ij}(\theta; n) \tag{7.37b}$$

$$u_i = \alpha \varepsilon_0 \left(\frac{J}{\alpha \sigma_0 \varepsilon_0 I_n} \right)^{1/(1+n)} r^{n/(1+n)} \tilde{u}_i(\theta; n)\,. \tag{7.37c}$$

In comparing (7.37a) with (7.6) we note that J plays the same role as that of K, i.e. J is a measure of intensity of the stress and strain fields ahead of the crack when the crack tip undergoes plastic deformation. A relationship between J and K can be found when the plastic zone in front of the crack is small (i.e. a condition of small-scale yielding prevails). The second observation is that the stress singularity in front of the crack is of the order $r^{-n/(1+n)}$ as compared to $r^{-1/2}$ for the linear elastic solution (note that $n = 1$ for a linear elastic material).

The dimensionless parameters $\tilde{\sigma}_{ij}(n, \theta)$, $\tilde{\varepsilon}_{ij}(n, \theta)$ and $\tilde{u}_i(n, \theta)$ in (7.37) depend on the mode of deformation, and the condition of plane problem (plane stress or plane strain) in addition to n and θ. In a similar manner, I_n is a non-dimensional parameter of exponent n. The variation of these parameters are given in [15, 23], among others.

7.3.3 RELATIONSHIP BETWEEN THE J-INTEGRAL AND ENERGY RELEASE RATE

Let us now consider the cracked body and spring shown in Fig. 7.7 with a nonlinear elastic behaviour where the total displacement Δ_T is prescribed. The strain energy in the system is the sum of the strain energy in the body and spring, i.e.

$$\mathrm{PE} = W(a, \Delta) + W(\Delta_M)\,, \tag{7.38}$$

where

$$W(a, \Delta) = \int_0^{\Delta} P(\Delta, a)\, d(\Delta) \quad \text{and} \quad W(\Delta_M) = \int_0^{\Delta_M} P d(\Delta_M)\,.$$

For the crack advancing along its line, J is defined as ([15, 22])

$$J = -\left[\frac{\partial PE}{\partial a} \right]_{\Delta_T} \tag{7.39}$$

where

$$\mathrm{dPE} = \left[\frac{\partial W(a,\Delta)}{\partial a}\right]_\Delta \mathrm{d}a + \left[\frac{\partial W(a,\Delta)}{\partial \Delta}\right]_a \mathrm{d}\Delta + W'\mathrm{d}\Delta_\mathrm{M}$$

$$= \left[\frac{\partial W(a,\Delta)}{\partial a}\right]_\Delta \mathrm{d}a + P(\mathrm{d}\Delta + \mathrm{d}\Delta_\mathrm{M}) = \left[\frac{\partial W(a,\Delta)}{\partial a}\right] \mathrm{d}a\,,$$

where the last step is obtained by noting that $\mathrm{d}\Delta_\mathrm{T} = 0$. Thus from (7.39) we get

$$J = -\left[\frac{\partial W(a,\Delta)}{\partial a}\right]_\Delta \mathrm{d}a = -\int_0^\Delta \frac{\partial P(\Delta,a)}{\partial a}\mathrm{d}\Delta\,. \tag{7.40}$$

We note that similar to the energy release rate in a linear elastic body, J in (7.40) is independent of the spring characteristic. In comparing (7.40) with (7.21) we note that J is a generalization of $\mathscr{G}$. Interpretation of (7.40) in a graphical form is shown in Fig. 7.12(a).

In the case of a load-controlled test, a similar derivation will show that J is independent of the spring characteristic [15], Fig. 7.12(b).

In the previous section J was defined in terms of a path-independent line integral (7.34). In the above J is defined in terms of the overall load displacement characteristic of the cracked body, and it is related to the energy release rate of the body at the current deformation state. For a more general derivation of the results, the reader is encouraged to consult Rice [12, 22].

It is worth mentioning that for a linear elastic body, J reduces to $\mathscr{G}$, i.e.

$$J = \mathscr{G}\,. \tag{7.41}$$

This is immediately verified by comparing Figs 7.8 and 7.12.

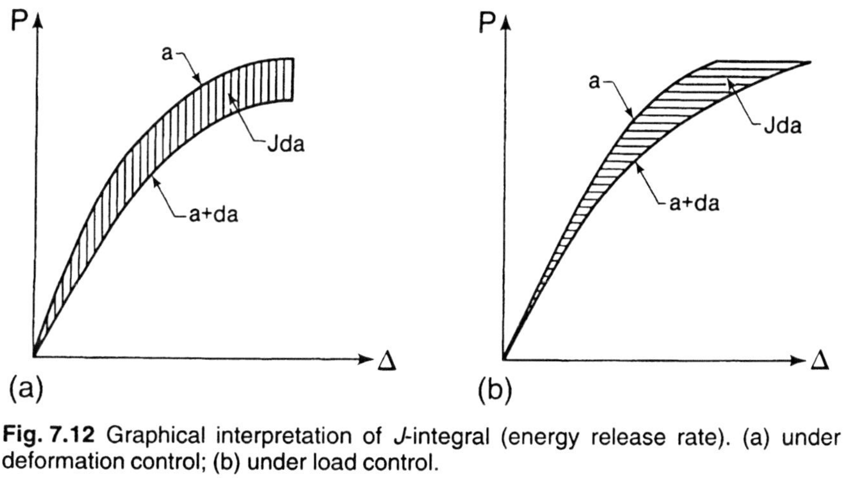

Fig. 7.12 Graphical interpretation of J-integral (energy release rate). (a) under deformation control; (b) under load control.

7.4 THE CONCEPT OF SMALL-SCALE YIELDING

In most metals, especially those with intermediate to low strength, yielding will take place ahead of the crack tip for a remotely applied stress $(\sigma_\infty/\sigma_0) \ll 1$. Yet extensive experimental and theoretical studies have shown that the predictions of linear elastic fracture mechanics (LEFM) provide a reasonable approximation to the observed behaviour up to a certain limit. In other words, linear elastic fracture mechanics is based on the concept of small-scale yielding (SSY) and we wish to investigate the condition which delineates SSY. A number of comparative studies have shown that SSY appears to be a reasonable assumption when the remotely applied load is below one half of the limit load for an elastic–perfectly plastic material [11, 12, 15].

In general, the plastic zone size establishes the condition under which elastic stress fields (*K*-fields) dominate. One would expect that when the size of the plastic zone is small compared to the relevant geometric properties (e.g. crack length, uncracked ligament width, thickness, etc.) then the condition of small-scale yielding would apply. Therefore, a study of the plastic zone size is necessary to ascertain whether a SSY condition exists or not. Strangely enough, we need to have a plastic solution of a given problem before we can definitely delineate the limit of SSY. For example, such a study was made by Rice [11, 12] for an anti-plane shear problem which showed that the predictions of the small-scale yielding solution for the plastic zone size is accurate up to about 50% of the limit load, i.e. τ_∞/τ_0 (net section) $\leqslant 0.5$. The effect of the plasticity was to shift the linear elastic stress distribution ahead of the crack by half the plastic zone size.

7.4.1 ELASTIC–PLASTIC SOLUTIONS IN SMALL-SCALE YIELDING

Rice [12] has discussed SSY elastic–plastic solutions. Closed form solutions for stresses and strains everywhere in the plastic zone have not yet been found for Modes I and II loading conditions. Hult and McClintock [25] were the first to provide such a solution for Mode III loading of an elastic–perfectly plastic material. For a strain-hardening material of the form

$$\frac{\gamma}{\gamma_0} = \frac{\tau}{\tau_0} \quad \text{for} \quad \tau \leqslant \tau_0, \tag{7.42a}$$

and

$$\left(\frac{\gamma}{\gamma_0}\right) = \left(\frac{\tau}{\tau_0}\right)^{1/n} \quad \text{for} \quad \tau > \tau_0, \tag{7.42b}$$

Rice [26] has given a solution of the form

$$\tau_\alpha = \tau_0 \left(\frac{r_p}{r}\right)^{n/(1+n)} \tilde{\tau}_\alpha(\theta), \quad (\alpha = r, \theta) \tag{7.43}$$

$$\gamma_\alpha = \gamma_0 \left(\frac{r_p}{r}\right)^{1/(1+n)} \bar{\gamma}_\alpha(\theta)$$

where

$$r_p = \frac{1}{\pi(1+n)}\left(\frac{K_{III}}{\tau_0}\right)^2 \tag{7.44a}$$

$$\tilde{\tau}(\theta) = \left[\frac{1+n}{2}\frac{\sin 2\phi}{\sin\theta}\right]^{n/(1+n)} \tag{7.44b}$$

$$2\phi = \theta + \arcsin\left[\frac{1-n}{1+n}\sin\theta\right] \tag{7.44c}$$

(note that $\tilde{\tau}(0) = 1$).

For example, the stress and strain ahead of the crack tip along the crack line ($y = 0$) are given by

$$\tau_{yz}(x,0) = \tau_0\left[\frac{K_{III}^2}{(1+n)\pi\tau_0^2 x}\right]^{n/(1+n)} \tag{7.45a}$$

and

$$\gamma_{yz}(x,0) = \gamma_0\left[\frac{K_{III}^2}{(1+n)\pi\tau_0^2 x}\right]^{1/(1+n)}. \tag{7.45b}$$

The plastic zone in this solution is circular of radius $R = (1+n)r_p/2$ where r_p, given by (7.44a), is the distance from the crack tip to the end of the plastic zone along the x-axis. Thus, the centre of the circular plastic zone is located at $\{(1-n)R/(1+n), 0\}$. When the small-scale yielding approximation is no longer appropriate, the solution given by Rice [26] shows the transition from the circular plastic zone to an elongated plastic zone. At the limit when $\tau_\infty \to \tau_0$, the plastic zone extends to infinity.

In comparing (7.7) and (7.9) we note that as long as the plastic zone size is small, the elastic crack tip stress fields are identical for Mode I and Mode III, if $K_I = K_{III}$ (see also McClintock [27]).

Using the above analogy, Kujawski and Ellyin [28] modified Rice's solution of Mode III for a crack loaded in Mode I for a Ramberg–Osgood stress–strain curve (Chapter 2, (2.7)). The stress and strain components normal to the crack line ($y = 0$) are given by

$$\sigma_y(x,0) = \sigma_0\left[\frac{K_\mathrm{I}^2}{(1+\bar{n})\pi\sigma_0^2 x}\right]^{\bar{n}/(1+\bar{n})} \tag{7.46a}$$

$$\varepsilon_y(x,0) = \frac{\sigma_0}{E}\left[\frac{K_\mathrm{I}^2}{(1+\bar{n})\pi\sigma_0^2 x}\right]^{\bar{n}/(1+\bar{n})} + \frac{\sigma_0}{E}\left[\frac{K_\mathrm{I}^2}{(1+\bar{n})\pi\sigma_0^2 x}\right]^{1/(1+\bar{n})} \tag{7.46b}$$

where

$$\bar{n} = \frac{1 + n(W^\mathrm{p}/W^\mathrm{e})}{1 + (W^\mathrm{p}/W^\mathrm{e})} \tag{7.47}$$

and W^p and W^e are the plastic and elastic components of the strain energy density, $W(\varepsilon) = \int_0^\varepsilon \sigma d\varepsilon$. Note that when $W^\mathrm{p} \gg W^\mathrm{e}$, $\bar{n} \approx n$ and for $W^\mathrm{p} \ll W^\mathrm{e}, \bar{n} = 1$. For the sake of brevity, (7.46) will be termed RKE fields.

Asymptotically, as the crack tip is approached, the contribution to the strain of the term which is linearly proportional to stress (first term on the RHS of (7.46b)) becomes small compared to the power law terms. Consequently, (7.45) and (7.46) (with K_III replaced by K_I) will give similar results. On the other hand, as the boundary of the plastic zone is approached, the linear term becomes much larger than the power law term. Therefore, the size of the elastic–plastic boundary as well as the stress and strain distribution near the boundary, depend on the type of stress–strain relationship employed. Figures 7.13 and 7.14 (taken from [29]) show the stress and

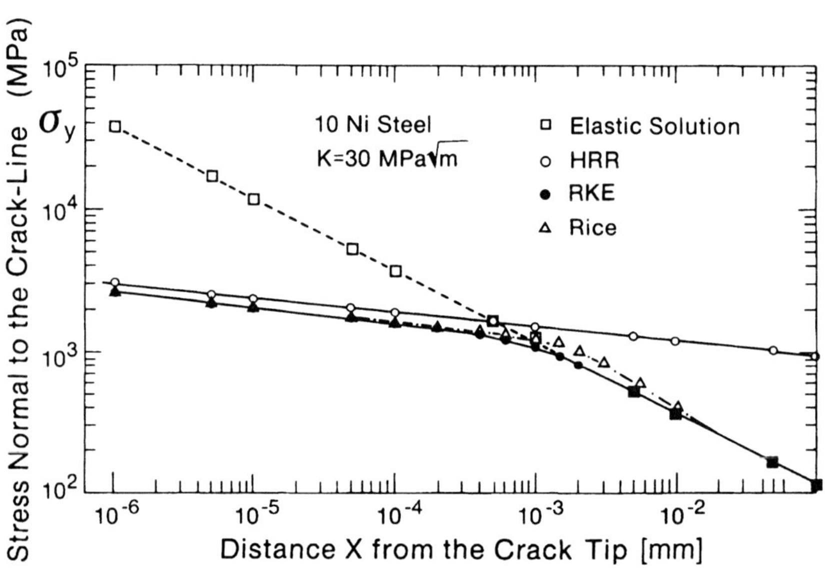

Fig. 7.13 Distribution of the stress component normal to the crack line, $\sigma_y(x,0)$ given by various singularity fields. Reprinted with kind permission from [29].

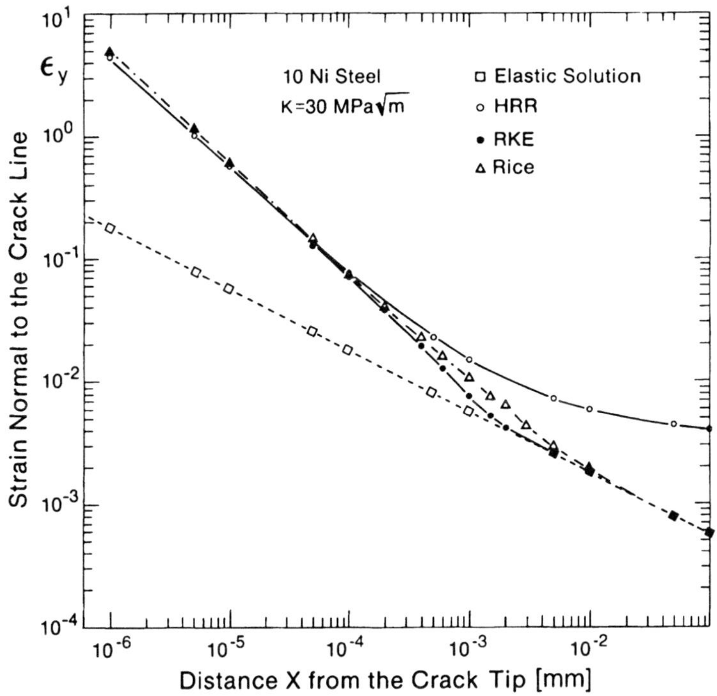

Fig. 7.14 Distribution of the strain component normal to the crack line, $\varepsilon_y\,(x,0)$ given by various singularity fields. Reprinted with kind permission from [29].

strain distribution along the crack line ($y = 0$). In these figures, the HRR singularity fields, (7.37), are compared with predictions of RKE fields, (7.46), Rice's solution, (7.45), and the elastic solution, (7.5) for the plane stress Mode I loading. It can be seen that near the crack tip, HRR, RKE and Rice's singularity fields predict similar values; however, HRR predictions deviate from those of RKE and Rice near the elastic–plastic boundary. Similar trends were observed by Shih and German [30] through finite element studies. For comparison with the experimental results, see [28].

It should be noted that in the case of small-scale yielding, the value of the J-integral (7.34) is known [22] and is given by

$$J = \frac{K^2}{\bar{E}}. \tag{7.48}$$

where $\bar{E} = E$ for the plane stress condition and $\bar{E} = E/(1 - \nu^2)$ when a plane strain condition prevails.

7.4.2 PLASTIC ZONE SIZE

The simplest estimate of the *plane stress* plastic zone size can be obtained from the LEFM solution (7.7) by equating the stress σ_y to σ_0, i.e.

$$X_0 = \frac{1}{2\pi}\left(\frac{K}{\sigma_0}\right)^2 . \tag{7.49}$$

Rice [11] proposed that stress redistribution will cause the yielded region to extend over roughly twice that estimated by (7.49), i.e.

$$R_p^R \simeq 2X_0 = \frac{1}{\pi}\left(\frac{K}{\sigma_0}\right)^2 . \tag{7.50}$$

In the case of small-scale yielding, Irwin [31] suggested that the elastic field beyond the plastic zone is the same as the one which would be present in the LEFM solution with the crack tip advanced by the plastic zone correction factor

$$\bar{r}_p = R_p^a - X_0 , \tag{7.51}$$

where R_p^a is the actual plastic zone size and X_0 is given by (7.49). He estimated that the plastic zone correction factor is in the range $0.3\,X_0 < \bar{r}_p < 0.5\,X_0$ for the *plane strain* condition. For computational purposes, it was suggested that

$$\bar{r}_p = \frac{1}{6\pi}\left(\frac{K}{\sigma_0}\right)^2 = \frac{1}{3}X_0 . \tag{7.52}$$

In the case of small-scale yielding, and when $\sigma_\infty/\sigma_0 \ll 1$, the Dugdale model, (7.26, 7.27 and 7.28), yields

$$R_p^D = \frac{\pi}{8}\left(\frac{K}{\sigma_0}\right)^2 . \tag{7.53}$$

which applies only for the plane stress condition.

From the above discussion, the plastic zone size in SSY can be written as [28]

$$R_p = \alpha\left(\frac{K}{\sigma_0}\right)^2 , \tag{7.54}$$

where $0.159 \leqslant \alpha \leqslant 0.393$, the lower limit is for plane strain and the upper one for the Dugdale plane stress. Schematic plastic zone sizes for plane strain and stress and the Dugdale model are shown in Fig. 7.15.

Using (7.46a), the plastic zone size is given by

$$R_0 = \frac{K^2}{(1+\bar{n}_0)\pi\sigma_0^2} , \tag{7.55}$$

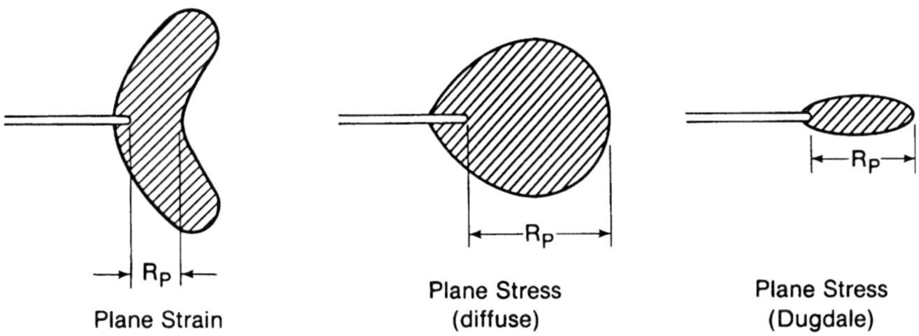

Fig. 7.15 Schematic plastic zone sizes for plane strain and plane stress.

with $\bar{n}_0$ calculated from (7.47) using $\sigma = \sigma_0$ and $\varepsilon^p = \varepsilon_0^p$. Analogous to (7.51) we define a plastic zone extension factor, $\bar{R}_p$, as

$$\bar{R}_p = R_0 - X_0 \,. \tag{7.56}$$

Substituting from (7.49) and (7.55) into the above, we get

$$\bar{R}_p = \frac{1 - \bar{n}_0}{1 + \bar{n}_0} X_0 \,. \tag{7.57}$$

The plastic zone extension, $\bar{R}_p$, is equal to the additional increase in the plastic zone size due to yielding of the material. From (7.47) and (7.57) we note that $\bar{R}_p$ depends not only on σ_0, but also on the strain hardening exponent of the material, n. The normalized plastic zone extension factor,

$$\frac{\bar{R}_p}{X_0} = \frac{1 - \bar{n}_0}{1 + \bar{n}_0} \,, \tag{7.58}$$

as a function of σ_0 and n, is shown in Fig. 7.16 (taken from [28]). The cross-hatched area in Fig. 7.16 corresponds to the plastic zone size correction range suggested by Irwin [31]. Predictions of (7.58) were found to be in good agreement with the experimental data [28]. The salient result of the above analysis is that the factor α in (7.54) is a function of the yield stress, σ_0, as well as the hardening exponent, n, of the material. Further studies have shown that (7.58) is a very good indication of the plastic zone size in metals [32].

Note that for the SSY condition to prevail, R_0 calculated from (7.55) or R_p determined from (7.54) must be larger than the specimen thickness in the plane stress condition, i.e. $R_p \gg t$, and smaller than t in the case of plane strain, $R_p \ll t$.

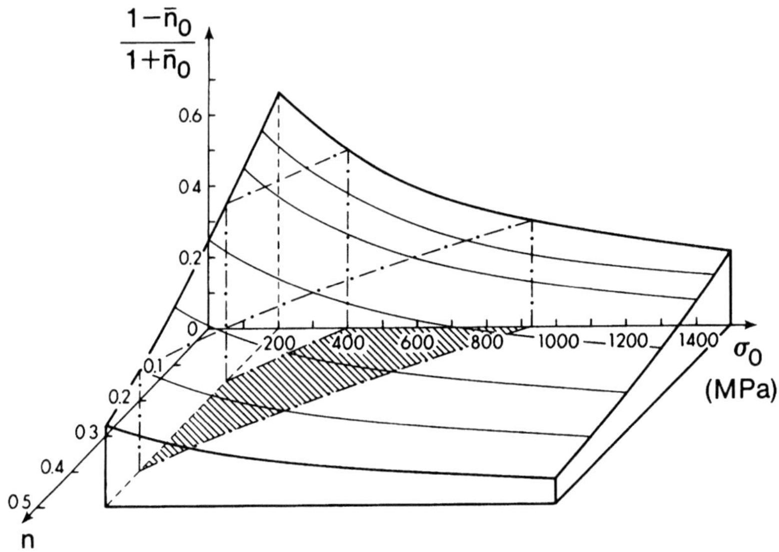

Fig. 7.16 Variation of the normalized plastic zone extension factor in terms of the yield stress σ_0 and hardening exponent *n*. Reprinted with kind permission from [28].

7.5 INITIATION OF CRACK GROWTH

For cases in which the small-scale yielding condition prevails, the stress intensity factor K for a given mode uniquely determines the intensity of stress/deformation independent of other geometric parameters or loading details (7.6). We could therefore postulate that a crack will propagate when

$$K = K_C , \tag{7.59}$$

where K_C is the critical stress intensity factor. Experiments have shown that K_C is not a constant value, except for very brittle materials. The value of K_C has to be obtained through experiments [33] and it is dependent on the mode of loading and constraint conditions, i.e. whether plane stress, plane strain or an intermediate condition prevails. In the case of plane strain and under Mode I loading, the critical stress intensity is denoted by K_{IC} and is termed the material fracture toughness at the given temperature. For continued crack growth the value of K has to be increased. Thus, for a given set of conditions, we obtain a curve

$$K = K_R(\Delta a) , \tag{7.60}$$

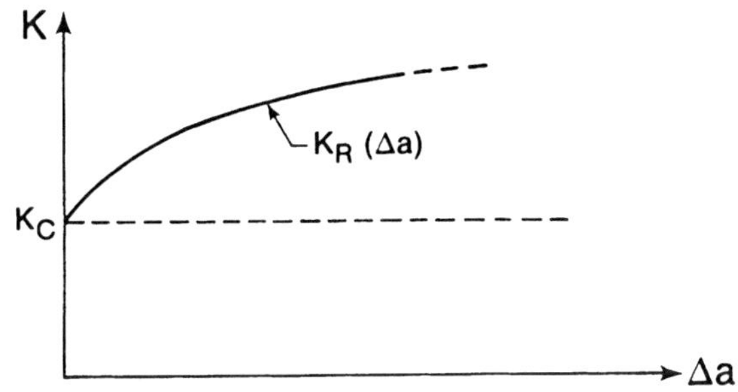

Fig. 7.17 Resistance of material to crack growth in terms of K, the R-curve.

which is called the resistance curve, or 'R-curve' as shown in Fig. 7.17. The condition for the stability of crack advance is then [14]

$$\left[\frac{\partial K}{\partial a}\right]_{\mathrm{L}} < \frac{\mathrm{d}K_{\mathrm{R}}}{\mathrm{d}(\Delta a)}, \tag{7.61}$$

where L denotes the prescribed loading condition which is to be kept constant. When the cracked specimen (or component) is loaded under compliance conditions (deformation control), the stability of crack growth $[\partial K/\partial a]_{\Delta_T}$ is strongly dependent on the compliance of the loading device. It can be shown that $[\partial K/\partial a]_{\Delta_T} < [\partial K/\partial a]_P$, e.g. Hutchinson [15].

In small-scale yielding, we have the following relationship:

$$J = \mathcal{G} = \frac{K^2}{\bar{E}}, \tag{7.62a}$$

where $\bar{E} = E$ for plane stress, and $\bar{E} = E/1 - \nu^2$ for plane strain. Thus there is a direct relationship between K_{IC} and J_{IC}, i.e.

$$J_{IC} = (1 - \nu^2)\frac{K_{IC}^2}{E}, \tag{7.62b}$$

which has been experimentally verified for a wide variety of metals and specimen geometries. Therefore we could express the initiation of crack growth in terms of the applied J in a manner similar to (7.60), i.e.

$$J = J_{\mathrm{R}}(\Delta a), \tag{7.63}$$

where $J_{\mathrm{R}}(\Delta a)$ is called the 'J-resistance curve' (Fig. 7.18).

The stability of crack growth is then expressed similarly to (7.61), i.e.

$$\left[\frac{\partial J}{\partial a}\right]_{\mathrm{L}} < \frac{\mathrm{d}J_{\mathrm{R}}}{\mathrm{d}(\Delta a)}, \tag{7.64a}$$

where $\mathrm{d}J_{\mathrm{R}}/\mathrm{d}(\Delta a)$ is referred to as 'tearing resistance' [13, 34]. Paris *et al.*

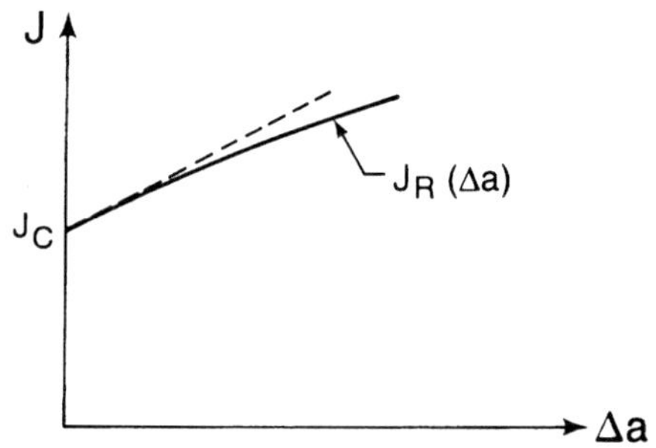

Fig. 7.18 Resistance of material to crack growth in terms of J, the J-resistance curve.

[34] have introduced a dimensionless 'tearing force' and tearing modulus defined as

$$T = \frac{E}{\sigma_0^2}\left(\frac{\partial J}{\partial a}\right) \quad \text{and} \quad T_{\mathrm{R}} = \frac{E}{\sigma_0^2}\left(\frac{\mathrm{d}J_{\mathrm{R}}}{\mathrm{d}a}\right). \tag{7.65}$$

The stability of crack growth (7.64a) is then expressed as

$$T < T_{\mathrm{R}} \tag{7.64b}$$

Similar to the K-field stability, it can be shown that $[\partial J/\partial a]_{\Delta_T} < [\partial J/\partial a]_{\mathrm{P}}$ (see Hutchinson [15] for the relationship between the two loading parameters).

7.5.1 J-DOMINANCE REGION

In the derivation of the HRR singularity fields, two basic assumptions are made. The first assumption is concerned with the use and applicability of the deformation theory of plasticity. The second implicit assumption involves the size of the region in which finite strain effects may be important and where the microscopic fracture process takes place. This region is now commonly referred to as the 'process zone'. The first assumption is valid (when compared to the more appropriate incremental theory of plasticity, Chapter 6) as long as the strains vary in a proportional manner. Several finite element studies using the incremental theory of plasticity, e.g. [9] have shown the J-values of the line integral (7.34) are approximately independent of the path, and are accurately predicted by the corresponding deformation theory.

In contrast, the size of the process zone depends on the fracture mechanism. For example, if the fracture mechanism is void nucleation, growth and coalescence (ductile fracture in monotonic loading), then a finite strain analysis would be required to delineate the J-dominance region. On the other hand, if the fracture mechanism is of a cleavage nature or micro-crack

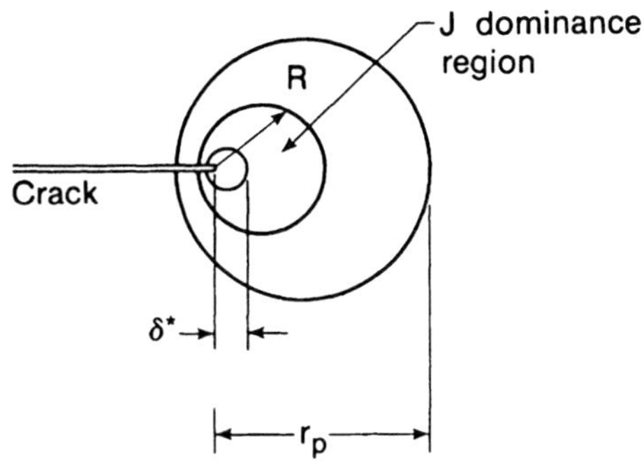

Fig. 7.19 The region of *J*-dominance.

formation as in fatigue loading, then the process zone size is of the order of grain size. Hutchinson [13–15] has fully discussed this issue. Let us denote by R the distance from the crack tip of the zone of dominance of the HRR singularity fields (Fig. 7.19). In the case of small-scale yielding, the J-dominance region is roughly

$$\delta^* < R < r_{\mathrm{p}}, \tag{7.66}$$

where δ^* is the 'radius' of the process zone and r_{p} is the plastic zone size. For large-scale yielding and ductile fracture

$$R > 3\,\delta_{\mathrm{t}}, \tag{7.67}$$

where δ_{t} is the crack opening displacement. The latter condition is established from finite element studies of a finite strain incremental formulation of the plane strain SSY condition [35].

It has been shown that in the case of an elastic–perfectly plastic material, (7.28) gives

$$\delta_{\mathrm{t}} = \frac{J}{\sigma_0}. \tag{7.68a}$$

For a strain-hardening material, Shih [36] has shown that

$$\delta_{\mathrm{t}} = d_n \frac{J}{\sigma_0}, \tag{7.68b}$$

where d_n is a function of α, ε_0 and n, $0.3 \leqslant d_n \leqslant 0.8$ for $0.08 \leqslant n \leqslant 0.33$.

7.6 MECHANICS OF FATIGUE CRACK GROWTH

Fatigue cracks may be classified according to their size, i.e. metallurgically small (or crystallographic) cracks, physically short cracks, and long cracks.

The metallurgically small cracks are defined by their crack size, which is small compared to metallurgical variables such as the grain size. In this regime the crack growth is strongly affected by the microstructure and its growth at stress levels below the fatigue limit of the material is generally stopped at microstructural barriers. The size of the crack would generally be less than 0.1 mm in this regime.

The physically small cracks are of such lengths that the resistance to crack growth by microstructural barriers is averaged out, but they are not long enough for crack closure, due to the plastic wake left behind the tip of a growing crack, to have reached a stable level. In this regime the cracks propagate at stress levels below the fatigue limit, σ_{FL}, and the long crack threshold stress, $\sigma_{th} = K_{th}/(F\sqrt{\pi a})$, where F is a parameter depending on the geometry (section 7.1). We may use the Kitagawa and Takahashi diagram [37], Fig. 7.20, to indicate the variation of the crack length and the stress level required to propagate the crack corresponding to the above-mentioned regimes.

In the following we will deal with 'long crack' problems and will leave the discussion of small cracks to Chapter 9. At this stage it suffices to mention that in-depth reviews on the mechanisms and mechanics of crack growth and the significance of crack closure are given, for example, by Miller [38], Schijve [39], and Topper, Yu and DuQuesnay [40].

Two different mechanisms have been proposed to describe the fatigue crack extension. One is based on a plastic sliding-off at the tip of an advancing crack, while the other is in terms of damage accumulation at the crack tip. In the literature there is a lack of agreement as to which mechanism controls the crack growth process. Proponents of each mechanism offer experimental evidence in support of their respective model.

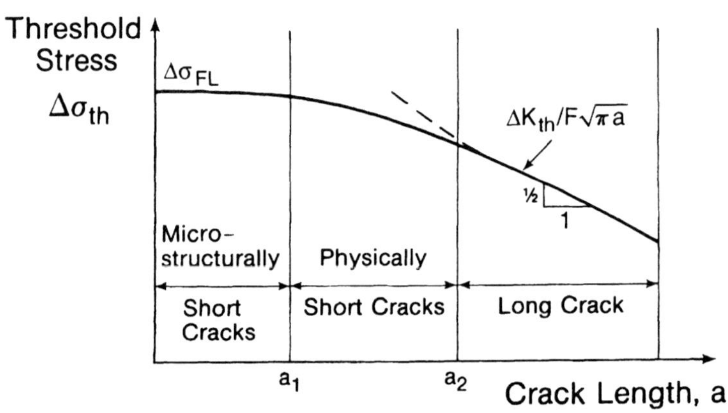

Fig. 7.20 Three regimes of crack growth.

The sliding-off mechanism of the crack extension process was initially proposed by Laird and Smith [41] and is now referred to as the plastic blunting model of fatigue crack extension. Figure 7.21 is a schematic representation of the crack tip opening, advance and blunting during the loading portion of a cycle (Fig. 7.21(a)–(c)). It can be seen that through the application of tensile load, highly localized plastic deformation takes place along the slip planes of maximum shear stress (45° to the applied uniaxial stress), Fig. 7.21(b). Upon further increase of the load, the width of the slip bands increases and the crack tip blunts to a semicircular shape, Fig. 7.21(c). As a result of this blunting the crack extends to about half the crack tip opening displacement. During compressive loading, the direction of slip is reversed and the distance between crack surfaces decreases.

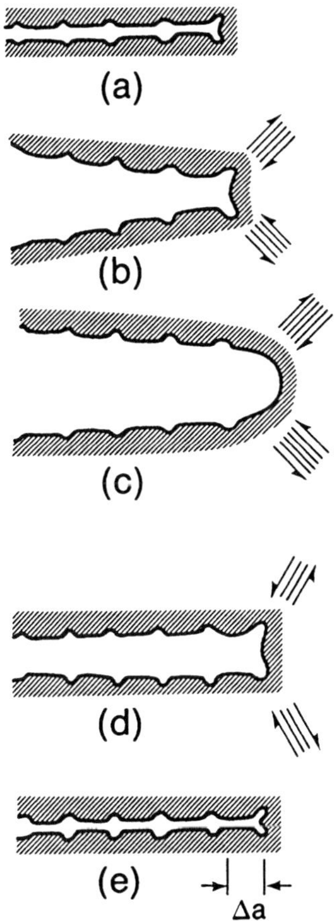

Fig. 7.21 Plastic blunting model of fatigue crack extension (Laird and Smith [41]). (a) zero load; (b) tensile load; (c) peak tensile load; (d) reversed loading; (e) peak compressive load. Slip direction is indicated by arrows.

The new surface created during tensile load is partly folded by buckling into a double notch at the tip, Fig. 7.21(d). At the maximum compressive stress, the crack tip is sharp again and the crack growth per cycle, Δa, is shown in Fig. 7.21(e). It should be noted that fatigue striations in this model are formed during the unloading part of the cycle.

Neumann's observation of the slip process in a single crystal copper [42] was an important contribution in advancing the sliding-off mechanism. In contrast to the Laird–Smith mechanism (Fig. 7.21), Neumann's observed crack tip was a V-groove with a constant angle. The 'blunting' process by plastic deformation takes place through widening of the 'V' at the constant crack tip angle, Fig. 7.22. During the compression part of the cycle, the crack closes by slip reversal. McEvily [43] has discussed various aspects of the sliding-off mechanics of the crack extension process, and the crack growth models based on the Laird–Smith–Neumann mechanism. Such models require cycle-by-cycle crack extension, and for the plastic deformation to occur in the crack tip, and the rate of crack growth to be of the order

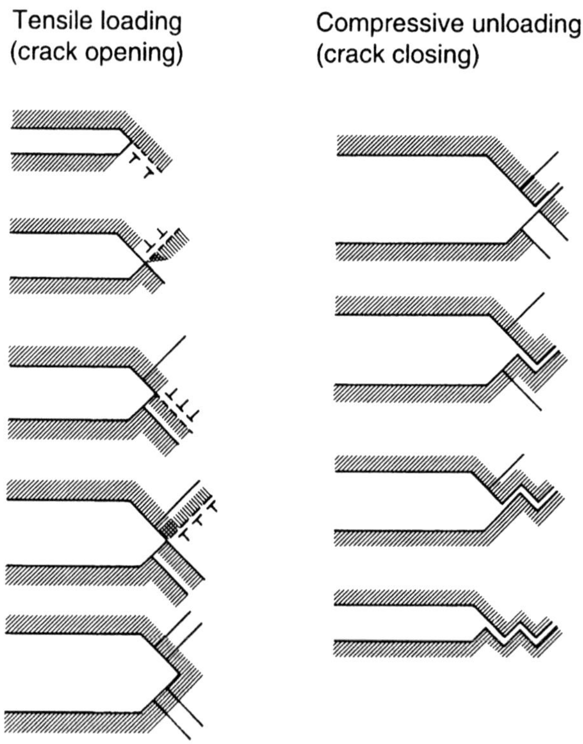

Fig. 7.22 Sliding-off mechanism of crack growth due to Neumann [42]. Alternating slip along two slip planes results in crack extension.

of one-half of the crack tip opening displacement (CTOD). There is experimental evidence to dispute this. Recent fatigue crack growth under high resolution conditions confirms that crack advance takes place intermittently. Only at high stress, where irreversible plastic deformation occurs on each loading cycle, may crack advance take place cycle-by-cycle. However, near the threshold region, plasticity is small and many cycles would be required for the crack to advance.

The crack growth rate for the sliding-off mechanism would require

$$\frac{da}{dN} \sim \text{CTOD} = \frac{\Delta K^2}{\sigma_0 E}. \tag{7.69}$$

The dependency on ΔK^2 is also not fully supported by the experimental evidence (e.g. Rice [11]). The use of the monotonic properties in (7.69) has also been criticized in the literature [44].

It was mentioned earlier that an alternative mechanism of fatigue crack growth is based on damage accumulation in a small element ahead of the crack. In this region reversed yielding occurs during unloading and reloading portions of a cycle. The crack advances through the region termed the process zone, when a sufficient amount of damage is accumulated. The 'failure' criteria are generally based on the low-cycle fatigue (LCF) process. A fatigue crack growth model based on the above mechanism will be presented in a later section. At this stage it suffices to mention that Baïlon and Antolovich [44], in reviewing various models of crack propagation, have identified micro crack tip LCF as the most appropriate mechanism.

The damage accumulation referred to in the above has been connected with the formation of dislocation subcell structures ahead of the crack. At low applied stresses, irreversibilities occur due to pinning of micro-cracks. Micro-cracks can also form at vacancy clusters caused by mobile intersecting of dislocations without formation of subcell structure (see Chapter 2 for a discussion of microstructure and cyclic loading).

Focusing now on the long crack regime (Fig. 7.20), the crack growth rate, da/dN, versus the range of stress intensity factor, ΔK, is shown in Fig 7.23 on a log–log scale. The curve typical of the fatigue growth rate at a prescribed condition (environment and stress ratio) is sigmoidal in shape and is bounded at the extremes by the range of stress intensity value at the threshold, ΔK_{th}, and the critical, ΔK_C. (The critical stress intensity range, ΔK_C, is also referred to as the cyclic fracture toughness, K_{CC}.) In the intermediate ΔK range, $\log(da/dN)$ is almost linearly related to $\log \Delta K$ as proposed by the Paris law [45]

$$\frac{da}{dN} = C(\Delta K)^m \tag{7.70}$$

where C and m are 'material' parameters and $\Delta K = K_{max} - K_{min}$.

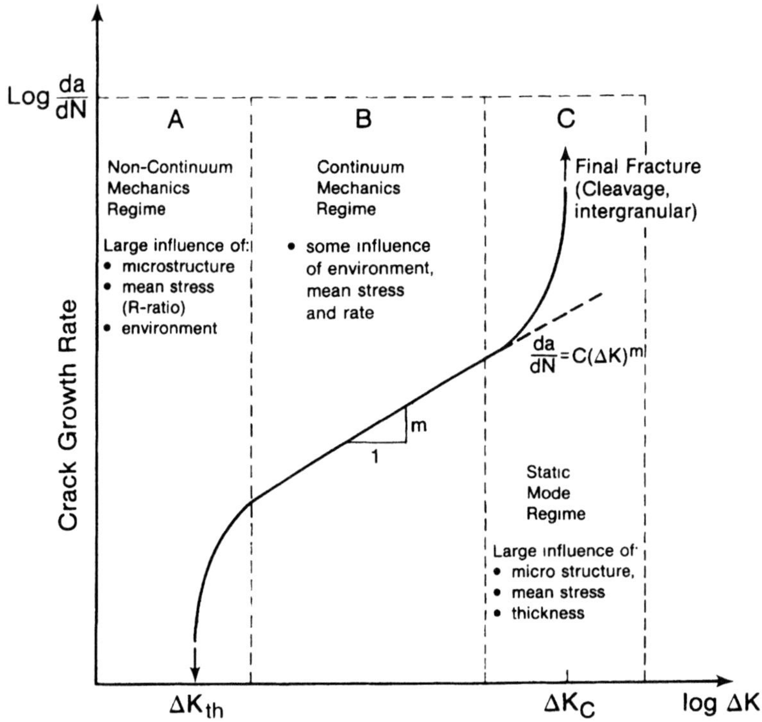

Fig. 7.23 The crack growth rate (da/dN) versus the range of stress intensity factor (ΔK) on a log–log scale showing three distinct regimes. (A) near threshold; (B) intermediate ΔK regime (Paris region); (C) static failure regime.

The threshold stress intensity, ΔK_{th}, is the stress intensity range below which a long fatigue crack will not grow. At the critical stress intensity, ΔK_C, the crack growth rate is fast and unstable.

To predict the crack growth rate for low and intermediate values of the stress intensity (Fig. 7.23) we require a crack growth model based on the principles of fracture mechanics and an appropriate microfracture mechanism. The first step in such an endeavour is to describe the stress and strain fields in front of a crack subject to cyclic loading.

7.6.1 ELASTIC–PLASTIC RESPONSE TO CYCLIC LOADING

A common feature of elastic–perfectly plastic solutions discussed by Rice [11, 12] is the proportional nature of the plastic flow, i.e. the plastic strain components remain in constant proportion to one another at each point of the plastic zone. In such circumstances, the response to unloading,

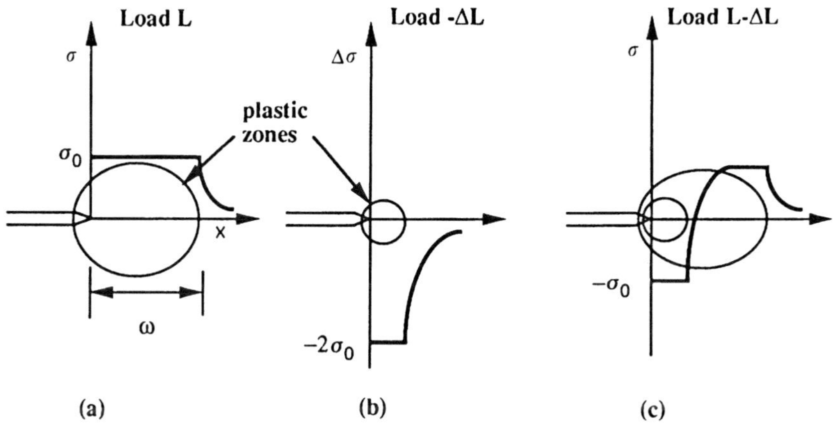

Fig. 7.24 Rigid plastic strip solution for (a) stresses for the applied load, L; (b) stresses for the reversed load, ΔL; (c) stresses from superposition [11].

reloading and cyclic loading is obtained through the plastic superposition method best illustrated in Fig. 7.24, from Rice [11].

The monotonic solution to the applied load L is shown in Fig. 7.24(a). Due to the very high strain concentration around the crack tip, a small amount of load reversal, ΔL, will cause a reversed plastic flow as shown in Fig. 7.24(b). The reversed plastic zone is smaller than that of the initial monotonic loading – in this case a quarter of monotonic loading, i.e. $\omega/4$. The response to unloading from L to $L - \Delta L$ is obtained by the superposition of Figs 7.24(a) and (b), as shown in Fig. 7.24(c). Reloading from $L - \Delta L$ to L leads to the initial configuration (Fig. 7.24(a)). Thus, the response to the unloading is obtained from the original monotonic loading by replacing loading parameters by ΔL and the yield stress and strain by **twice** the monotonic values. Except for the reversed direction, the distribution of the stress, strain and displacement remain the same, cf. Figs 7.24(a) and (b). The plastic superposition is valid as long as the reversed plastic zone is smaller or equal to the monotonic one, i.e. $\Delta L \leqslant 2L$. The equality sign applies for cases where the crack closure is negligible (crack closure will be discussed in section 7.8).

From the above, we see that the reversed plastic zone size and cyclic variation in stress, strain and displacements are obtained from the monotonic solution by replacing σ_∞ by $\Delta\sigma_\infty$ and σ_0 by $2\sigma_0$ and ε_0 by $2\varepsilon_0$, respectively.

We may similarly extend the monotonic work-hardening solution to the cyclic loading case by replacing the stress, strain and displacement components by their variations, $\Delta\sigma_{ij}$, $\Delta\varepsilon_{ij}$ and Δu_i and the monotonic stress–strain curve by the cyclic one, $\Delta\sigma_{ij} = f(\Delta\varepsilon_{ij})$, Chapter 2, (2.16) or (2.18).

For example, the HRR fields, (7.37), and RKE fields, (7.46), can be modified for a cyclically loaded cracked body by the following transformation:

$$K \to \Delta K\,, \quad \sigma \to \Delta\sigma\,, \quad \varepsilon \to \Delta\varepsilon\,, \quad n \to n'\,, \quad \sigma_0 \to \Delta\sigma'_y\,, \quad \alpha \to \alpha'\,, \tag{7.71a}$$

where n' is now the cyclic hardening component and σ'_y is the cyclic yield stress ($\Delta\sigma'_y \simeq 2\sigma'_y$), and

$$\alpha' = \frac{2E}{(2K')^{1/n'}\Delta\sigma_y'^{(n'-1)/n'}}\,. \tag{7.71b}$$

The above transformation is schematically shown in Fig. 7.25 where the origin of the loading portion of a cycle is placed at the lower tip of the stable hysteresis loop and that for the unloading at the upper tip. Since the loading and unloading branches are derived from a single curve, the material description corresponds to the Masing type behaviour (section 2.5). Therefore, within the framework of small-scale yielding, the deformation theory of plasticity and the Masing type cyclic stress–strain curve, the cyclic stress and plastic strain components of the HRR and RKE fields are given by

$$\Delta\sigma_{ij} = \Delta\sigma'_y \left(\frac{\Delta K^2}{\alpha'\Delta\sigma_y'^2 I_{n'} r}\right)^{n'/(1+n')} \tilde{\sigma}_{ij}(\theta; n')\,,$$

$$\Delta\varepsilon^p_{ij} = \frac{\alpha'\Delta\sigma'_y}{E}\left(\frac{\Delta K^2}{\alpha'\Delta\sigma_y'^2 I_{n'} r}\right)^{1/(1+n')} \tilde{\varepsilon}_{ij}(\theta; n') \tag{7.72}$$

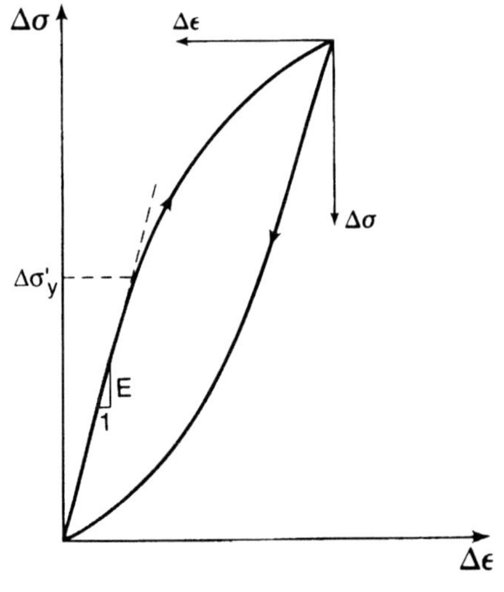

Fig. 7.25 Schematic representation of loading and unloading paths and the corresponding coordinate systems in terms of stress and strain ranges.

for the cyclic HRR fields, while the corresponding cyclic RKE fields are

$$\Delta\sigma_{ij} = \Delta\sigma'_y\left[\frac{\Delta K^2}{(1+\bar{n}')\pi\Delta\sigma'^2_y r}\right]^{\bar{n}'/(1+\bar{n}')}\tilde{\sigma}_{ij}(\theta)$$

$$\Delta\varepsilon^p_{ij} = \frac{\Delta\sigma'_y}{E}\left[\frac{\Delta K^2}{(1+\bar{n}')\pi\Delta\sigma'^2_y r}\right]^{1/(1+\bar{n}')}\tilde{\varepsilon}_{ij}(\theta)\,, \tag{7.73}$$

where

$$\bar{n}' = \frac{1+n'(\Delta W^p/\Delta W^e)}{1+(\Delta W^p/\Delta W^e)}\,.$$

7.6.2 FATIGUE CRACK PROPAGATION MODELS

In this section we are mainly concerned with the propagation of macro-cracks (long cracks). It is doubtful that the continuum plasticity solutions discussed earlier could be extrapolated to the microstructural scale (metallurgically short cracks), without due consideration of local properties.

A salient result of the elastic–plastic formulation for small-scale yielding described in the previous section is that the plastic deformation is entirely specified by the history of variation of the stress intensity factor, ΔK. The success of the Paris law, (7.70), to correlate data in the intermediate range of ΔK, is evidence of the aforementioned conclusion. It should also be noted that the range of small-scale yielding for cyclic loading is about twice that of the monotonic loading (Fig. 7.25). That is, the applied far field could be as high as $\Delta\sigma_\infty \approx \Delta\sigma'_y(A_{net}/A_{gross})$ where A is the cross-sectional area.

The fatigue crack growth rate, i.e. the extension of the crack length per cycle, da/dN, at a specific environment and temperature can be expressed in the form of

$$\frac{da}{dN} = f(\Delta K, R, C_i, \ell) \tag{7.74}$$

where $R = P_{min}/P_{max} = (\sigma_{min}/\sigma_{max})_\infty$ is the applied load (or stress) ratio, C_i's are material parameters such as cyclic yield stress, cyclic hardening exponent, etc. and ℓ is a characteristic length.

The earlier elastic–plastic crack growth models proposed by Head [46], McClintock [47], Liu [48], Frost and Dugdale [49], Rice [50], Weertman [51] and others, have been critically reviewed by Rice in [11]. To predict fatigue crack growth, one requires an expression for the stress, strain and displacement fields ahead of the crack, and a 'failure' criterion (separation of the surfaces). An averaging process over a characteristic length is required, since the stresses and strains are singular at the crack tip. Failure criteria can be classified into three categories, namely stress, strain or energy based

(Chapters 3 and 4). For example, in the aforementioned references, [46, 49] are stress-based leading to $da/dN \sim a^{3/2}$ or a, respectively; [47] is strain-based resulting in $da/dN \sim \Delta K^2$; and [11, 48, 50, 51] are energy-based predicting $da/dN \sim \Delta K^4$.

An alternative classification of fatigue crack propagation (FCP) models was given by Baïlon and Antolovich [44]. They divided the wide spectrum of models into three categories, namely phenomenological models, dislocation-based models and low-cycle fatigue-based models. The phenomenological models are variations of (7.70) to incorporate the influence of loading ratio, R, and environmental effects, etc. by modifying the parameters C and m to fit the experimental data, e.g. [52, 44], where further references to similar works are cited. These types of FCP equations are of limited usefulness as far as identifying intrinsic material properties which influence the rate of crack growth, and their predictive capabilities may not be valid for alloys different from that which the data was fitted to. Another major criticism of models of the form of (7.70) is that the constants C and m are not independent based on the dimensional analysis. That is, C and m are interrelated, [68].

Dislocation based models attempt to derive crack propagation by considering a physically sound, fundamental mechanism. However, due to highly idealized assumptions required to obtain a solution, some of the desired fundamental features are lost. Examples of this type of models are given by Weertman [53] and Yokobori *et al.* [54], where references to other similar works are listed.

Other review articles on various crack propagation models are also given by Yokobori [55], Kocanda [56], Knott [57] and Blom [58]. Explicit forms of (7.74) may be found in the above references as well as various textbooks on fatigue and fracture mechanics, [5–10].

A more recent review by Davidson and Lankford [59] which focuses on the physics of crack growth and the experimental micromechanics of crack advance, identified eight models which are physically based and consider damage accumulation in a zone ahead of the crack tip. A common feature of these models (see Table 6 in [59]) is that they assume damage to the microstructure to accumulate according to the low cycle fatigue properties of the material and that they include some definition of threshold condition (crack closure). One of the models belonging to this category will be discussed in the following section.

There has been so much work published on the fatigue crack growth rate that a complete review of all contributions will be well beyond the scope of this chapter. In the above we have attempted to identify the major approaches under which the various crack growth models can be classified. As such no attempt is made to include an exhaustive list of references, therefore omission from the limited reference list provided here is by no means a reflection on the quality of a particular contribution. The review articles cited above are to be consulted for the wealth of information they contain.

7.7 A LOW-CYCLE FATIGUE-BASED CRACK PROPAGATION MODEL

In this section we will describe a model for fatigue crack growth developed by Kujawski and Ellyin [60] and Ellyin [29]. In front of a stable growing crack, we distinguish three regions. A small region exists immediately ahead of the crack tip, denoted Region I in Fig. 7.26, whereby due to the crack tip blunting resulting from repeated loading and unloading cycles, the products of stress and strain have a finite magnitude as $r \to 0$. The extent of this zone, termed the 'fracture process zone', is denoted by δ^* in Fig. 7.26. Also, the stress and stress gradient in this region are much smaller than those predicted by (7.72) or (7.73). (Generally, singularity in the stationary problems is much stronger than that for a growing crack, e.g. Chitaley and McClintock [61]. At the same K or J, the strain is much larger near the tip in

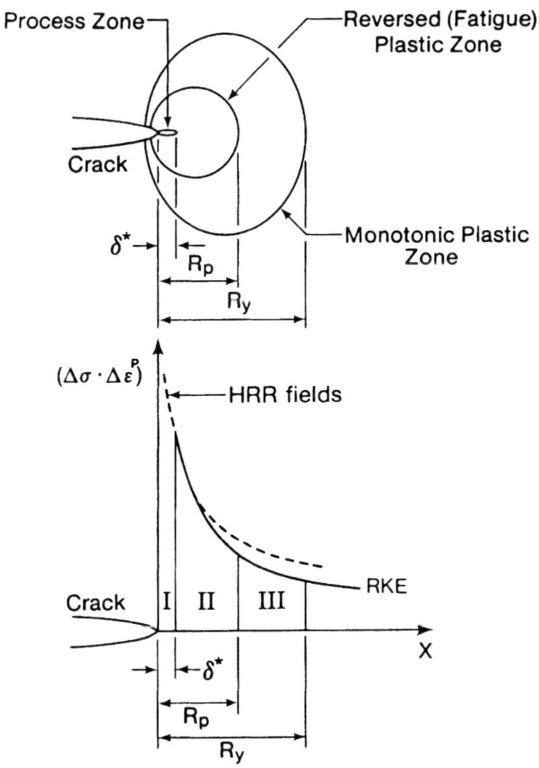

Fig. 7.26 Three regions in front of a fatigue crack (Kujawski and Ellyin [60]).

the stationary problem.) In the fracture process zone the plastic deformation is non-proportional and results based on the deformation theory of plasticity may not be accurate enough. We have already seen in Chapter 6 that in the case of non-proportional cyclic loading, there is an additional strain hardening compared to proportional cycling at the same deformation level. Thus for the same ΔK value, the material will exhibit lower straining. It is therefore fair to say that (7.72) or (7.73) are better representations of the stress and strain states beyond the process zone. Also, the effect of any crack closure at load ratio $R \geqslant 0$ may not be significant at $x \geqslant \delta^*$.

In the summary of their review, Davidson and Lankford [59] state that, 'the concept of a process zone in which damage is accumulating appears to be generally correct, as opposed to damage accumulating throughout the plastic zone'.

The second region in Fig. 7.26 is the reversed plastic zone embedded in the monotonic plastic zone and is denoted as Region II. We could expect a nearly proportional plastic deformation in this region. Indeed, the finite deformation elastic–plastic studies for the monotonic (stationary) crack have confirmed this type of behaviour (McMeeking and Parks [62]).

A crack growth law can be formulated by using (7.72) or (7.73) and a suitable failure (fracture) criterion. A critical plastic (or total) strain in front of the crack [47, 53, 61, 63–65] may not be appropriate for mode I loading (see McEvily [43] for a discussion). It has been demonstrated in Chapters 3 to 5 that an energy criterion is more suitable. A supply of energy is required to accomplish the processes which take place in front of a crack. Various methods of measuring the energy required to propagate a fatigue crack have been discussed by Fine and Davidson [66]. They divided the energy into two parts: hysteretic and permanent. As expected, experimental results indicate that the permanent (or damage) energy is negligible compared to that of the hysteresis energy (less than 1% of the latter). It was also concluded that fatigue crack propagation is not always a continuous process and one cannot ascribe a Δa for each cycle.

The fatigue crack growth model developed by Ellyin and co-workers [29, 60, 67] uses the hysteretic energy in front of the crack as a criterion for the crack advance. The stable hysteresis energy per cyclic and per unit volume (plastic strain energy per cycle and per unit volume) can be calculated from (Chapters 3–5)

$$\Delta W^{\mathrm{p}} = \left(\frac{1-n'}{1+n'}\right)(\Delta\sigma)_{\mathrm{eq}}(\Delta\varepsilon^{\mathrm{p}})_{\mathrm{eq}} \tag{7.75}$$

for a Masing material.

Using either HRR or RKE fields (7.72, 7.73) we get

$$\Delta\sigma_{\mathrm{eq}}\Delta\varepsilon^{\mathrm{p}}_{\mathrm{eq}} = \frac{\Delta K^2}{Er\psi(n',\theta)}, \tag{7.76}$$

where

$$\psi(n',\theta)=\frac{I_{n'}}{\tilde{\sigma}_{eq}(n',\theta)\tilde{\varepsilon}_{eq}(n',\theta)} \tag{7.77}$$

for the HRR fields, and

$$\psi(n',\theta)=\frac{(1+\bar{n}')\pi}{\tilde{\sigma}_{eq}(\theta)\tilde{\varepsilon}_{eq}(\theta)} \tag{7.78}$$

for the RKE fields.

The crack will propagate along the $y=0$ plane when (7.76) attains a critical value to be defined later. (On the crack plane $\theta=0$, and $r=x$). However, as mentioned earlier, we have to account for the crack blunting due to plastic deformation. The failure criterion is applied to a small material element defined by the process zone, i.e. for

$$x=\delta^*+\rho_c \tag{7.79}$$

where ρ_c is the critical blunting radius to be specified later. Evidence of crack tip blunting is clearly seen in the photomicrographs of [59].

For a specified stress and strain range, the cyclic plastic strain energy (area of the hysteresis loop per unit material volume) is given by (Chapter 4, (4.54) with $(\delta\sigma_0)_{eq}=0$ and $n^*=n'$)

$$\Delta W^p=\frac{1-n'}{1+n'}\Delta\sigma_{eq}\Delta\varepsilon^p_{eq}\,. \tag{7.80}$$

The number of cycles required for the material to 'fail'* cyclically at the specified stress and strain levels is obtained from (see (4.68a))

$$\Delta W^p_f=\kappa N_f^\alpha\,, \tag{7.81a}$$

and the total hysteretic energy to initiate a crack of a given size is

$$W_f=N_f\Delta W^p_f\,. \tag{7.81b}$$

For a cracked body, the RHS of (7.80) is obtained from (7.76) and (7.79), i.e.

$$\Delta W^p=\left(\frac{1-n'}{1+n'}\right)\frac{\Delta K^2}{E\,\psi(n')(\delta^*+\rho_c)}\,. \tag{7.82}$$

Let us denote the number of cycles required to penetrate δ^* (process zone) by ΔN. By equating (7.81a) and (7.82), we can then determine ΔN. However, material parameters κ and α are not generally known for different alloys. Therefore, as a first approximation, we would relate $\Delta\sigma$ and $\Delta\varepsilon^p$ to

* The word 'failure' is used here to signify initiation of a micro-crack of dimension $(\delta^*+\rho_c)$.

N_f, through the Basquin–Coffin–Manson relationships (3.5) and (3.9)

$$\begin{aligned}\Delta\sigma &= 2\sigma_f'(2N_f)^b \\ \Delta\varepsilon^p &= 2\varepsilon_f'(2N_f)^c ,\end{aligned} \tag{7.83}$$

where the symbols have already been defined in Chapter 3. Substituting from (7.83) into (7.80), we get

$$\Delta W^p = \frac{4(1-n')}{(1+n')}\sigma_f'\varepsilon_f'(2N_f)^{b+c} . \tag{7.84}$$

From equations (7.82) and (7.84) we obtain the conditions for crack advance for a distance δ^*. This could be interpreted as an averaged crack growth rate, $da/dN = \delta^*/\Delta N$, through the process zone. By setting

$$\frac{\Delta K^2}{E\,\psi(n')(\delta^* + \rho_c)} = 4\sigma_f'\varepsilon_f'(2\Delta N)^{-\beta} \tag{7.85}$$

where $\beta = -(b+c)$, we obtain

$$\delta^* = \frac{\Delta K^2}{4E\,\sigma_f'\varepsilon_f'\psi(n')}(2\Delta N)^{\beta} - \rho_c \tag{7.86}$$

or

$$\frac{da}{dN} = \frac{\delta^*}{\Delta N} = \frac{\Delta K^2}{4E\,\sigma_f'\varepsilon_f'\psi(n')}\frac{(2\Delta N)^{\beta}}{\Delta N} - \frac{\rho_c}{\Delta N} . \tag{7.87}$$

To find the critical blunting radius, ρ_c, we note that at the threshold stress intensity value, $\Delta K = \Delta K_{th}$, $da/dN \approx 0$, thus from (7.87) we get

$$\rho_c = \frac{\Delta K_{th}^2}{4E\,\sigma_f'\varepsilon_f'\psi(n')}(2\Delta N)^{\beta} . \tag{7.88}$$

From (7.86), (7.87) and (7.88) we obtain the crack growth rate relation

$$\frac{da}{dN} = 2\delta^*\left[\frac{\Delta K^2 - \Delta K_{th}^2}{4E\,\sigma_f'\varepsilon_f'\delta^*\psi(n')}\right]^{1/\beta} . \tag{7.89}$$

The above crack propagation rate is dimensionally correct independent of the value of β. For most metals β is in the range of 0.5 to 0.7 (Chapter 3). Equation (7.89) contains the following material properties: Young's modulus, E; cyclic hardening exponent, n'; and fatigue properties, σ_f', ε_f' and $\beta = -(b+c)$. In addition, it has a material length parameter, δ^*, associated with the process zone. We shall discuss the significance of this parameter later.

It may be of interest to note that Rice [11] proposed a somewhat similar crack growth law by considering that the material separation in front of the crack would occur when the total absorbed hysteresis energy equals a postulated critical value, U^*. Denoting the plastic displacement in front of the crack per load reversal by $\Delta u_y(x,0)$, the crack growth rate is given by

$$\frac{da}{dN}=\frac{4\sigma_0}{U^*}\int_0^{\omega^*}\Delta u_y(x,0)\,dx\,, \tag{7.90a}$$

where ω^* is the reversed plastic zone size (Fig. 7.26). In the case of small-scale yielding of the discrete surface model (section 7.3.1), we obtain

$$\frac{da}{dN}=\frac{5\pi(k+1)\sigma_0^2}{48G\,U^*}\left(\frac{\Delta K}{2\sigma_0}\right)^4, \tag{7.90b}$$

where k is the bulk modulus and G the shear modulus. However, Rice left the hysteresis energy required to create new fatigue crack surface undefined. The value of U^* can be obtained by equating (7.90b) with (7.89), but due to the rigid plastic material characterization, the physics of the solution will be lost.

Fine and Davidson [66] have written (7.90b) in the form of

$$\frac{da}{dN}=A\frac{\Delta K^4}{G(\sigma_y')^2U^*}, \tag{7.90c}$$

where A is a dimensionless universal constant, and σ_y' is now the cyclic yield stress. They then attempted to measure the plastic strain energy U^* integrated over the reversed plastic zone, ω^*. For the steel and aluminium alloys tested, the reported values were: $A=(2.9$ to $0.8)\times10^{-3}$ and $U^*=(0.17$ to $53)\times10^5\,\mathrm{J\,m^{-2}}$ [66].

7.7.1 PARTICULAR CASES

A number of proposed crack growth laws can be derived as a particular case of (7.89). For example, when ΔK_{th}^2 can be neglected in comparison to ΔK^2 (intermediate range of ΔK), then (7.89) can be written as

$$\frac{da}{dN}=2\delta^*\left[\frac{\Delta K^2}{4E\,\sigma_f'\,\varepsilon_f'\,\delta^*\,\psi(n')}\right]^{1/\beta}. \tag{7.91}$$

The Paris law (7.70) is immediately obtained with the coefficients

$$m=\frac{2}{\beta}\quad\text{and}\quad C=2\left[\frac{1}{4E\,\sigma_f'\varepsilon_f'\,\delta^{*(1-\beta)}\,\psi(n')}\right]^{1/\beta}. \tag{7.92}$$

The interrelation between m and C in the Paris law observed by several investigators is seen in (7.92) through the exponent β. In addition, the constants of the Paris law are fully determined in terms of the cyclic properties of the material, and the characteristic length (microstructural parameter) δ^*. Note that $\psi(n')$ depends on the singularity fields used, see (7.77) and (7.78).

The FCP model of Antolovich, Saxena and Chanani is discussed in [44], and is based on a plastic strain fatigue failure criterion. It can be written as

$$\frac{da}{dN} = 2\delta^* \left[\frac{C \Delta K^2}{E \sigma'_y \varepsilon'_f \delta^*} \right]^{-1/c}, \tag{7.93}$$

where C is a constant and c is the exponent in the low-cycle fatigue relation (7.83). Note that (7.93) could be obtained from (7.91) by setting $b = 0$ in $\beta = -(b + c)$, and $C = \sigma'_y / [\sigma'_f \psi(n')]$.

Equation (7.91) can also be written as

$$\frac{da}{dN} = \delta^* \left(\frac{\Delta K}{Z} \right)^m, \tag{7.94}$$

where $m = 2/\beta$ and $Z = 2^{-\beta/2} [4E \sigma'_f \varepsilon'_f \delta^* \psi(n')]^{1/2}$. Note that in the above equation, δ^* has the dimension of length, L, and Z has the dimension of $ML^{-1/2} T^{-2}$ (mass, length, time) which makes $\Delta K/Z$ dimensionless. The parameters δ^*, Z and m can be regarded as the material constants. A relationship of the form (7.94) was proposed by McCartney and Irving based on the dimensional analysis [68]. However, (7.94) contains specific relationships relating parameters δ^*, Z and m to fatigue and cyclic properties of the material.

Comparing (7.91) with (7.90c) we get $\beta = \frac{1}{2}$ and

$$\frac{\sigma_y'^2}{A} = \frac{4(1 + n')\psi^2(n')}{(1 - n')} (1 + \nu) E \varepsilon'_f \sigma'_f \tag{7.95a}$$

and

$$U^* = \frac{4(1 - n')}{1 + n'} \sigma'_f \varepsilon'_f \delta^*, \tag{7.95b}$$

which is the average cyclic (hysteretic) energy required for the crack to advance through the process zone, δ^*, in one cycle (cf. (7.84)). Note that A in (7.90c) depends on the cyclic hardening exponent n' and Poisson ratio and fatigue strength and ductility coefficients which could explain the variation in A reported in [66]. Thus, the undefined parameters in Rice's crack growth relation (7.90b) and those of Fine and Davidson (7.90c) are now fully determined in terms of the cyclic stress–strain exponent and the low-cycle fatigue properties of the material.

An alternative form of (7.91) is

$$\frac{da}{dN} = \gamma \left(\frac{\Delta K}{\sigma_y \sqrt{\delta^*}} \right)^m, \tag{7.96}$$

where

$$\gamma = 2\delta^* \left[\frac{\sigma_y^2}{4E \sigma'_f \varepsilon'_f \psi(n')} \right]^{1/\beta}.$$

A crack propagation law similar to (7.96) was proposed by Yokobori [55] after a critical review of the existing models up to 1979.

We noted earlier that the constants m and C in the Paris law (7.70) are interrelated. To discuss this further, let us equate (7.70) and (7.94) which results in

$$C = \frac{\delta^*}{Z^m} \tag{7.97a}$$

or

$$\log C = \log \delta^* - m \log Z\,. \tag{7.97b}$$

Such a relationship between C and m was observed experimentally by several investigators, e.g. see [55, 69, 70] among others.

From the foregoing discussion it is clear that the most common power law type relationship for the crack propagation rate can be shown to be a particular case of (7.89). The particular advantage of the relationship derived here, (7.89), is that it relates the coefficients to the basic fatigue properties which are readily available in the literature, see for example the compiled fatigue data in [71]. To illustrate this, let us study the exponent m in the Paris law (7.70). From (7.92) we get

$$m = \frac{2}{\beta} = \frac{2}{-(b+c)}\,, \tag{7.98}$$

where b and c are the fatigue exponents in the Basquin–Coffin–Manson relationships, (7.83). The value of exponent b varies from -0.05 to -0.15, and the exponent c varies from -0.4 to -0.8 for the majority of metals (see sections 3.2 and 3.3). The exponent β will thus vary between 0.55 and 0.85 making m vary between 2.35 and 3.64, which is in agreement with the slope found in [18].

7.7.2 COMPARISON WITH EXPERIMENTAL DATA

To compare predictions of the crack growth model, (7.89), with experimental results, we have to define the length parameter δ^* associated with the fracture process zone. This will be discussed later, and at this stage it suffices to mention that δ^* is of the order of the grain size of the material (10^{-5} to 10^{-6} m). It is in this zone that the microfailure mechanism takes place to allow for fatigue crack propagation.

An indication of the value of δ^* can be obtained by noting that when $\Delta K \to \Delta K_C$, experiments show an unstable crack growth rate. For the model developed here, this can be viewed as the crack growth through Region I of size δ^* (Fig. 7.26) at the instant of loading, i.e.

$$\Delta N = \tfrac{1}{2} \quad \text{for} \quad \Delta K = \Delta K_C\,. \tag{7.99}$$

Table 7.1 Material properties [60]

Material properties	10-Ni steel	Man-ten steel	2219-T851 Al alloy
σ'_y (MPa)	1106	331	334
E (GPa)	207	206	71
σ'_f (MPa)	2019	917	613
ε'_f	0.54	0.26	0.35
c	−0.645	−0.47	−0.55
b	−0.08	−0.095	−0.0756
n'	0.109	0.20	0.121
K' (MPa)	2177	1200	710
ΔK_{TH} (MPa m$^{1/2}$)	5	15	2.7
ΔK_c (MPa m$^{1/2}$)	150	95	30
$\delta^*_{(\Delta K = \Delta K_C)}$(m)	7.7×10^{-6}	1.12×10^{-5}	4.4×10^{-6}
$I_{n'}$	3.02	3.40	3.08

Substituting the above into (7.87) and using (7.88), we obtain the value of δ^* which is given in Table 7.1 for three different metals.

The function $\psi(n')$ at $\theta = 0$, in addition to the cyclic hardening exponent n', depends on the nature of the plane problem under consideration, i.e. plane stress or plane strain. For the HRR fields, $\psi(n')$ is given by (7.77) and it can be calculated from the curves given in [15, 23]. In the case of RKE fields $\psi(n')$ is given by (7.78) and can be determined analytically noting that $\tilde{\sigma}_{eq}(\theta = 0) = \tilde{\varepsilon}_{eq}(\theta = 0) = 1$.

Figures 7.27–7.29 show the comparison between the experimental crack growth rates and those predicted by (7.89). It can be seen that agreement between the predicted results and experimental data is fairly good. Also, the crack growth rate predictions based on the modified Rice solution, RKE fields (7.73), shown by dashed lines, yield results very close to those of HRR fields (7.72), shown by a solid line. In Fig. 7.29 the two lines coincide and this is shown by intermittent solid and broken lines. The advantage of using the expression for ψ from (7.78) is the availability of analytical expressions for all the parameters entering into the crack growth rate model, (7.89). In contrast, the parameters contained in (7.77) have to be evaluated numerically from the HRR singularity fields.

7.7.3 PROCESS ZONE SIZE

We will now comment on the material length parameter δ^* in Fig. 7.26. This parameter is associated with the microstructure and/or microfailure mechanism [72]. (Various proposed crack growth mechanisms were discussed in the first part of section 7.6.) The fatigue crack growth can be described as a local microfailure mechanism phenomenon ahead of the

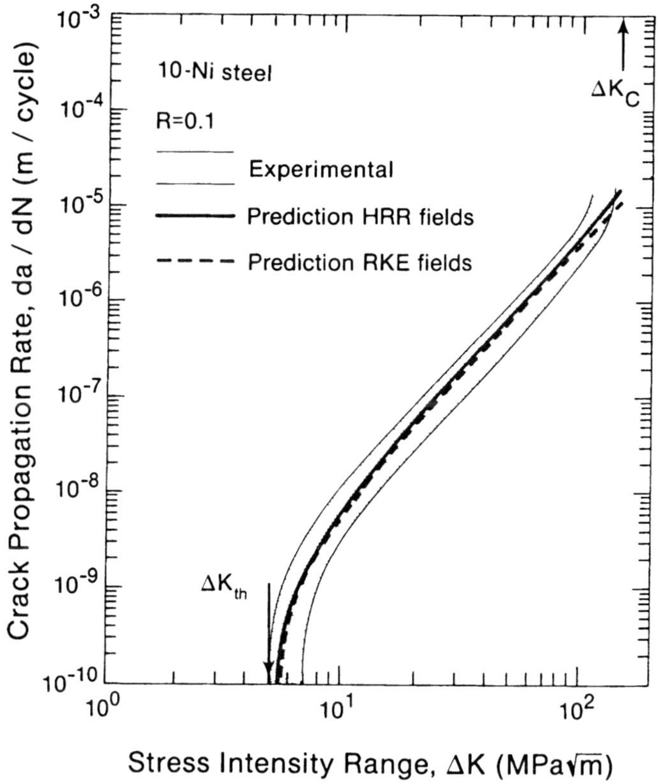

Fig. 7.27 Comparison between the theoretical and experimental data for 10-Ni steel. Reprinted with kind permission from [29].

crack tip. This mechanism is influenced by factors such as temperature, loading history, geometry and environmental conditions. In metals, the fatigue crack growth may be influenced by the grain size and other microstructural features such as distance between bands in dual phase steels, carbide spacing, mean cell diameter, etc. In metals and alloys with multiple slip capacity, the material ahead of the crack tip can be modelled as an aggregate of cells within the process zone δ^*. This zone can be defined as the set of cells that has reached the state of fatigue damage (7.81) which is proportional to the dissipated plastic energy. Outside the process zone, the damaging plastic strain energy decreases at a high rate due to the steep gradient of the stress and strain fields. The material properties within the process zone can be derived from the mechanical properties of the bulk material, if the size of the zone is large enough in comparison to the unit

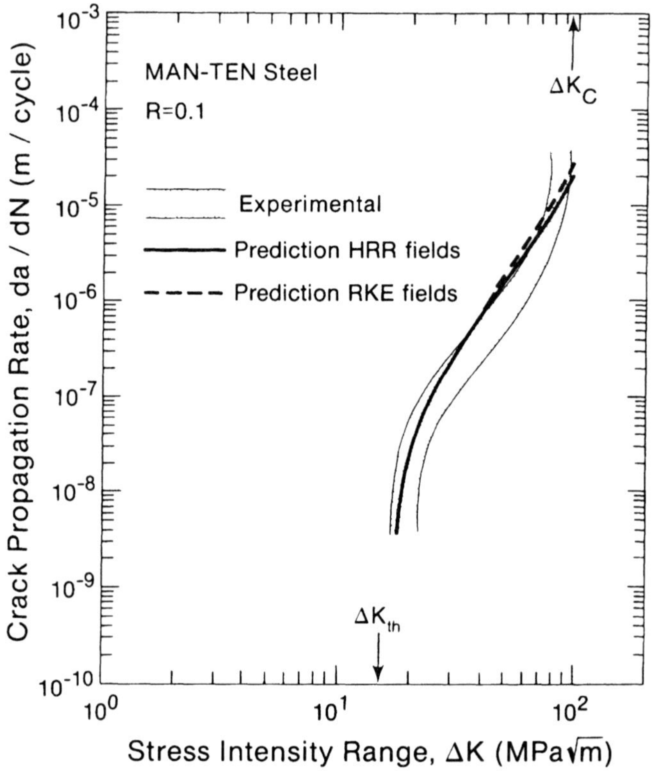

Fig. 7.28 Comparison between the theoretical and experimental data for Man-Ten steel. Reprinted with kind permission from [29].

cell dimensions. It has been reported that crack tip cell sizes for a variety of alloys range from 0.1–1.5 μm [59].

A discussion of the process zone can also be found in Baïlon and Antolovich [44]. They attribute the original notion to McClintock [47]. It was reported that for the low stacking fault energy of solid solution copper–aluminium alloys (higher aluminium), the process zone was about equal to the grain size. In the case of the higher stacking fault energy material (lower aluminium), the process zone was of the order of the cell size. The fatigue crack propagation rates decreased with the larger process zone size. The improved fatigue life was thus related to the size of the process zone, and planarity of the slip mode.

The grain size (or spacing parameter), λ, for different types of steels may be estimated from empirical relations proposed by various researchers, e.g. Gurland [73]. The ratio of δ^*/λ calculated for a number of steels was

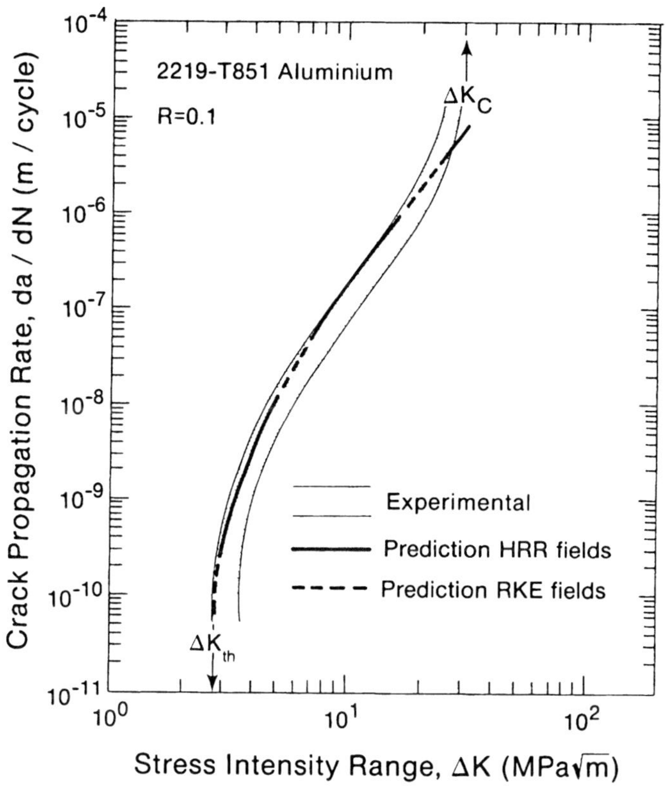

Fig. 7.29 Comparison between the theoretical and experimental data for 2219-T851 aluminium alloy. Reprinted with kind permission from [29].

always greater than one, thus indicating that δ^* is of the order of magnitude of λ. Therefore we can conclude that the size of the process zone, δ^*, is larger than the microstructural parameter λ, and the phenomenological approach based on the mechanical and cyclic properties of the bulk material can be applied to the process zone. In addition, Yokobori [55] has shown that for low carbon steel and 3% silicon iron, the coefficient C in the Paris law (7.70) is a function of the ferrite grain diameter. The crack propagation law (7.89) predicts Yokobori's observation, (7.97).

7.7.4 LOAD RATIO EFFECT

In the previous section, a crack growth model was developed for the case of load (stress) ratio $R = P_{min}/P_{max} = K_{min}/K_{max} \approx 0$. Experimental investigations have shown that the crack growth rate, da/dN, varies with R when all other conditions are kept constant. This section is concerned with the influence of the load ratio, R, on the fatigue crack growth behaviour at low and intermediate stress intensities.

The effect of the load ratio on the crack growth rate is depicted schematically in Fig. 7.30. This type of behaviour is observed in many metals and alloys including aluminium alloys [74], carbon and low carbon steels [52, 75–77], ferritic and perlitic steels [52, 78, 79], ultra-high strength steel [80, 81], cast steels [82, 83], stainless steel [84] and nickel alloys [85, 86]. As shown in Fig. 7.30, the shape of the curves when plotted on a logarithmic coordinate scale is sigmoidal with three distinct regions. It has been indicated earlier that the FCP in regions A and C is sensitive to the microstructure of the material, e.g. [80, 81]. The effect of the load ratio, R, on the fatigue crack propagation rates is also more pronounced in these two regions [80].

For some metals and alloys the R-ratio effect is less pronounced or non-existent in Region B [77–80, 85]. It has been suggested that the R-ratio and microstructure sensitive crack growth occurs when the size of the

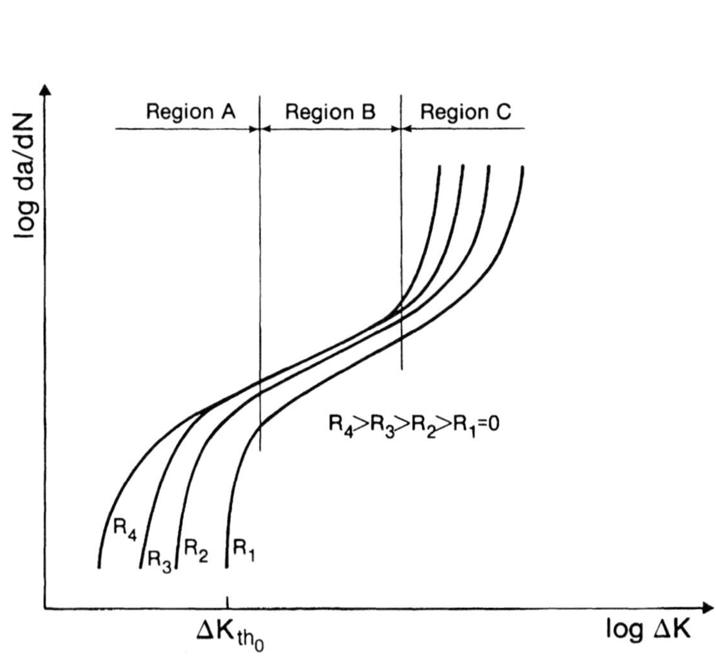

Fig. 7.30 Schematic illustration of the effect of load ratio, R, on the fatigue crack growth. Reprinted with kind permission from Kujawski and Ellyin [97].

cyclic plastic zone of the crack tip becomes equal to the grain size [80, 82, 85]. The quantitative influence of microstructure on the fatigue crack growth curve has yet to be determined.

Several concepts have been proposed to account for the R-ratio effect, namely, fatigue crack closure [87], residual compressive stresses [75], environmental influence by chemical reaction with the fresh surface at the crack tip when the crack opens [80, 85], etc. These concepts have been used to provide an explanation for the increased near-threshold growth rates at the higher load ratios. However, the validity of these explanations is restricted to the specimen type and the material used in experiments. For example, some materials tested in a vacuum [74, 85] exhibit no R-ratio dependence of the fatigue crack growth rate. This would then indicate that there is no closure or residual stress effect in these instances. Therefore, the above concepts do not entirely explain the reason why the R-ratio effect is less pronounced at the region B, and is diminished in a vacuum environment. The ability of the above parameters to rationalize the mean stress and environment effects have not been widely accepted, see for example [88]. Moreover, recent experimental measurements of crack closure indicate that the crack opening load, P_{op}, is not a unique value, and varies depending on the measurement location and the technique employed [39, 89, 90] (see also discussion in Chapter 9).

A number of investigators have studied the R-ratio effect on the near-threshold behaviour in laboratory air and vacuum in terms of the relation between ΔK_{th} and R [74–76, 78, 79, 91, 92] to mention a few. The general trends show that for many materials in air, ΔK_{th} tends to decrease with increasing R, but for some materials and $R > 0.5$, the sensitivity is less pronounced [93]. (See also [94, 95] for a general review of the threshold behaviour.) Most of the experimental results in a vacuum indicate that the R-ratio effect on ΔK_{th} is less pronounced or is diminished.

The influence of the R-ratio on ΔK_{th} can be described by a general equation of the form

$$\Delta K_{th} = \Delta K_{th_0} f(R)\,, \tag{7.100}$$

where ΔK_{th} is the threshold value for a given R-ratio and ΔK_{th_0} is the threshold value for $R = 0$. Several empirical and theoretical relations have been proposed for $f(R)$ in the above relation. The most popular empirical relationship between ΔK_{th} and R is the one proposed by Klesnil and Lukáš [75, 91]:

$$\Delta K_{th} = \Delta K_{th_0}(1 - R)^{\mu}\,, \tag{7.101}$$

where μ is a parameter which is chosen to fit the experimental data.

For various materials tested in laboratory air, a μ value in the range of 0.3 to 1 is obtained, but in a vacuum $\mu \approx 0$. Most of the experimental results in air appear to support the empirical relation (7.101) as a reasonable approximation of the threshold variation. It can thus be concluded that the influence of the R-ratio on ΔK_{th} and da/dN is also dependent on the environment and test conditions.

Following Elber's work [87, 96], the effective stress intensity factor ΔK_{eff} has been used to correlate crack growth under different cyclic loadings in region B of Fig. 7.30. The general form of this type of relationship is

$$\frac{da}{dN} = C(U\,\Delta K)^m\,, \tag{7.102}$$

where

$$U = \frac{\Delta K_{eff}}{\Delta K} = \frac{K_{max} - K_{op}}{K_{max} - K_{min}}\,.$$

A major difficulty with the application of (7.102) is the determination of the crack opening stress intensity factor, K_{op}. To alleviate this, attempts have been made to correlate U with the load ratio, R. Most functional relations have been of a linear form, except for Schijve [39] who introduced a quadratic form. Thus U can be represented as

$$U = C_1 + C_2 R + C_3 R^2 \tag{7.103}$$

where the C_i's are appropriate constants determined from fitting the experimental data (see [39] for the variation of constants).

Combining (7.102) and (7.103), the general form of the crack growth laws of the above type can be written as

$$\frac{da}{dN} = f_1(\Delta K, R)\,. \tag{7.104}$$

Although some experimental data, mostly for aluminium alloys, support the above form, a function of the form

$$\frac{da}{dN} = f_2(K_{max}, R) \tag{7.105}$$

would be more appropriate [52, 88]. The concept of crack closure will be discussed further in section 7.8.

In the following we will extend the crack propagation model given by (7.89) to include the R-ratio effect.

7.7.5 ANALYSIS OF THE STRESS RATIO IN THE CRACK TIP AREA

Let us now consider smooth and cracked fatigue specimens which are subjected to a constant amplitude cyclic loading such that the remotely applied load varies between P_{min} and P_{max}.

In the case of the smooth specimen, the distribution of stress $\underset{\sim}{\sigma}$ and strain $\underset{\sim}{\varepsilon}$ is homogeneous with the load ratio, $R = P_{min}/P_{max} = \sigma_{min}/\sigma_{max}$. In contrast, the distribution of the stress and strain at the crack tip area of the cracked specimen is highly inhomogeneous, and in this case, $R = P_{min}/P_{max} \neq \sigma_{min}/\sigma_{max}$ near the crack tip. If the specimen is unloaded, compressive stresses will be required to fit the plastic zone in the elastic surroundings. The stress distribution at the maximum load is the same as for monotonic loading, schematically shown by a full line in Fig. 7.31(a). Applying Rice's superposition argument of the reverse flow during unloading [11] (Fig. 7.24), the stress at the minimum load is shown schematically by the dashed line in Fig. 7.31(a). The cyclic stress history at the crack tip along the x-axis is depicted in Fig. 7.31(a)) by the zig-zag lines.

Due to the local elastic–plastic material behaviour, three different regions can be recognized (Fig. 7.31(a)), namely

- Region I or the cyclic plastic zone, r_c, where plastic deformation takes place during the loading and unloading half-cycles.
- Region II between the monotonic plastic zone, r_m, and the cyclic plastic zone, r_c, where plastic deformation occurs only during the loading part of the cycle, and is elastic during the unloading part.
- Region III or the elastic zone beyond the monotonic plastic zone, r_m, where cyclic strains are fully elastic during the loading and unloading.

It is evident that the stress ratio, $R_\sigma = \sigma_{min}/\sigma_{max}$, within these three regions is not constant and changes with the material element position along the x-axis. The stress ratio, R_σ, for a particular location is dependent upon ΔK and R. Schematic illustration of the stress ratio distribution in the crack tip area is shown in Fig. 7.31(b). It can be seen from this figure that in Region III where $x > r_m$, the stress ratio R_σ is approximately equal to the applied load ratio R. In Region II where $r_c < x < r_m$, the stress ratio, R_σ, will decrease (in an algebraic sense) as one moves towards the crack tip. In this region a steep gradient for the stress ratio is expected. The gradient of the stress ratio in Region II is dependent upon ΔK and R. The higher the value of ΔK and the lower the R, the steeper will be the stress ratio gradient.

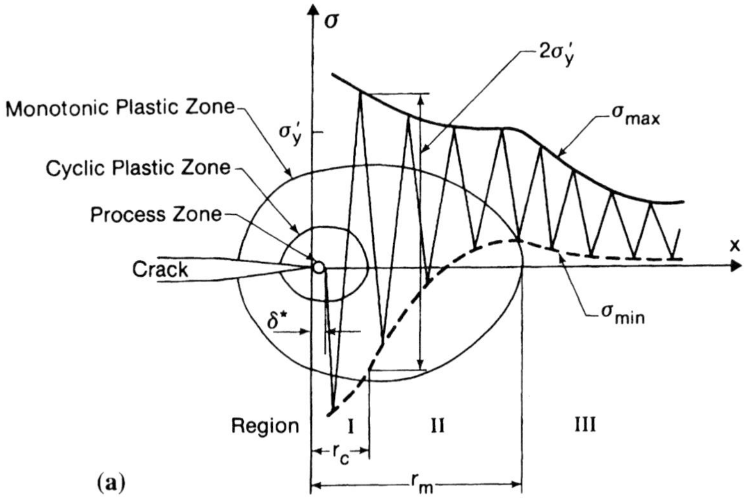

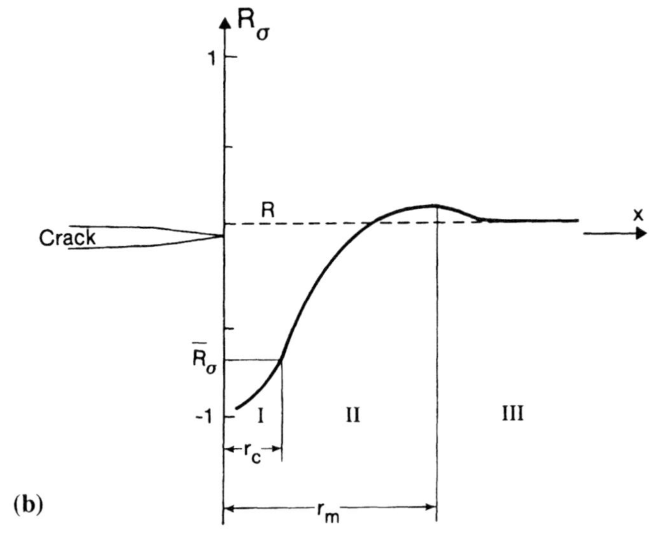

Fig. 7.31 Schematic plot of (a) the cyclic stress distribution, and (b) the stress ratio, R_σ, in the crack tip region. Reprinted with kind permission from [97].

For the material element at $x = r_c$, and assuming the kinematic hardening rule, the stress ratio, $\bar{R}_\sigma$, can be calculated as

$$\bar{R}_\sigma = \frac{\bar{\sigma}_{\min}}{\bar{\sigma}_{\max}} = \frac{(\bar{\sigma}_{\max} - \Delta\bar{\sigma})}{\bar{\sigma}_{\max}} = \frac{(\bar{\sigma}_{\max} - 2\sigma'_y)}{\bar{\sigma}_{\max}} = 1 - \frac{2\sigma'_y}{\bar{\sigma}_{\max}}, \tag{7.106}$$

where $\bar{\sigma}_{min}$, $\bar{\sigma}_{max}$ and $\Delta\bar{\sigma}$ are the minimum, maximum and range of the stress at $x = r_c$, respectively, and σ'_y is the cyclic yield stress of the material.

For a material element in Region I, the stress ratio R_σ will have a value between $\bar{R}_\sigma$ and -1. In this region the gradient of the stress ratio, R_σ, is expected to be smaller than that in Region II. For example, in the case of an elastic–perfectly plastic material, $\sigma_{max} = \sigma'_y$ for $x \leqslant r_c$ which implies a constant value of $R_\sigma = -1$ within Region I.

We have discussed in section 7.6.3 the existence of a small zone, termed the 'process zone' δ^*, immediately ahead of the crack tip. Within this zone, the stress has a finite magnitude with a small gradient. Consequently, as an approximation, a constant value of R_σ within the process zone δ^* can be assumed.

Based on the above description, a relationship will be derived later for the stress ratio, R_σ, depending on its location in one of the three regions mentioned above.

To obtain a crack growth law, we proceed in a manner similar to that in section 7.6.2, except that we have to account for the mean stress, σ_m, within the process zone. The mean stress effect was discussed in Chapter 3, and we use (3.7) to include the mean stress effect. (This is a fairly simple way of introducing the mean stress effect, however, this may be appropriate for small R values.) Thus (7.89) becomes [97]:

$$\frac{da}{dN} = 2\delta^* \left[\frac{\Delta K^2 - \Delta K_{th}^2}{4E(\sigma'_f - \sigma_m)\varepsilon'_f \delta^* \psi(n')} \right]^{1/\beta}. \tag{7.107}$$

Let us now introduce an 'effective' stress intensity range, ΔK_{eff}, defined as

$$(\Delta K_{eff})_0 = (\Delta K^2 - \Delta K_{th}^2)^{1/2}, \tag{7.108}$$

where subscript '0' refers to $R = 0$.

Therefore, the crack growth rate, (7.107), reduces to

$$\frac{da}{dN} = 2\delta^* \left(\frac{(\Delta K_{eff})_0}{[4E(\sigma'_f - \sigma_m)\varepsilon'_f \delta^* \psi(n')]^{1/2}} \right)^{2/\beta}. \tag{7.109}$$

It should be noted that when ΔK_{th}^2 is small in comparison to ΔK^2, then $(\Delta K_{eff})_0 \simeq \Delta K$. In comparing the above equation with that of the Paris equation, (7.70), we note that exponent m becomes a function of the fatigue exponents b and c only, and C is a function of the fatigue properties b, c, ε'_f, σ'_f, cyclic exponent n', modulus of elasticity E, mean stress σ_m and the material length parameter, δ^*.

As discussed earlier, empirical modifications of the power law relation (7.102) to account for the R-ratio effect have been proposed [52, 75, 91, 94–96, 98–101]. One of the contentious issues has been dependence of the da/dN on K_{max} which is not evident from (7.102) or (7.104). The relation derived in this section is of the functional form

$$\frac{da}{dN} = f\,[(\Delta K_{eff})_0, K_{max}, R, C_v, \ell\,], \tag{7.110}$$

which is a generalization of the earlier proposed relations (7.102)–(7.105) and is in general agreement with the observed experimental results. The dependence on K_{max} and R is through the mean stress, σ_m (see (7.111) in section 7.7.6).

A further advantage of the crack growth law (7.107) is that all the material parameters contained therein have well-defined physical meaning and are given in standard materials handbooks. In a similar manner to section 7.6.2, we can show that most of the power law crack growth models, e.g. (7.102), are special cases of (7.107).

It should be noted that in deriving relation (7.107) we have used the boundary condition $da/dN \approx 0$ at $\Delta K = \Delta K_{th}$. If ΔK associated with the crack opening is different from the familiar ΔK_{th}, then the appropriate stress intensity factor has to be substituted for ΔK_{th} (see discussion in section 7.8).

7.7.6 COMPARISON WITH EXPERIMENTAL DATA, $R \neq 0$

Six different types of steel were chosen for analysis and comparison with the predictions of (7.107). In addition to the crack propagation data, the fatigue and cyclic properties were also required in order to make such a comparison. All the relevant material data are given in Table 7.2. The values of δ^* were determined by using (7.107) and the experimental crack growth data at the region A and B of the da/dN vs. ΔK curve (Fig. 7.30). In the range B, due to the high plastic strain within the process zone and the resulting high mean-stress relaxation, the effect of the mean stress σ_m can be neglected. Using the above approximation and matching (7.107) with the experimental da/dN vs. ΔK curve at only one value of ΔK, the δ^* values were calculated. The values of δ^* thus determined are reported in Table 7.2.

To calculate the mean stress, σ_m, within δ^*, the following relation was used:

$$\sigma_m = \frac{\sigma_{max} + \sigma_{min}}{2} = \frac{\sigma_{max} + R_\sigma \sigma_{max}}{2} = \frac{1 + R_\sigma}{2}\sigma_{max}, \tag{7.111}$$

where σ_{max}, σ_{min} and R_σ are the maximum, minimum stresses and the stress ratio *at the end* of the process zone, respectively. The value of σ_{max} may be calculated from the RKE singularity fields, (7.46)

$$\sigma_{max} = \sigma'_y \left[\frac{K^2_{max}}{(1+\bar{n}')\pi\sigma'^2_y x}\right]^{\bar{n}'/(1+\bar{n}')} \tag{7.112}$$

Table 7.2 Mechanical and cyclic properties of various steel

Steels (Ref.)	E (GPa)	σ'_y (MPa)	n'	σ'_f (MPa)	ε'_f	c	b	ΔK_{TH} (MPa m$^{1/2}$)						δ^* (m)
								$R \approx 0$	$R = 0.1$	$R = 0.3$	$R = 0.5$	$R = 0.7$	$R = 0.8$	
0.050A [83]	209	400	0.171	1337	0.30	−0.569	−0.127	13			10			8.6×10^{-5}
C-Mn [83]	208	372	0.141	868	0.15	−0.514	−0.101	13			9			9.1×10^{-4}
M-Mo [83]	211	427	0.096	1116	0.78	−0.729	−0.101	14			11			7.6×10^{-4}
8630 [83]	207	682	0.122	1936	0.42	−0.693	−0.121	13			10			1.6×10^{-3}
A533-B1 [102]	200	345	0.165	869	0.32	−0.52	−0.085		7.7	5.4	4.5	2.9	2.8	8.5×10^{-5}
4340 [103]	209	724	0.146	1713	0.83	−0.65	−0.095		7.8		3.6			3×10^{-6}

The value of R_σ for a particular ΔK is calculated in the following manner [97], Fig. 7.31. Depending on the size of the process zone relative to the plastic zone, we have:

(i) when $\delta^* > r_m$, then $R_\sigma = R$. (7.113)

(ii) when $r_c \leqslant \delta^* \leqslant r_m$, due to the steep decrease of the stress ratio with the decreasing distance from the crack tip, the following semi-log relation is employed:

$$R_\sigma = \bar{R}_\sigma + \frac{\log \delta^* - \log r_c}{\log r_m - \log r_c}(R - \bar{R}_\sigma)\,. \tag{7.114}$$

(iii) when $\delta^* < r_c$, a linear relationship is used, i.e.

$$R_\sigma = \bar{R}_\sigma - \frac{r_c - \delta^*}{r_c}(1 - \bar{R}_\sigma)\,, \tag{7.115}$$

where $\bar{R}_\sigma$ is the stress ratio at the distance $x = r_c$ from the crack tip, see Fig. 7.31(b), (7.106).

Using the relation $K_{max} = \Delta K/(1 - R)$ and substituting (7.112) into (7.106), the value of $\bar{R}_\sigma$ can be calculated from

$$\bar{R}_\sigma = 1 - 2\left[\frac{(1-R)^2}{4}\right]^{\bar{n}'/1+\bar{n}'}\,. \tag{7.116}$$

It should be noted that near the crack tip $\Delta\varepsilon^p \gg \Delta\varepsilon^e$, and from the last of equations (7.73) it follows that $\bar{n}' \approx n'$.

Typical results of the fatigue crack growth data for the six types of steel investigated are shown in Figs 7.32–7.34 with R ranging from 0 to 0.8. Also included in these figures is the prediction of (7.107). It can be seen that the agreement between the predicted results and the experimental data is fairly good. The model predicts that the effect of the R-ratio for values greater than 0.6 is not significant (Fig. 7.33). This is to be expected since at these high stress ratios the crack will be open during the entire loading cycle. However, the agreement between the model predictions and experimental data for R-values greater than 0.5 appears to deteriorate, e.g. Fig. 7.33 for $R = 0.7$ and 0.8.

7.8 THE CRACK CLOSURE PHENOMENON

It has long been recognized that fatigue cracks do not open immediately upon load application. A fatigue crack remains 'closed' during part of the loading cycle. This then makes it difficult to determine the 'driving force', or the 'effective stress intensity factor', which causes crack propagation.

Crack closure occurs when the upper and lower surfaces of a crack come to contact and load is transmitted through the contact area. Elber [96]

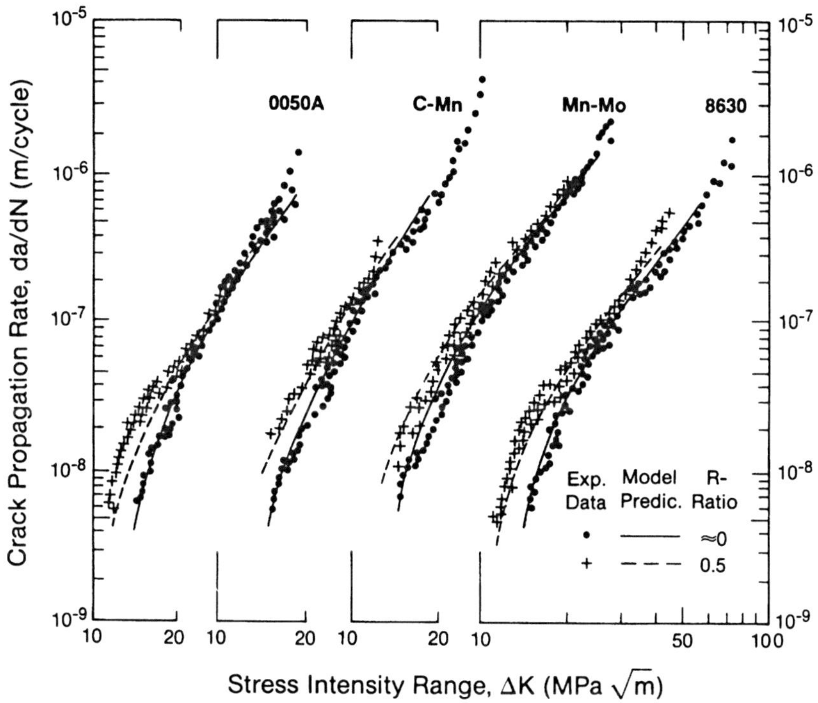

Fig. 7.32 Comparison of the predicted crack growth rate with the experimental data for four cast steels at room temperature. Reprinted with kind permission from [97].

was the first to report crack closure under cyclic tension by measuring crack opening behaviour. Closure mechanisms are rather complicated, varying from plasticity induced closure to closure due to roughness of separated surfaces, surface oxidation, bridging, crack filling, and asymmetric (tortuous) crack path. For a fixed loading and environment, more than one of the aforementioned mechanisms is operative. Consequently, it has proved difficult to predict crack closure based on a continuum modelling.

One of the extensively studied mechanisms is the plasticity induced closure. Even for this case, it has not been clearly established whether closure is caused by a band of stretched material in the wake of the growing crack, or by the compressive residual stresses ahead of the crack tip due to slip irreversibility. McEvily [104] has reviewed the relative importance of several closure mechanisms and concluded that the effect of the plasticity induced closure on crack growth is not as important as previously thought, except during overloads.

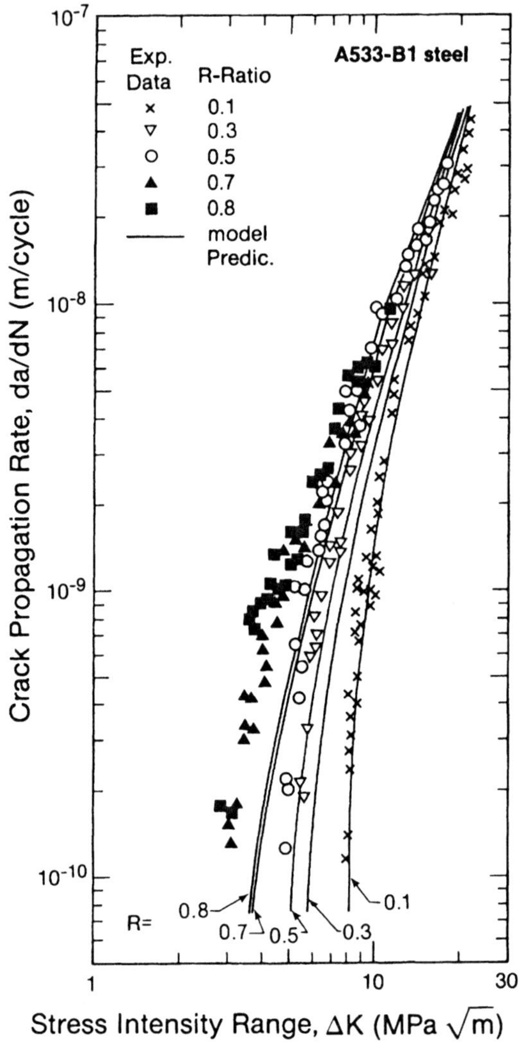

Fig. 7.33 Comparison of the predicted crack growth rate with the experimental data for A533-B1 steel. Reprinted with kind permission from [97].

It is worth mentioning that due to the irregular crack path topology, a perfect matching of two separated surfaces cannot be achieved in general. Bowles and Schijve [105] reported that cracks remain open on a microscopic scale. In a survey of experimental methods of crack closure measurements, Schijve [39] concluded that an 'accurate and unambiguous' indication of the stress level at which full crack tip opening or closure occurs is difficult to identify.

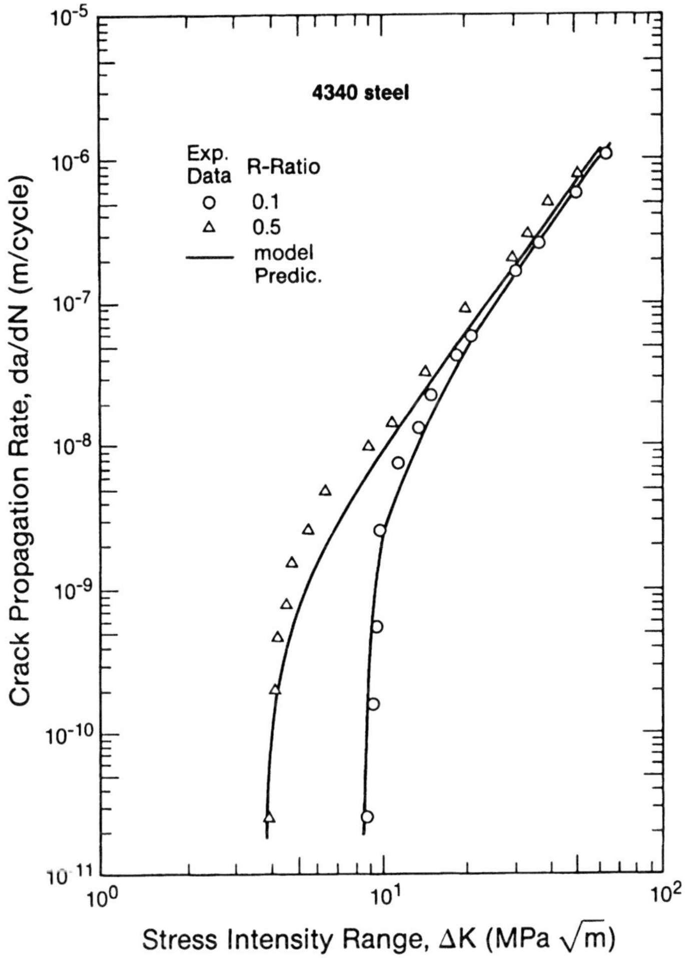

Fig. 7.34 Comparison of the predicted crack growth rate with the experimental data for 4340 steel. Reprinted with kind permission from [97].

Davidson and Lankford [59] have used high resolution stereoimaging photographs made at the minimum load and at progressively higher loads, to measure crack opening profiles. They have reported that at the minimum load most fatigue cracks are closed at the tip and for 'hundreds of micrometres' behind it, at least for $0 < R < 0.5$. The most important causes of crack closure are attributed to: (i) plastic deformation ahead and that left behind the crack tip, (ii) roughness of the crack surface, and (iii) growth of oxide on the crack surfaces (environmental interaction).

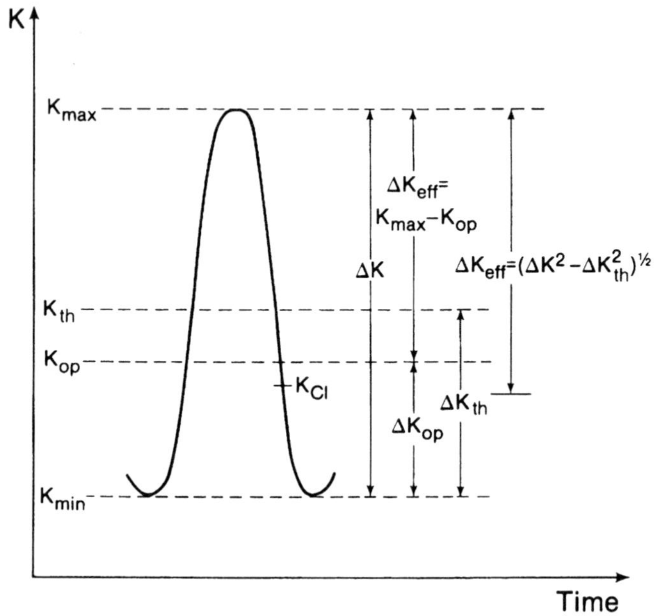

Fig. 7.35 Definition of various stress intensity factors and ranges.

Figure 7.35 shows various definitions of the stress intensity factor and its range associated with constant amplitude loading. As shown in the figure, generally the crack opening stress intensity $K_{op}(\Delta K_{op})$ is smaller than the threshold one, $K_{th}(\Delta K_{th})$, i.e. $\Delta K_{op} < \Delta K_{th}$. Furthermore, the crack opening or closure stress (or stress intensities) are not the same, i.e. $\Delta K_{cl} < \Delta K_{op} < \Delta K_{th}$. The effective stress intensity defined by Elber [96] and that defined by Kujawski and Ellyin [97] are also shown in the figure. Note that from (7.108) for $R \approx 0$,

$$(\Delta K_{eff})_0 = (\Delta K^2 - \Delta K_{th}^2)^{1/2} \approx \Delta K\left[1 - \frac{1}{2}\left(\frac{\Delta K_{th}}{\Delta K}\right)^2\right] \qquad (7.117)$$

$$> \Delta K - \Delta K_{th}\,.$$

Hudak and Davidson [106] have conducted extensive experimental measurements on crack closure on 7091 aluminium alloy and 304 stainless steel under constant amplitude loading. From the experimental results, they deduced that

$$\Delta K_{eff} = \Delta K\left(1 - \frac{\Delta K_{th}}{\Delta K}\right). \qquad (7.118)$$

Comparing (7.117) with the above, we see that the term $(\Delta K_{th}/\Delta K)$ appears in squared form with a 0.5 multiplier in our definition. Figure 7.36 shows the relationship between the two effective stress intensity factors for $R=0$.

For $R \neq 0$, the dependence on K_{max} and R is through the mean stress, σ_m (7.111). We note that while for $R \approx 0$, ΔK_{eff} defined by (7.117) is comparable to other definitions (Figs 7.35, 7.36), in the case of $R \neq 0$, the functional form is more complicated, i.e.

$$\Delta K_{eff} = \frac{(\Delta K_{eff})_0}{[1-(\sigma_m/\sigma_f')]} = \frac{(\Delta K_{eff})_0}{[1-(1+R_\sigma)\sigma_{max}/2\sigma_f')]}, \tag{7.119}$$

which shows dependence on σ_{max} (or K_{max}) and R and material cyclic properties and microstructure (cf. (7.113)–(7.116)). This type of dependence is observed experimentally, e.g. [39, 105, 106]. Note that for $R_\sigma = -1$, $\Delta K_{eff} = (\Delta K_{eff})_0$ which correspond to $R=0$ at the far field (Fig. 7.31). When the K_{op}/K_{max} is traced against the load ratio, R, we note a wide variation for different aluminium alloys and titanium alloys (Fig. 7.37). The K_{op}/K_{max} ratio varies from about 0.2 to 0.6 at $R \approx 0$. Therefore it is clear that simple definitions of ΔK_{eff}, although desirable, could not reflect different facets of crack closure phenomena.

An alternative form of representing the data in Fig. 7.37 is to plot crack opening stress (or stress intensity) in terms of the minimum stress (or stress

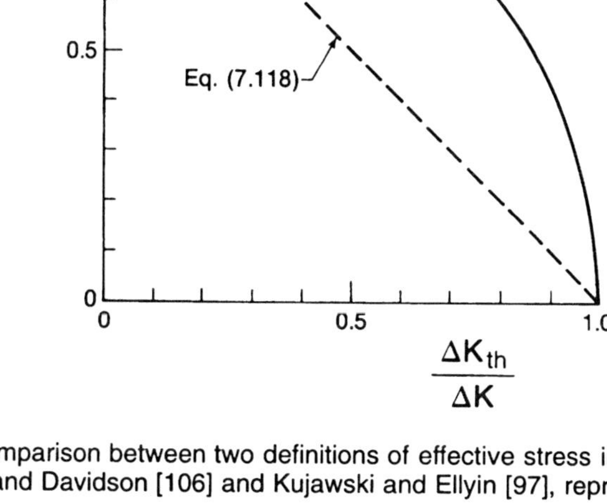

Fig. 7.36 Comparison between two definitions of effective stress intensity factor, from Hudak and Davidson [106] and Kujawski and Ellyin [97], reprinted with kind permission.

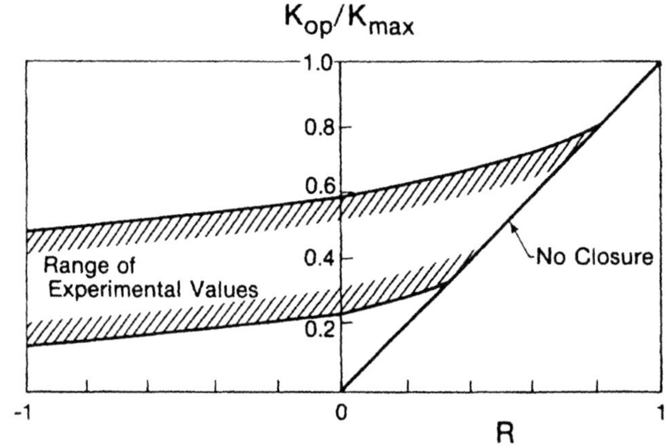

Fig. 7.37 Variation of K_{op}/K_{max} versus R constructed from experimental observation of various aluminium alloys and titanium alloys.

intensity) while keeping the maximum stress (or K_{max}) constant. (This type of test is preferable to that of the conventional ΔK decreasing test procedure, R = constant, when determining ΔK_{th} values [107].) Figure 7.38, taken from Topper, Yu and DuQuesnay [40], shows that the crack opening stress increases with the increase in the minimum stress for a load-controlled test with constant amplitude. The crack will open at zero stress when the minimum stress reaches S'', i.e. $K_{op}/K_{max} \approx 0$ at $R \approx -2$, and $K_{op}/K_{max} \approx 0.5$ at $R \approx 0$. At about $R \approx 0.6$, the crack remains open during the entire load cycle. From Fig. 7.38, it can be seen that $\Delta K_{eff}/\Delta K$ is minimum at

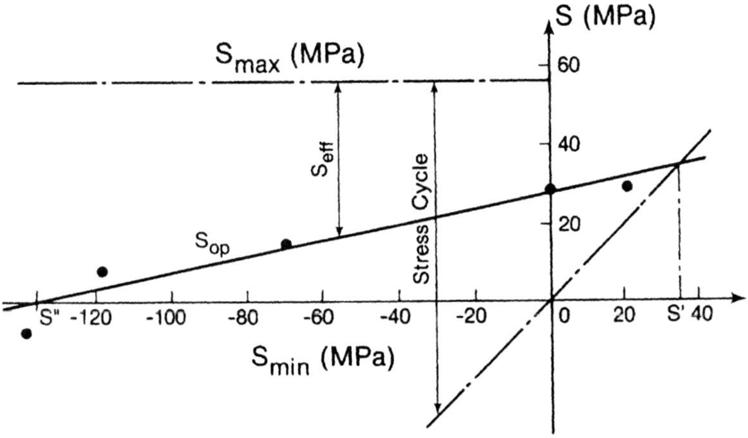

Fig. 7.38 Crack opening stress versus minimum stress for an SAE 1010 steel, from [40].

$R \approx 0$, and it increases with positive and negative R values. The effective stress intensity factor defined by (7.119) follows this trend.

Williams and Lam [108] have proposed an alternative effective stress intensity factor range, ΔK_{eff}, in terms of the contact stress intensity factor, K_{cont}, arising from the crack closure, i.e.

$$\Delta K_{eff} = \Delta K - (K_{cont,max} + K^*) , \tag{7.120}$$

where K^* is a 'material constant'. From the crack arrest experiments, the values of $K_{cont,max}$ and K^* were determined. They subsequently showed that for two aluminium alloys, 2024-T3 and 6061-T6, when the crack growth rate, da/dN, for constant amplitude, was plotted against ΔK_{eff} (7.120), the data collapsed to a single curve for three R ratios (0, 0.3 and 0.5).

Most crack closure models are based on the plastic induced closure and a few typical ones will be described in a later section. It will suffice here to mention that there have not been successful models to predict roughness induced closure or oxide induced closure, although both make important contributions to the closure phenomenon.

7.8.1 EFFECT OF VARIABLE AMPLITUDE LOADING

Crack growth under variable amplitude loading is a rather complex process and it can partly be explained through crack closure. For example, crack growth retardation and acceleration following an overload observed in experiments, was explained to a certain extent by Elber in terms of the plastic deformation left in the wake of a growing crack [87]. Schijve, in his comprehensive review article [39], discusses the problems involved with variable loading, e.g. the extension of the ΔK_{eff} concept to variable loading requires the capability of predicting the crack opening stress intensity, K_{op}, under the prescribed load history, accounting for the sequence effects (load interaction). The problem is compounded by the three-dimensional closure aspects, namely plane stress/plane strain transition, shear lips, and the resulting curved crack front (section 1.3.3). In addition, the plastic response is a history-dependent phenomenon which will affect the deformation behaviour in the crack front for a variable loading history compared to a constant amplitude one (see Chapter 6 for the transient effects depending on the strain history).

The aforementioned effects are of concern when attempting to apply the similitude concept, i.e. the crack extension for a given ΔK_{eff} in an individual load cycle of a variable history loading is the same as that of a constant amplitude loading. However, some experimental results seem to indicate that for a random loading history, the crack opening stress, σ_{op}, may reach an almost constant value [109, 110]. A critical issue in predicting the crack growth under variable loading is the contribution of a large number of small amplitude cycles. A slight change in the crack opening stress, σ_{op}, could have a significant effect on the damage caused by the small amplitude cycles, e.g. in vibratory systems.

The effect of an overload can be illustrated by examining a stationary crack subjected to a constant amplitude cyclic loading with a superimposed overload cycle. The results of an elastic–plastic finite element study by Ellyin and Wu [111] indicate that an overload ratio of 1.5 (50% increase in the peak stress) causes a three-fold increase in the plastic zone size ahead of the crack, i.e. $r_p \approx 0.307\ (K_I/\sigma_0)^2$ vs. $r_p \approx 0.932\ (K_I/\sigma_0)^2$ (Fig. 7.39). (The size of the plastic zone will increase with increasing overload ratio.) No plastic zone is observed in front of the crack at the peak of the next baseline cycle following the overload (point 7). This is due to the rather high value of the compressive residual stresses left after the unloading from the peak overload [111]. The diminished plastic zone at the peak applied load following an overload is an indication of the reduction of the crack driving force. Further evidence of this is the reduction of the maximum tensile stress near the crack tip following an overload [111].

The crack opening displacement corresponding to the reversal points of the ending cycles are shown in Fig. 7.40. It can be seen that while the crack opens with the increased load, no crack closure is observed following the overload cycle. This is because the material near the crack tip is stretched during the overload cycle and the crack tip is blunted (profile 5). An

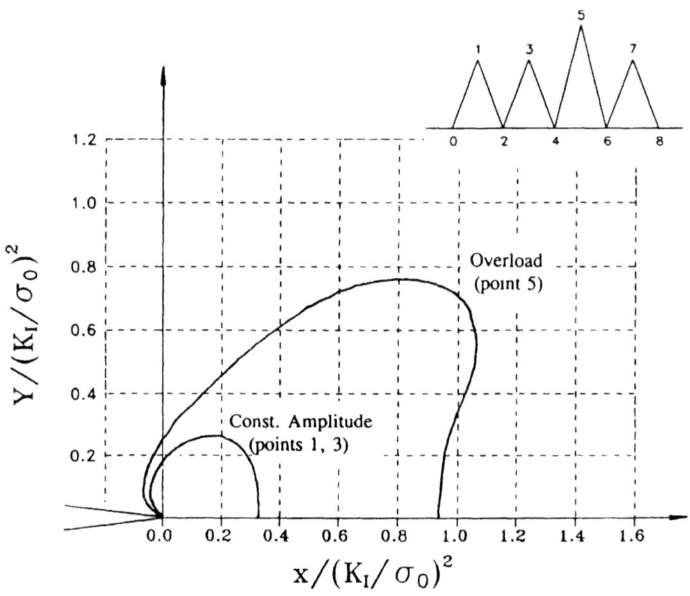

Fig. 7.39 A comparison of the plastic zone size of the constant amplitude base cyclic loading and that due to a 50% overload (see inset). Reproduced with permission from [111].

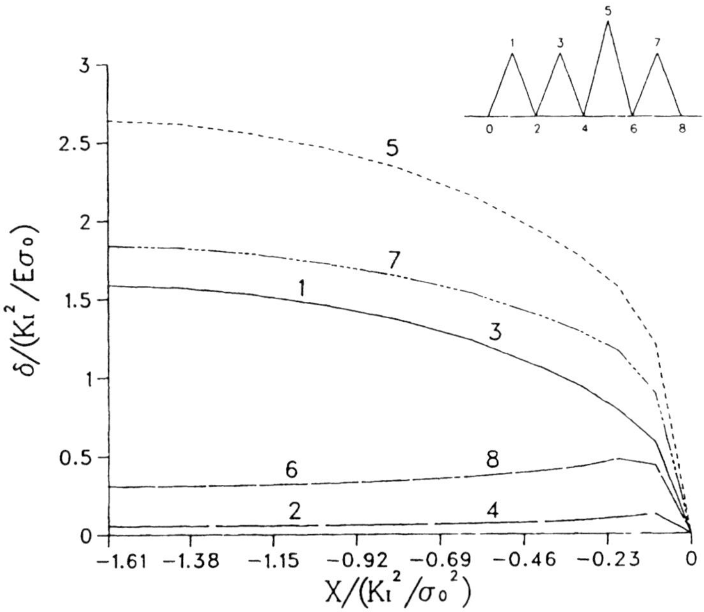

Fig. 7.40 Crack opening displacement at each reversal point of the loading history (see inset). Reproduced with permission from [111].

analysis of the crack opening profile indicates that the cyclic crack opening displacement range, Δu_y, is essentially unaffected as a result of an overload, but the details of the crack tip cyclic stress and strain fields are significantly altered [111]. This then indicates that global measures such as crack opening displacement are not good measures to indicate the influence of an overload in a stationary crack. While such a measurement will show an increased displacement due to the overload (profile 5 in Fig. 7.40) the subsequent measurement will indicate no appreciable change in the cyclic displacement range, i.e. the difference between profiles 7–8 vs. 3–4, is not noticeable. However, experiments show that an overload has a measurable influence on the crack growth. From the results of this analysis, it would seem that a clear indication of changes caused by a cyclic overload can be obtained by concentrating on the crack tip plastic zone, and the residual strain/stress patterns.

In the following we will discuss the changes in the crack growth rate following an overload. As discussed above, a tensile overload in a constant amplitude fatigue test will result in an increase of the plastic zone size and the tensile stretch in front of the crack tip as compared to the baseline cyclic loading. The plastically deformed material ahead of the crack tip will tend to keep the crack open causing a decrease in the opening stress magnitude,

σ_{op}. This will then result in an increased crack growth rate. However, as the crack grows into the overload plastic zone it will encounter higher residual compressive stresses which are produced by the overload residual strain upon unloading. These stresses will effectively reduce ΔK_{eff} in the crack tip, and thus the crack growth rate will decrease. In the meantime, the size and magnitude of the compressive residual stress behind the crack will increase, further decreasing the crack growth. This is in agreement with the delayed retardation following an overload. Figure 7.41 (taken from Ward-Close and Ritchie [112]) shows the effect of a single overload on the fatigue crack growth in IMI 550 titanium alloy with two different microstructural conditions. They concluded that the retardation phenomenon is the result of a sequence of several mechanisms. The application of an overload causes crack blunting and crack opening for some distance behind the crack tip which, coupled with the actual crack advance by the overload, results in an initial increase of the crack growth rate. When the crack growth reaches the overload plastic zone, the higher compressive residual stresses cause crack closure behind the crack tip, leading to progressively increased crack tip closure and subsequent delayed growth retardation. As the crack grows out of the overload zone, the above effects diminish and the baseline growth rate is reached (Fig. 7.41). The effect of the coarser-grained

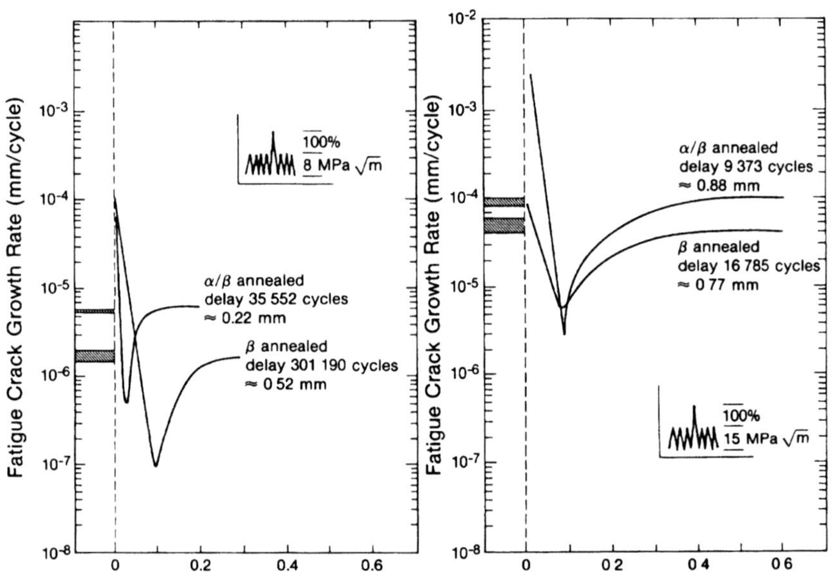

Fig. 7.41 Variation in fatigue crack growth rate for two different microstructure titanium alloys following a single 100% overload. (a) base $\Delta K = 8$ MPa m$^{1/2}$; (b) base $\Delta K = 15$ MPa m$^{1/2}$. From Ward-Close and Ritchie [112].

microstructure was a larger retardation period (Fig. 7.41). The closure mechanism at low ΔK (near threshold) levels appears to be controlled by the wedging mechanism (asperity-induced), see also [113].

At this stage of crack closure development, the consensus appears to be that the main reasons for the crack growth retardation following a single tensile overload are plasticity induced closure and irregular (rough) fractured surfaces including crack tip branching [39, 40, 105, 107, 112–117]. In particular, Fleck's paper [117] contains a critical review of the literature and test results, strongly supporting plasticity induced crack closure as the main mechanism of crack retardation following a tensile overload. Residual humps of stretched material were observed on the crack surfaces at the overload location behind the advancing crack.

Crack growth retardation due to an overload is most effective at low stress intensity ranges and high overload ratios (greater than 1.5), e.g. the experimental investigation by Kim and Tai [118].

From the foregoing discussion it is clear that several competing and cooperating events take place following an overload. First, there is a crack growth jump and then a stage of crack growth retardation. Crack tip blunting and compressive residual stresses result in immediate crack growth retardation while crack front irregularities and plasticity induced closure require a certain amount of crack growth after the overload. Furthermore, these mechanisms can also be divided into those that act in front of the crack tip and those that act behind and influence crack closure.

Multiple tensile overloads are known to cause additional retardation, e.g. [119, 120]. Topper and Yu [121] performed tests on a centre notched specimen of annealed SAE 1010 steel which was first cycled at a base load of 0–45 MPa, followed by three repeated tensile overloads (1.5 times base load) and subsequently cycled at the previous load level. The effect of the repeated tensile overloads was a significant retardation and even crack growth arrest for the near threshold base loading. Figure 7.42(a) shows crack length versus applied cycles which indicates significant crack growth before the application of overload and crack growth arrest sometime after the overload application. The variation of the opening stress following the application of the overloads is shown in Fig. 7.42 (b). It can be seen that the crack opening stress increases as the crack enters the overload plastic zone. This then leads to decreased ΔK_{eff} and crack arrest.

A compressive load following a tensile overload would partly negate the residual stresses built up by the tensile overload and could cause an accelerated crack growth in comparison to one with tensile overload only [122]. However, for the above to happen, the magnitude of the compressive stresses should be a high fraction of the compressive yield stress.

It is worthwhile noting that a number of earlier investigators have reported that the effect of compression excursions on the fatigue response and crack growth ($R < 0$) is negligible (e.g. [7]). The reason for this type of behaviour

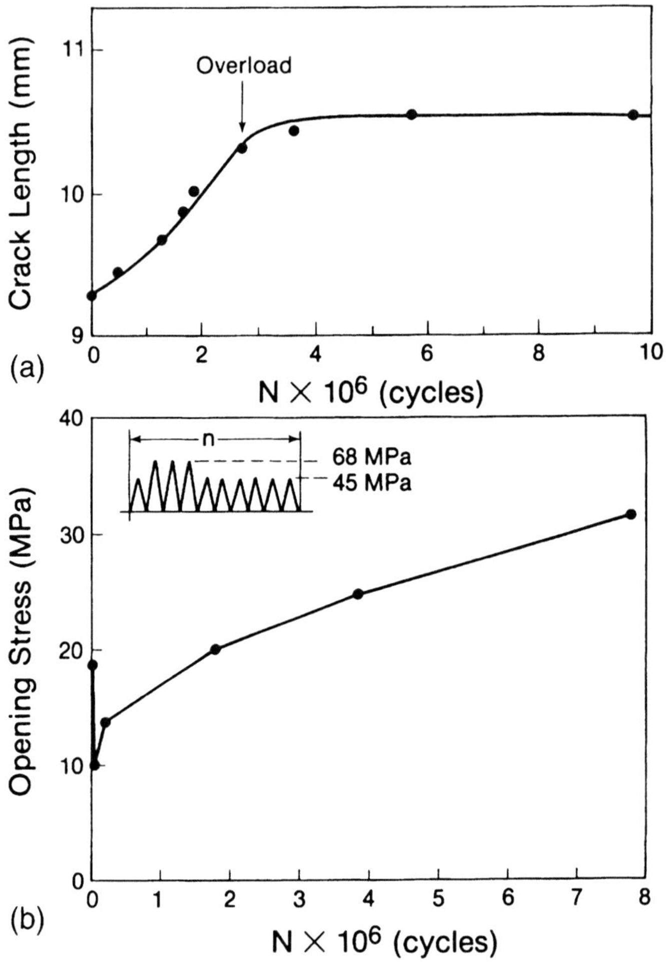

Fig. 7.42 Effect of multiple tensile overloads. (a) crack length versus number of cycles; (b) opening stress versus number of cycles (Reprinted with kind permission from Topper and Yu [121]).

is that the magnitude of the compressive stress compared to the yield stress was small. Stephens *et al.* [123] reported that cyclic loading with a negative R-ratio drastically reduced crack growth retardation. Tension–compression or compression overloads were detrimental.

The effect of multiple compressive underloads may be seen in Fig. 7.43, taken from Topper *et al.* [40, 121]. In Fig. 7.43(a) the reduction in the crack opening stress and its recovery following the compressive underloads is depicted. A comparison between the crack growth rate with and without (open circles) the compressive overloads is shown in Fig. 7.43(b). The crack

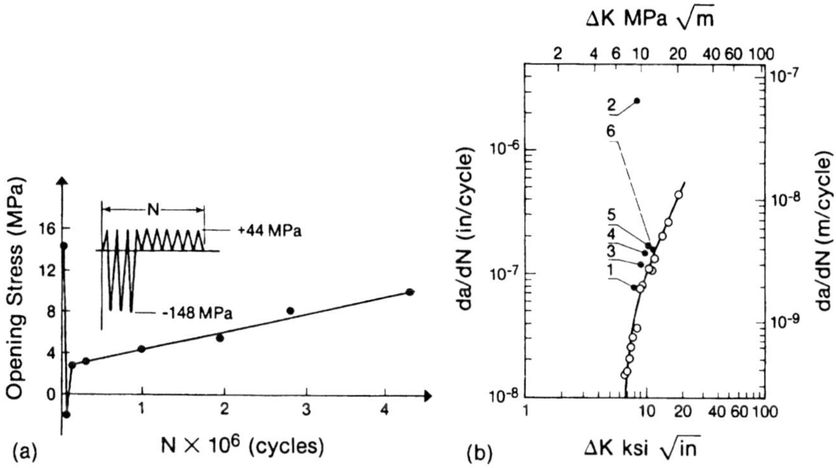

Fig. 7.43 Effect of compressive underloads on crack growth. (a) opening stress versus number of cycles following application of the compressive overloads; (b) crack growth rate versus stress intensity factor. Open circles are baseline data and filled ones are average growth after certain number of cycles (see text). From Topper and Yu [121] with kind permission.

growth rates after three compressive overloads are shown in the figure with dark circles, dot 2 shows the average crack growth rate 500 cycles after the compressive overload. The rate of crack growth increased by a factor of 33 from 1.98×10^{-9} to 6.6×10^{-8} m/cycle. Dots 3–6 are the crack growth rates after further cycling. The effect of the intermittent compressive underloads is a significant decrease in the threshold stress intensity and increased crack growth rate [121].

A comparison of the crack opening stress in Figs 7.42(b) and 7.43(a) shows that the opening stress increased with the tensile overload and decreased with the compressive underload in comparison to the steady-state opening stress before the overload application. However, immediately following a tensile or compressive overload, there is a decrease in the crack opening stress.

In a subsequent paper, Yu *et al.* [124] reported results on the effect of compressive peak stress on the fatigue crack growth. They concluded that in compression–tension tests the crack propagation increased and threshold stress intensity decreased linearly with the increasing magnitude of the compressive peak stress. It was also noted that during compression–compression cycling, the crack in a centre notched aluminium alloy 2024-T351 was not fully closed. Indeed cracks were initiated in the notch of the aluminium alloy which gradually became non-propagating. The extent of the crack growth was a function of the maximum and minimum peak stresses. (For crack initiation and growth in notches under compression–compression loading see section 8.5.)

Papers on crack growth under simple and complex/random load spectra, and further references on the influence of service/test environments may be found in ASTM Symposia [119, 125] and in more recent specialized conferences e.g. [126].

In a random spectrum loading such as the flight spectrum [127], various load interactions, e.g. tensile overloads, compressive underloads, preceded or followed by tension–tension, tension–compression, compression–compression, take place. The results of a round-robin analysis to predict the fatigue crack growth behaviour and lives for a specified random flight spectrum test case are given in [125]. The objective was to assess whether data from constant amplitude fatigue crack growth tests could be used to predict crack growth under random load sequence. A summary paper by Chang concludes that reasonably accurate predictions were achieved and that there is a trade-off between computational costs and prediction accuracy.

In the following section we will discuss some of the analytical and numerical methods used to predict crack closure and load interaction effects.

7.9 CRACK CLOSURE MODELS

Various attempts have been made to model crack closure phenomena. They can be grouped into two categories, namely analytical models, e.g. [127–132] and finite element models [133–144], among others.

Most of the analytical models are based on the extension of the Dugdale–Barenblatt model which was discussed in section 7.3.1. It would be instructive to review briefly the stationary crack solution for the case of small-scale yielding for loading and unloading cases. The crack opening displacement with the condition $\sigma_\infty/\sigma_0 \ll 1$, is plotted in Fig. 7.44(a) for $K = K_{max}$. Using Rice's plastic superposition method (section 7.6.1) we obtain the unloaded deformed shape at $K = K_{min} = 0$ as shown in Fig. 7.44(b). Under constant amplitude cyclic loading between K_{max} and K_{min}, the stationary crack changes between the two configurations shown in Figs 7.44(a) and (b). When this figure is compared with the finite element results presented in Fig. 7.40, profiles 1 and 2, one notes the idealized nature of the Dugdale–Barenblatt model. Due to the applied cohesive forces along the plastic zone (Fig. 7.9) the opened crack profile takes a concave shape rather than the convex profile resulting from the crack tip blunting.

In the case of a propagating crack, the crack grows through the plastically stretched material in front of it (Figs 7.15 and 7.39) which alters the displacement profile behind the crack due to plastically stretched material

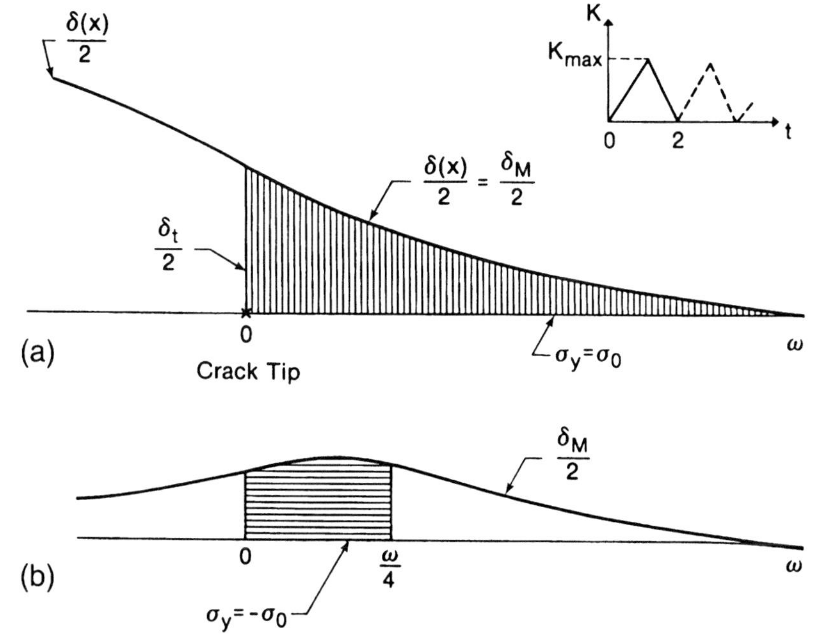

Fig. 7.44 Displacement profile of a stationary Dugdale–Barenblatt model. (a) $K = K_{max}$; (b) unloading to $K_{min} = 0$. Reproduced with permission from ASME from [128].

left in its wake. We will now discuss some of the techniques used to model the aforementioned process.

7.9.1 BUDIANSKY AND HUTCHINSON MODEL

Budiansky and Hutchinson [128] modified the Dugdale–Barenblatt model shown in Fig. 7.44 by assuming (i) that the crack grows always under constant K_{max}, and (ii) that the upper and lower crack surfaces are in contact all along their length. The thickness of residually stretched material, δ_R, and the length of the reversed plastic zone, R_p, were obtained from solving the boundary value problem shown in Fig. 7.45. δ_M in this figure refers to the stationary profile shown in Fig. 7.44. The results indicated that the reversed yielding zone, R_p, was about 0.1ω in contrast to 0.25ω in the stationary crack; and that the residual stretch $\delta_R \approx 0.86\,\delta_t$, where δ_t is the crack tip displacement shown in Fig. 7.44(a).

For a positive load ratio, $R = K_{min}/K_{max} > 0$, the results of the analysis are shown in Fig. 7.46 which indicates the variation of the opening or contact values of the stress intensity factor in terms of the load ratio R. A comparison between Figs 7.46 and 7.37 indicates that the predicted results

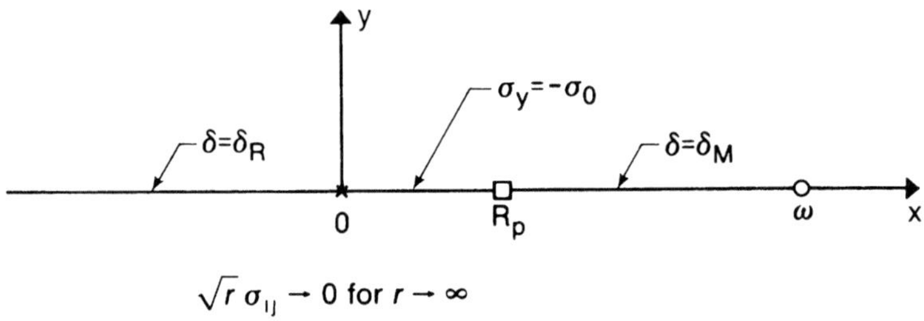

Fig. 7.45 Formulation of a steady-state growing crack, from Budiansky and Hutchinson [128] with permission from ASME.

constitute an almost upper bound to the experimentally observed curves for aluminium and titanium alloys. However, when the effect of cyclic hardening is incorporated by increasing or decreasing the yield stress, the curves shown in Fig. 7.46 shift upwards for hardened material, and downwards for a softening material. Despite its simplifying assumptions,

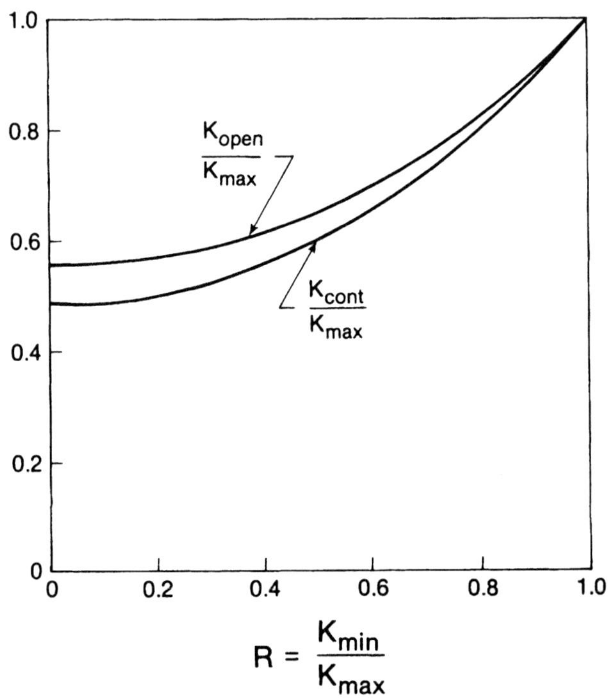

Fig. 7.46 Opening and contact stress intensity ratios versus the applied load ratio, *R*, from Budiansky and Hutchinson [128] with permission from ASME.

the prediction of this model is in fairly good agreement with the experimental observations.

7.9.2 OTHER MODELS BASED ON YIELDING ON A DISCRETE SURFACE

There are other crack closure analyses based on the extension of the Dugdale–Barenblatt model, e.g. Führing and Seeger [129], Newman [130], de Koning and Liefting [131], Ibrahim, Thompson and Topper [132], among others. These models differ in the way in which the residual displacement left in the wake of an advancing crack is incorporated in the Dugdale–Barenblatt model. For example, in the Führing and Seeger model the residual displacement distribution was obtained from assuming a crack extension through an imaginary cut while the load is kept constant at its maximum level. The crack opening displacement is then determined for the 'cut model' and is subsequently used as a residual displacement for the unloading calculations of the physical crack of length $2a$. Crack closure occurs when the upper and lower crack surfaces come into contact, and the contact stress depends on the crack tip opening displacement and the unloading displacement.

Newman [130] modelled the crack region by rigid–perfectly plastic bar elements with a flow stress $\sigma_{flow} \approx (\sigma_0 + \sigma_{ult})/2$. He assumed that in the wake of the crack the broken elements (Fig. 7.47(a)) carry compressive stresses only when in contact and yield at a stress level of $-\sigma_{flow}$. At the front of the crack (plastic zone) an element yields at a tensile stress level of $\alpha\,\sigma_{flow}$, where α is a constraint factor to simulate plane stress ($\alpha = 1$) and plane strain ($\alpha = 3$) conditions. It should be noted that the Dugdale–Barenblatt model is not suited for a plane strain analysis (cf. plastic zone shapes in Fig. 7.15), and the use of modified flow stress is an approximate way to simulate plane strain conditions. The shaded regions in Figs 7.47(a) and (b) indicate the material which is stressed in the plastic state.

This model essentially shows the trends predicted by Budiansky and Hutchinson's analysis. The crack opening stresses were then used as an input to an empirical crack growth law to predict crack extension due to constant and variable amplitude loads [130].

Newman has extended the above model to a crack emanating from a circular hole in a finite width specimen [127]. The modified model was used to calculate the applied stress load, s_0', at which the crack surfaces fully open as a function of crack length and load history. Figure 7.48 shows a typical result of calculated s_0', normalized by the maximum applied stress, s_{max}, as a function of crack length, a for a constant amplitude loading with $R = -1$. The crack configuration depicted in the upper part of the figure is to simulate an inclusion or void. It can be seen that as the crack grows, the crack opening stress rises rapidly and reaches a steady state at

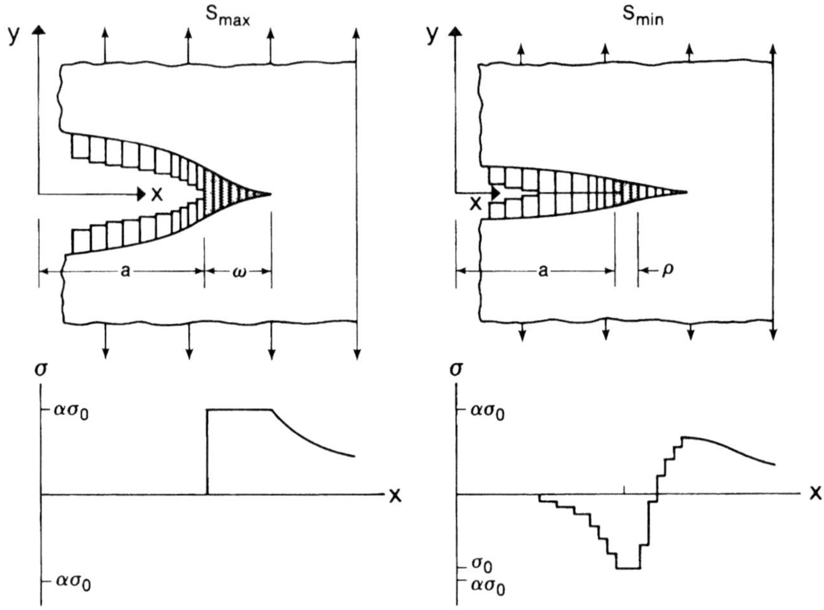

(a) Maximum stress (b) Minimum stress

Fig. 7.47 Crack surface displacement and stress distribution at maximum and minimum applied stress, from Newman [130].

$a > 20 - 40\,\mu$m. Crack opening equations found from fitting to the calculated results are given in [127]. Experimental data for fatigue crack growth rates for two high strength aluminium alloys for three R-ratios of -1, 0 and 0.5 correlated fairly well with the ΔK_{eff} calculated from

$$\Delta K_{\text{eff}} = K_{\max} - K_0' = \Delta K \frac{1 - s_0'}{s_{\max}(1 - R)} \tag{7.121}$$

except near the threshold region. Near the threshold crack growth region, closure in Mode II may dominate crack growth rate e.g. [59].

The de Koning and Liefting model [131] is essentially a combination of the two approaches by Führing and Seeger [129] and Newman [130]. The plastically deformed region in the wake of the crack is treated similarly to [129] and numerical treatment is by discretization into uniaxial elements of constant stress, similar to [130]. The effect of a single overload or a single underload and variable loading are then investigated.

In summary, an analytical crack closure model can give a better insight into the mechanism of plastically induced crack closure than a limited experimental investigation. It also enables one to study systematically the effect of various factors on the crack growth. However, due to inherent

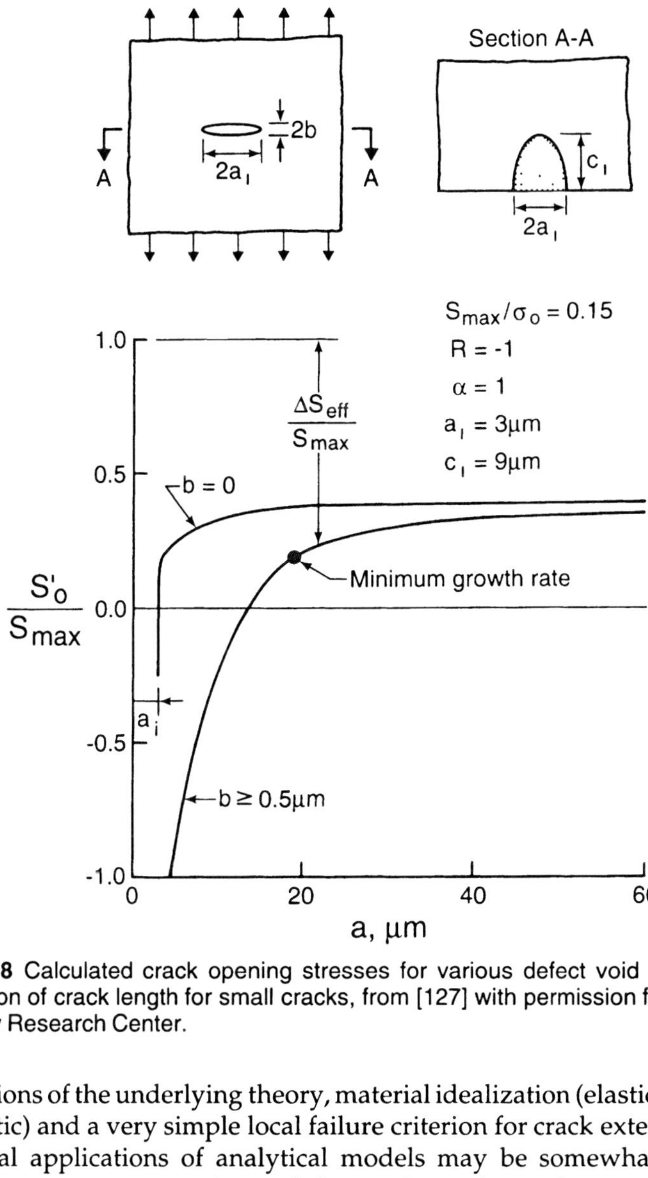

Fig. 7.48 Calculated crack opening stresses for various defect void heights as a function of crack length for small cracks, from [127] with permission from NASA Lougley Research Center.

limitations of the underlying theory, material idealization (elastic–perfectly plastic) and a very simple local failure criterion for crack extension, the practical applications of analytical models may be somewhat limited, without recourse to experimental data to determine various parameters.

7.9.3 FINITE ELEMENT STUDIES OF CRACK CLOSURE

Numerical studies of crack closure have been reported by a number of investigators, e.g. [133–144]. Most of these investigations have been based on two-dimensional finite element models, and a few three-dimensional models e.g. [143, 144] have been used to investigate the constraint effects. A brief outline of these models will be given below.

To carry out nonlinear analysis using the finite element method, it is necessary to model: (a) the crack tip singularity, (b) an iterative scheme to solve the elastic–plastic response, (c) a technique to change boundary conditions during a load history, and (d) a local failure criteria for the crack growth.

There are two ways to deal with the crack tip singularity. Either one uses very small elements and shifts the midside nodes to the quarter point, or a special element is employed whereby the singularity fields are embedded in it, e.g. [135, 137, 145].

Procedures to solve nonlinear problems which are formulated incrementally are well documented in most finite element texts, e.g. [146].

As a crack grows the internal nodes become boundary nodes and the boundary nodes may change from a free to a fixed condition, or from a fixed to a free condition as the crack closes or opens. A technique to modify the boundary conditions is therefore an important requirement in any finite element analysis employed for crack growth. Generally, fictitious spring elements are used to change boundary conditions. For free nodes along the crack surfaces the spring stiffness is set to zero, whereas for the fixed nodes the spring stiffness is set to a very large number, e.g. [135].

Recently Wu [147] has used a number of truss elements with varying stiffness to incrementally release a node at the peak of a cycle. Rigid contact elements were also used to prevent overlapping of material into the opposite free surface during unloading (Fig. 7.49). This node release scheme can avoid possible numerical difficulties.

As mentioned earlier, the crack advance is simulated by a node release technique. The level of load at which the node is released affects the residual stress and strain distributions in the wake of the crack tip as well as the crack closure and opening stresses. After a comparison between the

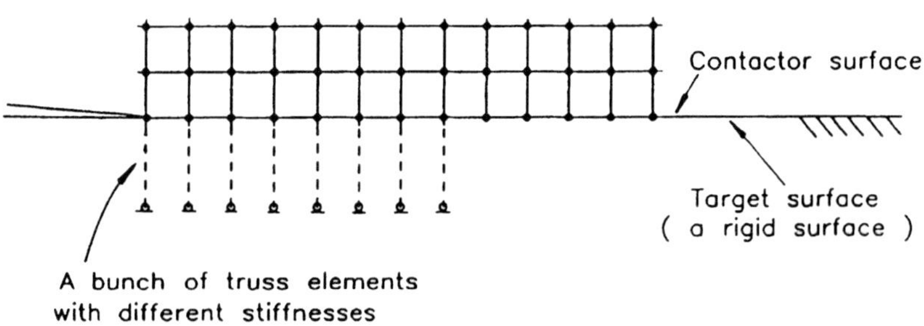

Fig. 7.49 A schematic of truss elements connected to boundary nodes and a pair of contact surfaces along the crack plane. Crack advance is achieved by gradual cut of the truss elements at the peak of the loading cycle [147].

node release at the top and bottom of a cycle, Wu [147] has pointed out that calculating the crack opening stress in a finite element analysis by releasing the crack tip node at the bottom of a cycle has major drawbacks.

An example of crack opening and closing stress levels is shown in Fig. 7.50 for a constant amplitude loading. Figure 7.51 indicates the effect of a single overload superimposed on a constant amplitude loading. It can be seen from Fig. 7.50 that there is a relatively stable response for the first few cycles, following which the crack opening stress rises steadily. This is due to the build-up of residual stress and strain in the wake and front of the crack which play a dominant role in the crack opening and closing stresses. Note that the crack opening stress is slightly lower than the crack closing stress in this analysis (see [147] for definitions).

Experimental observations of the effect of a single tensile overload has been discussed earlier, noting that its primary effect is the retardation of crack growth rate.

To simulate the crack growth following a 100% overload, five nodes were released at the maximum overload cycle, in contrast to a single node released per cycle before and after the overload event. Figure 7.51 shows that before the overload cycle, the crack opening and closing stresses had already reached a stable level. During the overload cycle, only the closing stress rises sharply. Following the overload cycle both the opening and closing stress levels are higher than those of the preceding base load with the opening stress being higher than the closing one. However, both

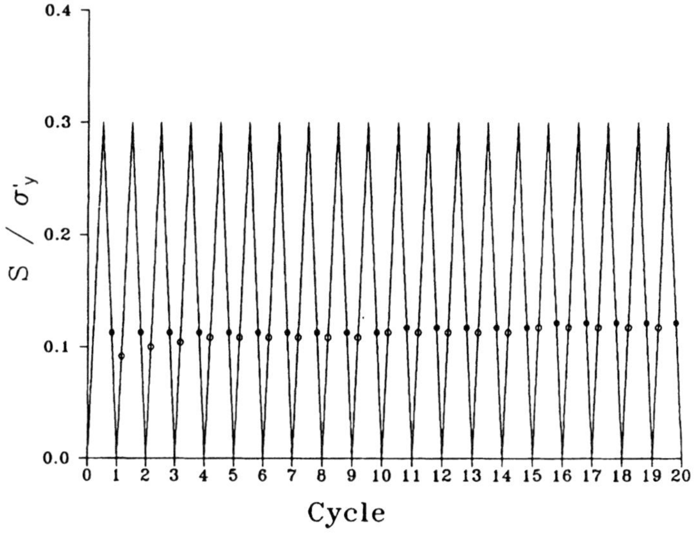

Fig. 7.50 Crack opening stresses (open circles) and crack closing stresses (solid circles) stresses for a constant amplitude cyclic loading with $S_{max} = 0.3\,\sigma_y'$ and $R = 0$ [147].

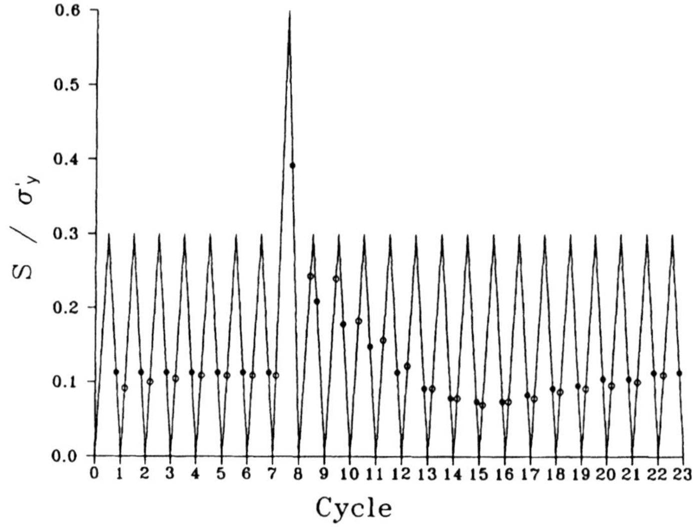

Fig. 7.51 Crack opening stresses (open circles) and crack closing stresses (solid circles) for a 100% tensile overload superimposed on a constant amplitude cyclic loading [147].

opening and closing stresses decrease gradually with further cycling. They decrease to a minimum value which is less than the previous stabilized base load level. Finally, both of these stresses gradually increase to the pre-overload level.

The results shown in Fig. 7.51 essentially capture the experimental observations. The reduction in the effective stress intensity range (ΔK_{eff}) implies a crack growth retardation. Following the crack growth retardation, the numerical analysis indicates a slight growth acceleration which corresponds to the crack opening stresses being less than the stabilized base load level.

Due to cyclic loading damage accumulates in front of a crack and failure will occur when the accumulated damage reaches a critical value (Chapters 3 and 4). Unfortunately, in most finite element analyses the local failure criteria used are rather simplistic. The crack tip node is generally released at each load cycle, regardless of the magnitude of the load or prior history.

Chermahini *et al.* [143] developed a three-dimensional finite element model incorporating an eight-noded hexahedron element. The material response was assumed to be elastic–perfectly plastic. The results of this study on a centre-cracked tension specimen showed that, as to be expected, the crack front closes first on the exterior (free surface) planes and closes last in the mid-plane. The sequence is then reversed during the loading portion of the cycle. For the interior the results showed similar characteristics to those of the plane strain analysis, whereas results on the exterior surfaces were similar to the plane stress calculations.

The finite element modelling of cracked bodies enables one to study the effect of constraint (plane stress versus plane strain) and various load histories and material properties. The mesh size in front of the crack is found to be important because of the characteristics of stress and strain fields at the crack tip with very steep gradients. The crack extension model based on releasing either the crack tip node or the element at the maximum applied load, which is used by most investigators, makes the crack growth per cycle dependent on the size of the element ahead of the crack tip. It is found that the crack closure becomes insensitive to the mesh size only if the mesh size is an order of magnitude smaller than the plastic zone ahead of the crack tip. Finally, the implementation of the elastic–plastic finite element analyses requires a fair amount of resources both in time and cost.

7.10 TIME-DEPENDENT CRACK GROWTH – TEMPERATURE EFFECTS

Fatigue crack growth, in addition to the factors already discussed, is affected by temperature, loading rate and environment. For example, an increase in temperature generally results in an increase in the crack propagation rate. At low temperatures (room temperature and below), crack growth is frequency independent, i.e. crack growth is time independent to a certain extent. For example, see [5], where the effect of cyclic frequency on the crack growth is shown to be negligible for frequencies ranging from 6 to 3000 cycles per minute for an A36 steel in air. As the temperature is increased, changes in plastic flow occur, thereby altering the crack front plastic zone and stress and strain distribution. The crack growth rate thus becomes frequency/time dependent.

Cyclic loading at elevated temperatures engenders an interactive mechanism combining rate-independent fatigue and rate-dependent creep phenomena. The dominance of one mechanism over the other depends upon temperature, material microstructure, and loading rate and cyclic wave form (Chapter 3). For example, an asymmetric wave form with a slower tensile loading rate, a hold-time and an increasing load ratio, R, all enhance creep cavity growth resulting in an increased crack growth rate. In general, lower strain rates (or reduced wave form frequencies) may result in an increase in the crack growth rate. It should be noted, however, that low frequencies, slow–fast wave forms, and long hold-times do not always increase crack growth rate. The material type (microstructure) and environment have an important role in the relationship between crack growth and loading. It has been mentioned in Chapters 1 and 3 that oxidation results in weakening and embrittlement of grain boundaries, especially at the crack tip region near the surface. Cyclic action and increased temperature enhance the oxidation effect and this may have a greater influence on crack growth than the hold-time effect [148].

Creep crack growth kinetics is influenced by two opposing processes. Creep deformation causes crack tip stress relaxation and blunting, and creep damage manifests itself in the formation and growth of cavities. Thus, the relative rate of stress relaxation and cavity growth rate control the growth of a crack in the creep range of a material. The increased crack growth rate may therefore be limited to a certain range of temperatures and strain-rates.

In the following we will first discuss the influence of the high temperature environment (in the range of temperature in which a material exhibits creep deformation) on the stress and strain fields in front of a crack. The fracture driving force parameters used to correlate the crack growth data will then be outlined. We will conclude this section with the combined effect of cyclic loading and creep on the crack growth rate.

7.10.1 TIME-DEPENDENT STATIONARY CRACK TIP FIELDS

In this section we will briefly describe the stress and strain fields in front of a crack tip under two different creep regimes. In steady-state or secondary creep deformation, the creep strain-rate in front of the crack will dominate. In contrast, the elastic strain-rate may not be negligible under a primary creep regime until sufficient time has elapsed to allow for the development of extensive primary creep in front of the crack. We will study these two cases.

(a) Elastic–secondary creep

An elastic–secondary creep material may be represented by a power law relationship of the type*

$$\dot{\varepsilon} = \left(\frac{\dot{\sigma}}{E}\right) + B\,\sigma^{1/n}, \tag{7.122}$$

for a uniaxial stress state. The above equation is similar to the uniaxial elastic–plastic material description (2.7) when ε is replaced by $\dot{\varepsilon}$ and $B = \alpha\varepsilon_0\sigma_0^{-1/n}$. The stationary crack tip fields for such a material may be obtained from the HRR fields (7.37) by replacing $\mathbf{u}$ and $\boldsymbol{\varepsilon}$ by $\dot{\mathbf{u}}$ and $\dot{\boldsymbol{\varepsilon}}$, J by $C(t)$, and noting that elastic strains are negligible compared to creep

* For the sake of consistency throughout, here we have adopted a notation for the material exponent n which is the inverse of that used in the creep literature.

strains near the crack tip, i.e.

$$\begin{aligned}\sigma_{ij} &= \left[\frac{C(t)}{BI_n r}\right]^{n/(1+n)} \tilde{\sigma}_{ij}(\theta; n)\\ \dot{\varepsilon}_{ij} &= B\left[\frac{C(t)}{BI_n r}\right]^{1/(1+n)} \tilde{\varepsilon}_{ij}(\theta; n)\\ \dot{u}_i &= B\left[\frac{C(t)}{BI_n r}\right]^{1/(1+n)} r^{n/(1+n)}\tilde{u}_i(\theta; n)\,,\end{aligned} \tag{7.123}$$

where the amplitude factor $C(t)$ depends on the magnitude and history of the applied load, geometry of the cracked body, the crack length, material properties, and time. The time-dependent amplitude factor $C(t)$ is determined from far-field loading conditions. In the case of small-scale creep where the creep dominated region is very small compared to the crack length, $C(t)$ is given by [149]

$$C(t) = \int_\Gamma \frac{1}{1+n}\sigma_{ij}\dot{\varepsilon}_{ij}\,\mathrm{d}y - \sigma_{ij}n_j\frac{\partial \dot{\mathbf{u}}_i}{\partial x}\mathrm{d}s \tag{7.124a}$$

where Γ is a contour that encircles the crack tip in regions where $\dot{\varepsilon}^c\,(=B\sigma^{1/n}) \gg \dot{\varepsilon}^c(=\dot{\sigma}/E)$.

Under extensive creep conditions, with $\dot{\sigma}=0$ everywere, the above integral is path independent, i.e. $C(t)\rightarrow C^*$ with C^* given by

$$C^* = \int_\Gamma W(\dot{\varepsilon}_{ij})\,\mathrm{d}y - \sigma_{ij}n_j\frac{\partial \dot{u}_i}{\partial x}\mathrm{d}s\,. \tag{7.124b}$$

By analogy to the elastic–plastic crack tip fields, we may view the creep zone as a time dependent plastic zone, and thus define a small-scale yielding domain whereby the creep zone is small with respect to the crack size and other characteristic dimensions of the specimen. Small-scale creep solutions have been presented by Riedel [150] for Mode III and Riedel and Rice [151] for Mode I. They determined $C(t)$ in (7.123) by assuming path independence of the J-integral and equated the asymptotic value of $J\,(r\rightarrow 0)$ to that calculated at the far-field, J_∞, thus obtaining $C(t)$ in (7.124) in the format

$$C(t) = \frac{nJ_\infty}{(1+n)t} = \frac{n(1-\nu^2)K^2}{E(1+n)t}\,, \tag{7.125}$$

for the plane strain condition. For a plane stress case, replace $(1-\nu^2)$ by unity.

From (7.123), (7.124b) and (7.125) we see that K and C^* are loading parameters for the extreme cases of small-scale yielding and large-scale creep regimes, respectively. Finite element analysis by Bassani and McClintock [149] and Ehler and Riedel [152] have shown a good agreement with the short and long term asymptotic behaviour of the stress and strain fields,

(7.123). For the intermediate range, there is a smooth transition from K dominated fields to C^* controlled response proportional to $1/t$. A good approximation for the transient behaviour is obtained when $C(t)$ in (7.123) is replaced by [152]

$$C(t) \simeq C^*\left(1 + \frac{t_T}{t}\right), \tag{7.126}$$

where the transition time t_T is obtained from (7.125) by replacing $C(t)$ by C^*, i.e.

$$t_T = \frac{nJ_\infty}{(1+n)C^*}. \tag{7.127}$$

A thorough evaluation of the $C(t)$ parameter is given in a paper by Bassani, Hawk and Saxena [153] encompassing both experimental and finite element investigations. They concluded that $C(t)$ correlated fairly well with the creep growth rate, da/dt, for a variety of transient regions.

(b) Elastic–primary creep

For an elastic–primary creep, the material law may be represented by a Bailey–Norton form

$$\dot{\varepsilon} = B\varepsilon^{-p}\sigma^{1/n}. \tag{7.128}$$

In the case of a time-dependent proportional loading, we may express

$$\sigma_{ij}(r,\theta,t) = P(t)\tilde{\sigma}_{ij}(r,\theta) \tag{7.129}$$

where $P(t)$ is the time-dependent loading parameter and $\tilde{\sigma}_{ij}(r,\theta)$ is a time-dependent function of coordinates r and θ. Substitution from (7.129) into (7.128) and integrating with respect to time, yields

$$\varepsilon = \left\{[P(t)]^{-1/n}\int_0^t [P(\tau)]^{1/n}\mathrm{d}\tau\right\}^{1/(p+1)} [(p+1)B]^{1/(p+1)}\sigma^{1/n(p+1)} \quad \text{for} \quad p > -1. \tag{7.130}$$

We may introduce a path-independent integral analogous to the J-integral which is time dependent such as

$$C_h^*(t) = J(t)\left\{\frac{P(t)^{1/n}}{\int_0^t [P(\tau)]^{1/n}\mathrm{d}\tau}\right\}^{1/(p+1)}. \tag{7.131}$$

The C_h^*-integral depends on the current load, $P(t)$. Note that for $p = 0$, (7.128) and (7.122) are identical, and thus $C_h^* = C^*$. The C_h^*-integral may be measured similarly to the J-integral.

The analogous HRR fields are given in e.g. [8] as $r \to 0$:

$$\sigma_{ij} = \left(\frac{C_h^*(t)}{[B(1+p)]^{1/(p+1)} I_m r} \right)^{m/(m+1)} \tilde{\sigma}_{ij}(\theta; m)$$
$$\varepsilon_{ij} = \left(\frac{C_h^*(t)}{[B(1+p)]^{1/(p+1)} I_m r} \right)^{1/(m+1)} \tilde{\varepsilon}_{ij}(\theta; m) \,, \tag{7.132}$$

where $m = n(p+1)$. Note that the primary creep stress field has a singularity of $r^{-m/(m+1)}$ at the crack tip. For a constant load, $P(t) = \text{constant}$,

$$C_h^* = t^{-1/(p+1)} J_\infty . \tag{7.133}$$

Thus in a cracked body under a constant load, the development of the primary creep is controlled by the C_h^*-integral, (7.131), and the corresponding stress and strain fields are given by (7.132). Upon development of the secondary creep, the C_h^*-integral will still be the controlling parameter, until the secondary creep supersedes the primary creep at which time the C^*-integral, (7.124b), will control the creep process.

A critical assessment of the creep intensity parameters was reported by Bensussan, Piques and Pineau [154]. They found that C^* correlated with the creep crack initiation time for the large-scale secondary creep and C_h^* for the large-scale primary creep conditions.

7.10.2 CREEP CRACK GROWTH FIELDS

Creep crack growth, similar to the time-independent fatigue crack propagation, may require a period for damage accumulation, following which crack growth takes place. To predict crack growth, we would need a failure criterion. In most creep analyses, the failure criterion has been a strain-based one – a critical strain value. However, unlike the time-independent fatigue crack growth or stationary creep crack, the HRR type fields cannot represent the asymptotic behaviour of a growing crack for an elastic–power law type material, (7.122).

The asymptotic creep crack tip fields for a growing crack have been determined by Hui and Riedel [155] in the form of

$$\sigma_{ij} = \alpha_n \left[\frac{\dot{a}(t)}{BEr} \right]^{n/(1-n)} \tilde{\sigma}_{ij}(\theta; n)$$
$$\varepsilon_{ij} = \frac{\alpha_n}{E} \left[\frac{\dot{a}(t)}{BEr} \right]^{n/(1-n)} \tilde{\varepsilon}_{ij}(\theta; n) \tag{7.134}$$

where α_n is a numerical factor that depends on the exponent n. It is to be noted that contrary to the time-independent crack growth fields and the stationary time-dependent fields (7.37) and (7.123), (7.134) indicate that the stress and strain singularities for a growing creep crack are of the **same** order, i.e. $r^{-n/(1-n)}$, and are independent of the applied loading or the specimen geometry. Furthermore, the creep crack driving force is a function only of the material properties B and E, and the current speed of crack extension, $\dot{a}(t)$. These observations are not supported by the experimental data, and could not be supported on physical grounds either. Bassani *et al.* [156] have shown that for a small-scale creep condition, the stress field evolves in a complex manner as the crack propagates. This is due to the competing effects of stress relaxation because of constrained creep, and stress elevation caused by the instantaneous elastic response. Their analysis shows that the stress field can be approximated by matching three asymptotic singular crack tip solutions, namely Hui and Riedel growing crack, (7.134), the relaxing stress field, Riedel and Rice, (7.123) with $C(t)$ given by (7.125) and the remote elastic singularity (K) field, (7.6).

It should be noted that Bensussan *et al.* [154] reported that they were unable to find a correlation between the creep crack growth rates of 316L steel and any usual crack tip parameter such as C^* or K.

7.10.3 CORRELATION WITH EXPERIMENTAL DATA

Time-dependent crack growth experiments performed at high temperatures combine two effects, namely creep and environmental processes. For a given material and temperature, one process may dominate over the other. This is one of the reasons why several parameters such as stress intensity factor, K, the C^*-integral, crack opening displacement, δ_t, net section stress, σ_{net}, and reference or equivalent stress, σ_{ref}, have been shown to correlate with the experimental data. Sadananda and Shahinian [157, 158] have reviewed some of the available experimental data on structural alloys at high temperatures. They concluded that an appropriate parameter to characterize crack growth would depend on the environmental sensitivity of the material and its creep ductility. It has been suggested that time-dependent crack growth may be grouped into two categories in which growth occurs either by a brittle or ductile mode.

In a brittle mode, the crack growth takes place essentially in a continuous manner along the grain boundary through the grain sliding mechanism. An aggressive environment would accelerate the crack growth rate in this case. If the speed of the crack growth is high enough, then the effect of crack tip stress relaxation due to the creep of the matrix may not be significant. Thus, the conditions of small-scale creep could apply and a linear fracture mechanics parameter such as K may be used to correlate experimental data. This appears to be the case observed with high strength materials [157–159]. However, if the creep zone is outside the K-field, then the parameter to be used would probably be C^* as long as a steady-state creep condition prevails [158, 159]. In this case one would expect a ductile mode of creep crack growth where micro-cracks (or voids) initiate in front of the crack and creep crack growth occurs through the micro-crack (or void) growth and coalescence of cracks. Creep deformation ahead of a crack would cause crack tip stress relaxation and blunting. Therefore, crack growth occurs due to the accumulation of creep damage resulting in micro-crack or cavity formation and their subsequent growth and coalescence with the main crack.

A further critical assessment of various proposed models for creep crack growth is given by Riedel [160].

7.10.4 COMBINED CYCLE AND TIME-DEPENDENT CRACK GROWTH

Time-dependent effects will be superimposed on the cycle-dependent fatigue process when the strain-rate or frequency is decreased and/or a hold-time is added to the wave form. Depending on the process which dominates the crack growth, one could expect a correlation with the cycle-dependent or time-dependent parameters which have been discussed in the earlier parts of this chapter. However, there is a range of temperatures and strain-rates where a strong interaction takes place causing a higher rate of crack growth. As shown in Fig. 7.52, the rate of

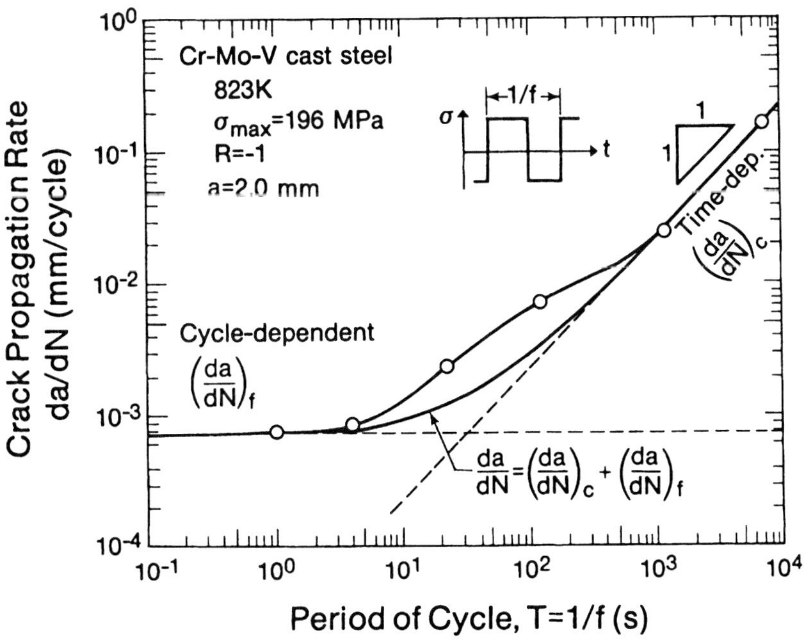

Fig. 7.52 Crack growth rate versus frequency for a Cr-Mo-V cast steel at high temperature, from Ohtani [161].

crack growth is higher than the linear summation of the time-independent $(da/dN)_f$ and time-dependent $(da/dN)_c$ growth rates [161].

We have already discussed in Chapter 3 the mechanism of crack initiation for the creep–fatigue interaction. Here we will confine our discussion to the fracture parameters which are used to correlate the experimental data in the creep–fatigue interaction range. An aggressive environment may have a strong influence on the time-dependent part of the crack growth process, and in any fundamental study an attempt should be made to determine the effect of each process separately and in combination. For the creep–fatigue crack growth study this would necessitate crack growth measurements in high vacuum and air or other corrosive environments.

In Chapter 6 we have seen how a prior plastic strain could influence the subsequent creep deformation and vice versa. Ohtani [161] has attempted to divide the crack growth into two regimes correlating it with either the time-dependent creep J-integral range, akin to the $C_h^*(t)$ equation (7.131) or the cycle-dependent fatigue J-integral range.† Although a certain success is obtained by this procedure, a proper method would be to use a procedure similar to that given in Chapter 3 for time-dependent fatigue.

Figure 7.53, taken from Ohtani's paper [161], is a good example of the strong interactive effect showing a gross underestimation of the linear summation.

Jaske has recently summarized a creep–fatigue model developed by him and his collaborators [163]. The creep–fatigue interaction is taken into account through consideration of a crack tip zone model where the extent of each process (creep or fatigue) is determined. He concludes that for most practical situations, the fatigue crack growth rate is significantly affected by the prior creep growth. This conforms with the prediction of the constitutive model and experiments presented in section 6.9.

The majority of creep–fatigue experiments have been conducted under isothermal conditions, i.e. constant temperature. There may be conditions where the temperature may fluctuate. James [164] has conducted fatigue crack growth experiments under non-isothermal conditions where the temperature cycled between 149°C and 243°C, and was not in-phase with the fatigue cycle. Initially the crack growth rate was high – approximately equal to that at 243°C. It then steadily decreased and reached a steady-state condition equivalent to that expected at the lower temperature. These results illustrate the complex nature of creep–fatigue interaction.

† The definition of the J-integral range, ΔJ, requires careful consideration of the unloading process. Some definitions used in the literature do not preserve the path-independent properties of this integral. An interested reader is referred to a paper by Ellyin and Fakinlede [162] where an appropriate method of evaluation is given.

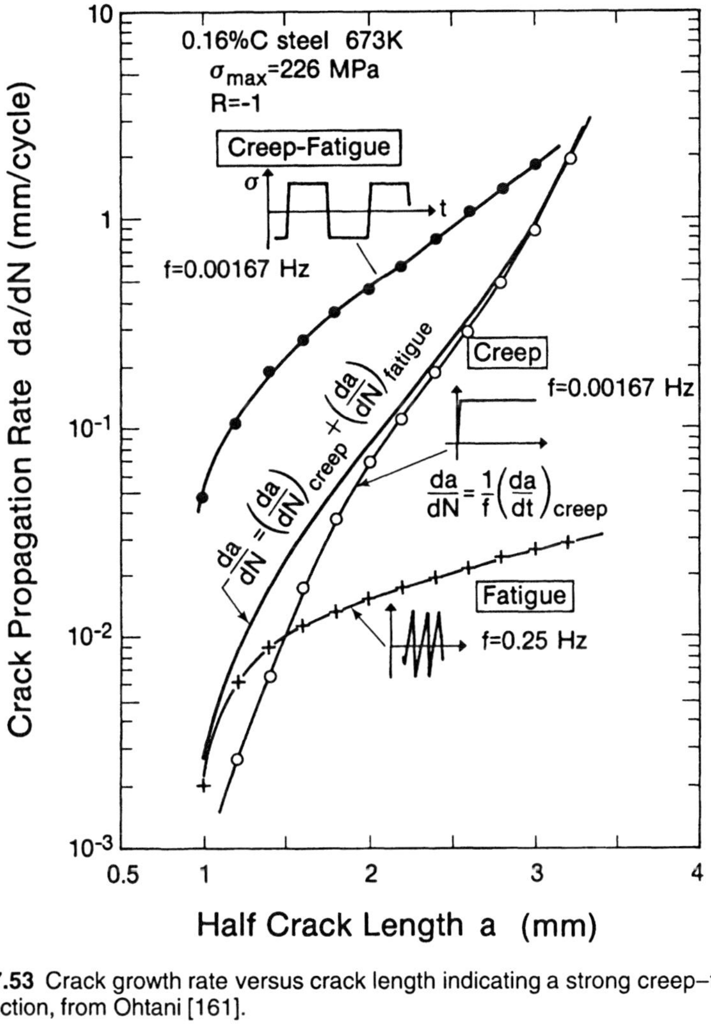

Fig. 7.53 Crack growth rate versus crack length indicating a strong creep–fatigue interaction, from Ohtani [161].

The subject of crack growth in the creep–fatigue interaction range is presently under active investigation and hopefully more definitive results will emerge in the near future.

7.11 TIME-DEPENDENT CRACK GROWTH – ENVIRONMENTAL EFFECTS

The mechanism by which oxygen (or any other aggressive medium) promotes an early crack nucleation and growth was discussed in section

1.3.2. It was pointed out that the environment strongly influences slip irreversibility and thus affects the life to initiate a fatigue crack.

Experimental data at high temperatures which indicate the effect of surface oxidation on fatigue life were presented in section 3.6. Further data on the influence of oxygen in promoting early crack initiation and accelerated crack growth at room temperature were discussed in section 3.8.

In this section we will restrict our attention to the effects of an aggressive environment on the crack growth behaviour. The majority of previously referred to experimental studies cited in this chapter were concerned with crack propagation in air. However, it is well-known that crack growth behaviour is dependent on both the environment around the crack tip and the material's susceptibility to environmental attack.

It was pointed out in both Chapters 1 and 3 that to understand the role of an environment on fatigue crack growth, the effect of the environment must be separated from the other fatigue growth parameters such as loading history, processing history, temperature, etc. However, this may not always be possible since there are synergistic effects between mechanical and environmental conditions. This is well documented by the observation that the environmental effect on crack propagation varies with the change in loading frequency. Microscopic examinations have revealed that a corrosive environment alters the metallurgical and mechanical properties both on the crack surfaces and ahead of the crack tip. The extent of this change depends on the material–environment system.

There is extensive literature on various aspects of environmentally-assisted crack growth behaviour, e.g. review papers by Wei and Gangloff [165], Gangloff [166], Garud [167], and a recent symposium [168] where numerous references on the subject are cited. It is not our intention here to discuss all aspects of corrosion fatigue, rather to focus on those elements which influence crack growth behaviour beyond what has been discussed in earlier sections. In a strict sense, any test conducted outside an inert environment, for example in laboratory air, involves certain environmental effects, e.g. section 3.8. However, the words aggressive environment are sometimes used in a relative sense, e.g. compared to air, for example, the environmentally-assisted corrosion fatigue crack propagation curves in the ASME Boiler and Pressure Vessel Code, Section XI [169].

7.11.1 PROPOSED MECHANISMS

A number of mechanisms have been proposed to explain how a corrosive environment promotes or retards crack growth. These mechanisms can generally be grouped into three categories: (a) formation of oxide products in the crack, (b) hydrogen embrittlement, and (c) the time for which a crack is exposed to the corrosive environment.

(a) Oxidation of the crack surfaces

Oxidation at the crack tip plays two important roles in fatigue crack growth. First, it alters the material's properties around the crack tip, and second it promotes development of closure. The freshly exposed surfaces of a propagating crack develop a hard but brittle oxide layer which through subsequent crushing and reforming, can eventually lead to a wedge-like shape at the crack tip. This wedge of oxide debris may locally reduce the crack driving force (Suresh and Ritchie [170]). Zhiqiang *et al.* [171] pointed out that the closure promotes the movement of the corrosive solution out of the crack tip and may subsequently reduce exposure to the corrosive medium. Thus with the reduction in the driving force induced by the oxide products and the shielding properties that the closure generates, the crack growth rate may decrease. However, the effect of the oxide induced closure is restricted to the near threshold propagation regime, region A in Fig. 7.23, where the size scale of the oxide debris is of the same order of magnitude as the crack tip opening displacement. Beyond this regime, experiments indicate that the effect of oxide induced closure rapidly decreases and disappears resulting in increased crack propagation.

The process of repeated oxide film rupture on a local scale is thought to be the main mechanism of stress corrosion cracking (SCC) in an aqueous environment [167]. The presence of debris and corrosion products near the crack tip may cause rupture of the protective oxide layer during the compression part of the cyclic load. They may also contribute to an increased growth by keeping the crack tip open during the compression part of the cycle.

(b) Hydrogen embrittlement

Hydrogen embrittlement is due to chemical absorption and molecular dissociation of H_2O which results in the formation of atomic hydrogen, H^+. These atomic hydrogens migrate to regions of high stress concentration (crack tip). The absorption of the hydrogen ions into the crack tip, through dislocation motion [172] or transport up the hydrostatic stress gradients [173], promotes embrittlement of the crack tip material. A micro-crack may initiate within the embrittled area, grow and join the already existing dominant crack. This type of accelerated crack growth has been observed in high strength steels. In lower strength steels, the hydrogen effect is different. It may enhance the nucleation and growth of micro-voids at the crack tip thereby accelerating crack growth.

In certain zirconium, titanium and niobium alloys, hydride platelets are formed which migrate up the hydrostatic stress gradient and align themselves with the crack plane. The hydride precipitates near the crack tip then causes a redistribution of stress. The brittle hydride subsequently cracks and the main crack grows and the process repeats itself [173].

(c) Exposure time

The fatigue crack growth rate is affected by the time available for chemisorption at the crack tip during each cycle. Interesting results on the effect of frequency on the corrosion fatigue growth were reported by Bartlett and Hudak [174]. At high values of ΔK, a higher frequency resulted in lower fatigue crack growth rates than those at reduced frequencies. The higher growth rate result at lower frequencies occurs because the corrosive environment has more contact time with the crack surfaces, while at higher frequencies, the corrosive environment cannot fully penetrate the crack tip before the crack begins to close again. However, the reverse trend occurs at lower values of ΔK (near threshold region), i.e. reduced frequencies tend to decrease the crack growth in a corrosive environment. The explanation for this is that at lower frequencies, the higher exposure time produces larger amounts of corrosion products which promote early crack closure, thus reducing the effective driving force. This has already been discussed in section 7.11.1(a).

The effect of exposure time on fatigue crack growth reaches a plateau. This is achieved when the exposure time is long enough for the crack to be saturated with the corrosive product, allowing the chemisorption process to take place.

7.11.2 CRACK GROWTH MODELS

It has been indicated in section 7.2 that stationary cracks begin to grow once the monotonically applied load exceeds a certain value, i.e. when $K_{appl} > K_C$ (Fig. 7.17). In a corrosive environment crack growth initiates when $K_{appl} \geqslant K_{scc}$, where K_{scc} is the stress intensity factor for stress corrosion cracking (in general $K_{scc} < K_C$).

Various attempts have been made to model environmentally-assisted crack growth under cyclic loading by superposing different mechanisms. This is essentially a variation of the superposition equation contained in Fig. 7.52. The nature of the cyclic load–environment interaction is described by three components: (a) A mechanical component is associated with an inert environment, i.e. test results similar to Fig. 7.23 with the experiment conducted in an inert environment, (ii) A so-called 'true corrosion fatigue' where the environment by itself is assumed to have no effect on the crack growth behaviour. The crack growth rate is affected due to the degradation of the crack tip material by any of the mechanisms discussed in section 7.11.1. Thus, when the cyclic load is applied, the crack growth rate is accelerated which is schematically shown in Fig. 7.54(a). It can be seen that there is a reduction of the threshold stress intensity range, ΔK_{th}, due to the presence of a corrosive environment and a general upward shift of the 'Paris regime'. (iii) Finally there is a so-called 'stress

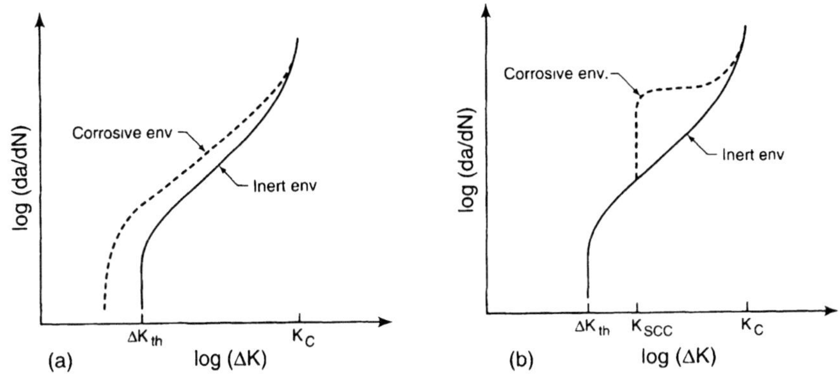

Fig. 7.54 A schematic of the corrosion fatigue process. (a) 'true corrosion fatigue' indicating accelerated crack growth due to cyclic loading in an aggressive environment; (b) 'stress corrosion cracking' a time-dependent component.

corrosion cracking', a time-dependent environmental component which incorporates the effect of frequency into the environment. This effect is shown in Fig. 7.54(b), where there is a plateau-like behaviour above K_{scc} which was defined earlier.

When a material is exposed to an aggressive environment, it will react to a combination of the above three phenomena. Often it is assumed that each of the above effects is mechanistically independent from each other, and the total damage is the sum of the individual contributions [99]. For example, Chalant, Petit and Suyitno [175] have proposed an environmental fatigue crack propagation model for structural steels. In this model, the crack growth in a corrosive environment is the sum of the crack growth rate in an inert environment plus the additional crack growth provided by the hydrogen embrittlement mechanism, all adjusted for the effect of frequency, i.e.

$$\left(\frac{da}{dN}\right)_{env} = \left[\left(\frac{da}{dN}\right)_{inert} + \left(\frac{da}{dN}\right)_{hydrogen}\right] \times f. \tag{7.135}$$

In the above 'f' stands for the correction due to the frequency effect, and the (da/dN) terms are all based on the effective stress intensity range, ΔK_{eff}, so that the effect of closure in an inert environment and oxygen-assisted closure are accounted for. Generally (da/dN) in an inert environment is not sensitive to the frequency of loading (section 3.6). Therefore, by multiplying the first term on the RHS of (7.135) by f the crack growth rate in an inert environment may be overestimated.

Further attempts at modelling the crack growth rate in a corrosive environment by the superposition type of equations are listed in review papers [166–168]. Essentially these models attempt to describe experimentally observed trends through a set of fracture mechanics parameters

(ΔK, K_{max}, $\dot{K}$) with adjustable coefficients. Using the afore-mentioned stress-based global parameters to describe the crack driving force in a corrosive environment may not be sufficient, since the corrosion process mainly takes place near the crack tip. An appropriate model would have to take into account effects such as the altered crack tip material properties, crack tip opening profile, plastic zone size, and fatigue damage accumulation at the crack tip.

To illustrate a few of the above-mentioned problems, Fig. 7.55 shows the stress distribution near the crack tip as the hydride precipitates [173]. It can be seen that the peak stress near the crack tip decreases, and shifts towards the front end of the hydride. There is also a considerable unloading near the crack tip and crack tip blunting. These changes occur while the far-field stress or K, is kept constant. Thus, a stress intensity factor calculated based on the applied far-field is not representative of the stress state in the presence of hydride formation and precipitation. The corresponding strain distribution is shown in Fig. 7.56. It should be noted that the crack tip strain is less affected by the hydride precipitation, and thus a strain-based model may be more appropriate in the case of environmentally-assisted cracking. Indeed Garud [167] advocated such an approach following his survey of

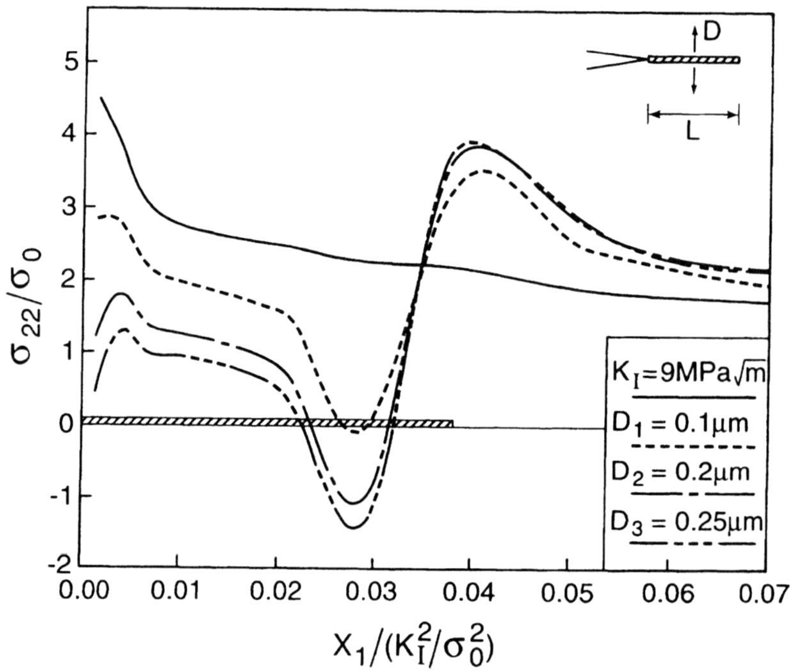

Fig. 7.55 Distribution of stress component normal to the crack plane due to hydride expansion (length of hydride platelet $L = 10\ \mu m$, $X_1 = 0$ denotes the crack tip) [173].

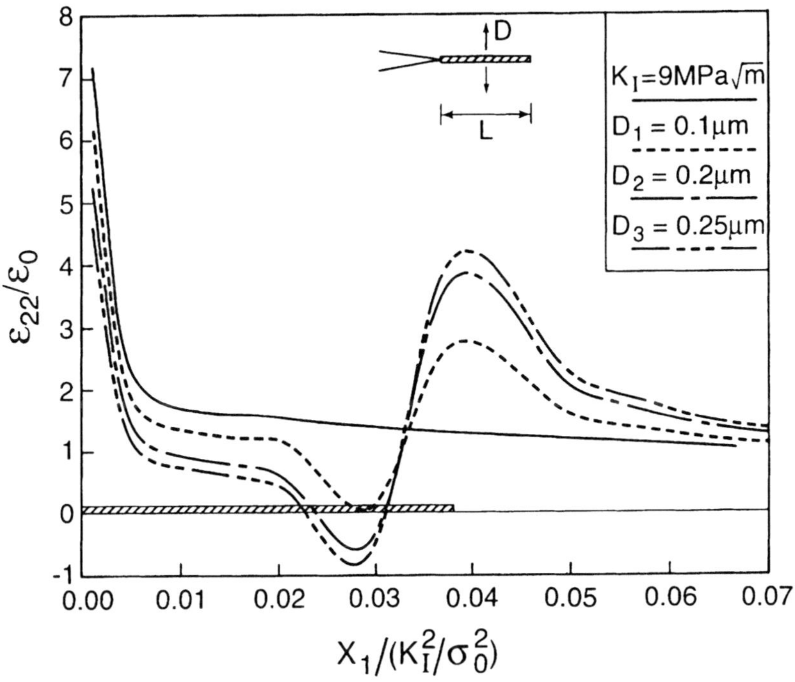

Fig. 7.56 Distribution of strain component normal to the crack plane due to hydride expansion (length of hydride platelet $L = 10\,\mu$m, $X_1 = 0$ denotes the crack tip) [173].

developments in the field. Similarly James [164] states that the crack tip strain rate, $\dot{\varepsilon}_{ct}$, is acknowledged to be 'the fundamentally correct mechanical crack-driving parameter for describing environmentally assisted cracking'. However, he points out that as yet there is no consensus for its definition. Numerous models have been suggested, e.g. Turnbull [176].

One of the methods that has become popular in the pressure vessel industry is mainly due to Shoji and his collaborators [177, 178]. They proposed that the time-based crack growth rate, $(\mathrm{d}a_b/\mathrm{d}t)$ is proportional to the crack tip strain-rate, $(\mathrm{d}\varepsilon_{ct}/\mathrm{d}t)$, i.e.

$$\left(\frac{\mathrm{d}a_b}{\mathrm{d}t}\right) \propto \left(\frac{\mathrm{d}\varepsilon_{ct}}{\mathrm{d}t}\right). \tag{7.136}$$

However, the nature of the proportionality is unknown, but for practical purposes, the use of $(\mathrm{d}a_b/\mathrm{d}t)$ instead of $\dot{\varepsilon}_{ct}$ was suggested (see (7.134) for a relationship in the case of creep crack growth). The $(\mathrm{d}a_b/\mathrm{d}t)$ is taken to be the mean crack growth rate in air divided by the load rise time, t_r, i.e.

$$\left(\frac{\mathrm{d}a_b}{\mathrm{d}t}\right) \simeq \frac{(\mathrm{d}a/\mathrm{d}N)_{air}}{t_r}. \tag{7.137}$$

Similarly, a time-based environmental rate is defined by

$$\left(\frac{da_e}{dt}\right) \simeq \frac{(da/dN)_{env}}{t_r}. \tag{7.138}$$

When (da_e/dt) is plotted against (da_b/dt), an environmentally-affected crack growth manifests itself by being significantly above a 1:1 line. An attractive feature of this comparative model is that $\dot{a}_b$ in (7.137), through its RHS, can combine ΔK, the R-ratio, and a time-dependent variable (frequency), in a single parameter. However, by simply comparing da/dN curves for different environments, the true mechanisms of the fatigue crack growth enhanced by corrosion are not properly modelled.

The subject of corrosion fatigue crack growth is of considerable practical interest. In particular, components in nuclear power plant, gas pipelines, and chemical process plants, are experiencing environmentally-assisted cracking as the service life increases. Consequently, intensive research is being undertaken in connection with materials used in the above applications, e.g. [168]. As more discriminating experimental data becomes available, the detailed contribution of each mechanism will be better understood, and a complex model of corrosion fatigue crack growth will emerge. Through the use of computers, mechanical and chemical interactions can be simulated to arrive at models for a particular material–environment–loading system.

REFERENCES

1. Liebowitz, H. (ed.) (1968) *Fracture*–Advanced Treatise in 7 volumes, Academic Press, New York.
2. Tada, H., Paris, P. and Irwin, G. (1973) *The Stress Analysis of Cracks Handbook*, Del Research Corporation, Hellertown, PA.
3. Sih, G.C. (1973) *Handbook of Stress-Intensity Factors for Researchers and Engineers*, Institute of Fracture and Solid Mechanics, Lehigh University, Bethlehem, PA.
4. Erdogan, F. (1983) Stress intensity factors. *J. Appl. Mech., Trans. ASME*, 50th anniversary issue, **103**, 992–1002.
5. Barsom, J.M. and Rolfe, S.T. (1987) *Fracture and Fatigue Control in Structures*, 2nd edn, Prentice-Hall, Englewood Cliffs, NJ.
6. Knott, J.F. (1973) *Fundamentals of Fracture Mechanics*, Butterworths, London.
7. Broek, D. (1986) *Elementary Engineering Fracture Mechanics*, 4th revised edn, Martinus Nijhoff, Dordrecht, Netherlands.
8. Kanninen, M.F. and Popelar, C.H. (1985) *Advanced Fracture Mechanics*, Oxford University Press, New York.
9. Shih, C.F. and Gudas, J.P. (eds) (1983) *Elastic-Plastic Fracture: Second Symposium*, in two volumes, ASTM STP 803, American Society for Testing and Materials, Philadelphia, PA.
10. Anderson, T.L. (1991) *Fracture Mechanics: Fundamentals and Applications*, CRC Press, Boca Raton, FL.

11. Rice, J.R. (1967) Mechanics of crack tip deformation and extension by fatigue, in *Fatigue Crack Propagation*, ASTM STP 415, American Society for Testing and Materials, Philadelphia, PA, pp. 247–311.
12. Rice, J.R. (1968) Mathematical analysis in the mechanics of fracture, in *Fracture*, vol. 2, Mechanical Fundamentals (ed. H. Liebowitz), Academic Press, New York, pp. 191–311.
13. Hutchinson, J.W. (1979) *Recent Developments in Nonlinear Fracture Mechanics*. Proc. 7th Canadian Cong. of Appl. Mech., Sherbrooke, PQ, vol. 1 (eds F. Ellyin and K.W. Neale), pp. 24–36.
14. Hutchinson, J.W. (1983) Fundamentals of the phenomenological theory of nonlinear fracture mechanics. *J. Appl. Mech., Trans. ASME*, 50th anniversary issue, **103**, 1042–51.
15. Hutchinson, J.W. (1979) *Nonlinear Fracture Mechanics*, Dept. of Solid Mechanics, Tech. Univ. of Denmark, Lyngby, Denmark .
16. Muskhelishvili, N.I. (1953) *Some Basic Problems of the Mathematical Theory of Elasticity*, P. Noordhoff, Holland.
17. Irwin, G.R. (1960) Fracture Mechanics, in *Structural Mechanics* (Proc. 1st Symp. Naval Struct. Mech., 1958) (eds J.N. Goodier and N.J. Hoff), Pergamon Press, New York, pp. 557–91.
18. Paris, P.C. and Erdogan, F. (1963) A critical analysis of crack propagation laws. *J. Basic Engng., Trans. ASME*, **85**, 528–34.
19. Barenblatt, G.I. (1962) Mathematical theory of equilibrium cracks in brittle fracture, in *Advances in Applied Mathematics*, Vol. VII, Academic Press, New York, pp. 55–129.
20. Dugdale, D.S. (1960) Yielding of steel sheets containing slits. *J. Mech. Phys. Solids*, **8**, 100–4.
21. Goodier, J.N. (1968) Mathematical theory of equilibrium cracks, in *Fracture* vol. 2 (ed. H. Liebowitz), Academic Press, New York, pp. 1–66.
22. Rice, J.R. (1968) A path independent integral and the approximate analysis of strain concentration by notches and cracks. *J. Appl. Mech., Trans. ASME*, **35**, 379–86.
23. Hutchinson, J.W. (1968) Singular behaviour at the end of a tensile crack in a hardening material. *J. Mech. Phys. Solids*, **16**, 13–31.
24. Rice, J.R. and Rosengren, G.F. (1968) Plain-strain deformation near a crack tip in a power-law hardening material. *J. Mech. Phys. Solids*, **16**, 1–12.
25. Hult, J.A. and McClintock, F.A. (1956) *Elastic-Plastic Stress and Strain Distribution Around Sharp Notches Under Repeated Shear*. Proc. 9th Int. Congr. of Appl. Mech., Brussels, Vol. 8, pp. 51–8.
26. Rice, J.R. (1967) Stresses due to a sharp notch in a work-hardening elastic-plastic material loaded by longitudinal shear. *J. Appl. Mech., Trans. ASME*, **34**, 287–98.
27. McClintock, F.A. (1961) Discussion of fracture testing of high strength sheet materials. *Mater. Res. Standards*, **1**, 277–9.
28. Kujawski, D. and Ellyin, F. (1986) On the size of plastic zone ahead of a crack-tip. *Engng. Fract. Mech.*, **25**, Elsevier Science Ltd, Oxford, 229–36.
29. Ellyin, F. (1986) Crack growth rate under cyclic loading and effect of different singularity fields. *Engng. Fract. Mech.*, **25**, Elsevier Science Ltd, Oxford, 463–73.
30. Shih, C.F. and German, M.D. (1981) Requirements for a one parameter characterization of crack-tip field by HRR singularity. *Int. J. Fract.*, **17**, 27–43.
31. Irwin, G.R. (1960) *Plastic Zone Near a Crack and Fracture Toughness*. Proc. 7th

Sagamore Ordnance Materials Conf., Syracuse Univ., New York, Vol. IV, pp. 63–78.
32. Lies, B.N., Brust, F.W. and Scott, P.M. (1991) Development and Validation of a Ductile Flaw Growth Analysis for Gas Transmission Line Pipe. Report to the American Gas Association (AGA) NG-18 Report No. 193, Bettle Memorial Institute, Columbus, OH, June 1991.
33. American Society for Testing and Materials (1973) Fracture Toughness Evaluation by R-Curve Methods, ASTM STP 527, Philadelphia, PA.
34. Paris, P.C., Tada, H., Zahoor, A. and Ernst, H. (1979) Instability of the Tearing Modulus of Elastic-Plastic Crack Growth in *Elastic-Plastic Fracture*, ASTM STP 668 (eds J.D. Landes *et al.*), American Society for Testing and Materials, Philadelphia, PA, pp. 5–36.
35. Rice, J.R., McMeeking, R.M., Parks, D.M. and Sorensen, E.P. (1979) Recent finite element studies in plasticity and fracture mechanics. *Computer Methods in Appl. Mech. and Engng.*, **17/18**, 411–42.
36. Shih, C.F. (1981) Relationships between the J-integral and the crack opening displacement for stationary and extending cracks. *J. Mech. Phys. Solids*, **29**, 305–26.
37. Kitagawa, H. and Takahashi, S. (1976) *Applicability of Fracture Mechanics to Very Small Cracks*. Proc. Int. Conf. Mech. Behaviour of Materials (ICM2), Amer. Soc. Metals, pp. 627–31.
38. Miller, K.J. (1987) The behaviour of short fatigue cracks and their initiation, Part I – A review of two recent books, and Part II – A general summary. *Fatigue Fract. Engng. Mater. Struct.*, **10**, 73–113.
39. Schijve, J. (1988) Fatigue crack closure: Observations and technical significance, in *Mechanics of Fatigue Crack Closure*, ASTM STP 982 (eds J.C. Newman and W. Elber), American Society for Testing and Materials, Philadelphia, PA, pp. 5–34.
40. Topper, T.H., Yu, M.T. and DuQuesnay, D.L. (1988) *Mechanics and Mechanisms of Fatigue Crack Initiation and Growth*. Proc. Int. Symp. on Fracture Mechanics (eds W.R. Tyson and M. Mukherjee), Pergamon Press, pp. 81–94.
41. Laird, C. and Smith, G.C. (1963) Initial stages of damage in high stress fatigue in some pure metals. *Phil. Mag.*, **8**, 1945–63.
42. Neumann, P. (1974) New experiments concerning the slip processes at propagating fatigue cracks. *Acta Metall.*, **22**, 1155–65.
43. McEvily, A.J. (1983) On the quantitative analysis of fatigue crack propagation, in *Fatigue Mechanisms; Advances in Quantitative Measurement of Physical Damage*, ASTM STP 811 (eds J. Lankford *et al.*), American Society for Testing and Materials, Philadelphia, PA, pp. 283–312.
44. Baïlon, J.-P. and Antolovich, S.D. (1983) Effect of microstructure on fatigue crack propagation: A review of existing models and suggestions for further research, ASTM STP 811 (eds J. Lankford *et al.*), American Society for Testing and Materials, Philadelphia, PA, pp. 313–49.
45. Paris, P.C., Gomez, M.P. and Anderson, W.E. (1961) A rational analytical theory of fatigue, *The Trend in Engineering*, **13** (Univ. of Washington, Seattle, WA), 9–14.
46. Head, A.K. (1956) The propagation of fatigue cracks. *J. Appl. Mech., Trans. ASME*, **23**, 407–10.
47. McClintock, F.A. (1963) On the plasticity of the growth of fatigue cracks, in *Fracture of Solids* (eds D.C. Drucker and J.J. Gilman). Interscience, New York, pp. 65–102.

48. Liu, H.W. (1964) Fatigue crack propagation and the stress and strain in vicinity of a crack. *Appl. Mat. Res.*, **3** (4), 229–37.
49. Frost, N.E. and Dugdale, D.S. (1958) The propagation of fatigue cracks in sheet specimens. *J. Mech. Phys. Solids*, **6** (2), 92–110.
50. Rice, J.R. (1965) *Plastic Yielding at a Crack Tip*. Proc. 1st Int. Conf. on Fracture, Sendai, Japan, part A-17, Japan Soc. for Strength and Fract. Mater., pp. 283–308.
51. Weertman, J. (1966) Rate of growth of fatigue cracks as calculated from the theory of dislocations distributed on a plane. *Int. J. Fract.*, **2**, 460–7.
52. Ellyin, F. and Li, H.-P. (1984) Fatigue crack growth in large specimens with various stress ratios. *J. Pressure Vessel Technol., Trans. ASME*, **106** (3), 255–60.
53. Weertman, J. (1979) Fatigue crack propagation theories, in *Fatigue and Microstructure*, Amer. Soc. Met., Metals Park, OH, pp. 279–306.
54. Yokobori, T., Konosu, S. and Yokobori, A.T. Jr. (1977) *Micro and Macro Fracture Mechanics Approach to Brittle Fracture and Fatigue Crack Growth*. In Fracture 1977, Proc. ICF4, Waterloo, Canada, Vol.1 (ed. D.M.R. Taplin), Pergamon, pp. 665–81.
55. Yokobori, T. (1979) A critical evaluation of mathematical equations for fatigue crack growth with special reference to ferrite grain size and monotonic yield strength dependence, in *Fatigue Mechanisms*, ASTM STP 675 (ed. J.T. Fong), American Society for Testing and Materials, Philadelphia, PA, pp. 683–706.
56. Kocanda, S. (1978) *Fatigue Failure of Metals*, Sijthoff and Noordhoff, Amsterdam.
57. Knott, J.F. (1986) Models of fatigue crack growth, in *Fatigue Crack Growth – 30 Years of Progress* (ed. R.A. Smith), Pergamon Press, Oxford, pp. 31–52.
58. Blom, A.F. (1989) Modelling of fatigue crack growth, in *Advances in Fatigue Science and Technology* (eds C. Moura Branco and L. Guerra Rosa), Kluwer Academic, Dordrecht, Netherlands, pp. 77–110.
59. Davidson, D.L. and Lankford, J. (1992) Fatigue crack growth in metals and alloys: Mechanisms and micromechanics. *Int. Mater. Rev.*, **37**, 45–76.
60. Kujawski, D. and Ellyin, F. (1984) A fatigue crack propagation model. *Engng. Fract. Mech.*, **20**, 695–704.
61. Chitaley, A.D. and McClintock, F.A. (1971) Elastic-plastic mechanics of steady crack growth under anti-plane shear. *J. Mech. Phys. Solids*, **14**, 147–63.
62. McMeeking, R.M. and Parks, D.M. (1979) On criteria for J-dominance of crack-tip fields in L.S.Y., in *Elastic-Plastic Fracture*, ASTM STP 668 (eds J.D. Landes, J.A. Begley and G.A. Clarke), American Society for Testing and Materials, Philadelphia, PA, pp. 175–94.
63. Lanteigne, J. and Baïlon, J.-P. (1981) Theoretical model for FCGR near the threshold. *Metall. Trans.*, **12A**, 459–66.
64. Glinka, G. (1982) Cumulative model of fatigue crack growth. *Int. J. Fatigue*, **4**, 59–67.
65. Radon, J.C. (1982) A model for fatigue crack growth in a threshold region. *Int. J. Fatigue*, **4**, 161–6.
66. Fine, M.E. and Davidson, D.L. (1983) Quantitative measurement of energy associated with a moving fatigue crack, in *Fatigue mechanisms: Advances in Quantitative Measurement of Physical Damage*, ASTM STP 811 (eds. D.L. Davidson *et al.*), American Society for Testing and Materials, Philadelphia, PA, pp. 350–70.
67. Ellyin, F. and Fakinlede, C.O.A. (1984) Crack-tip growth rate model for cyclic loading, in *Modelling Problems in Crack Tip Mechanics* (ed. J.T. Pindera), Martinus Nijhoff, Dordrecht, pp. 224–37.

68. McCartney, L.N. and Irving, P.E. (1977) Comment on a correlation for fatigue crack growth rate. *Scripta Metall.*, **11**, 181–3.
69. Niccols, E.H. (1976) A correlation for fatigue crack growth rate. *Scripta Metall.*, **10**, 295–8.
70. Tanaka, S., Ichikawa, M. and Akita, S. (1981) Variability of m and C in fatigue crack propagation law. *Int. J. Fract.*, **17**, R121–4.
71. Boller, C.H.R. and Seeger, T. (1987) *Materials Data for Cyclic Loading*, Elsevier, Amsterdam.
72. Broberg, K.B. (1982) The foundation of fracture mechanics. *Engng. Fract. Mech.*, **16**, 497–515.
73. Gurland, J. (1972) *Correlation Between Yield Strength and Microstructure of some Carbon Steels* in *Stereology and Quantitative Metallography*, ASTM STP 504, American Society for Testing and Materials, Philadelphia, PA, pp. 108–18.
74. Kirby, B.R. and Beevers, C.J. (1979) Slow fatigue crack growth and threshold behaviour in air and vacuum of commercial aluminum alloys. *Fatigue Engng. Mater. Struct.*, **1**, 203–15.
75. Klesnil, M. and Lukáš, P. (1972) Effect of stress cycle asymmetry on fatigue crack growth. *Mater. Sci. Engng.*, **9**, 231–40.
76. Masounave, J. and Baïlon, J.P. (1976) Effect of grain size on the threshold stress intensity factor in fatigue of a ferritic steel. *Scripta Metall.*, **10**, 165–70.
77. Tanaka, K., Nakai, Y. and Yamashita, M. (1981) Fatigue growth threshold of small cracks. *Int. J. Fract.*, **17**, 519–33.
78. Masounave, J. and Baïlon, J.P. (1975) The dependence of the threshold stress intensity factor on the cyclic stress ratio of the fatigue of ferritic-perlitic steels. *Scripta Metall.*, **9**, pp. 723–30.
79. Beevers, C.J. (1977) Fatigue crack growth characteristics at low stress intensities of metals and alloys. *Metal Sci.*, **11**, 362–7.
80. Ritchie, R.O. (1977) Influence of microstructure on near-threshold fatigue-crack propagation in ultra-high strength steel. *Metal Sci.*, **11**, 368–81.
81. Ritchie, R.O. (1977) Near-threshold fatigue crack propagation in ultra-high strength steel: Influence of load ratio and cyclic strength. *J. Engng. Mater. Technol., Trans. ASME*, **99**, 195–204.
82. Usami, S. and Shida, S. (1979) Elastic-plastic analysis of the fatigue limit for a material with small flaws. *Fatigue Engng. Mater. Struct.*, **1**, 471–81.
83. Stephens, R.I. *et al.* (1984) Constant and variable amplitude fatigue behaviour of five cast steels at room temperature and −45°C. *J. Engng. Mater. Technol., Trans. ASME*, **106**, 25–37.
84. Staal, H.V. and Elen, J.D. (1979) Crack closure and influence of cyclic ratio R on fatigue crack growth in type 304 stainless steel at room temperature. *Engng. Fract. Mech.*, **11**, 275–83.
85. King, J.E. (1982) Surface damage and near threshold fatigue crack growth in a Ni-base superalloy in vacuum. *Fatigue Engng. Mater. Struct.*, **5**, 177–88.
86. Taylor, D. and Knott, J.F. (1982) Growth of fatigue cracks from casting defects in nickel-aluminum bronze. *Metals Tech.*, **9**, 221–8.
87. Elber, W. (1971) The significance of fatigue crack closure, in *Damage Tolerance in Aircraft Structures*, ASTM STP 486, American Society for Testing and Materials, Philadelphia, PA, pp. 230–42.
88. Urangst, K.D., Shih, T.T. and Wei, R.P. (1977) Crack closure in 2219-T851 aluminum alloy. *Engng. Fract. Mech.*, **9**, 725–34.

89. Macha, D.E., Corbly, D.M. and Jones, J.W. (1979) On the variation of fatigue-crack-opening load with measurement location. *Proc. Soc. Exp. Stress Anal.*, **36**, 207–13.
90. Shin, C.S. and Smith, R.A. (1985) Fatigue crack growth from sharp notches. *Int. J. Fatigue*, **7**, 87–93.
91. Klesnil, M. and Lukáš, P. (1972) Influence of strength and stress history on growth and stabilization of fatigue cracks. *Engng. Fract. Mech.*, **4**, 77–92.
92. Davenport, R.T. and Brook, R. (1979) The threshold stress intensity range in fatigue. *Fatigue Engng. Mater. Struct.*, **1**, 151–8.
93. Radon, J.C. (1982) A model for fatigue crack growth in a threshold region. *Int. J. Fatigue*, **4**, 161–6.
94. Liaw, P.K. (1988) Overview of crack closure at near-threshold fatigue crack growth levels, in *Mechanics of Fatigue Crack Closure*, ASTM STP 982 (eds J.C. Newman Jr. and W. Elber), American Society for Testing and Materials, Philadelphia, PA, pp. 62–92.
95. Radon, J.C. and Guerra Rosa, L. (1989) Fatigue threshold behaviour, Parts I and II, in *Advances in Fatigue Science and Technology* (eds C. Moura Branco and L. Guerra Rosa), Kluwer Academic, Dordrecht, Netherlands, pp. 129–56.
96. Elber, W. (1970) Fatigue crack closure under cyclic tension. *Engng. Fract. Mech.*, **2**, 37–45.
97. Kujawski, D. and Ellyin, F. (1987) A fatigue crack growth model with load ratio effects. *Engng. Fract. Mech.*, **28** (4), Elsevier Science Ltd, Oxford, 367–78.
98. Forman, R.G. (1972) Study of fatigue crack initiation from flaws using fracture mechanics theory. *Engng. Fract. Mech.*, **4**, 333–45.
99. McEvily, A.J. and Wei, R.P. (1972) Fracture mechanics and corrosion fatigue, in *Corrosion-Fatigue: Chemistry, Mechanics and Microstructure*, Vol. NACE-2 (eds O.F. Devereux, A.J. McEvily and R.W. Staehle), National Association of Corrosion Engineers, Houston, TX, pp. 381–95.
100. Nicholson, C.E. (1973) *Influence of Mean Stress and Environment on Crack Growth*. Proc. Conf. on Mechanics and Mechanisms of Crack Growth, Cambridge, UK, British Steel Corp., pp. 226–43.
101. Hoeppener, D.W. and Krupp, W.E. (1974) Prediction of component life by application of fatigue crack growth knowledge. *Engng. Fract. Mech.*, **6**, 47–70.
102. Paris, P.C. *et al.* (1972) Extensive study of low fatigue crack growth rates in A533 and A508 steels, in *Stress Analysis and Growth of Cracks, Part I*, ASTM STP 513, American Society for Testing and Materials, pp. 141–76.
103. Glinka, G. (1985) A notch stress-strain analysis approach to fatigue crack growth. *Engng. Fract. Mech.*, **21**, 245–61.
104. McEvily, A.J. (1988) On crack closure in fatigue crack growth, in *Mechanics of Fatigue Crack Closure*, ASTM STP 982 (eds J.C. Newman Jr. and W. Elber), American Society for Testing and Materials, Philadelphia, PA, pp. 35–43.
105. Bowles, C.Q. and Schijve, J. (1983) Crack tip geometry for fatigue cracks grown in air and vacuum, in *Fatigue Mechanisms: Advances in Qualitative Measurement of Physical Damage*, ASTM STP 811 (eds J. Lankford *et al.*), American Society for Testing and Materials, Philadelphia, PA, pp. 400–26.
106. Hudak, S.J. and Davidson, D.L. (1988) The dependence of crack closure on fatigue loading variables, in *Mechanics of Fatigue Crack Closure*, ASTM STP 982 (eds J.C. Newman Jr. and W. Elber), American Society for Testing and Materials, Philadelphia, PA, pp. 121–38.

107. Hertzberg, R.W., Newton, C.H. and Jaccard, R. (1983) Crack closure: Correlation and confusion, in *Fatigue Mechanisms: Advances in Qualitative Measurement of Physical Damage,* ASTM STP 811 (eds J. Lankford *et al.*), American Society for Testing and Materials, Philadelphia, PA, pp. 139–48.
108. Williams, J.F. and Lam, Y.C. (1986) The quantification of crack closure based on a contact stress intensity factor. *Theor. Appl. Fract. Mech.*, **6**, 21–8.
109. Elber, W. (1976) Equivalent constant-amplitude concept for crack growth under spectrum loading, in: *Fatigue Crack Growth Under Spectrum Loading,* ASTM STP 595, American Society for Testing and Materials, Philadelphia, PA, pp. 236–47.
110. Schijve, J. (1980) Prediction methods for fatigue crack growth in aircraft material, in *Fracture Mechanics,* Twelfth Conf., ASTM STP 700, American Society for Testing and Materials, Philadelphia, PA, pp. 3–34.
111. Ellyin, F. and Wu, J. (1992) Elastic-plastic analysis of a stationary crack under cyclic loading and effect of overload. *Int. J. Fract.*, **56**, Kluwer Academic Publishers, Dordrecht, 182–208.
112. Ward-Close and Ritchie, R.O. (1988) On the effect of crack closure mechanisms in influencing fatigue crack closure following tensile overloads in a titanium alloy: near threshold versus higher ΔK behaviour, in *Mechanics of Fatigue Crack Closure,* ASTM STP 982 (eds J.C. Newman Jr. and W. Elber), American Society for Testing and Materials, Philadelphia, PA, pp. 93–111.
113. Allison, J.E., Ku, R.C. and Pompetzki, M.A. (1988) A comparison of measurement methods and numerical procedures for the experimental characterization of fatigue crack closure, in *Mechanics of Fatigue Crack Closure,* ASTM STP 982 (eds J.C. Newman Jr. and W. Elber), American Society for Testing and Materials, Philadelphia, PA, pp. 171–85.
114. Williams, J.F. and Stouffer, D.C. (1979) An estimate of the residual stress distribution in the vicinity of a propagating fatigue crack. *Engng. Fract. Mech.*, **11**, 547–57.
115. Suresh, S. (1983) Micromechanisms of fatigue crack retardation following overloads. *Engng. Fract. Mech.*, **18**, 577–93.
116. Tokaji, K., Ando, Z. and Kojima, T. (1984) Fatigue crack retardation of low carbon steel in salt water. *J. Engng. Mater. Technol., Trans. ASME,* **106**, 38–42.
117. Fleck, N.A. (1988) Influence of stress state on crack growth retardation, in *Basic Questions in Fatigue,* Vol. 1, ASTM STP 924 (eds J.T. Fong and R.J. Fields), American Society for Testing and Materials, Philadelphia, PA, pp. 157–83.
118. Kim, S. and Tai, W. (1992) Retardation and arrest of fatigue crack growth in A151 4140 steel by introducing rest periods and overloads. *Fatigue Fract. Engng. Mater. Struct.*, **15**, 519–30.
119. *Fatigue Crack Growth Under Spectrum Loads* (1996), ASTM STP 595, American Society for Testing and Materials, Philadelphia, PA.
120. Vandar, O. and Yildirim, N. (1990) Crack growth retardation due to intermittent overloads. *Int. J. Fatigue*, **12**, 283–7.
121. Topper, T.H. and Yu, M.T. (1985) The effect of overloads on the threshold and crack closure. *Int. J. Fatigue,* **7**(3), Elsevier Science Ltd, Oxford, 159–64.
122. Marissen, R., Trautmann, K.H. and Nowack, H. (1984) The influence of compression loads and of dK/da on the crack propagation under variable amplitude loading. *Engng. Fract. Mech.*, **19**, 863–79.
123. Stephens, R.I., Chen, D.K. and Hom, B.W. (1976) Fatigue crack growth with negative stress ratio following single overloads in 2024-T3 and 7075-T6 aluminum alloys, in *Fatigue Crack Growth Under Spectrum Loads,* ASTM STP 595, American Society for Testing and Materials, Philadelphia, PA, pp. 27–40.

124. Yu, M.T., Topper, T.H., DuQuesnay, D.L. and Levin, M.S. (1986) The effect of compressive peak stress on fatigue behaviour. *Int. J. Fatigue*, **8**, 9–15.
125. Chang, J.B. and Hudson, C.M. (eds) (1981) *Methods and Models for Predicting Fatigue Crack Growth Under Random Loading*, ASTM STP 748, American Society for Testing and Materials, Philadelphia, PA.
126. Petit, J., Davidson, D.L., Suresh, S. and Robbe, P. (eds) (1988) *Fatigue Crack Growth Under Variable Amplitude Loading*, Elsevier, London.
127. Newman, J.C., Jr., Wu, X.R., Venneri, J.L. and Li, C.G. (1994) Small Crack Effects in High-Strength Aluminum Alloys. NASA Reference Publication 1309, NASA Langley Research Center, Hampton, VA, May 1994, 118 pp.
128. Budiansky, B. and Hutchinson, J.W. (1978) Analysis of closure in fatigue crack growth. *J. Appl. Mech., Trans. ASME*, **45**, 267–76.
129. Führing, H. and Seeger, T. (1979) Dugdale crack closure analysis of fatigue cracks under constant amplitude loading. *Engng. Fract. Mech.*, **11**, 99–122.
130. Newman, J.C., Jr. (1981) A crack-closure model for predicting fatigue crack growth under aircraft spectrum loading, in *Methods and Models for Predicting Fatigue Crack Growth under Random Loading*, ASTM STP 748 (eds J.B. Chang and C.M. Hudson), American Society for Testing and Materials, Philadelphia, PA, pp. 53–84.
131. de Koning, A.V. and Liefting, G. (1988) Analysis of crack opening behaviour by application of a discretized strip yield model, in *Mechanics of Fatigue Crack Closure*, ASTM STP 982 (eds J.C. Newman Jr. and W. Elber), American Society for Testing and Materials, Philadelphia, PA, pp. 437–59.
132. Ibrahim, F.K., Thompson, J.C. and Topper, T.H. (1986) A study of the effect of mechanical variables on fatigue crack closure and propagation. *Int. J. Fatigue*, **8**, 135–42.
133. Newman, J.C., Jr. and Armen, H., Jr. (1975) Elastic-plastic analysis of a propagating crack under cyclic loading. *AIAA Journal*, **13**, 1017–23.
134. Ohji, K., Ogura, K. and Ohkubo, Y. (1975) Cyclic analysis of a propagating crack and its correlation with fatigue crack growth. *Engng. Fract, Mech.*, **7**, 457–64.
135. Newman, J.C., Jr. (1976) A finite-element analysis of fatigue crack closure, in *Mechanics of Crack Growth*, ASTM STP 590, American Society for Testing and Materials, Philadelphia, PA, pp. 281–301.
136. Ogura, K. and Ohji, K. (1977) FEM analysis of crack closure and delay effect in fatigue crack growth under variable loading. *Engng. Fract. Mech.*, **9**, 471–80.
137. Nakagaki, M. and Atluri, S.N. (1979) Fatigue crack closure and delay effect under Mode I spectrum loading: An efficient elastic-plastic analysis procedure. *Fatigue Engng. Mater. Struct.*, **1**, 421–9.
138. Blom, A.F. and Holm, D.K. (1985) An experimental and numerical study of crack closure. *Engng. Fract. Mech.*, **22**, 997–1011.
139. Kobayashi, H. and Nakamura, H. (1987) Investigation of fatigue crack closure (Analysis of plasticity induced crack closure), in *Current Research on Fatigue Crack*, The Society of Material Science, Kyoto, Japan, pp. 229–47.
140. Fleck, N.A. and Newman, J.C., Jr. (1988) Analysis of crack closure under plane strain condition, in *Mechanics of Fatigue Crack Closure*, ASTM STP 982 (eds J.C. Newman Jr. and W. Elber), American Society for Testing and Materials, Philadelphia, PA, pp. 319–41.
141. Lalor, P.L. and Sehitoglu, H. (1988) Fatigue crack closure outside a small-scale yielding regime, in *Mechanics of Fatigue Crack Closure*, ASTM STP 982 (eds J.C.

Newman Jr. and W. Elber), American Society for Testing and Materials, Philadelphia, PA, pp. 342–60.

142. Anquez, L. and Baudin, G. (1988) Correlation between numerically predicted crack opening load and measured load history dependent crack growth threshold, in *Mechanics of Fatigue Crack Closure*, ASTM STP 982 (eds J.C. Newman Jr. and W. Elber), American Society for Testing and Materials, Philadelphia, PA, pp. 380–97.
143. Chermahini, R.G., Shivakumar, K.N. and Newman, J.C., Jr. (1988) Three-dimensional finite-element simulation of fatigue crack growth and closure, in *Mechanics of Fatigue Crack Closure*, ASTM STP 982 (eds J.C. Newman Jr. and W. Elber), American Society for Testing and Materials, Philadelphia, PA, pp. 398–413.
144. Chermahini, R.G., Palmberg, B. and Blom, A.F. (1993) Fatigue crack growth and closure behaviour of semicircular and semielliptical surface flaws. *Int. J. Fatigue*, **15**, 259–63.
145. Fawkes, A.J., Owen, D.R.Y. and Luxmore, A.R. (1979) An assessment of crack tip singularity models for use with isoparametric elements. *Engng. Fract. Mech.*, **11**, 143–59.
146. Bathe, K.-J. (1996) *Finite Element Procedures in Engineering Analysis*, Prentice-Hall, Englewood Cliffs, NJ.
147. Wu, J. (1995) Nonlinear analysis of cracked bodies. Doctoral dissertation, University of Alberta.
148. Michel, D. (1987) *Elevated Temperature Fatigue Crack Propagation*. Fatigue 87, Proc. 3rd Int. Conf. on Fatigue and Fatigue Threshold, Virginia, vol. II (ed. R.O. Ritchie), EMAS, London, pp. 1077–85.
149. Bassani, J.L. and McClintock, F.A. (1981) Creep relaxation of stress around a crack tip. *Int. J. Solids Struct.*, **17**, 479–92.
150. Riedel, H. (1981) Creep deformation at crack tips in elastic-viscoplastic solids. *J. Mech. Phys. Solids*, **29**, 35–44.
151. Riedel, H. and Rice, J.R. (1980) Tensile cracks in creeping solids, in *Fracture Mechanics*, Twelfth Conf., ASTM STP 700, American Society for Testing and Materials, Philadelphia, PA, pp. 112–30.
152. Ehler, R. and Riedel, H. (1981) A finite element analysis of creep deformation in a specimen containing a macroscopic crack, in *Advances in Fracture Research*, Proc. 5th Int. Conf. on Fracture, Cannes, France, Vol. 2 (ed. D. Francois), Pergamon, Oxford, pp. 691–8.
153. Bassani, J.L., Hawk, D.E. and Saxena, A. (1989) Evaluation of the C_t parameter for characterizing creep crack growth rate in the transient regime, in *Nonlinear Fracture Mechanics: Vol. 1–Time-dependent Fracture*, ASTM STP 995 (eds A. Saxena, J.D. Landes and J.L. Bassani), American Society for Testing and Materials, Philadelphia, PA, pp. 7–26.
154. Bensussan, P., Piques, R. and Pineau, A. (1989) A critical assessment of the global mechanical approach to crack initiation and creep crack growth in 316L steel, in *Nonlinear Fracture Mechanics: Vol. 1–Time-dependent Fracture*, ASTM STP 995 (eds A. Saxena, J.D, Landes and J.L. Bassani), American Society for Testing and Materials, Philadelphia, PA, pp. 27–54.
155. Hui, C.Y. and Riedel, H. (1981) The asymptotic stress and strain field near the tip of a growing crack under creep conditions. *Int. J. Fract.*, **17**, 409–25.
156. Bassani, J.L., Hawk, D.E. and Wu, F.-H. (1989) Crack growth in small-scale creep, in *Nonlinear Fracture Mechanics, Vol. I: Time-Dependent Fracture*, ASTM

STP 995 (eds A. Saxena, J.D. Landes and J.L. Bassani), American Society of Testing and Materials, Philadelphia, PA, pp. 68–95.

157. Sadananda, K. and Shahinian, P. (1981) Review of the fracture mechanics approach to creep crack growth in structural alloys. *Engng. Fract. Mech.*, **15**, 327–42.

158. Sadananda, K. and Shahinian, P. (1981) Creep-fatigue crack growth, in *Cavities and Cracks in Creep and Fatigue* (ed. J. Gittus), Elsevier, London, pp. 109–95.

159. Koterazawa, R. and Mori, T. (1977) Applicability of fracture mechanics parameters of crack propagation under creep condition. *J. Engng. Mater. Technol., Trans. ASME*, **99**, 298–305.

160. Riedel, H. (1987) *Fracture at High Temperature*, Springer-Verlag, Berlin.

161. Ohtani, R. (1987) Substance of creep-fatigue interaction examined from the point of view of crack propagation mechanics, in *Low Cycle Fatigue and Elasto-Plastic Behaviour of Materials* (ed. K.-T. Rie), Elsevier, London, pp. 211–22.

162. Ellyin, F. and Fakinlede, C.O.A. (1987) A dislocation model for work-hardening material and cyclic J-integral. *Int. J. Fract.*, **33**, 95–110.

163. Jaske, C.E. (1995) Life prediction in high-temperature structural materials. *J. Pressure Vessel Technol., Trans. ASME*, **117**, 1–6.

164. James, L.A. (1995) Environmentally-assisted cracking behaviour of a low-alloy steel under non-isothermal conditions, in *Fatigue and Crack Growth: Environmental Effects, Modeling Studies, and Design Considerations*, PVP 306 (eds S. Yukawa, D.P. Jones and H.S. Mehta), American Society of Mechanical Engineers, New York, pp. 19–27.

165. Wei, R.P. and Gangloff, R.P. (1989) Environmentally assisted crack growth in structural alloys: Perspectives and new directions, in *Fracture Mechanics: Perspectives and Directions* (20th Symp.), ASTM STP 1020 (eds R.P. Wei and R.P. Gangloff), American Society for Testing and Materials, Philadelphia, PA, pp. 233–64.

166. Gangloff, R.P. (1990) *Corrosion Fatigue Crack Propagation in Metals.* Proc. Environment-Induced Cracking of Metals (eds R.P. Gangloff and M.B. Ives), NACE-10, National Association of Corrosion Engineers. Houston, Texas, pp. 55–109.

167. Garud, Y.S. (1991) Quantitative evaluation of environmentally assisted cracking: A survey of developments and application of modeling concepts. *J. Pressure Vessel Technol., Trans. ASME*, **113**, 1–9.

168. Yukawa, S., Jones, D.P. and Mehta, H.S. (eds) (1995) *Fatigue and Crack Growth: Environmental Effects, Modeling Studies, and Design Considerations*, PVP 306, American Society of Mechanical Engineers, New York.

169. American Society of Mechanical Engineers (1993) *ASME Boiler and Pressure Vessel Code*, Section XI: Rules for Inervice Inspection of Nuclear Power Plant Components.

170. Suresh, S. and Ritchie, R.O. (1982) Mechanistic dissimilarities between environmentally influenced fatigue-crack rates in lower strength steels. *Metal Sci.*, **16**, 529–38.

171. Zhiqiang, X., Yujiu, S. and Mingjing, T. (1991) Crack closure induced by corrosion products and its effect in corrosion fatigue. *Int. J. Fatigue*, **13**, 69–72.

172. Henaff, G., Petit, J. and Bouchet, B. (1992) Environmental influence on the near-threshold fatigue crack propagation behaviour of a high strength steel. *Int. J. Fatigue*, **14**, 211–18.

173. Ellyin, F. and Wu, J. (1994) Effect of hydride precipitation on the elastoplastic stress field near a crack tip. *Acta Metallurgica et Materialia*, **42**, 2709–17.
174. Bartlett, M.L. and Hudak, S.J., Jr. (1990) The influence of frequency-dependent crack closure on corrosion fatigue crack growth, in *Fatigue 90*, Vol. III (eds H. Kitagawa and T. Tanaka), Materials and Components Engineering . Publications Ltd., UK, pp. 1783–8.
175. Chalant, G., Petit, J. and Suyitno, B. (1990) An environmental fatigue crack propagation model for structural steels, in *Fatigue 90*, Vol. III (eds H. Kitagawa and T. Tanaka), Materials and Components Engineering Publications Ltd., Birmingham, UK, pp. 1771–6.
176. Turnbull, A. (1993) Modelling of environment assisted cracking. *Corrosion Sci.*, **34**, 921–60.
177. Shoji, T., Takahashi, H., Suzuki, M. and Kondo, T. (1981) A new parameter for characterizing corrosion fatigue crack growth. *J. Engng. Mater. Technol., Trans. ASME*, **103**, 298–304.
178. Shoji, T. and Takahashi, H. (1983) *Role of Loading Variables in Environmental Enhanced Crack Growth for Water-Cooled Nuclear Reactor Pressure Vessel Steels*. Proc. IAEA Specialists' Meeting on Subcritical Crack Growth, NUREG/CP-0044, May 1983, pp. 143–71.

Fatigue of notched members 8

8.1 INTRODUCTION

Most structural/mechanical elements contain notches of various forms. For example, to connect structural elements holes are drilled and bolts are used. The bolts themselves have threaded roots and head to shank geometric discontinuity. Pressure vessels always have openings for functional requirements. Shafts contain keyways and steps. At these geometric discontinuities, stresses are generally higher than the nominal values, and if proper care is not taken, notches could be sites of crack initiation and the subsequent crack propagation may result in failure of a component/ structural element. Small fatigue cracks which start from a smooth surface in blunt notches cannot be classified as crack-like defects which can be treated by the classical fracture mechanics approach as discussed in Chapter 7. The word 'notch' is used here in a generic sense, to imply geometric discontinuities of all shapes, e.g. holes, grooves, shoulders (with or without fillets), keyways, etc. Notches act as stress raisers and the degree of stress concentration depends on the geometry of the notch. A concentration factor is defined which indicates the severity of the notch,

$$K_\sigma = \frac{\sigma_{max}}{\sigma_{nom}}, \quad K_\varepsilon = \frac{\varepsilon_{max}}{\varepsilon_{nom}}, \tag{8.1}$$

where K_σ, K_ε are stress and strain concentration factors and σ_{nom} and ε_{nom} are nominal stress and strain at sections removed from the notch (in some fatigue literature a notation of s and e is used to indicate nominal values). In some instances, σ_{nom} or ε_{nom} are defined with respect to a reduced (net) cross-section at the notch location. Although there is a straightforward relationship between the two methods, the definition based on gross cross-section, i.e. stress and strain calculated at a region remote from the effect of stress concentration, is preferable on physical grounds. When the material behaviour is elastic, then

$$K_\sigma = K_\varepsilon \approx K_t \tag{8.2}$$

where K_t is often referred to as a 'theoretical' stress concentration factor determined from an elastic analysis of a notched element. This will be demonstrated in section 8.2.1 for an elliptical notch in a flat plate.

For an increased nominal stress, the stress and strain values at a notch root may exceed a material's elastic limit. In the plastic regime, the stress concentration relaxes with increased plastic deformation whereas the strain concentration increases. Methods used to determine the maximum stress and strain concentrations in the elastic–plastic range of a material's response will be discussed in sections 8.2.2 to 8.2.4.

A combination of a high stress or strain concentration and cyclic loading may lead to initiation of a crack at a notch root and its subsequent propagation. These will be the subject of sections 8.3 and 8.4, respectively.

8.2 NOTCH ANALYSIS

Analytical studies on the effect of notches date back to Kirsch (1898) who examined the problem of a circular hole in a thin, elastic-sheet. The stress around elliptical holes was subsequently studied by Kolosov about 1909, who showed that the stress concentration factor for elliptical holes depends upon the rounding-off radius (increasing with decreasing radii) and the location of the hole relative to the external tensile load. This problem will be discussed in detail in the following section.

Neuber's book, '*Kerbspannungslehre*', first published in 1934 [1], was an important contribution in advancing theoretical solutions of some stress concentration problem. Solutions of a large class of plane elasticity problems were made possible mostly due to the work of Muskhelishvili and his school in the former Soviet Union, through the application of conformal mapping techniques [2]. Savin [3] has detailed the problems of stress concentration in plates, by presenting the results in the form of simple formulae, graphs and tables which are useful for designers. Another source for obtaining the stress concentration factors for a large number of problems of practical significance is Peterson's book [4]. Most exact solutions of plane elasticity problems were obtained for notches in infinite bodies. However, due to the very localized effect of the stress concentration, they are applicable to finite bodies when the notch to width ratio exceeds a certain value.

A treatment based on a three-dimensional stress analysis becomes necessary when the radius of the notch and the half thickness of the plate are of about the same magnitude. Solutions of this kind have been given by Green [5], Sternberg and Sadowsky [6], Reiss [7] and Youngdahl and Sternberg [8], among others. These investigations have shown that the circumferential stress remains the largest component, and its value is close to that given by plane theory. A more general problem of oblique holes in flat plates was attempted by Ellyin *et al.* [9, 10] who proposed a method whereby the stress field for an oblique hole could be determined from the plane stress field of an equivalent straight hole. The results indicated a steep rise in the maximum stress in terms of the angle of skewness, sensitivity to Poisson's ratio, and shift of the location of the maximum stress from the midsurface for thick plates towards the edge for thin ones. The theoretical predictions were found to be in good agreement with the experimental results, e.g. [10].

A review of mainly plane stress concentration problems investigated up to 1965 are given by Neuber and Hahn [11]. The application of the complex potential functions has played a very important role in the solution of plane stress concentration problems. In the following an example of an elliptical hole in a flat plate will be given. This problem is of practical significance since as the minor axis of the ellipse tends towards zero ($b \to 0$), the elliptical hole tends towards a narrow slit (crack-like). Thus, the transition from a notch with finite radius of curvature to a slit with zero radius of curvature could be studied. Indeed, this was the manner in which Griffith derived his theory of brittle fracture of solids.

8.2.1 ELASTIC ANALYSIS OF AN ELLIPTICAL HOLE IN A PLATE

The stress field in the plane theory of elasticity may be expressed by means of a single auxiliary function which is called the stress function or Airy function, $\phi = \phi(x, y)$. It may be shown that the single auxiliary function can itself be expressed in a simple manner with the help of two analytical functions of the complex variable $z = x + iy$, such that

$$\phi = \mathrm{Re}[\bar{z}\psi(z) + \chi(z)], \tag{8.3}$$

where a bar denotes a conjugate quantity, i.e. $\bar{z} = x - iy$, and Re means the 'real part of'.

For an elliptical hole with semi-major axis, a, and semi-minor axis, b, subjected to a far-field tensile stress of N, normal to the major axis, the complex potentials are given by [12]

$$\begin{aligned} 4\psi(z) &= Nc[-e^{2\alpha}\cosh z + (1 + e^{2\alpha})\sinh z] \\ 4\chi(z) &= -Nc^2[(\cosh 2\alpha + 1)z - \tfrac{1}{2}e^{2\alpha}\cosh 2(z - \alpha)] \end{aligned} \tag{8.4}$$

where $c^2 = a^2 - b^2$, $a = c\cosh\alpha$, $b = c\sinh\alpha$ and $e^{2\alpha} = (a + \mathrm{b})/(a - b)$.
With reference to the elliptical coordinates ξ, η shown in Fig. 8.1, the stress components $\sigma_{\xi\xi}, \sigma_{\eta\eta}$ and $\sigma_{\xi\eta}$ are found from substituting (8.4) into (7.1) of Chapter 7 and changing θ to η, r to ξ and noting that $e^{2i\theta} = \sinh z/\sinh\bar{z}$.

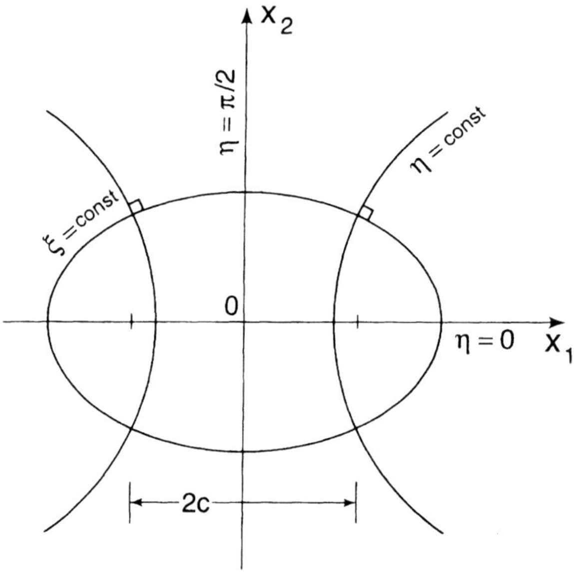

Fig. 8.1 An elliptical hole with curvilinear and Cartesian coordinates.

The result after certain algebraic manipulations is

$$\sigma_{\eta\eta} = \frac{N}{4}\mathrm{Re}(2[-e^{2\alpha} + (1+e^{2\alpha})\coth z] - \mathrm{csch}\, z\, \mathrm{csch}\, \bar{z}\, \{(1+e^{2\alpha})\mathrm{csch}\, z \cosh \bar{z}$$
$$- e^{2\alpha}[2\cosh 2(z-\alpha) - \coth z \sinh 2(z-\alpha)] - 2\cosh^2 \alpha \coth z\}) \quad (8.5a)$$

$$\sigma_{\xi\xi} = \frac{N}{4}\mathrm{Re}(2[-e^{2\alpha} + (1+e^{2\alpha})\coth z] + \mathrm{csch}\, z\, \mathrm{csch}\, \bar{z}\, \{(1+e^{2\alpha})\mathrm{csch}\, z \cosh \bar{z}$$
$$- e^{2\alpha}[2\cosh 2(z-\alpha) - \coth z \sinh 2(z-\alpha)] - 2\cosh^2 \alpha \coth z\}) \quad (8.5b)$$

$$\sigma_{\xi\eta} = \frac{N}{4}\mathrm{Im}(-\mathrm{csch}\, z\, \mathrm{csch}\, \bar{z}\, \{(1+e^{2\alpha})\mathrm{csch}\, z \cosh \bar{z} - e^{2\alpha}[2\cosh 2(z-\alpha)$$
$$- \coth z \sinh 2(z-\alpha)] - 2\cosh^2 \alpha \coth z\}) \quad (8.5c)$$

It could easily be verified that at the tips of the major and minor axes, the circumferential stress, (8.5a), reduces to

$$\sigma_{\eta\eta}/N = (1 + 2a/b) \quad \text{for} \quad \xi = \alpha, \eta = 0 \quad (8.6a)$$

$$\sigma_{\eta\eta}/N = -1 \quad \text{for} \quad \xi = \alpha, \eta = \pi/2 \quad (8.6b)$$

$$\sigma_{\xi\xi} = \sigma_{\xi\eta} = 0 \quad \text{for} \quad \xi = \alpha \quad (8.6c)$$

Figure 8.2 shows the distribution of the circumferential stress for two major to minor axes ratios, $a/b = 4$ and $a/b = 10$. It is seen that as a/b

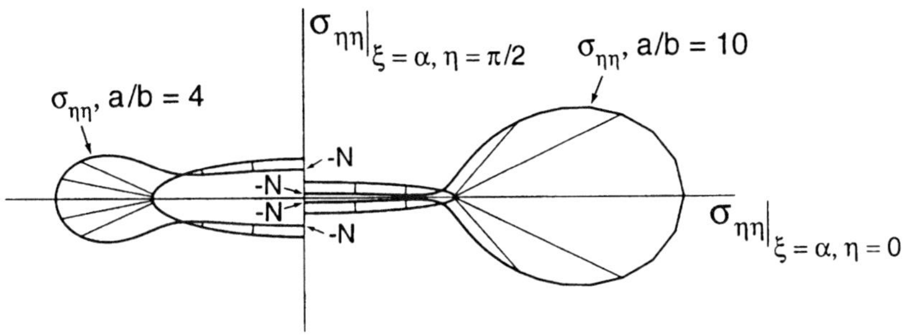

Fig. 8.2 Distribution of the circumferential stress, $\sigma_{\eta\eta}$, around an elliptical notch for two values of *a/b*.

increases, the tensile stress concentrates around the major axes and its magnitude increases. The highest stress is at the apex of the major axis of the ellipse where the radius of curvature,

$$\rho = b^2/a\,, \tag{8.7}$$

is minimum. Note that for a slender ellipse with $a/b \gg 1$, the maximum stress is given by

$$\sigma_{\max} = 2N\left(\frac{a}{\rho}\right)^{1/2} \tag{8.8}$$

where $\rho \ll a$.

From the foregoing analysis it is clear that a crack may initiate at the apex of the major axis of a slender ellipse, for either a high value of N (monotonic loading) or a lower alternating N value (cyclic loading). Initially the crack would be very small, and its propagation will depend on the stress and strain state of the notch. Thus, the study of stress distribution beyond the notch tip may prove instructive. Figure 8.3 shows the stress distribution along the major axis of the ellipse. The abscissa is made dimensionless by dividing distance from the centre of the notch by the semi-major axis of the ellipse, a. The ordinate is also non-dimensional, indicating stress magnification due to the notch. Three ratios of a/b are considered, $a/b = 2, 4$ and 10. The normal stress $\sigma_{\eta\eta}$, and the axial stress $\sigma_{\xi\xi}$ at $\eta = 0$, are plotted in the figure. Note that the shear stress $\sigma_{\eta\xi} = 0$ on the plane of symmetry, $\eta = 0$. It can be seen that for $a/b > 4$ the stress concentration factor, $K_\sigma = \sigma_{\eta\eta}/N$ has a very steep gradient along the major axis. Also note that the axial stress, $\sigma_{\xi\xi}$, which was zero at the notch tip, increases rapidly and then decays as x/a increases. The effect of the notch is restricted to an area about $x/a < 1.0$, i.e. the notch effect is negligible beyond a distance of one semi-major axis from the notch tip. The biaxial/triaxial stress state and a steep stress gradient in the vicinity of a notch are characteristic notch effects and have to be considered in a fatigue analysis. It follows that a stress concentration

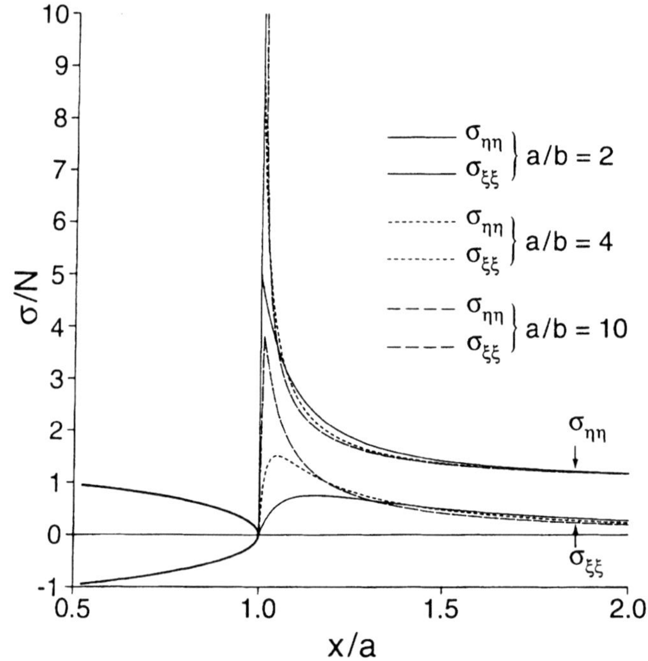

Fig. 8.3 Distribution of normal and axial stress components, $\sigma_{\eta\eta}$ and $\sigma_{\xi\xi}$, along the major notch axis, $\eta = 0$, for three ratios of $a/b = 2,4$ and 10.

factor by itself may not be a good indicator of the stress state near a notch root as far as fatigue analysis is concerned.

In Fig. 8.3, it was noted that the maximum stress increases rapidly with the decrease in the radius of curvature at the notch root, cf. (8.8). Therefore, for ductile materials yielding will generally take place around a notch root. To obtain an indication of the plastic zone size and its shape, contours of equivalent stress

$$\sigma_{\mathrm{eq}} = \frac{1}{\sqrt{2}}[\sigma_{\eta\eta}^2 + \sigma_{\xi\xi}^2 + 3\tau_{\xi\eta}^2]^{1/2}, \tag{8.9}$$

are traced in Fig. 8.4. The abscissa is the dimensionless length, x/a, and the ordinate is similarly non-dimensionalized, as y/a. Contours of σ_{eq} ranging from $\sigma_{\mathrm{eq}} = 2N$ to $\sigma_{\mathrm{eq}} = 7N$ are plotted in Fig. 8.4, for an elliptical notch with the major to minor axes ratio, $a/b = 4$. It can be seen that for low nominal applied stresses, the plastic zone would be very small and will be confined to a region around the apex of the major axis. However, as the far-field stresses are increased, the plastic zone expands in an almost self-similar manner. Note that for $\sigma_{\mathrm{nom}}/\sigma_0 < 0.5$, the plastic zone is confined to a region

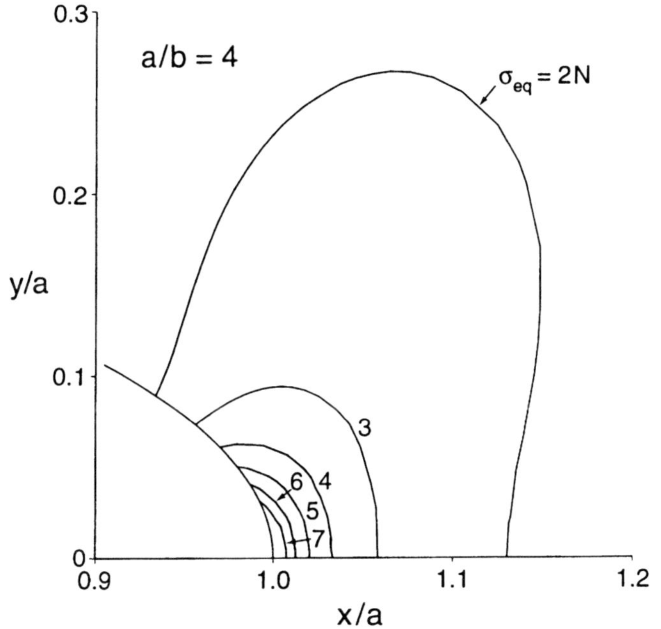

Fig. 8.4 Contours of equivalent stress for an elliptical notch loaded perpendicular to its major axis, with $a/b = 4$, and $\rho = a/16$.

of about $0.15a$. However, as the a/b ratio increases, the plastic zone at initial stages of loading will be restricted to a region around the apex of the major axis. With increased loading it will extend outward in a manner similar to that shown in Fig. 8.4.

8.2.2 STRESS DISTRIBUTION AROUND SLENDER NOTCHES

The stress distribution near a 'sharp' notch may be calculated from the stress intensity factor as described in section 7.2.1 (7.10–7.13). For a deep notch with the origin of cylindrical coordinates placed one half of the radius from the notch apex, as shown in Fig. 8.5, the stress distribution for mode I loading was derived by Creager and Paris [13] as follows:

$$\begin{Bmatrix} \sigma_{xx} \\ \tau_{xy} \\ \sigma_{yy} \end{Bmatrix} = \frac{K_{\mathrm{I}}}{\sqrt{2\pi r}} \frac{\rho}{2r} \begin{Bmatrix} -\cos\dfrac{3\theta}{2} \\ -\sin\dfrac{3\theta}{2} \\ \cos\dfrac{3\theta}{2} \end{Bmatrix} + \frac{K_{\mathrm{I}}}{\sqrt{2\pi r}} \cos\frac{\theta}{2} \begin{Bmatrix} 1 - \sin\dfrac{\theta}{2}\sin\dfrac{3\theta}{2} \\ \sin\dfrac{\theta}{2}\cos\dfrac{3\theta}{2} \\ 1 + \sin\dfrac{\theta}{2}\sin\dfrac{3\theta}{2} \end{Bmatrix}. \tag{8.10}$$

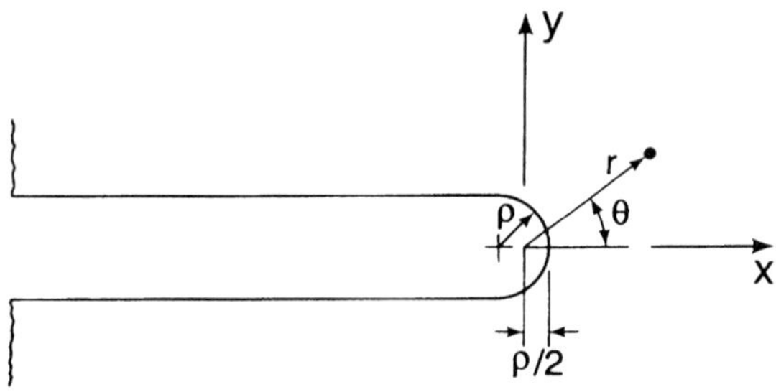

Fig. 8.5 A slender notch and origin of coordinate systems.

Comparing the above with (7.5), it can be seen that the second term on the RHS of (8.10) is the same as that of (7.5). Thus, in the case of a deep notch, apart from the shift of the coordinate system to a point $\rho/2$ from the notch tip, there is an additional term which is the first term on the RHS of (8.10) when compared to a sharp crack. The effect of the first term on the RHS of (8.10) is significant where $r \rightarrow \rho/2$, whereas for $r \gg \rho$ but small compared to other planar dimensions, its effect becomes small. Therefore, the first term on the RHS of (8.10) is the local (perturbation) effect of a blunted crack within the original sharp crack stress field (7.5).

An advantage of (8.10) and similar ones given in [14] for modes II and III loading, is that the formulae given for stress intensity factors available in handbooks can be used to compute stress fields for sharp notches.

To compare the predictions of (8.10) with those of a narrow elliptical notch, (8.8), we compute σ_{yy} from the last equation of (8.10) at the notch tip, i.e.

$$\sigma_{\max} = (\sigma_{yy})_{\substack{r=\rho/2 \\ \theta=0}} = \frac{2K_{\mathrm{I}}}{\sqrt{\pi\rho}}. \tag{8.11}$$

For a mode I central crack in a panel, $K_{\mathrm{I}} = \sigma_{\infty}\sqrt{\pi a}$, (7.10), and substitution in (8.11) yields

$$\sigma_{\max} = 2\sigma_{\infty}\left(\frac{a}{\rho}\right)^{1/2} \tag{8.12}$$

which is identical to (8.8) with $\sigma_{\infty} \equiv N$, for the far-field applied stress.

8.2.3 NONLINEAR ANALYSIS

It has been mentioned in earlier chapters that the fatigue phenomenon of metals can be explained in terms of irrecoverable energy due to plastic deformation. The presence of stress raisers can magnify the effect of

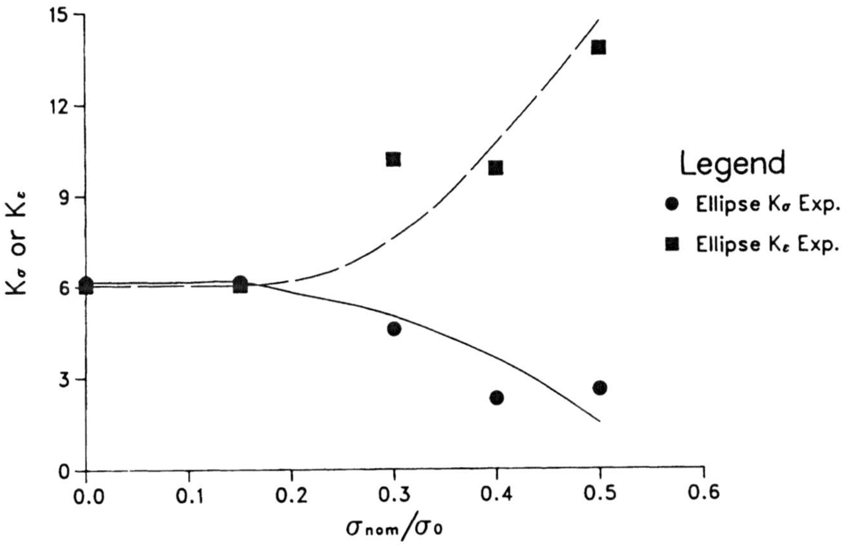

Fig. 8.6 Variation of the stress and strain concentration factors with the increasing nominal stress of an elliptical notch with $a/b = 2.58$ in a plate.

plasticity and therefore reduce the fatigue life. The occurrence of plastic strain at the notch root causes stress relief and thus, a decrease in the stress concentration. Figure 8.6 shows the variation of stress and strain concentrations with increasing nominal to yield stress ratio, σ_{nom}/σ_0, for an elliptical hole in a flat plate with $a/b = 2.58$ [15]. It can be seen that for $\sigma_{nom}/\sigma_0 > 0.17$, plastic deformation occurs at the root of the notch and the strain concentration K_ε increases, while the stress concentration K_σ decreases.

There are only a few analytical solutions which are concerned with plastic deformation around notches. Earlier attempts are documented in a book by Sokolovskii [16]. An approximate solution for a circular hole in an infinite plate subject to a tensile load was given by Stowell [17], which related the stress concentration factor to the secant modulus of a material. A numerical trial and error procedure is necessary to obtain the stress concentration factor.

The case of a circular hole in an infinite plate subjected to equal biaxial tension was investigated by Budiansky and Mangasarian [18]. They used the deformation theory of plasticity with the Ramberg–Osgood stress–strain relationship.

Neuber [19] derived an expression for notched prismatic bodies under anti-plane shear. He showed that the geometrical mean value of the maximum stress and strain concentration factors is a constant independent of the stress–strain curve, i.e.

$$K_\sigma K_\varepsilon = K_t^2 , \tag{8.13}$$

where K_t is the elastic stress concentration factor. The above simple relation has proved to be very popular and been used beyond its original intent. For example, Rice [20] has demonstrated that (8.13) is valid for small-scale yielding, i.e. when $\sigma_{nom}/\sigma_0 < 0.5$ (see Chapter 7 for the definition of small-scale yielding). It should be noted that all the references cited above are strictly valid only for monotonically increasing loads. If unloading takes place, as is the case for fatigue loading, the analysis becomes very complicated. As yet no complete solution has been given for cyclic loading, hence approximate solutions have been derived and they will be discussed below.

8.2.4 APPROXIMATE METHODS

Following the works of Stowell and Neuber, most investigators have attempted to correlate the stress concentration in the inelastic domain to that of the elastic stress concentration factor, K_t. Among these are Hardrath and Ohman's modification of Stowell's formula in the form of [21]

$$K_\sigma = 1 + (K_t - 1)\frac{E_s}{E_{ns}} \tag{8.14}$$

where $E_s = \sigma_{max}/\varepsilon_{max}$ is the secant modulus at the point of maximum local notch deformation, and $E_{ns} = \sigma_{nom}/\varepsilon_{nom}$ is the nominal secant modulus. From basic definitions (8.1) it is clear that $E_s/E_{ns} = K_\sigma/K_\varepsilon$.

It has been shown that both Neuber's relation (8.13) and that of Hardrath and Ohman (8.14) tend to over-predict notch root strains, e.g. [15, 22]. Thus they may not be accurate enough for predicting crack initiation life.

Based on an energy interpretation of Neuber's rule, Molski and Glinka [23] proposed a relationship for the elastic–plastic notch root stress and strain in the form of

$$W_n K_t^2 = W_a^{max} \tag{8.15}$$

where W_n is the nominal elastic strain energy, and W_a^{max} is the 'actual' maximum strain energy density at the notch root. For nominal stresses which are relatively small compared to the yield stress (8.15) in certain instances gives a more accurate prediction than that of Neuber's rule (8.13). However, when σ_{nom} approaches or exceeds yield stress, (8.15) tends to underestimate the notch root strain [24].

Using Rice's J-integral definition [25], Ellyin and Kujawski [24] obtained a similarity measure of the strain energy density along a smooth notch boundary. The maximum stress and strain were related to the nominal ones through

$$\frac{\sigma_{\max}^2}{2E} + \frac{1}{1+n}\sigma_{\max}\varepsilon_{\max}^{\mathrm{p}} = K_{\mathrm{t}}^2\left[\frac{\sigma_{\mathrm{nom}}^2}{2E}\right] + \frac{1}{1+n}\sigma_{\mathrm{nom}}\varepsilon_{\mathrm{nom}}^{\mathrm{p}}, \tag{8.16}$$

where n is the monotonic strain hardening exponent of the material. Note that when the nominal stress/strain values are below the yield stress/strain of the material, i.e. $\varepsilon_{\mathrm{nom}}^{\mathrm{p}} = 0$, then (8.16) reduces to

$$\frac{\sigma_{\max}^2}{2E} + \frac{1}{1+n}\sigma_{\max}\varepsilon_{\max}^2 = K_{\mathrm{t}}^2\left(\frac{\sigma_{\mathrm{nom}}^2}{2E}\right) \tag{8.17}$$

which is that proposed by Molski and Glinka [23], cf. (8.15). On the other hand, if the material has a high strain hardening slope, i.e. $2/(1+n) \approx 1$, then (8.16) reduces to

$$K_\sigma K_\varepsilon = K_{\mathrm{t}}^2\,,$$

which is Neuber's product formula. Therefore, (8.16) is of a more general nature, and the proposed formulae in [19] and [23] can be derived as particular cases of it.

8.2.5 A GENERAL APPROACH

In this section we will present a more general approach to the notch analysis of bodies undergoing elastic–plastic deformation near a notch root. This analysis is applicable to both monotonic and constant amplitude cyclic loading in the case of uniaxial as well as multiaxial states of stress. A detailed derivation is given in a paper by Ellyin and Kujawski [26], and here we will only give an outline of the approach and the final results.

Let us consider a notched body as depicted in Fig. 8.7, showing yielding in a small region near the smooth-ended notch, i.e. the condition of small-scale yielding as described in Chapter 7, is assumed to prevail. Under these conditions, it can be shown that there exists a relationship between the stress and strain fields in the notched body for two types of materials, the actual elastic–plastic material and a linear elastic one, i.e.

$$\int_{V_{R_e}} \sigma_{ij}^{\mathrm{a}}\varepsilon_{ij}^{\mathrm{a}}\,\mathrm{d}V = \int_{V_{R_e}} \sigma_{ij}^{\mathrm{e}}\varepsilon_{ij}^{\mathrm{e}}\,\mathrm{d}V \tag{8.18}$$

where σ_{ij}^{a}, $\varepsilon_{ij}^{\mathrm{a}}$ denote the actual values of the stress and strain fields, and σ_{ij}^{e}, $\varepsilon_{ij}^{\mathrm{e}}$ the stress and strain values obtained from a linear elastic solution, and R_{e} is the far-field boundary where the effect of the plastic zone R_{p} is negligible, i.e. $R_{\mathrm{e}} \gg R_{\mathrm{p}}$, and $V_{R_{\mathrm{e}}}$ is the volume enclosed by R_{e}.

It should be noted that

$$\int \sigma_{ij}\varepsilon_{ij}\,\mathrm{d}V = \int_V [W(\varepsilon) + W(\sigma)]\,\mathrm{d}V\,, \tag{8.19}$$

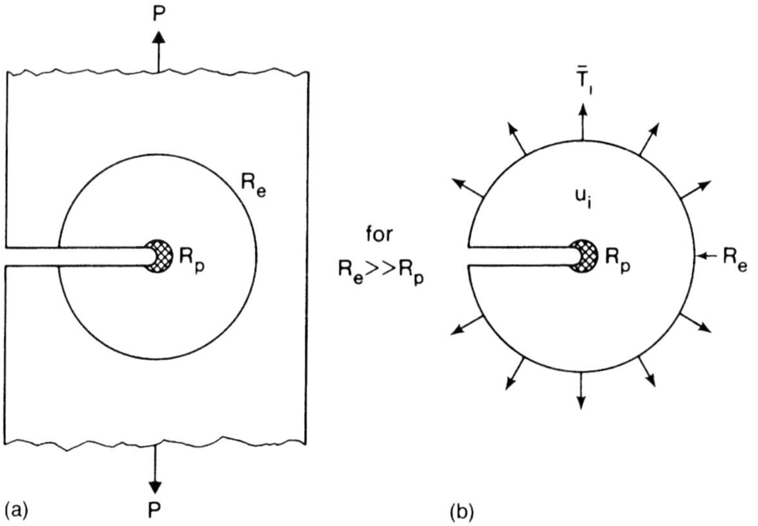

Fig. 8.7 A smooth-ended notch with small-scale yielding at its root. (a) plastic zone size; (b) tractions at a radius R_e away from the notch root.

where $W(\varepsilon) = \int \sigma_{ij} \, d\varepsilon_{ij}$ is the strain energy density and $W(\sigma) = \int \varepsilon_{ij} \, d\sigma_{ij}$ is the complementary strain energy density. For a nonlinear material, the equivalent stress–strain curve may be represented by a power law relationship:

$$\varepsilon_{eq} = (\sigma_{eq}/K)^{1/n} \tag{8.20a}$$

and then the strain energy density is given by

$$W(\varepsilon) = \int \sigma_{eq} \, d\varepsilon_{eq} = \frac{1}{1+n} \sigma_{eq} \varepsilon_{eq} \tag{8.20b}$$

and the complementary strain energy density is found from

$$W(\sigma) = n \, W(\varepsilon) , \tag{8.20c}$$

where σ_{eq} and ε_{eq} are equivalent stress and strain defined earlier (e.g. Chapter 2, (2.28, 2.29)) and n is the strain hardening exponent of the σ_{eq} versus ε_{eq} curve. Substituting from (8.20c), (8.19) into (8.18), we obtain

$$\int_{V_{R_e}} [qW^a(\varepsilon) - W^e(\varepsilon)] dV = 0 \tag{8.21}$$

where $q = (1+n)/2$ and is bounded by $\frac{1}{2} \leqslant q \leqslant 1$. The principle of minimum strain energy states that of all sets admissible stress components, σ_{ij}, the actual one (σ_{ij}^a) would make $W^a(\varepsilon)$ a minimum. Thus, for any other stress state we will have $W^e(\varepsilon) \geqslant qW^a(\varepsilon)$. At the far-field boundary we have

$W^{e}(\varepsilon) = qW^{a}(\varepsilon)$ and from (8.21) the integral over the volume, $\int qW^{a}(\varepsilon)\,dV = \int W^{e}(\varepsilon)\,dV$. Thus it follows that

$$qW^{a}(\varepsilon) = W^{e}(\varepsilon) \tag{8.22a}$$

or

$$\sigma_{ij}^{a}\,\varepsilon_{ij}^{a} = \sigma_{ij}^{e}\,\varepsilon_{ij}^{e}\,. \tag{8.22b}$$

In the case of uniaxial stress states, (8.22b) reduces to Neuber's relation (8.7).

It may be of interest to note that Glinka and co-workers [27] have recently adopted relationship (8.22b) as a starting point for the extension to multiaxial loading. However, no statement is made regarding the underlying restrictions as explained above and given in our earlier work, [26].

(a) Constant amplitude cyclic loading

We will now examine the difference between a nonlinear elastic and an elastic–plastic material when subjected to monotonic and cyclic loading. In the case of a nonlinear elastic material the loading and unloading branches are the same (Fig. 8.8), whereas for an elastic–plastic material they are different. The loading path 0AB in Fig. 8.8 is followed by the unloading path BD0 for a uniaxial stress state. If the branches 0AB and BD0 are similar (see Chapter 2 for conditions for which this applies), then the origin for the unloading case can be placed at B (Fig. 8.8) and we can use equality

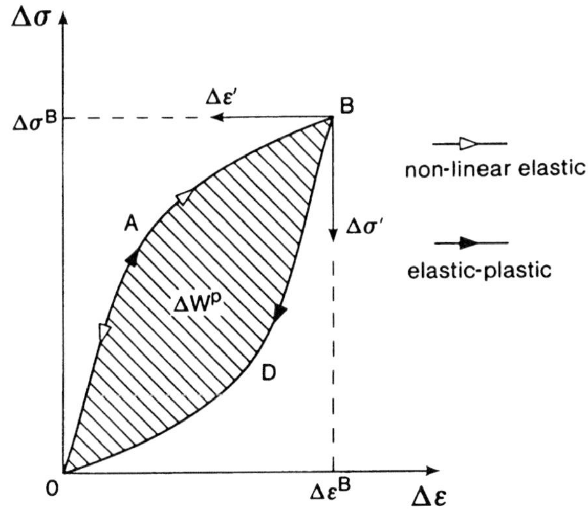

Fig. 8.8 Cyclic stress and strain variation for loading and unloading of a nonlinear elastic and an elasto-plastic material.

condition (8.19) for the unloading case, provided the strain components remain proportional through the unloading process.

For a nonlinear elastic material, the range of stress and strain during cyclic loading will depend on the fluctuation of remotely applied loads and (8.22b) can be written as

$$\Delta\sigma^{e}\Delta\varepsilon^{e} = \Delta\sigma^{a}\Delta\varepsilon^{a} \,. \tag{8.23}$$

Using the above energy interpretation for an elastic–plastic material as depicted in Fig. 8.8, we get [26]

$$\Delta\sigma^{e}\Delta\varepsilon^{e} = \Delta W_{L} + \Delta W_{RL} = \Delta\sigma^{a}\Delta\varepsilon^{a} + \Delta W^{p} \,, \tag{8.24}$$

where ΔW_L and ΔW_{RL} are the energies for the loading path 0AB, and the reversed path BD0 with the origin placed at B, and ΔW^p is the absorbed strain energy per cycle (area of the hysteresis loop 0ABD0). Note that for a nonlinear elastic material, $\Delta W^p = 0$. It could easily be shown that (8.24) reduces to (8.17) when the applied nominal stresses are within the elastic range of a material. However, it should be noted that (8.22b) applies to multiaxial loading cases whereas the special cases discussed above are for a uniaxial loading condition.

Predictions of (8.24) are compared with the experimental results of a circular notched plate loaded under stress-controlled conditions [28]. Figure 8.9 shows plots of the theoretical notch stress amplitude, $\sigma_{nom}K_t$ versus the maximum notch strain amplitude, predicted by (8.24) and Neuber's rule (8.13) along with the experimental data. The stable cyclic stress–strain curve of the material is also depicted in the figure. The experimental data, with the exception of one point, lie close to the prediction of (8.24) or Neuber's rule provides a lower bound. Further comparisons with experimental and numerical results are given in [24, 26].

8.2.6 THE FINITE ELEMENT METHOD

In the previous sections we have seen that the extension of the analytical method to the plastic range of a material's response is a difficult task. It has been shown thus far that if an elastic solution exists, then an 'equivalent' elastic–plastic stress distribution may be found under certain restrictive conditions, e.g. small-scale yielding and proportional loading. With the advance of numerical techniques, especially the finite element method and the increased capacity of digital computers, the elastic–plastic stress and strain distribution around a notch can now be determined with relative ease. Examples of the earlier attempts for isotropic and orthotropic elastic materials can be found in Zienkiewicz's book [29] among others.

Extension into the elastic–plastic regime requires the specification of an appropriate constitutive relation. It has been shown in Chapter 6 that the

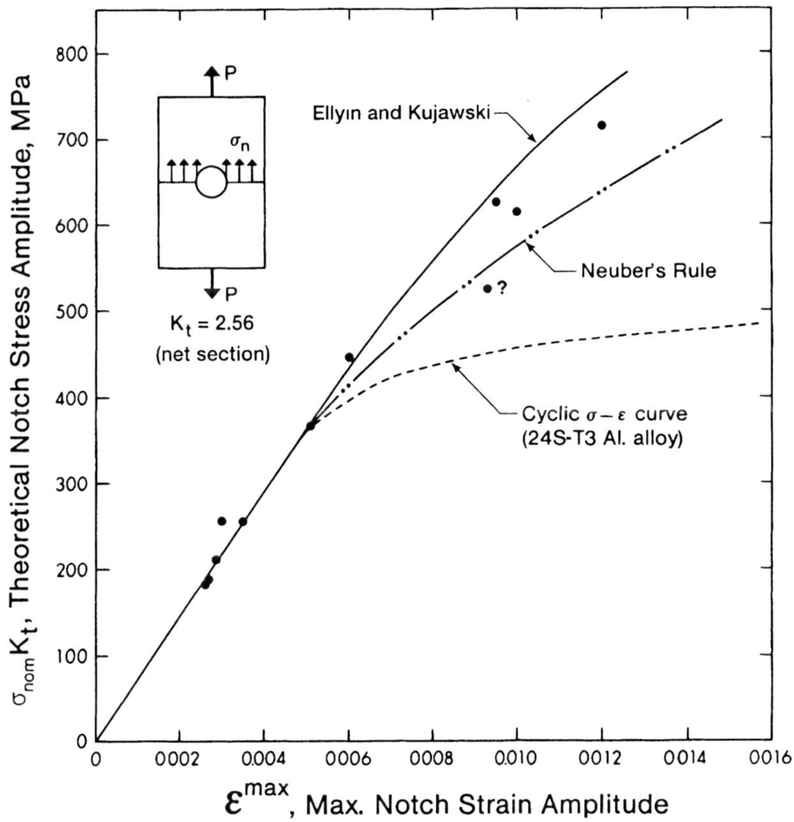

Fig. 8.9 Theoretical notch stress amplitude versus maximum notch strain amplitude for a centrally notched circular hole in a plate under stress-controlled conditions. Reprinted with kind permission from [26].

classical isotropic or kinematic hardening rules are not adequate for cyclic loading. However, once a proper elastic–plastic constitutive relation is incorporated into a finite element code, the determination of the stress and strain distribution around a notch becomes an easy task, albeit a time-consuming operation. Two major problems requiring attention are appropriate element sizes in the high stress gradient regions and the load increments during plastic deformation to ensure proper convergence and accuracy of results.

To demonstrate the accuracy of the finite element analysis, a plate containing an elliptical central hole with $a/b = 4$ and width $W = 2.5$ in (63.5 mm) will be analysed. The plate was made of aluminium alloy 2024-T351, and was tested under cyclic strain-controlled conditions [30]. Due to the symmetry of the geometry and load, only one quadrant of the plate had to be analysed. The solution was obtained for a monotonically increasing load using an elastic–plastic constitutive relation described in [31]. The stable cyclic stress–strain relation was used in the model, as depicted in Fig. 8.10 by dashed lines. A comparison of the calculated and measured results is shown in Fig. 8.10, in which the maximum strain amplitude at the notch root is plotted against the theoretical (elastic) notch strain, $K_t\varepsilon_n$ ($K_t = 4.6$ net section). The agreement between the calculated and measured values is very good. Similar results were obtained for other notch geometries [31].

In summary, it has been shown in this section that several methods are available by which the notch root stress and strain distributions can be estimated, once the nominal (far-field) loading conditions are specified. Depending on the desired accuracy, one can use available methods ranging from analytical solutions to approximate methods to the finite element discretization, to obtain stress and strain values around the notch. The required level of effort increases with the degree of accuracy demanded.

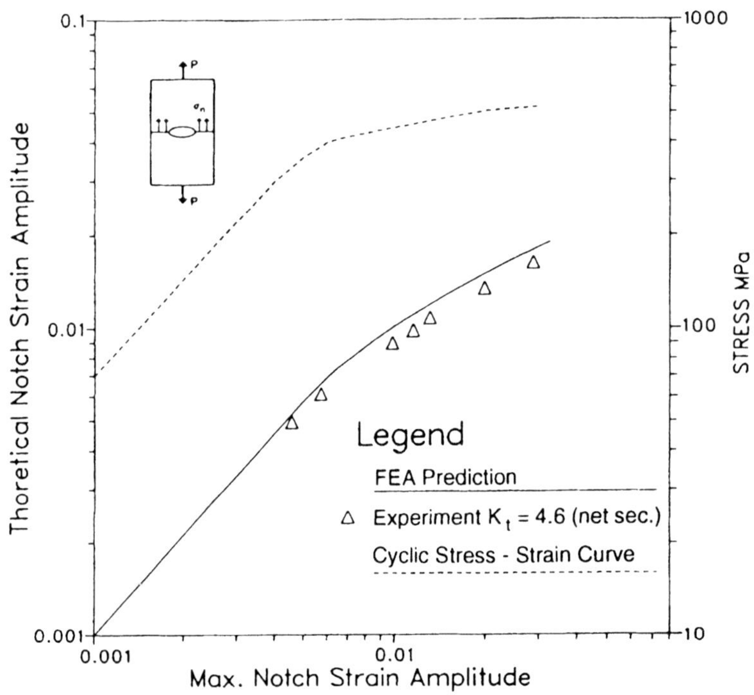

Fig. 8.10 Theoretical notch strain amplitude versus maximum strain amplitude, for an elliptically notched plate under strain-controlled conditions [31].

8.3 LIFE TO CRACK INITIATION

The number of cycles required to initiate a crack at a notch root may be estimated by several methods. The first step is to determine the stress and strain at the critical region. This has been dealt with in the previous section. The next step is to account for the damage accumulation at the critical location by means of a damage criterion. In some earlier investigations, attempts were made to determine damage at a notch root indirectly by using the nominal (remote) applied stress or strain and a fatigue concentration factor. More recent approaches have been to use the local stress and strain to determine the accumulation of damage. An underlying motivation has been to relate the crack initiation life at the notch root to that of a smooth uniaxial specimen subjected to an equivalent cyclic load.

The above methods may not yield satisfactory results because they do not take into account two major characteristics of the stress/strain distribution at a notch root. The first characteristic is the steep stress/strain gradient near the notch root (Fig. 8.3) and the second is the multiaxial nature of the stress state near the notch root. In the following we will briefly review the methods used to predict the crack initiation life under constant amplitude cyclic loading.

8.3.1 FATIGUE NOTCH FACTOR APPROACH

This method is a stress-based approach and it can be explained with reference to Neuber's rule, (8.13). Let us assume that the applied alternating nominal stresses are within the elastic range of a material's response. Replacing the stress and strain in Neuber's relation by their cycling range, (8.13) becomes

$$K_t \Delta\sigma_{nom} = (\Delta\sigma_{max} \Delta\varepsilon_{max} E)^{1/2} \tag{8.25}$$

where the nominal stress range at the remotely applied region is related to the range of the maximum stress, maximum strain and Young's modulus at the notch root. It will therefore suffice to relate fatigue crack initiation life to the applied nominal stress range through the LHS of (8.25) provided K_t is known. A number of earlier investigators, e.g. [30], have shown that the life to crack initiation of a notched specimen can be predicted from (8.25) if K_t is replaced by a fatigue notch factor, K_f, defined as

$$K_f = \frac{\text{Nominal alternating stress in a smooth specimen}}{\text{Nominal alternating stress in a notched specimen}} \tag{8.26}$$

at the same fatigue life. The fatigue notch factor, K_f, is generally determined from long life (high cycle) fatigue tests, e.g. at $N \geqslant 10^6$. The essence of this approach is that the stress in a smooth specimen is scaled by a factor K_f, to represent that prevailing in a notched specimen. Thus, the determination of K_f is the central issue in this approach. In addition to the experimental determination of K_f for various notch geometries, empirical relations have been proposed for its estimation.

It has been shown earlier, (8.8), that the stress concentration factor is related to the notch radius ρ, in the following manner:

$$K_t \propto \left(\frac{1}{\rho}\right)^{1/2}. \tag{8.27}$$

Neuber [1] and Peterson [4] have proposed formulae relating K_f to K_t which can be written as

$$K_f = 1 + \frac{(K_t - 1)}{\left[1 + \left(\frac{\lambda}{\rho}\right)^{\gamma}\right]} \tag{8.28}$$

where λ is a length parameter which has to be determined for a given material, and $\gamma = \frac{1}{2}$ in the case of Neuber, and $\gamma = 1$ in Peterson's formula. Note that as $\rho \to \infty$, $K_f \to K_t$, i.e. for notches with a large radius of curvature, K_f approaches K_t. The inverse of the quantity in the square bracket in (8.28) is called the 'notch sensitivity' factor.

Experiments have shown that K_f, unlike K_t, depends on material properties, e.g. strength and ductility, but also on the level of applied strain [32]. Generally the ratio of K_f/K_t decreases with an increase in the strain or stress level. Also, for a given K_t, the fatigue notch factor decreases with decreasing notch size. This is known as the 'notch size' effect. It has been shown that below a certain size, the effect of a notch on the fatigue strength is negligible [33].

From the above discussion it is clear that the fatigue notch factor approach is limited to the low strain, high-cycle regime and notches of a certain physical size.

8.3.2 LOCAL STRAIN APPROACH

In this approach an attempt is made to account directly for plasticity at the notch root (Fig. 8.11(a)). Generally an assumption is made in the sense that both smooth and notched samples will accumulate the same amount of damage and have the same life to fatigue crack initiation, if stress–strain histories are imposed identical to those prevailing at the crack initiation site (Fig. 8.11(b)). Topper and co-workers have shown that the above assumption is a reasonable one, e.g. [30]. The accuracy of this approach then depends upon the method used to determine the local stress and strain fields at the notch root. The available methods have already been discussed in the previous section. It suffices here to mention that (8.25) does not hold since, at the local level (Fig. 8.11(b)), $\Delta\sigma_{nom} \neq E\Delta\varepsilon_{nom}$. The

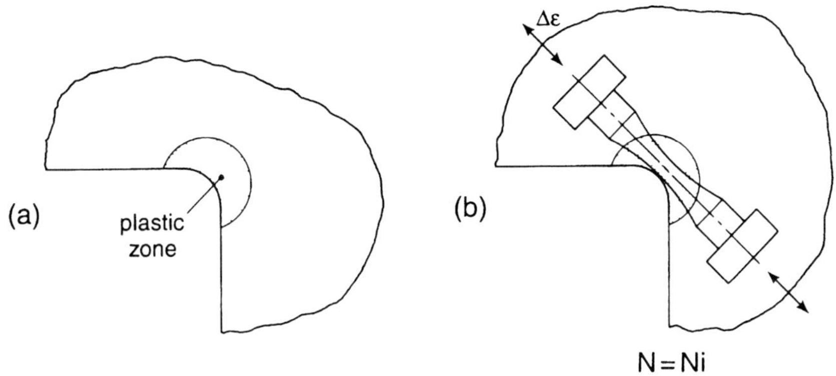

Fig. 8.11 (a) Small-scale yielding at a notch root; (b) schematic of the local strain approach by subjecting a smooth specimen to the same strain as that prevailing at the notch root.

plastic zone at the notch root is constrained by the elastic material surrounding it; thus, a strain-controlled test is a good representation of the condition to which the notch root material is subjected.

A major advantage of this method is that an estimation of the life to crack initiation of a notched member can be made from the widely available uniaxial data (Chapter 3). Dowling [34] has reported that the estimated lives are not very sensitive to estimated strains, and advocates local strain and crack growth approaches to determine the total fatigue life for initiation plus propagation, respectively.

8.3.3 ENERGY APPROACH

In section 8.2.4, an energy relation was derived relating the energy based on a linear elastic response to that of an elastic–plastic material undergoing cyclic loading, (8.24). This equation, which holds for any point in the notch vicinity, may be written as

$$\Delta\sigma^{e}_{max}\Delta\varepsilon^{e}_{max} = \Delta\sigma^{a}_{max}\Delta\varepsilon^{a}_{max} + \Delta W^{p}_{max} \tag{8.29}$$

or

$$K_t^2\left(\frac{1}{2}\Delta\sigma_{nom}\Delta\varepsilon_{nom}\right) = \frac{1}{2}\Delta\sigma^{a}_{max}\Delta\varepsilon^{a}_{max} + \frac{1}{2}\Delta W^{p}_{max} \tag{8.30}$$

where definition (8.1) were used in the LHS of (8.29). In (8.30) $\frac{1}{2}\Delta\sigma_{nom}\Delta\varepsilon_{nom}$ is the nominal strain energy density at the far-field, ΔW_{nom}, which is known. The RHS of (8.30) is the cyclic strain energy density represented by the loading curve 0AB in Fig. 8.8. Thus (8.30) can be written as

$$K_t^2\left(\frac{1}{2}\Delta\sigma_{nom}\Delta\varepsilon_{nom}\right) = \Delta W^p + \Delta W^e. \tag{8.31}$$

In section 3.4.3, the RHS of (8.31) was related to the number of cycles to failure, N_f, through the following relationship:

$$\Delta W^p + \Delta W^e = \kappa(2N_f)^\alpha + \Delta W_0. \tag{8.32}$$

Thus, using the local approach, the number of cycles to initiate a crack at the notch root could be determined from (8.31) and (8.32) if the material data is available (see Chapters 3 and 4 regarding the required test data). An advantage of the above relation is that it applies to both uniaxial and multiaxial states of stress under a constant amplitude cyclic loading. Therefore, the effect of multiaxiality at the notch root can be incorporated in this approach.

Recently Du-yi and De-jun [35] have used (8.30) along with a modified definition of (8.13), $K_f = (K_\sigma K_\varepsilon)^{1/2}$, to derive a fatigue notch factor of the form $K_f = K_f(n', \Delta\varepsilon^e/\Delta\varepsilon^p, K_t)$. A comparison between a wide range of experimental data and predicted results indicates a good agreement.

8.4 GROWTH OF CRACKS INITIATED FROM NOTCHES

In the previous section we discussed the requirements for crack initiation from notches. The growth of these initiated cracks will be the subject of this section. A crack initiated from a notch may grow under a 'driving force' which is influenced by the notch stress–strain field. After reaching a certain length, ℓ_0 (a fraction of the notch radius), the effect of the notch field is diminished and the continuation of growth will depend on the crack tip field as discussed in Chapter 7. In general, a crack initiated from a **blunt** notch will propagate to failure, similarly to a smooth specimen, since the required bulk stresses will be high. In contrast, a crack initiated from a sharp notch may subsequently become non-propagating. The reason for this is that once a crack leaves the notch influence zone, the applied bulk stresses may not be high enough for the crack to propagate.

To illustrate this, let us consider a small through thickness crack initiated at the ends of the major axis of an elliptical notch in an infinite plate (Fig. 8.12). The applied far-field stress is perpendicular to the crack plane, and its magnitude is such that the local response at the crack tip is either elastic or the small-scale yielding condition discussed in section 7.4 prevails.

The stress intensity factor, $(K_I)_s$, for this problem is given in [14],

$$(K_I)_s = 1.122\, K_t \sigma_{nom}\sqrt{\pi\ell}, \tag{8.33}$$

where the subscript S refers to a short crack, $\ell/(a+\ell) \to 0$, a being the semi-major axis of the ellipse (Fig. 8.12).

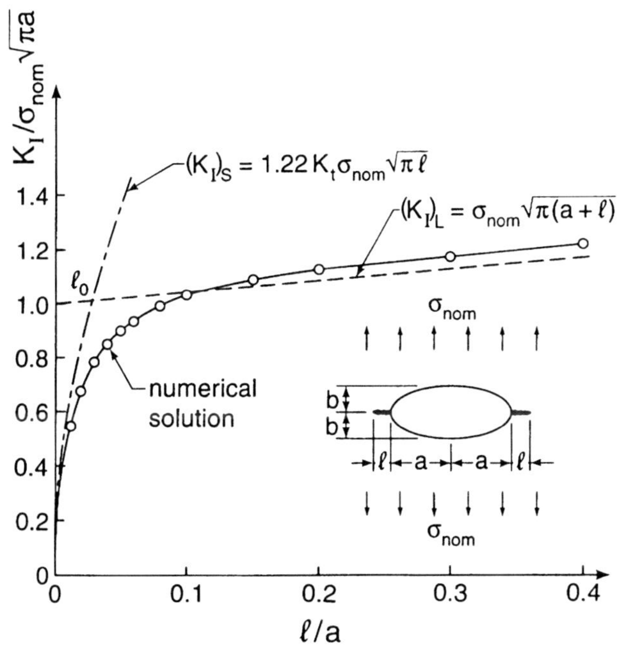

Fig. 8.12 Variation of the stress intensity factor K_I non-dimensionalized by dividing through by $\sigma_{nom}\sqrt{\pi a}$ as the crack length ℓ emanating from an elliptical notch increases. Two cases of a small crack, and one long crack, are compared with the numerical solution of [36].

For a long crack we have

$$(K_I)_L = \sigma_{nom}\sqrt{\pi(a+\ell)}\,, \tag{8.34}$$

where the subscript L denotes a long crack where the influence of the notch is negligible.

Equating (8.33) and (8.34) we find the intersection point denoted by ℓ_0,

$$\ell_0 = \frac{a}{1+(1.122\ K_t)^2}\,, \tag{8.35}$$

which is a transition crack length beyond which the effect of the notch field is negligible and the nominal stress intensity factor for a crack length $a+\ell$, can be used.

A graphical representation of (8.33) and (8.34) for an ellipse with $K_t = 5$ is shown in Fig. 8.12. The abscissa is normalized with respect to the semi-major axis of the ellipse, and the ordinate is also normalized with respect to the stress intensity of a crack with length $2a$, $\sigma_{nom}\sqrt{\pi a}$. The numerical prediction by Newman [36] is also shown in the figure.

In general, for cracks emanating from notches, the stress intensity factor has a form

$$K = F\,\sigma_{\text{nom}}\sqrt{\pi\ell}\,, \tag{8.36}$$

where function F is determined for each configuration through analytical or numerical means. For example, in the case of a crack initiated from an elliptical notch (Fig. 8.12), a comparison of (8.33) and (8.36) results in

$$F = 1.122\,K_t\,. \tag{8.37}$$

For an elliptical notch, $K_t = 2(a/\rho)^{1/2}$, (8.8). Therefore, attempts have been made to correlate the stress intensity factors found through numerical results with a function of ℓ/ρ. Lukáš and Klesnil [37] have shown that

$$F = 1.122\,K_t[1 + 4.5(\ell/\rho)]^{1/2} \tag{8.38}$$

provides a good estimate for relatively small ratios of ℓ/ρ in the case of blunt notches, $K_t < 3$.

A more general expression is given by Kujawski [38] with a wider range of application:

$$F = \frac{1}{2} g f K_t \left[\left(1 + 2\frac{\ell}{\rho}\right)^{-1/2} + \left(1 + 2\frac{\ell}{\rho}\right)^{-3/2}\right], \tag{8.39}$$

where g is a geometric shape factor, e.g. $g = 1.122$ for an edge through thickness crack, and f is a correction factor,

$$\begin{aligned} f &= 1 && \text{for} \quad \frac{\ell}{\rho} < 0.2 \\ f &= 1 + \frac{\tan(\pi/2\,K_t)}{2.8}\left(\frac{\ell}{\rho} - 0.2\right) && \text{for} \quad \frac{\ell}{\rho} \geqslant 0.2\,. \end{aligned} \tag{8.40}$$

The predictions of (8.39) are in close agreement with the numerical results [38].

At this stage, it is worthwhile noting that the stress intensity factor, K, cannot strictly characterize crack initiation from a notch, since $K \equiv 0$ in the absence of a crack, i.e. when $\ell = 0$. Thus, as indicated in Chapter 7, the stress intensity factor can only be used in connection with initiated cracks of a certain length.

8.4.1 CRACK GROWTH RATE

It was mentioned earlier that crack initiation and early crack growth are controlled by the notch stress and strain fields which decay rapidly in a sharp notch, e.g. Fig. 8.3. Therefore, the early growth rate of an initiated crack may decrease with the increase in crack length, and it may stop unless the level of the applied bulk stress is large enough. For long cracks it has been shown that the condition for continued growth is $\Delta K \geqslant \Delta K_{\text{th}}$.

However, short cracks grow at stress intensity values smaller than ΔK_{th} (see the Kitagawa–Takahashi diagram shown in Fig. 7.20).

For cracks initiated from notches, the condition for continued crack growth may be found as follows. In the case of cyclic loading, we use the stress intensity range, $\Delta K = K_{max} - K_{min}$; thus (8.36) becomes

$$\Delta K = F\,\Delta\sigma_{nom}\sqrt{\pi\ell}\,. \tag{8.41}$$

For a given ΔK and notch geometry, we can plot the variation of $\Delta\sigma = F\,\Delta\sigma_{nom}$ versus ℓ. Figure 8.13 shows the variation of $\Delta\sigma$ versus ℓ for cracks initiated from two types of notches, a sharp and a blunt one. The tangent point between these curves and the corresponding Kitagawa–Takahashi diagram indicates the required condition for the crack growth. It can be seen that while the required $\Delta\sigma_{nom}$ for a sharp notch is smaller than that of a blunt one, the reverse condition holds for the crack lengths. That is, a crack initiated from a sharp notch has to reach a certain length before it can grow under its own driving force, whereas the required crack length for a blunt notch is smaller than that of a sharp notch. The experimental evidence of the above can be found in Frost and Dugdale [39], Frost [40], Smith and Miller [41], Topper and co-investigators [41, 42], among others.

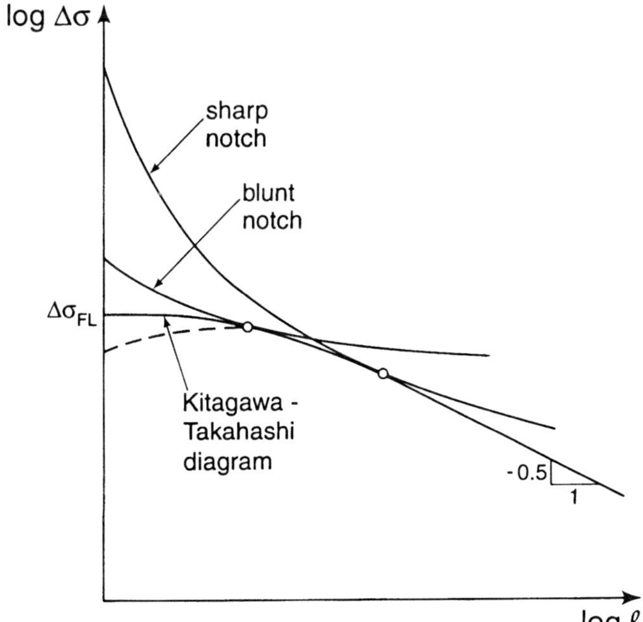

Fig. 8.13 Requirement for the growth of an initiated crack from two types of notches: a sharp and a blunt one. For explanation of dashed line see section 8.4.1.

In principle, once the stress intensity range, ΔK, for an initiated crack from a notch is determined, then its growth rate can be obtained from the crack growth laws described in Chapter 7. However, this will only apply if $\Delta K \geqslant \Delta K_{th}$, i.e. the applied bulk stress is high enough for a crack of any size to grow (Fig. 8.13), and the small-scale yielding at the crack tip prevails. However, predicting early crack growth from sharp notches by the LEFM approach is problematic. It was mentioned earlier that these cracks grow at a higher rate than those in plain specimens with the same crack size and under the same applied ΔK. However, this initial high growth rate decreases with the increased crack length resulting in crack arrest, or it reaches a minimum rate and then increases and approaches that of a long crack (e.g. Fig. 9.1). However, no dip is observed under a high stress ratio, R, where crack closure does not occur. This short crack behaviour will be discussed in the next chapter. It will suffice here to mention that the long crack fracture mechanics discussed in Chapter 7 cannot predict short crack behaviour. Consequently, a number of investigators [42, 44–46] have proposed various parameters to describe the transition of fatigue crack growth rate from that of notched to notch-free long crack specimens. These models are reviewed in [43, 47], and the reader is encouraged to consult these papers. A major conclusion reached by Shin and Smith [47] is that neither the stress intensity factor range, ΔK, nor the effective stress intensity range, $\Delta K_{eff} = K_{max} - K_{open}$, nor a modified strain-based intensity factor range, $\Delta K_{\varepsilon} = E\,\Delta\varepsilon\,[\pi(\ell + \ell_0)]^{1/2}$, where $\ell_0 = (1/\pi)(\Delta K_{th}/\Delta\sigma_{FL})^2$ [45], nor the total shear deformation parameter, ϕ_{total} [44], correlate all the notch fatigue crack data with that of a long crack growth from a notch.

The basic problem is that the above-mentioned parameters are essentially empirical in nature, and as such are unable to describe the physical phenomenon associated with the growth of a short crack. A microstructural model based on the surface grain behaviour and crack closure has been put forward by Abdel-Raouf *et al.* [48], which predicts the fatigue limit stress range for a range of notch sizes and stress ratios. It is also capable of predicting the non-propagating crack depth in a notched specimen. A brief description of the model will follow.

8.4.2 FATIGUE LIMIT STRESS RANGE IN NOTCHES WITH SHORT CRACKS

It has been mentioned in Chapter 1 and in other chapters that surface grains can deform more easily than those in the interior of a material which are supported and constrained by the surrounding grains. These highly strained surface grains are preferred sites for crack initiation (section 1.3.1). Even if the applied bulk strain is in the elastic range, plastic deformation can occur in large surface grains. In grains favourably oriented for slip, the surface strain concentration, here denoted by Q_{ε}, was estimated by

Abdel-Raouf, Topper and Plumtree [49] to be 6.3. It drops to unity at a distance of about three grains from the free surface, according to

$$Q_i = \frac{\Delta \varepsilon_{max}}{\Delta \varepsilon_{nom}} = 1 + q \exp\left(\frac{-a}{D}\right) \tag{8.42}$$

where q is a constant equal to 5.3, a is the crack depth, and D is the average grain size in the crack growth direction. Figures 8.14(a) and (b) illustrate the local strain variation and the extent of the surface affected zone. Within this surface zone, initiated short cracks experience local strains well above

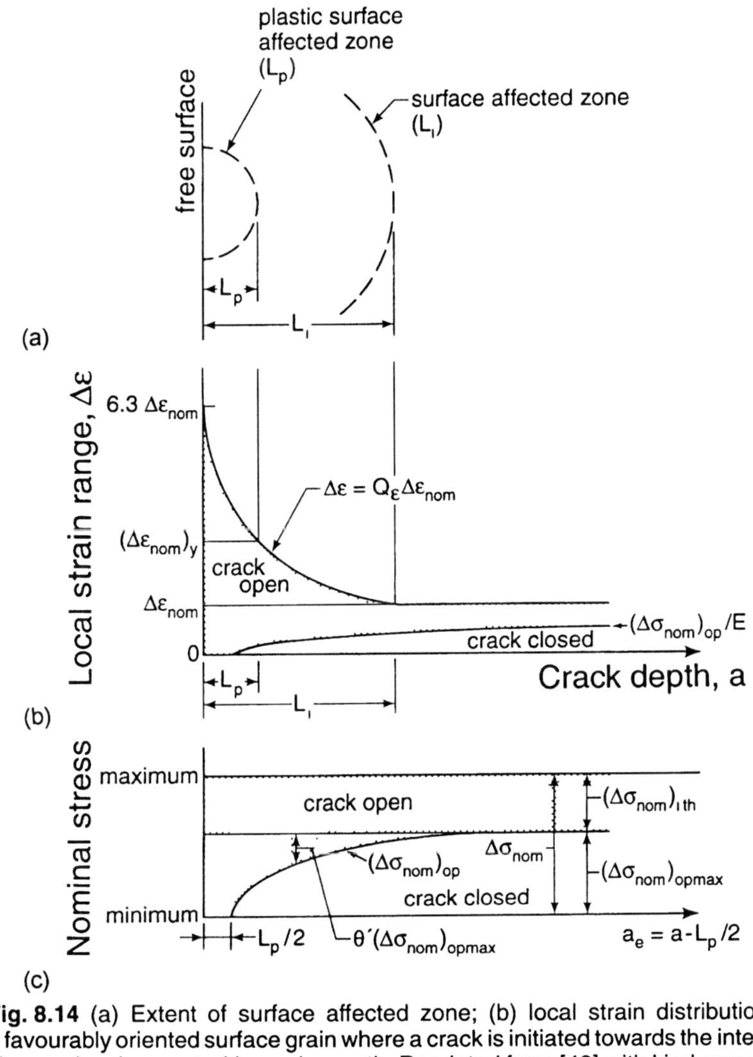

Fig. 8.14 (a) Extent of surface affected zone; (b) local strain distribution from a favourably oriented surface grain where a crack is initiated towards the interior; (c) closure development with crack growth. Reprinted from [48] with kind permission.

the nominal value, i.e. $\Delta\varepsilon > \Delta\varepsilon_{nom}$, which increases the local crack driving force well beyond those predicted by long crack fracture mechanics.

In a notched specimen, in addition to the above discussed inherent surface strain concentration, Q_ι, there is a geometric concentration, K_ε, which depends on the notch geometry and loading. The latter has been amply discussed in section 8.2. Therefore, the net strain concentration acting at a notch root could be the product of $K_\iota Q_\iota$. The thershold stress intensity range, ΔK_{th}, can be determined from (8.41) by considering the local stress field in the surface affected zone, i.e.

$$\Delta K_{th} = FK_\iota Q_\iota (\Delta\sigma_{nom})_{th}\sqrt{\pi a}\,. \tag{8.43}$$

An implicit asumption in using (8.43) is that the surface affected zone depicted in Fig. 8.14(a) is large compared to the short crack, so that the far-field stress for the initiated crack can be taken as $K_\varepsilon Q_\iota \sigma_{nom}$. Obviously this is an upper bound, and it requires that the notch radius be large enough to contain several grains. As the notch root radius becomes smaller, the net concentration factor will fall below $K_\iota Q_\iota$. The inherent strain concentration reaches that of the smooth specimen, Fig. 8.14(b), when $\rho \gg D$. In this model, crack closure develops as the crack tip approaches the centre of the reversed plastic zone, $L_p/2$ (Fig. 8.14(c)). Denoting by $\Delta K_{\iota th}$ the part of the threshold stress intensity factor range for which the crack is open, then the corresponding nominal intrinsic threshold stress range, $(\Delta\sigma_{nom})_{\iota th}$, is obtained from (8.43)

$$(\Delta\sigma_{nom})_{\iota th} = \frac{\Delta K_{\iota th}}{FK_\iota Q_\iota\sqrt{\pi a}}\,. \tag{8.44}$$

For a nominal applied stress in the elastic range, we may write (Fig. 8.14(c))

$$\Delta\sigma_{nom} = (\Delta\sigma_{nom})_{\iota th} + (\Delta\sigma_{nom})_{opmax}\,. \tag{8.45}$$

The crack closure development is described by a relation [48]

$$H_{cl} = \frac{\Delta\sigma_{nom}}{(\Delta\sigma_{nom})_{\iota th} + \theta'(\Delta\sigma_{nom})_{opmax}} \tag{8.46}$$

where

$$\theta' = \exp(-ka_e), \quad \text{for} \quad a_e > 0$$

and

$$\theta' = 1 \qquad\qquad \text{for} \quad a_e < 0$$

where $a_e = a - L_p/2$ is an effective crack depth and k is a material constant which reflects the rate of closure development. For a very small crack $\theta' = 1$, and for long cracks $\theta' \to 0$, as depicted in Fig. 8.14(c). Thus, the nominal threshold stress intensity range can be expressed as

$$(\Delta\sigma_{nom})_{th} = (\Delta\sigma_{nom})_{ith} H_{cl} = \frac{\Delta K_{ith} H_{cl}}{F K_i Q_i \sqrt{\pi a}} \tag{8.47}$$

where the last step is obtained by substituting for $(\Delta\sigma_{nom})_{ith}$ from (8.44). Figure 8.15 shows the variation of the threshold stress range with crack depth a (8.47) for a circular notch with a notch radius $\rho = 0.12$ mm [48]. It can be seen that the threshold stress range, $(\Delta\sigma_{nom})_{th}$, decreases from its crack initiation value $(\Delta\sigma_{nom})_C$ for the smallest possible crack depth of 3 μm [49] to a minimum with increasing crack depth. The threshold stress range then increases to a maximum value at a crack depth which increases with the notch size [48]. Both the minimum and maximum values of $(\Delta\sigma_{nom})_{th}$ decrease with the increasing value of the stress ratio R and the notch size. The model correctly predicts the initial decrease in the stress intensity factor, ΔK_{th}, with increasing crack depth, a, due to combined reduction of the inherent and geometric strain concentrations. The condition for the continued crack growth is $\Delta\sigma_{nom} \geqslant (\Delta\sigma_{nom})_{th\,max}$; thus, the maximum value of $(\Delta\sigma_{nom})_t$ defines the fatigue limit of the notched specimen. For any applied nominal stress $(\Delta\sigma_{nom})_c < \Delta\sigma_{nom} \leqslant (\Delta\sigma_{nom})_{th\,max}$ an initiated crack may grow but will self-arrest. Further discussion regarding the effect of notch radius, non-propagating cracks, and a comparison with the experimental data can be found in [48]. The key equation in this model is (8.47) which requires a knowledge of crack closure development through the specification of the parameter k in (8.46), and the determination of the

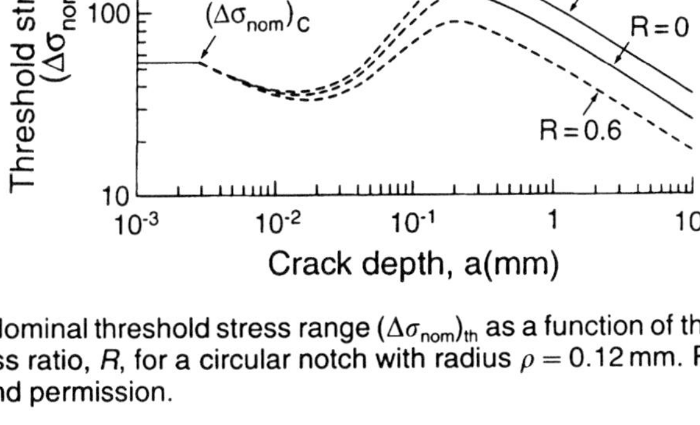

Fig. 8.15 Nominal threshold stress range $(\Delta\sigma_{nom})_{th}$ as a function of the crack depth, a, and stress ratio, R, for a circular notch with radius $\rho = 0.12$ mm. Reprinted from [48] with kind permission.

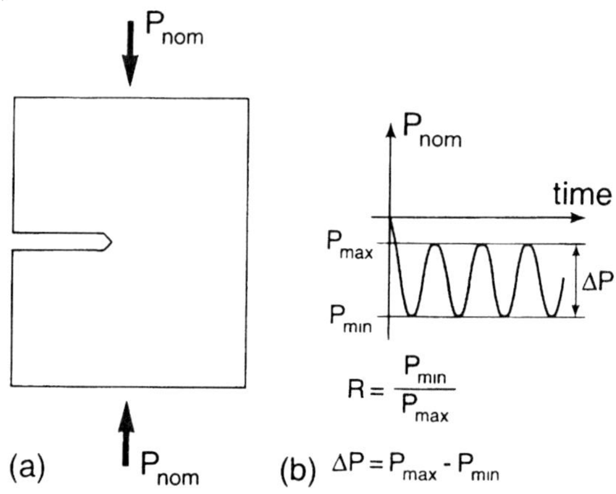

Fig. 8.16 (a) Schematic of a notched specimen subjected to a cyclic nominal load; (b) variation of applied load with time.

intrinsic threshold stress intensity factor range, ΔK_{ith}, through tests with a high R ratio, e.g. $R \geqslant 0.6$.

8.5 INITIATION AND GROWTH OF CRACKS FROM NOTCHES SUBJECT TO FAR-FIELD CYCLIC COMPRESSIVE LOAD

Consider a notched specimen subject to a far-field cyclic compressive load as shown in Fig. 8.16. Upon the first compressive load cycle, a residual tensile plastic zone is created at the notch root. This can be illustrated with

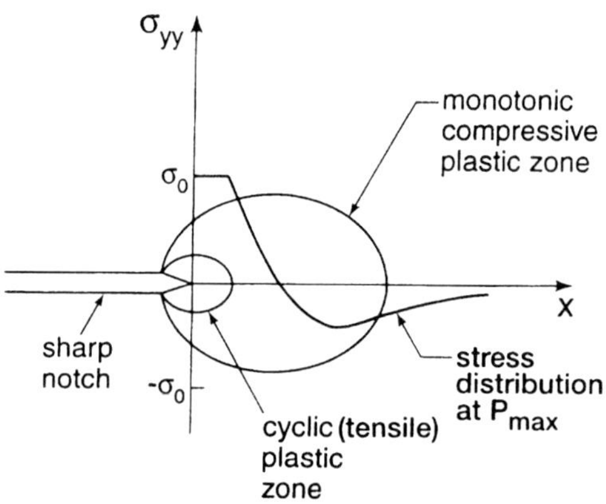

Fig. 8.17 Schematic of the residual tensile stresses, monotonic compressive plastic zone and the cyclic plastic zone in a sharp notch subjected to a nominal compressive cyclic load.

reference to Fig. 7.24 in Chapter 7, but with an opposite applied load sign. That is, for an applied compressive load during the first half cycle there will be a monotonic compressive plastic zone at the notch root followed by a smaller (about a quarter in the case of an elastic–perfectly plastic material) cyclic (tensile) plastic zone upon the application of the next half cycle (Fig. 8.17). Therefore, a crack may initiate and grow inside the residual plastic zone, provided that the applied far-field stress is large enough to cause plastic deformation at the notch root, and that enough cycles are applied for the accumulated damage to cause crack initiation and subsequent growth in the notched specimen. Crack growth under a compressive cyclic loading will self-arrest due to the build-up of crack closure stresses to be explained later.

Experimental observations on crack initiation under cyclic compressive loads are reported by Hubbard [50], Reid, Williams and Hermann [51], Chu *et al.* [52], Holm, Blom and Suresh [53] and Pippan [54], among others. They have all reported crack growth normal to the applied compressive load over a distance of a few millimetres from the notch tip.

A series of notched ASTM compact tension type specimens with sharp notches were tested by Craig *et al.* [55]. The material was a low carbon (0.25% C) ferritic–pearlitic steel conforming to ASTM A-516 Gr. 70 specification. The mechanical, chemical and microstructural properties are given in [55].

The extent of the monotonic plastic zone emanating from the notch can be estimated using any of the expressions given in section 7.4.2. None of the equations are strictly applicable since the far-field loading is entirely compressive. However, since the geometry of the specimen tested was such that the faces of the notch did not touch during loading, the expressions in section 7.4.2 provide a reasonable estimate of the monotonic plastic zone. Using the analogy of an elastic–prefectly plastic material, the size of the cyclic plastic zone can be estimated to be one quarter of the size of the monotonic plastic zone. Figure 8.18 shows the estimated plastic zone sizes using the Dugdale model (7.53), Irwin's plane stress estimate (7.50), and the Kujawski–Ellyin elastic–plastic equation (7.57). The final crack sizes on both surfaces of the notched specimen (CT type) subject to a cyclic compressive load with $R = P_{min}/P_{max} = 10$, are also shown in the figure. It can be seen that with the exception of specimen B8, the resulting cracks emanate from the notch self-arrest outside the tensile cyclic zone and within the monotonic compressive plastic zone.

The reason for crack arrest is the build-up of closure as the initiated crack grows from the notch tip. The faces of the growing crack come into contact during most of the compressive load cycle, and the effective stress, $\sigma_{eff} = \sigma_{max} - \sigma_{cl}$ decreases with increasing crack length until it reaches

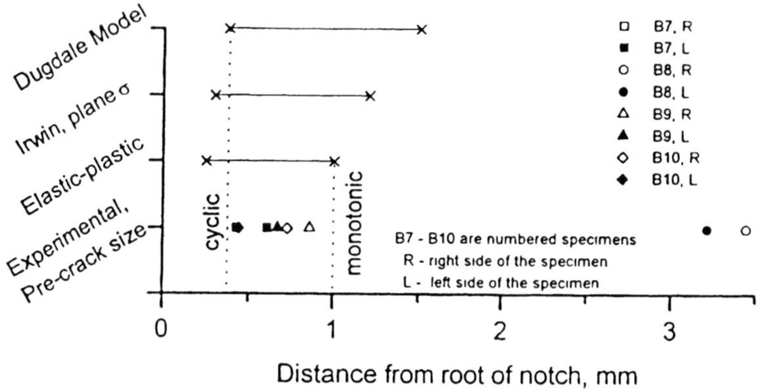

Fig. 8.18 Estimated monotonic compressive plastic zone, and cyclic plastic zone, and the resulting final crack sizes emanating from the notch. Reproduced with kind permission from [55].

a level which is below the intrinsic threshold stress range $\Delta\sigma_{ith}$. As a result, the driving force falls below the threshold value and the crack ceases to grow.

Crack closure measurements and numerical analysis are reported by Suresh and co-investigators [53, 56] which support the above-mentioned mechanism of crack initiation, closure build-up and crack arrest under far-field cyclic compressive loads. The role of the first cycle in producing a tensile plastic zone was discussed at the beginning of this section. Aswarth *et al.* [57] have shown the importance of the amplitude of the first cycle on the crack growth rate and the final crack length. They reported that a three-fold increase in the amplitude of the first cycle resulted in ten-fold increase in the final crack length. This can be explained by noting that the plastic zone size $R_p \propto \sigma^2$ (section 7.4.2). Thus, an increase in the applied load amplitude will cause an increase in the tensile plastic zone in Fig. 8.17, in a squared proportion. Perhaps the results of specimen B8 in Fig. 8.18 may be explained by an inadvertent first cycle overload. No other explanation was found for a crack four times longer than the others in similar specimens.

Recently Ellyin, Xia and Wu [58] have used their comprehensive elasto-plastic constitutive model together with the node release scheme described in section 7.9.4 to determine crack closure development under far-field cyclic compressive loading. Figure 8.19 shows the crack closure stress levels as the crack propagates. It can be seen that the effective stress $\sigma_{eff} = \sigma_{max} - \sigma_{cl}$ diminishes with the crack extension. Thus, after a small amount of crack growth, the effective stress falls below the intrinsic threshold stress range and the crack cannot grow further with the cyclic loading unless the applied load level is increased.

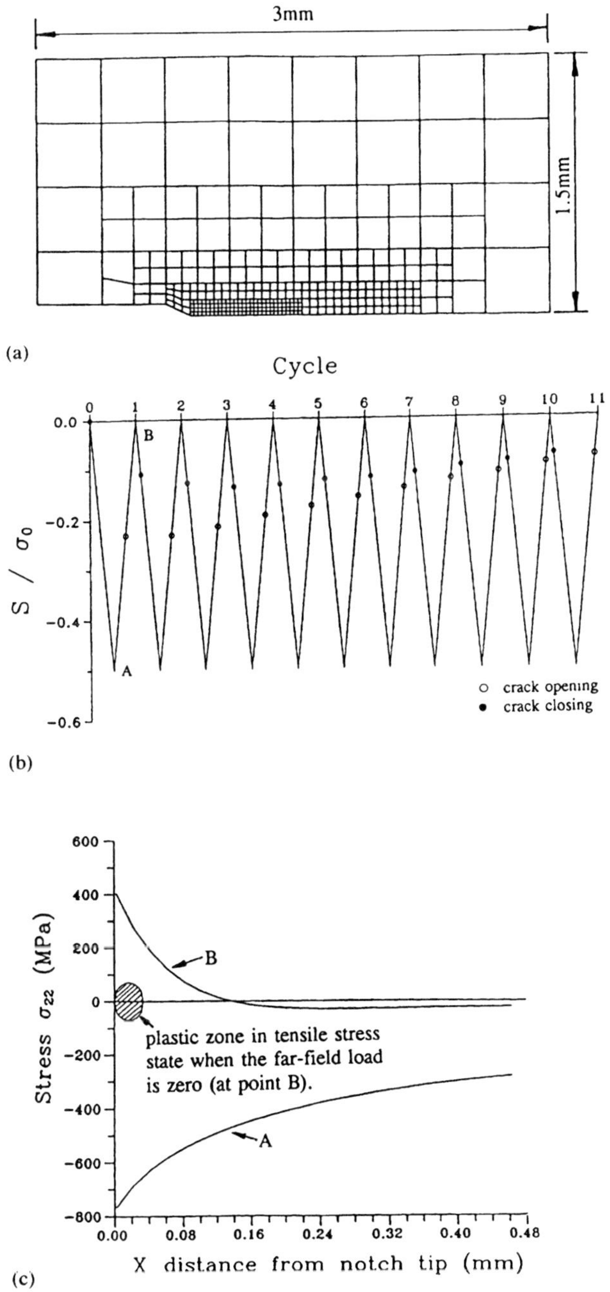

Fig. 8.19 Crack closure development with crack extension under applied far-field compressive stresses. Reprinted with kind permission from [58]. (a) Finite element mesh of a quarter of a centre cracked panel; (b) development of crack opening and closing stress with applied cycles; (c) normal to crack plane stress distribution during the first cycle and occurrence of a tensile plastic zone.

REFERENCES

1. Neuber, H. (1958) *Kerbspannungslehre*, Springer, Berlin. Translation: *Theory of Notch Stresses*, U.S. Office of Technical Services, Washington, DC, 1961.
2. Muskhelishvili, N.I. (1953) *Some Basic Problems of Mathematical Theory of Elasticity*, P. Noordhoff Ltd., Groningen, The Netherlands.
3. Savin, G.N. (1961) *Stress Concentration Around Holes*, Pergamon Press, New York.
4. Peterson, R.E. (1974) *Stress Concentration Factors*, John Wiley, New York.
5. Green A.E. (1948) Three-dimensional stress systems in isotropic plates. *Phil. Trans. R. Soc. London* A, **11**, 561–97.
6. Sternberg, E. and Sadowsky, M.A. (1948) Three-dimensional solution for the stress concentration around a circular hole in a plate of arbitrary thickness. *J. Appl. Mech., Trans. ASME*, **16**, 27–38.
7. Reiss, E.L. (1963) Extension of an infinite plate with a circular hole. *J. Soc. Ind. App. Math. (SIAM)*, **11**, 840–54.
8. Youngdahl, C.K. and Sternberg, E. (1966) Three-dimensional stress concentration around a cylindrical hole in a semi-infinite elasticity body. *J. Appl. Mech., Trans. ASME*, **33**, 855–65.
9. Ellyin, F. and Sherbourne, A.N. (1968) Effect of skew penetration on stress concentration. *J. Engng. Mech. Div., Proc. ASCE*, **94**, 1317–36.
10. Ellyin, F. (1970) Elastic Stresses Near a Skewed Hole in a Flat Plate and Application to Oblique Nozzle Attachments in Shells. *Welding Research Council Bulletin*, No. 153, Aug. 1970, pp. 32–51.
11. Neuber, H. and Hahn, H.G. (1966) Stress concentration in scientific research and engineering. *Appl. Mech. Rev.*, **19**, 187–99.
12. Timoshenko, S. and Goodier, J.N.(1970) *Theory of Elasticity*, 3rd edn, McGraw-Hill, New York.
13. Creager, M. and Paris, P.C. (1967) Elastic field equations for blunt cracks with reference to stress corrosion cracking. *Int. J. Fract. Mech.*, **3**, 247–52.
14. Tada, H., Paris, P.C. and Irwin, G.R. (1973) *The Stress Analysis of Cracks Handbook*, Del Research Corporation, Hellertown, PA.
15. Ellyin, F. (1973) A class of stress concentration problems in plastic range, in *Experimental Mechanics in Research and Development* (eds J.T. Pindera *et al.*), SM Study No. 9, University of Waterloo, Waterloo, ON, pp. 205–24.
16. Sokolovskii, V.V. (1950) *Theory of Plasticity*, 2nd edn, Academic Press, Moscow.
17. Stowell, E.Z. (1950) Stress and Strain Concentration at a Circular Hole in an Infinite Plate. NACA Technical Note 2073, National Advisory Committee for Aeronautics.
18. Budiansky, B. and Mangasarian, O.L. (1960) Plastic stress concentration at a circular hole in an infinite sheet subjected to equal biaxial tension. *J. Appl. Mech., Trans. ASME*, **27**, 59–64.
19. Neuber, H. (1961) Theory of stress concentration for shear strained presmatical bodies with arbitrary nonlinear stress-strain law. *J. Appl. Mech., Trans. ASME*, **28**, 544–50.
20. Rice, J.R. (1967) Mechanics of crack tip deformation and extension by fatigue, in *Fatigue Crack Propagation*, ASTM STP 415, American Society for Testing and Materials, Philadelphia, PA, pp. 247–311.
21. Hardrath, H.F. and Ohman, L. (1953) A Study of Elastic and Plastic Stress Concentration Factors Due to Notches and Fillets in Flat Plates. NACA TN 117, National Advisory Committee for Aeronautics.

22. Morrow, J.D., Wetzel, R.M. and Topper, T.H. (1970) Laboratory simulation of structural fatigue behaviour, in *Effect of Environment and Complex Load History on Fatigue Life*, ASMT STP 462, American Society for Testing and Materials, Philadelphia, PA, pp. 74–91.
23. Molski, K. and Glinka, G. (1981) A method of elastic-plastic stress and strain calculation at a notch root. *Mater. Sci. Engng.*, **50**, 93–100.
24. Ellyin, F. and Kujawski, D. (1987) Notch root stress/strain prediction for elastic-plastic loading. *Res. Mechanica*, **20**, 177–90.
25. Rice, J.R. (1968) A path independent integral and the approximate analysis of strain concentration by notches and cracks. *J. Appl. Mech., Trans. ASME*, **35**, 379–86.
26. Ellyin, F. and Kujawski, D. (1989) Generalization of notch analysis and its extension to cyclic loading. *Engng. Fract. Mech.*, **32**, Elsevier Science Ltd, Oxford, 819–26.
27. Moftakhar, A., Buczynski, A. and Glinka, G. (1995) Calculation of elasto-plastic strains and stresses in notches under multiaxial loading. *Int. J. Fract.*, **70**, 357–73.
28. Leis, B.N. and Frey, N.D. (1982) Cyclic-inelastic deformation and fatigue resistance of notched-thin aluminum plates. *Proc. Soc. for Experimental Stress Analysis (SESA)*, **39**, 287–95.
29. Zienkiewicz, O.C. (1971) *The Finite Element Method in Engineering Science*, McGraw-Hill, London.
30. Leis, B.N., Gowda, C.V.B. and Topper, T.H. (1973) Some studies of the influence of localized and grow plasticity on the monotonic and cyclic concentration factors. *J. Testing Eval.*, **1**, 341–8.
31. Wu, J. (1989) An elasto-plastic material model and applications. M.Sc. Thesis, Department of Mechanical Engineering, University of Alberta, Edmonton.
32. Leis, B.N., Gowda, C.V.B. and Topper, T.H. (1973) Cyclic inelastic deformation and the fatigue notch factor, in *Cyclic Stress-Strain Behaviour*, ASTM STP 519, American Society for Testing and Materials, Philadelphia, PA, pp. 133–50.
33. Murakami, Y., Tazunoki, Y. and Endo, T. (1984) Existence of the coaxing effect and effects of small artificial holes on fatigue strength of an aluminum alloy and 70–30 brass. *Metall. Trans.*, **15A**, 2029–38.
34. Dowling, N.E. (1982) *A Discussion of Methods for Estimating Fatigue Life*. Proc. SAE Fatigue Conf. P-109, April 1982, Society of Automotive Engineers, Warrendale, PA, pp. 161–74.
35. Du-yi, Y. and De-jun, W. (1996) A new approach to the prediction of fatigue notch reduction factor K_f. *Int. J. Fatigue*, **18**, 105–9.
36. Newman, J.C. (1971) An Improved Method of Collocation for the Stress Analysis of Cracked Plates with Various Shaped Boundaries. *National Aeronautics and Space Administration (NASA)*, Technical Note D-6376, Langley, Virginia.
37. Lukáš, P. and Klesnil, M. (1978) Fatigue limit of notched bodies. *Mater. Sci. Engng.*, **34**, 61–6.
38. Kujawski, D. (1991) Estimation of stress intensity factors for small cracks at notches. *Fatigue Fract. Engng. Mater. Struct.*, **14**, 953–65.
39. Frost, N.E. and Dugdale, D.S. (1957) Fatigue tests on notched mild steel plates with measurements of fatigue cracks. *J. Mech. Phys. Solids*, **5**, 182–92.
40. Frost, N.E. (1960) Notch effect and the critical alternating stress required to propagate a crack in an aluminum alloy subject to fatigue loading. *J. Mech. Engng. Sci.*, **2**, 109–19.

41. Smith, R.A. and Miller, K.J. (1977) Fatigue cracks at notches. *J. Mech. Engng. Sci.*, **19**, 11–22.
42. El Haddad, M.H., Topper, T.H. and Smith, K.N. (1979) Prediction of non-propagating cracks. *Engng. Fract. Mech.*, **11**, 573–84.
43. Du Quesnay, D.L., Yu, M.T. and Topper, T.H. (1988) An analysis of notch-size effects at the fatigue limit. *J. Testing Eval.*, **16**, 375–85.
44. Hammouda, M.M., Smith, R.A. and Miller, K.J. (1979) Elastic-plastic fracture mechanics for initiation and propagation of notch fatigue cracks. *Fatigue Engng. Mater. Struct.*, **2**,139–54.
45. El Haddad, M.H., Smith, K.N. and Topper, T.H. (1979) A strain based intensity factor solution for short fatigue cracks initiating from notches, in *Fracture Mechanics*, Eleventh Conf., ASTM STP 677, American Society for Testing and Materials, Philadelphia, PA, pp. 274–89.
46. El Haddad, M.H., Dowling, N.E., Topper, T.H. and Smith, K.N. (1980) J-integral application to short fatigue cracks at notches. *Int. J. Fract.*, **16**, 15–30.
47. Shin, C.S. and Smith, R.A. (1985) Fatigue crack growth from sharp notches. *Int. J. Fatigue*, **7**, 87–93.
48. Abdel-Raouf, H., Du Quesnay, D.L., Topper, T.H. and Plumtree, A. (1992) Notch-size effect in fatigue based on surface strain redistribution of crack closure. *Int. J. Fatigue*, **14**, Elsevier Science Ltd, Oxford, 57–62.
49. Abdel-Raouf, Topper, T.H. and Plumtree, A. (1991) A short fatigue crack model based on the nature of the free surface and its microstructure. *Scripta Metallurgica et Materialia*, **25**, 597–602.
50. Hubbard, R.P. (1969) Crack growth under cyclic compression. *J. Basic Engng., Trans. ASME*, **91**, 625–31.
51. Reid, C.N., Williams, K. and Hermann, R. (1979) Fatigue in compression. *Fatigue Engng. Mater. Struct.*, **1**, 267–70.
52. Chu, W., Hsiao, C.M., Jin, L.J. and Liu, T.H. (1983) Fatigue cracks initiation from a notch tip under a cyclic compressive load. *Scripta Metall.*, **17**, 993–6.
53. Holm, D.K., Blom, A.F. and Suresh, S. (1986) Growth of cracks under far-field cyclic compressive loads: Numerical and experimental results. *Engng. Fract. Mech.*, **23**, 1097–1106.
54. Pippan, R. (1988) The sensitivity to measure crack closure with strain gauges near the crack tip. *Engng. Fract. Mech.*, **31**, 867–71.
55. Craig, D.F., Kujawski, D. and Ellyin, F. (1995) An experimental technique to study the behaviour of small corner cracks. *Int. J. Fatigue*, **17**, Elsevier Science Ltd, Oxford, 253–9.
56. Suresh, S., Christman, T. and Bull, C. (1986) Crack initiation and growth under far-field cyclic compression: Theory, experiments and application, in *Small Fatigue Cracks* (eds R.O. Ritchie and J. Lankford), The Metallurgical Society of the American Institute of Mining, Mineral and Petroleum Engineers, Warrendale, PA, pp. 513–40.
57. Aswarth, P.B., Suresh, S., Holm, D.K. and Blom, A.F. (1988) Load interaction effects on compression fatigue crack growth in ductile solids. *J. Engng. Mater. Technol., Trans. ASME*, **110**, 278–85.
58. Ellyin, F., Xia, Z. and Wu, J. (1995) A new elasto-plastic constitutive model inserted into the user-supplied material model of ADINA. *Computers & Structures*, **56**, Elsevier Science Ltd, Oxford, 283–94.

Growth and behaviour of small cracks

9

9.1 INTRODUCTION

It has been pointed out in sections 7.6 and 8.4.1 that small cracks behave differently from their long counterparts. The similitude concept of fracture mechanics, i.e. that cracks with the same crack tip condition (e.g. ΔK) will propagate at the same rate, does not hold for a physically short crack. There is no unique definition as to what crack size constitutes departure from the long crack fracture mechanics approach described in Chapter 7, nevertheless some boundaries will be specified in the next section. An understanding of the behaviour of small cracks is important from a practical viewpoint. This knowledge can be used to evaluate the size of a flaw which can be tolerated in a production process. In Chapters 3 to 5 we have discussed the initiation of cracks whereas in Chapter 7 we were concerned with the growth of initiated cracks. Small crack modelling in a sense attempts to bridge these two seemingly different approaches. An example of this type was provided in section 8.4.2.

The two main characteristics of a physically small crack are that it grows at a higher rate than a long crack for the same value of applied stress intensity factor range, ΔK, and it also grows at values of ΔK, below that of the threshold value, ΔK_{th}, of the long cracks.

This behaviour is illustrated in Fig. 9.1 where the crack growth rate, da/dN, is plotted against the crack length, a on a log–log coordinate system. The figure shows typical results for a given material and environment during constant amplitude cyclic loading. It indicates that the growth rate depends on the applied stress level. Of particular interest are cracks which grow below a material's fatigue limit, σ_{FL}, but self-arrest due to microstructural barriers. The reduction in crack growth rate with increased crack length at stress levels above the fatigue limit is attributed to the development of crack closure (Fig. 8.14). Obviously, the crack tip driving force in a physically small crack is different from that of a long crack, and linear elastic fracture mechanics parameters such as ΔK which

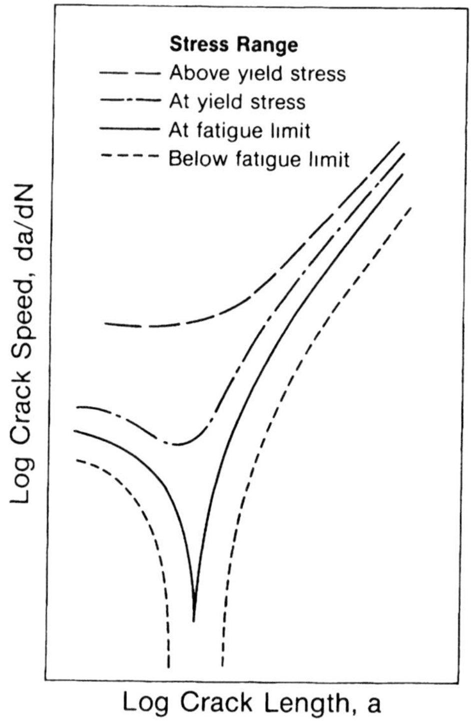

Fig. 9.1 A schematic representation of small crack behaviour as a function of applied cyclic stress.

relate the crack tip stress/strain field to the applied far-field stress are inadequate to describe the small crack driving force at higher stress levels, $\Delta\sigma > \Delta\sigma_{th}$. This is partly due to violation of the requirement of the small-scale yielding (SSY) condition at the crack tip. Expressed another way, the ratio of the plastic zone size, R_p, to the crack length, a, is different for the two cases. For long cracks under $\Delta\sigma < \Delta\sigma_{FL}$ loading, $R_p \ll a$, whereas for physically short cracks, R_p and a are of the same order of magnitude.

Pearson [1] was the first to report on the anomalous growth of small cracks in an aluminium alloy. Since then numerous investigators have reported on the growth characteristics of small cracks in a variety of metals and alloys, for example [2–9], and review articles [10–13], where further references are cited. The literacture on the behaviour of small cracks is extensive. At least three conferences have been held on the subject [14–16]. Two of these conferences were reviewed by Miller in [17], which contains 69 papers reporting on a variety of topics ranging from the definition of an initiated crack to microstructural effects, experimental techniques and devices, environmental aspects, analytical and numerical models, small cracks initiated from notches, and engineering design applications. In addition to the above-mentioned special conferences, the subject of small crack behaviour has been covered in general fatigue conferences, e.g. [18, 19], and in numerous journal articles. What has emerged from the study of these contributions is that an understanding of small crack behaviour will require an interdisciplinary effort in a number of areas.

In this chapter we will discuss a few topics which are essential to the understanding of small crack behaviour, namely, the mechanisms involved and modelling of small crack growth. A mathematical model based on physical and microstructural observation is essential for the implementation of a damage-tolerance design procedure, and life estimation for critical industrial components.

9.2 SMALL CRACK REGIMES

Figure 9.1 shows that crack growth behaviour is a function of physical crack size, a, as well as the applied stress. For example at an applied stress above the yield stress, no dip in the crack growth rate is observed. Therefore, the problem of physically small and long cracks subject to nominal stresses at or above the yield stress of the material can, in principle, be dealt with by the elastic–plastic fracture mechanics methodology described in Chapter 7. However, large-scale yielding can occur in this case, which will affect the development of closure stresses differently in the small and long cracks, thus resulting in dissimilar growth rates. The so-called short crack anomaly is, however, more pronounced when the crack is physically small and it is subjected to a nominal stress below the yield stress of the material.

Typical behaviour of a small initiated surface crack, $a > 3\,\mu m$, stressed at a nominal value slightly above the fatigue limit of a material in a fully-reversed condition, is shown schematically in Fig. 9.2. It can be seen that the initiated crack first grows in a shear mode which is termed Stage I after Forsyth [20]. This growth pattern continues through one to three surface grains at an angle of about 45 degrees, following which the crack plane rotates to become normal to the maximum principal stress. The latter is termed Stage II cracking, or Mode I in fracture mechanics terminology (Fig. 7.3).

The growth of a Stage I crack and its transition from Stage I to Stage II are governed by the crystallographic slip system and constraint of adjacent grains. Thus, crack growth in this stage is highly influenced by the microstructure, and the first grain boundary and other microstructural features. (For example, pearlite bands in dual-phase steels constitute barriers to crack growth, Fig. 9.7). Stage I cracks of the order of one or a few grains will be termed **microstructural** or **crystallographic small cracks**. The extent of the maximum size of microstructural small cracks is dependent on material

(a)

(b)

(c)

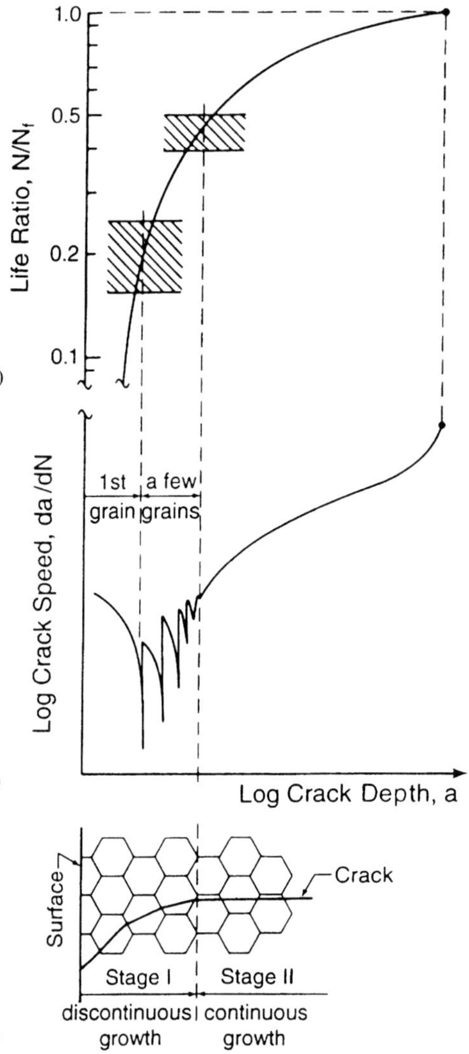

Fig. 9.2 A schematic illustration of small crack growth and transition to long crack. (a) relative life ratio spent at each stage; (b) crack growth rate versus crack depth; (c) transition from a microstructurally small crack to a physically small crack and eventually a long crack.

anisotropy and texture. The fatigue limit of a material is thus dictated by the microstructural barriers which arrest the growth of initiated crystallographic cracks.

For smooth specimens tested at constant stress amplitude in the high-cycle fatigue regime (elastic bulk response), the early crack growth (say, up to the depth of a few grain diameters) may occupy more than 50% of the total fatigue life. This is shown in Fig. 9.2(a), where the abscissa is the crack depth and the ordinate is the fatigue life ratio, N/N_f. The fatigue limit of a smooth specimen is defined either when the crack growth through a few grains takes 10^6–10^7 cycles, or a small initiated crack is arrested (or trapped) at microstructural barriers (features). The strongest barrier for a given material and stress range defines the extent of the microstructural regime. Thus, as rightly pointed out by Miller [17], the grain size alone does not define the microstructural small crack regime.

The physically small crack (PSC) regime is bounded by the upper limit of the microstructurally small crack regime discussed above, and the lower limit of the long crack regime. The latter can be defined with respect to the threshold stress intensity range, ΔK_{th}. Referring to Fig. 9.3, the upper limit of this regime is defined by the straight line on the RHS which is obtained at stress levels below the fatigue limit of a smooth specimen (see also Fig. 8.15). In relation to Fig. 9.2, the PSC regime covers the latter part of the transition from Stage I to Stage II and the early part of the Stage II crack growth regime where the crack size and consequently the driving force is large enough that microstructural barriers can decelerate crack growth but are unable to arrest it. Although the upper limit of the PSC regime, i.e. the boundary with long crack growth, is stress and crack depth dependent, generally for a variety of industrial metals and alloys, the crack depth, a, is of the order of a millimetre or more, i.e. $a \geqslant 1$ mm. Identification of a lower limit is more difficult, since it depends on the dominant microstructural barrier. For example, in a dual-phase ferrite/pearlite steel (ASTM A-516 Gr. 70), with a banded structure perpendicular to the rolling direction, investigated in [21], the ferrite/pearlite boundaries were irregularly

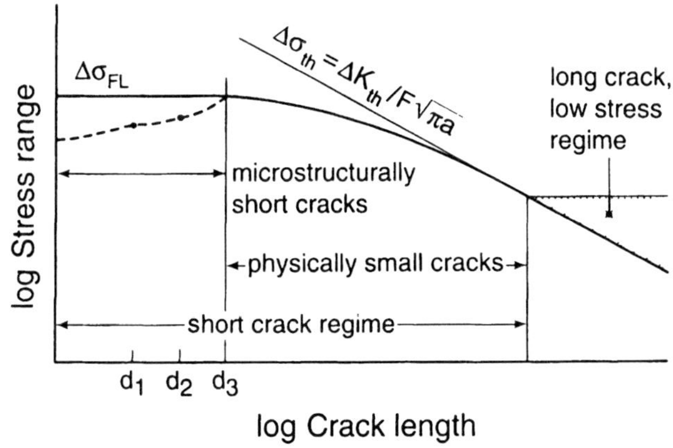

Fig. 9.3 A modified Kitagawa–Takahashi diagram showing various small crack regimes.

spaced, 0.1–0.3 mm apart, and the average grain size was 0.013 mm (Fig. 9.7). Thus, for this steel, the microstructural small crack regime is given by $a \leqslant 0.3$ mm, whereas the physically small crack regime can be bounded by $2\,\text{mm} \geqslant a > 0.1$ mm. Referring to Fig. 9.3, a modified Kitagawa–Takashashi diagram [17], the microstructural barriers for this steel are: $d_1 = 0.013$ mm, and $d_3 = 0.3$ mm.

Other definitions have been proposed, e.g. [10], which distinguishes four categories, namely (i) microstructurally small flaws, (ii) mechanically small flaws, (iii) physically small flaws, and (iv) chemically small cracks. In the above discussion, we have combined definitions (ii) and (iii) together under the physically small crack regime. Even so, there is an overlap when the material is microstructurally anisotropic, e.g. the distance between pearlite bands and pearlite colonies in the dual-phase steel mentioned earlier. Thus depending upon the crack growth direction, the distance between dominant microstructural features will differ (see also discussion by Miller [17] on this topic).

Apparent anomalies exhibited by physically long cracks in an aggressive environment have to be discussed separately since the operative mechanism in this case is quite different. The growth of cracks in an aggressive environment is affected by the rate of loading, the kinetics of a chemical reaction, and the embrittlement of the material due to the chemical environment (see discussions in Chapters 3 and 7).

9.3 MECHANISMS OF SMALL CRACK GROWTH

Microstructurally small (or crystallographic) crack growth (or Stage I growth) occurs in a slip plane and along a slip direction. These characteristics have been found for small cracks in many polycrystals [22], copper single crystals [23], and aluminium bicrystals [24], to mention but a few.

Stage I crack growth results from irreversible plastic deformation at the crack tip. As shown in Figs 9.4(a) and (b) [25], an intrusion forms due to the relative displacement of a slip band CC′ and the one above it under shear stress reversal. In the next applied positive shear deformation (Fig. 9.4(c)) dislocations on the CC′ plane will produce a greater offset. An opposite deformation (Fig. 9.4(d)) will cause the slip planes above and below CC′ to act and the intrusion will eventually grow and form a crack of the order of a few μm. Crack propagation along the PSBs depends on irreversible dislocation movement which results in crack growth in small integral multiples of the Burgers vector. This process has been fully discussed in sections 1.3.1 to 1.3.3 and we will not delve into it any further. Here it will suffice to mention that the local cyclic plastic strain range in persistent slip bands ahead of a Stage I crack in single copper crystals was found to be constant, e.g. [23]. The magnitude of the cyclic plastic strain is

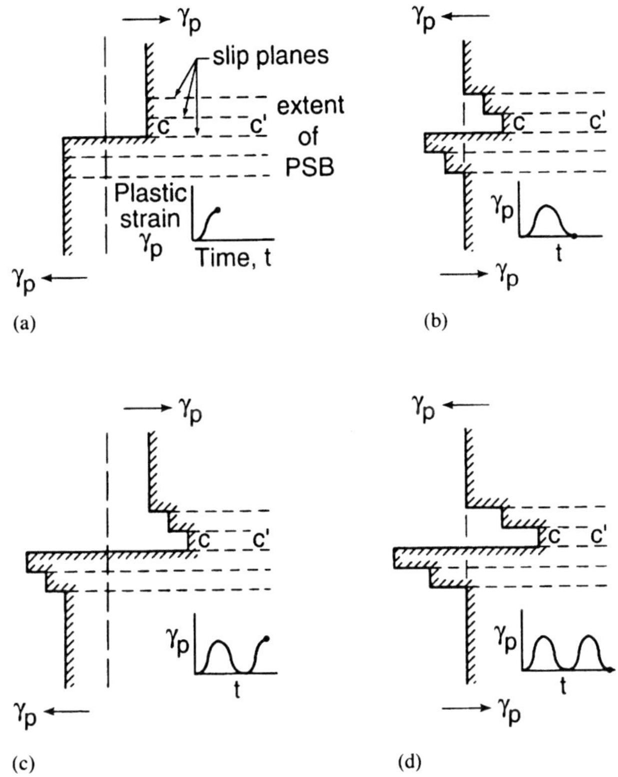

Fig. 9.4 Schematic illustration of the early stages of a crystallographic crack growth in a PSB on a surface normal to the primary Burgers vector. The inserted strain–time plot indicates the loading sequence corresponding to the slip behaviour, taken from [25].

a measure of the driving force Stage I crack growth [22]. This then suggests that the Stage I growth rate is not appreciably affected by the applied stress unless a crack shielding mechanism or a barrier is present. A crack may be shielded from the full effect of applied loads by altering the cyclic plastic deformation ahead of it. This can then lead to a reduction in the effective driving force experienced at the crack tip. Crystallographic crack tip shielding due to microstructural barriers causes a transition from Stage I growth to Stage II growth in addition to crack deceleration during constant cyclic loading (Fig. 9.2). Stage II growth in FCC single crystals arises from the irreversibility of shear deformation in two or more slip bands ahead of the crack tip. This has been described and experimentally verified by Neumann [26] (see also section 7.6 and Fig. 7.22).

Fatigue crack shielding by microstructural barriers leads to a breakdown of the similitude concept of linear elastic fracture mechanics. Consequently the stress intensity factor range, ΔK, is not an appropriate field parameter for small cracks. Microstructural barriers impede the slip band ahead of the crack tip resulting in crack deflection and roughness induced closure. Microstructurally short cracks which initially grow in a shear mode are subject to mixed mode deformation during the transition to Stage II growth. The size of the cyclic plastic zone ahead of a crack is an indicator of the crack driving force. It was mentioned earlier that the ratio of the plastic zone size to the crack size for a short crack is much larger than that for a long crack under the same nominal stress intensity loading [27]. As a crack approaches a microstructural barrier, secondary slip systems form and the primary plastic zone size decreases (Li [28]). Thus, the observed crack growth deceleration is primarily due to a microstructural barrier inducing crack tip shielding. Crack tip shielding in this case is due to the effect of primary slip incompatibility with the adjacent grain. As a result secondary slip occurs and the crack path deviates from its original direction, i.e. crack branching takes place. This process results in a crack tip cyclic plasticity reduction and redistribution. The secondary slip system plays a key role in increasing crack opening and driving the crack across the microstructural boundary [28]. The effect of the microstructural barrier is usually averaged out when the crack travels a length of about three times the distance between the strongest barriers, e.g. three times the grain diameter or three times the pearlite band spacing, etc.

In summary, the growth of a small crack is governed by the cyclic plasticity ahead of the crack which is constrained and changed by the microstructural barriers in front of it. Once the path of the crack is altered, roughness induced closure takes place which further reduces the crack driving force. The enhanced crack closure is mainly due to the mixed mode sliding and mismatch between crack face asperities. The height of the fracture surface asperities can be of comparable magnitude to the crack tip opening displacement. Thus, for small cracks in Stage I the crack closure mechanism is mainly due to crack branching and development of rough surface asperities rather than residual plasticity in the wake of the crack (plasticity induced closure). The effect of crack tip cyclic plasticity on the retardation of crystallographic (Stage I) crack growth in metals and alloys has been reviewed by Ellyin and Li [29]. It is shown that changes in the crack growth direction result from a redistribution of the crack tip plastic strain. Similarly, the crack growth rate deceleration is related to a reduction in the cyclic plastic strain range due to crack tip shielding. Thus, the growth of small cracks is strongly influenced by the change in cyclic plasticity ahead of the cracks.

In the following section experimental data will be presented to support some of the points discussed above, in particular the role of microstructural barriers and development of crack closure.

9.4 EXPERIMENTAL DATA ON SMALL CRACK BEHAVIOUR

There are a number of techniques available for the measurement of crack length and calculation of crack growth rate. They range from the potential drop method [30], to the use of a travelling microscope [31], scanning electron microscope measurement [32], the use of a plastic replica technique [33], ultrasonic methods [34], and acoustic microscopy [35]. No single method is ideal and different problems arise with each technique. The plastic replica approach is often a favoured technique because it gives good results and does not involve expensive and specialized equipment. A further benefit of the replica approach is that despite being labour intensive, it provides a permanent record.

Newman and Edwards [36] have compared several methods for measuring the growth of naturally initiated small cracks of the order of 10 μm and larger. They reported that only the plastic replica method was accurate enough to monitor the growth of very small cracks.

9.4.1 INITIATION OF SMALL CRACKS

To investigate early crack growth behaviour, a small crack has to be initiated. A number of methods have been explored to initiate small cracks. They include: (i) initiation from natural flaws in smooth surfaces under loads slightly above the fatigue limit, (ii) the introduction of a small notch into the test region, (iii) the use of large notch specimens with a tensile loading, (iv) the use of large notch specimens with compressive loading. None of the above methods of initiating small cracks is entirely satisfactory in producing damage free, quick and cheap specimens with very small cracks. In the first two methods one has no prior information on the depth of the initiated crack, while in the last two methods, the specimen preparation involves two stages of machining which may in turn introduce residual stresses. However, the technique of beginning with a notched specimens, applying a compressive load and subsequently machining away the notch and most of the self-arrested crack (section 8.5) comes close to producing a specimen with the desired attributes, albeit that it is one of the most labour-intensive procedures.

The technique is described by Kujawski and Ellyin in [37,38]. A pre-crack specimen, having a large sharp notch, is loaded in cyclic compression. The crack which grows from the notch is permitted to grow until it self-arrests. The crack front in this case is concave, i.e. the maximum crack growth occurs at the outside surface. A fatigue tension specimen is subsequently cut from this pre-cracked specimen so as to include just the tip of the self-arrested crack. This results in two small corner cracks which are naturally open with minimal damage ahead of the tip. Figure 9.5(b) shows the two resulting corner cracks and the approximation which

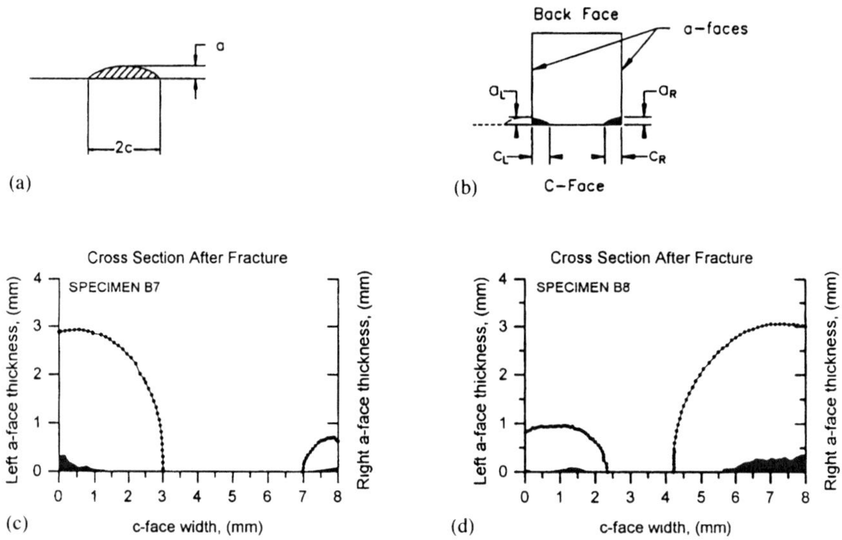

Fig. 9.5 (a) A surface crack; (b) corner cracks obtained by the procedure described in [37, 38]; (c) and (d) initial and final crack fronts based on after-fracture measurements, from [39], reproduced with kind permission.

a corner crack affords with respect to a surface crack, Fig. 9.5(a). The nomenclature for both surface and corner cracks is depicted in Fig. 9.5. An advantage of corner cracks is that both crack length, c, and depth, a, can be monitored.

9.4.2 GROWTH OF CORNER CRACKS

The results of a systematic study of the growth of small corner cracks in a ferritic/pearlitic steel (ASTM A-516 Gr. 70) are reported in [37, 39] for Mode I cracks and in [38] for the inclined (mixed mode) corner cracks. Here, we will present a few representative results to illustrate the behaviour of small corner cracks. Readers are encouraged to consult [37–39] for more details.

Crack growth rates were obtained from crack length measurements made by the plastic replica technique using a secant method described in [39]. Figure 9.6 shows a typical growth rate of both a- and c-faces versus crack depth, a. Note that once the crack is well-developed (reaching the long crack regime) the slopes of the growth rate curves for the crack length and depth converge towards a common value.

The microstructural features of the tested material are shown in Fig. 9.7. It can be seen that the c-face reveals a banded ferritic/pearlitic structure whereas the a-face contains irregularly dispersed pearlite regions (colonies)

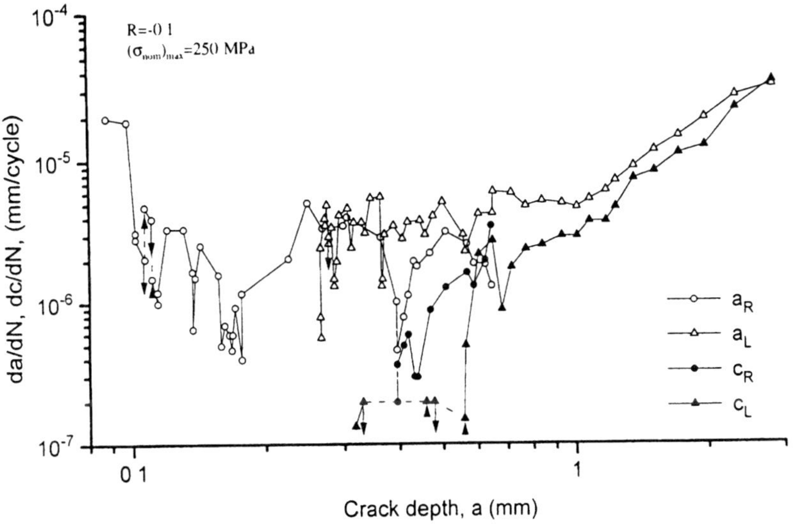

Fig. 9.6 Growth rate da/dN and dc/dN versus crack depth, a, of a ferritic/pearlitic steel. A vertical downward arrow indicates a growth rate below the scale of the graph while an upward arrow shows the subsequent occurrence of growth. A dashed line is placed between the two events to aid in showing the trend, from [39], reproduced with kind permission.

within the ferrite. The early growth of the crack is in the depth direction (a-face) as indicated in Fig. 9.6. The growth rate versus crack depth data is aligned with the microstructure photomicrographs of the crack in Fig. 9.8. When the crack is small, the crack growth rate shows a drastic deceleration and acceleration when it passes through a pearlite colony. The influence of microstructure on the growth of a small crack is clearly seen in Fig. 9.8.

In a study of small cracks in single-edge-notch tension specimens of LC9cs clad aluminium alloy (2 mm thick) extensively used in the aircraft industry, Newman *et al.* [40] found that cracks initiated in the cladding material (90 μm thick on each side) and grew as corner cracks in the core alloy. They observed that at crack lengths up to about 0.3 mm, corner cracks grew at a much lower rate than the surface cracks. After this length the growth rates were the same. Their observation appears to be similar to that shown in Fig. 9.6 which indicates that beyond a microstructural zone influence, the growth on the surface (c-face) is at a faster rate than that in depth, and the growth rates converge to the same rate as the crack becomes long.

The evolution of the a/c ratio throughout the test is described in Fig. 4 of [39]. The maximum aspect ratio (saturation) is usually less than 1.0, and it is reached at a crack depth of about 2 mm. The shape of the initial and final crack front measured after fracture of the specimens is shown in Figs 9.5(c) and (d).

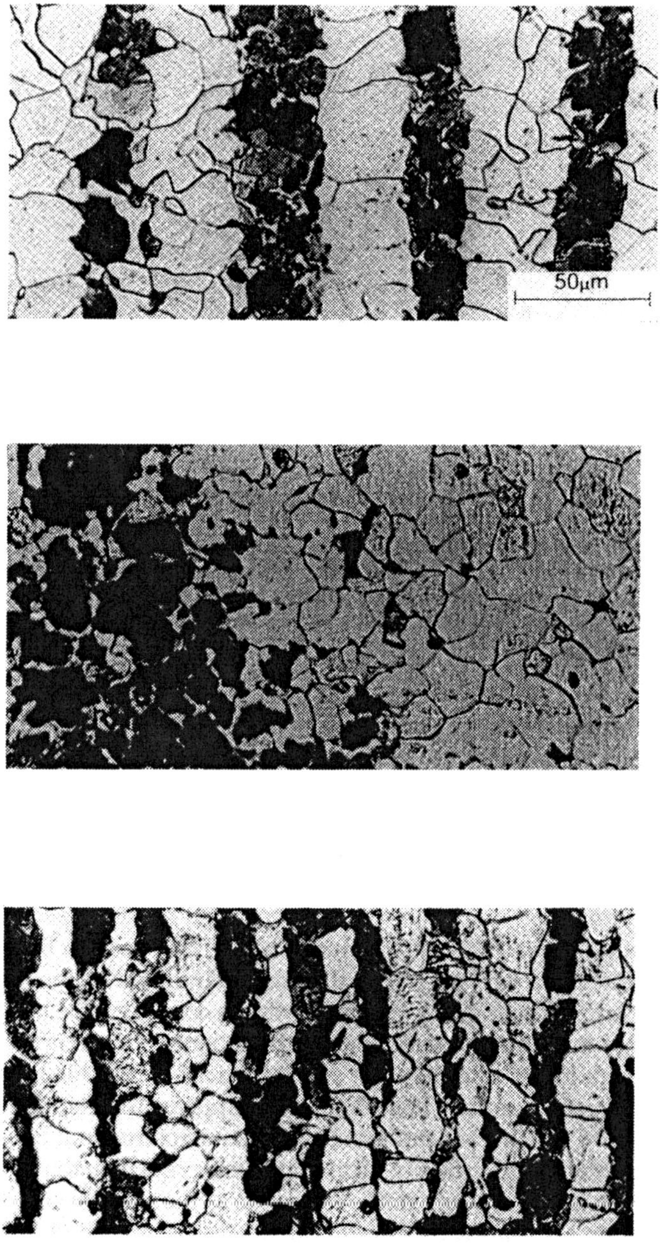

(a)

(b)

(c)

Fig. 9.7 Photomicrographs of a ferritic/pearlitic steel showing: (a) a banded pearlite structure in the *c*-face (see Fig. 9.5(b) for nomenclature); (b) irregular pearlite colonies in the *a*-face; (c) an irregularly spaced banded structure perpendicular to the rolling direction (magnification 425 ×).

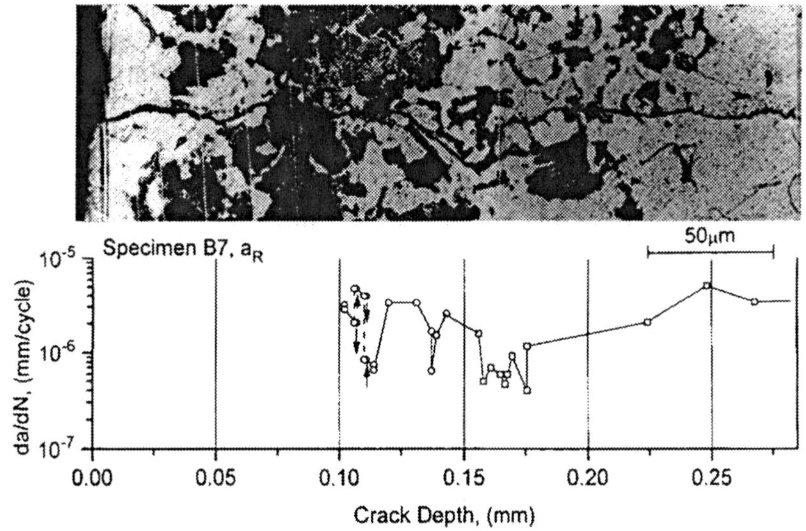

Fig. 9.8 Crack growth rate in the depth direction, d*a*/d*N*, aligned with the corresponding microstructural features of ASTM A-516 Gr. 70 ferritic/pearlitic steel. Reproduced with kind permission from [39].

9.4.3 CRACK CLOSURE DEVELOPMENT

A crack is said to be fully open when there is no contact point between its surfaces. Measuring the load at which the crack is fully open is complicated because the crack does not open instantaneously. Typically, a crack begins to open at its mouth, and gradually proceeds to open right up to the crack tip with the increasing load.

Several techniques have been employed to measure crack opening load, the most common being to use strain gauges adjacent to or on the crack to record change in strain while the crack opens. Due to the small plastic zone in front of a small crack, placing a simple strain gauge adjacent to the crack is not sensitive enough to record crack tip strain changes. A modification used by various investigators, e.g. Ogura, Miyoshi and Nishikawa [41], Chen, Weiss and Stickler [42] and DuQuesnay [43], is to place an active gauge on top of the crack. A second gauge is mounted away from the crack to record the nominal applied strain. The difference between the active and the far-field reading provides a signal dependent on the crack behaviour. The differential compliance curves obtained by this technique are examined to determine the opening load. Since crack opening is a gradual process, the differential compliance curve has a gradual change in slope. A common procedure is to fit a strainght line to the upper part of the compliance curve and a straight line to its lower part. The point where the two lines intersect is taken as the opening load. It should be noted that at the fully opened and fully closed crack configurations, the corresponding compliance curves exhibit a linear behaviour. The intersection of these two lines is an idealized crack opening/closing load level. However, as mentioned earlier, a crack opens gradually in a 'peel open' process starting from its mouth towards the tip.

A new differential compliance method for corner cracks is described in [44]. The crack opening load is determined through a curvilinear fit of a Ramberg–Osgood type relation. The fitted curve, which is representative of a fully open crack, is compared with the test data, and the opening load is determined from the loading branch of the compliance curve where the two become tangent. This is illustrated in Fig. 9.9 where the fitted curve is compared to the data. This curvilinear fit was found to be more sensitive, since it takes into account the nonlinearity of the differential displacement [44].

Figure 9.10 shows the opening load data versus number of cycles for a corner crack, obtained by the linear and curvilinear methods described above. The graphs are plotted using linear scales, with an inset to show in greater detail the evolution of the crack opening load during the first 2000 cycles. The data points are connected by straight lines. The trend in the crack opening data is shown by dashed lines which are fitted to the average of each cluster in Fig. 9.10. Further data is provided in [21] and [39]. The curvilinear fit indicates a more reasonable trend, an increasing opening load with increasing number of cycles, reaching a steady state at about 20%

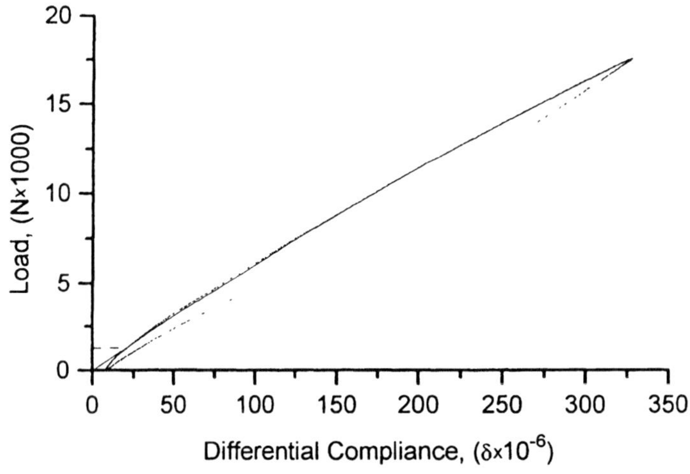

Fig. 9.9 Curvilinear fit method: dotted curve is the differential compliance curve whereas the solid curve represents a fully open crack. The intersection point is the opening stress level. Reprinted with kind permission from [44].

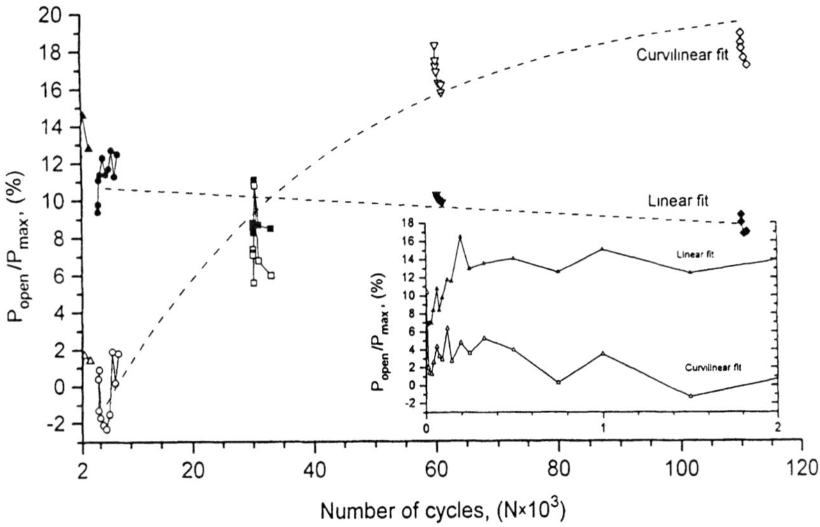

Fig. 9.10 Relative crack opening versus number of applied cycles. Data obtained by the curvilinear fit method are represented by open symbols whereas those obtained by the linear intercept approach are indicated by solid symbols. Replacement of the strain gauges is shown by using a symbol of a different shape. Reprinted with kind permission from [39].

of the maximum applied load. The linear intercept method in this case shows a slightly decreasing opening load. Intuitively, one would expect a build-up of the closure stress as a crack grows into the crack tip plastic zone.

9.5 MODELS DESCRIBING SMALL CRACK BEHAVIOUR

From the discussion of mechanisms of small crack growth and experimental data in the previous sections, it is clear that any model attempting to predict this behaviour must be microstructurally-based. A purely continuum model would be of limited use, though it may be shown to correlate with a certain set of experimental data. However, when the applied far-field stress is above the yield stress of the material, large-scale yielding takes place and microstructural barriers are not very effective in impeding the crack advance [29]. A mathematical treatment of the physical processes occurring during small crack growth is of importance for several reasons. The availability of such a model permits the evaluation of small defect sizes in fatigue and the time required for a small flaw to grow to a size detectable by non-destructive methods. It also helps in the design of new materials or in the alteration of the grain size of existing materials, to increase the fatigue limit. Finally, it assists in delineating the importance of the small crack regime with respect to the applied stress, through construction of a phase diagram, e.g. Fig. 1.12.

In the following we will describe only a few recent models. It is not our intention here to review various proposed models, but rather to concentrate on a few microstructurally-based models. An interested reader may consult [14–19] for other contributions.

9.5.1 SLIP BAND MODELS

The threshold stress for short cracks was defined by Tanaka *et al.* [45] in terms of slip band–grain boundary interactions. In their model, they investigated the required condition for coplanar slip bands emanating from a crack tip to be blocked by a grain boundary. Figure 9.11 shows a schematic of their model in which a slip band originating from a crack tip is blocked by a grain boundary. An analysis of this condition yields

$$\sigma_{th} = \frac{K_c^m}{\sqrt{\pi b}} + \frac{2}{\pi}\sigma_{fr}^* \operatorname{arc\,cos}\left(\frac{a}{b}\right) \tag{9.1}$$

and

$$K_{th} = \sigma_{th}\sqrt{\pi a}\,, \tag{9.2}$$

where $2a$ is the crack length and $b = a + \omega_0$, ω_0 being the width of the blocked slip band, σ_{fr}^* is the friction stress for dislocation motion, and K_c^m is termed the microscopic stress intensity factor at the tip of the slip band.

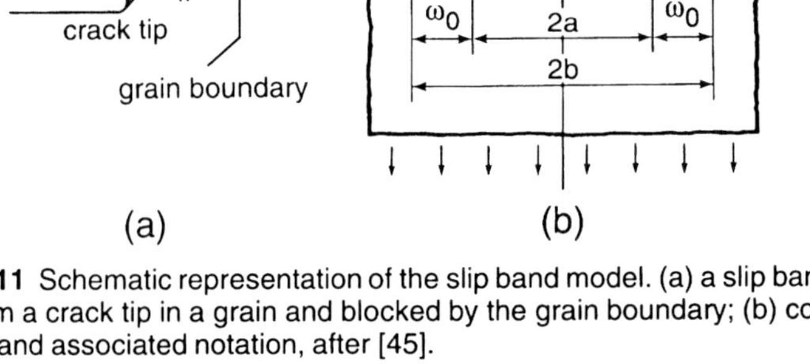

Fig. 9.11 Schematic representation of the slip band model. (a) a slip band originating from a crack tip in a grain and blocked by the grain boundary; (b) coplanar slip bands and associated notation, after [45].

The fatigue limit is obtained from (9.1) by letting $a \to 0$, i.e. very small flaws, which results in

$$\sigma_{FL} = \frac{K_c^m}{\sqrt{\pi \omega_0}} + \sigma_{fr}^* . \tag{9.3}$$

For long cracks, $a \gg \omega_0$, and from (9.1) one gets

$$(\sigma_{th})_{long} = \frac{K_c^m}{\sqrt{\pi a}} + \frac{2}{\pi}\left(\frac{2}{a}\right)^{1/2} \sigma_{fr}^* \sqrt{\omega_0} , \tag{9.4}$$

and substituting in (9.2) yields

$$(K_{th})_{long} = K_c^m + 2\left(\frac{2}{\pi}\right)^{1/2} (\omega_0)^{1/2} \sigma_{fr}^* . \tag{9.5}$$

Another special case is obtained by setting $K_c^m = 0$ and $\sigma_{fr}^* = \sigma_0$ in (9.1), which gives

$$\sigma_{th} = \frac{2}{\pi} \sigma_y \text{ arc cos}\left(\frac{a}{a+\omega_0}\right) . \tag{9.6}$$

Solving for ω_0 gives

$$\omega_0 = a\left[\sec\left(\frac{\pi \sigma_{th}}{2\sigma_o}\right) - 1\right] \tag{9.7}$$

which is the Dugdale plastic zone, (7.26).

The single slip pattern ahead of short cracks for stresses under the fatigue limit is similar to that of long cracks for stress intensities less than the threshold stress intensity range, ΔK_{th}. Usami and co-investigators [46, 47] have suggested a single criterion for both long and short cracks based on the similitude of the cyclic plastic zone at the crack tip. They postulated that the cyclic plastic zone size under the threshold stress range, $\Delta\sigma_{th}$, is a material property. This critical cyclic plastic zone size is obtained from (9.7) by replacing σ_{th} by $\Delta\sigma_{th}$ and σ_o by $\Delta\sigma_y' \approx 2\sigma_0$, i.e.

$$r_c^* = a\left(\sec\frac{\pi \Delta\sigma_{th}}{4\sigma_o} - 1\right) . \tag{9.8}$$

Equation (9.8) successfully simulates the Kitagawa–Takahashi plot in Fig. 9.3. However, the assertion that r_c^* is a material property requires further studies. A single slip band of the characteristic length of the microstructure is the minimum required cyclic plastic zone size for the crack to grow continuously. In other words, a crack tip planar slip band of such a length determines the ability of a crack to cross the microstructural barriers. However, the threshold of short crack growth is different from that of a long crack. Apart from their physical sizes, numerous grain boundaries along the long crack front block single slip bands in individual grains

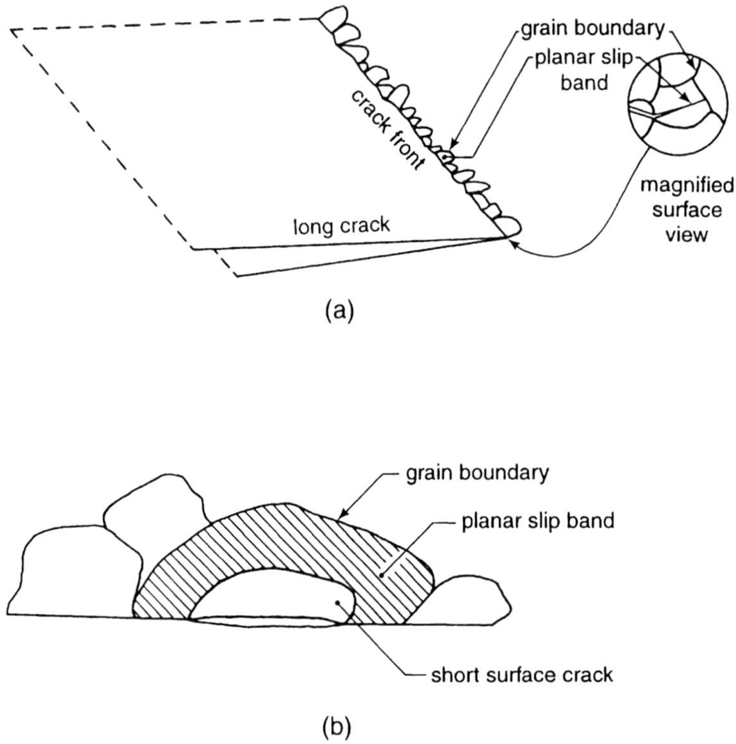

Fig. 9.12 Schematic illustration of planar slip bands. (a) in front of a long crack under the threshold stress intensity range, ΔK_{th}; (b) in front of a surface short crack front under the short crack threshold stress (fatigue limit), $\Delta\sigma_{FL}$ [29].

(Fig. 9.12(a)). Therefore, ΔK_{th} is an averaged resistance of the material to single slip related growth of long cracks. In contrast, only one or a few barriers lie in front of a short crack, which is frequently a penny-shaped crack or a corner crack (Fig. 9.12(b)). Therefore, the growth rate of a short crack is dependent on its approach to individual barriers, such as misorientation of the adjacent grains. Finally, the threshold stress for short cracks is much larger than the applied stress corresponding to ΔK_{th} for long cracks. The effect of cyclic plasticity in front of a crack is discussed in more detail in a recent review paper [29].

9.5.2 SURFACE STRAIN REDISTRIBUTION MODEL

This model has been described in section 8.4.2 in connection with notched specimens. It is based on the non-uniformity of strains in the surface layer and the development of crack closure [48]. Surface grains oriented for easy slip are subject to an inherent microstructurally-dependent strain concentration, Q_ι, which decays with depth into the material (Fig. 8.14(b) and (8.42)). For small cracks in smooth specimens, the threshold stress range $(\Delta\sigma_{nom})_{th}$ can be obtained from (8.47) by setting the notch geometry strain concentration factor, $K_\iota = 1$, i.e.

$$(\Delta\sigma_{nom})_{th} = \frac{\Delta K_{ith} H_{cl}}{F Q_\iota \sqrt{\pi a}} \tag{9.9}$$

where ΔK_{ith} is the intrinsic threshold stress intensity factor range (fully open crack, $R \geqslant 0.6$), H_{cl} is a crack closure development parameter described by (8.46), F is a shape factor with a value of approximately 0.72 for a semicircular surface crack in an infinite plate, and a is the crack depth.

The prediction of (9.9) is shown in Fig. 9.13 which shows the relationship between the nominal threshold stress range, $(\Delta\sigma_{nom})_{th}$, and crack depth for three stress ratios. It can be seen that the threshold stress range has a local maximum, representing the stress range required for continuous crack growth, that defines the fatigue limit. This was the point raised with respect to the slip band model described in section 9.5.1, that the definition of the threshold condition must be such that a crack cannot be stopped by the major microstructural barriers. This condition has been further discussed in [29].

9.5.3 SURFACE LAYER YIELD STRESS REDISTRIBUTION MODEL

In contrast to the previous model, which considers the strain redistribution at the surface layers, this model is concerned with the distribution of the yield stress through the surface layer [49]. There is ample experimental

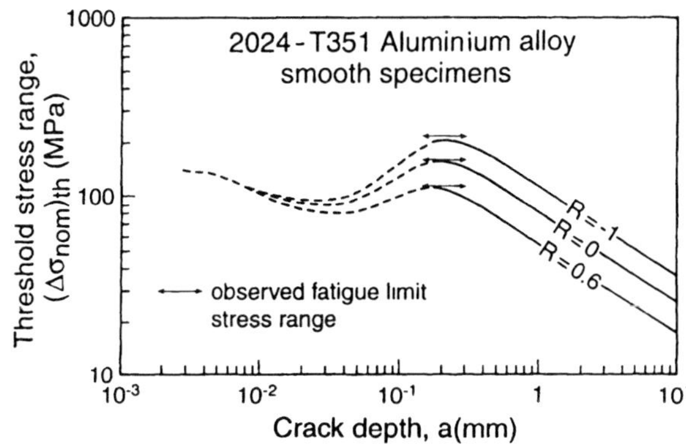

Fig. 9.13 Nominal threshold stress range as a function of crack depth and stress ratio for smooth specimens, after [48].

evidence which indicates that the surface layer is much softer than that of the bulk material, and the yield stress can be significantly lower in larger grains than that of the bulk material. A steep yield strength gradient extending a fraction of a millimetre has been found [50, 51]. In this model the material thickness is subdivided into a series of strips whose thickness is equal to the process zone. (The process zone has been defined in section 7.7.) Figure 9.14(a) is a schematic representation of the material modelling which indicates that the surface layer thickness is about three grain diameters, after which the yield strength approaches that of the bulk material, σ_{yb}. For an applied elastic bulk strain amplitude of $\Delta\varepsilon_b/2$, each material layer of thickness δ^* exhibits a different hysteresis loop, as depicted in Fig. 9.14(b).

The plastic strain energy density per cycle, ΔW^p, for the m-th layer is given by

$$\Delta W_m^p = \Delta\sigma_{ym}\Delta\varepsilon_m^p\,. \tag{9.10}$$

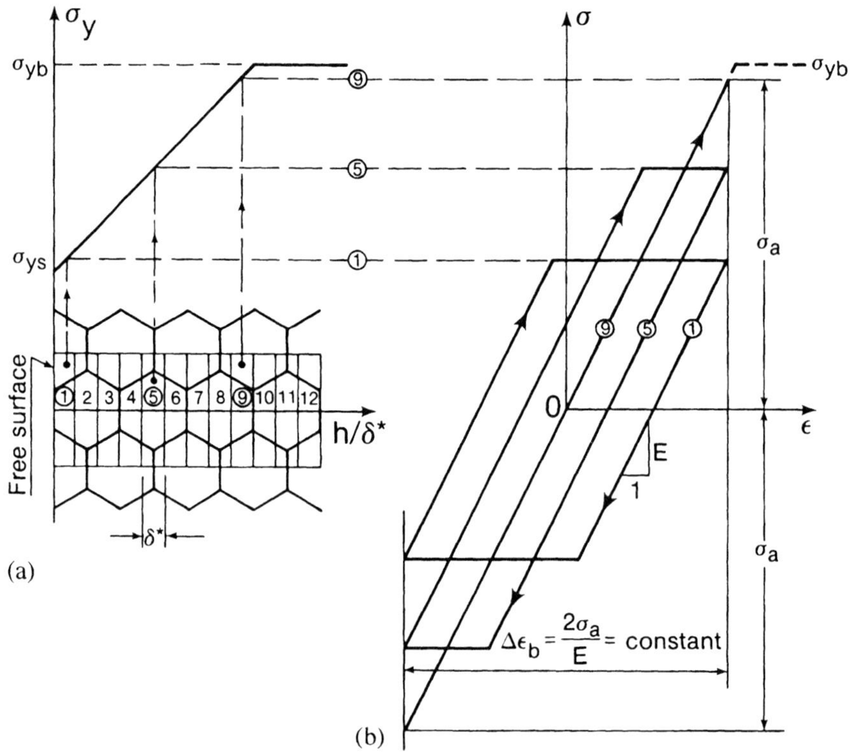

Fig. 9.14 Schematic representation of soft surface layer. (a) material modelling by assuming the process zone to be a third of a grain diameter; (b) three cyclic stress–strain loops near the free surface region. Reproduced with permission from [49].

A crack initiates at the surface layer due to microplastic deformation of the surface layer, e.g. Fig. 9.4. The propagation of the initiated small crack is based on the local stress–strain state within the process zone (see development in section 7.7). The stress/strain range ahead of a crack tip within the process zone can be determined from (7.76) by taking $n' = 0$ (elastic–perfectly plastic material) and $x = \delta^*$, i.e.

$$\Delta\sigma\Delta\varepsilon^{p} = \frac{\Delta K^2}{\pi E\,\delta^*}. \tag{9.11}$$

The use of ΔK is justified since the initiated crack in the surface layer is constrained by the bulk elastic matrix.

A failure criterion for the process zone is required for the crack to advance. As in section 7.7, we may use the Basquin and Coffin–Manson relationships (3.5) and (3.9)

$$\Delta W^{p} = \Delta\sigma\Delta\varepsilon^{p} = 4\sigma_f'\varepsilon_f'(2N_f)^{b+c} \tag{9.12}$$

where ε_f' and σ_f' are fatigue ductility and strength coefficients; exponents b and c are material properties, and $2N_f$ is the number of reversals to failure.

The plastic energy density at the m-th layer is the sum of (9.10) and (9.11). (The summation is valid because energy is a scalar quantity.) Thus

$$(\Delta\sigma\Delta\varepsilon^{p})_{mt} = \Delta\sigma_{ym}\Delta\varepsilon_m^{p} + \frac{\Delta K^2}{\pi E\delta^*} = \Delta\sigma_{ym}\left(\Delta\varepsilon_b - \frac{\Delta\sigma_{ym}}{E}\right) + \frac{\Delta K^2}{\pi E\delta^*}. \tag{9.13}$$

The number of cycles, N^*, required for the crack to advance through the process zone, δ^*, can be determined from (9.12) and (9.13), i.e.

$$N^* = \frac{1}{2}\left[\frac{(\Delta\sigma\Delta\varepsilon^{p})_{mt}}{4\sigma_f'\varepsilon_f'}\right]^{1/(b+c)}. \tag{9.14}$$

Therefore, the crack extension per cycle, da/dN, is given by

$$\frac{da}{dN} = \frac{\delta^*}{N^*} = 2\delta^*\left[\frac{\Delta\sigma_{ym}(\Delta\varepsilon_b - \Delta\sigma_{ym}/E) + (\Delta K^2/\pi E\delta^*)}{4\sigma_f'\varepsilon_f'}\right]^{1/(b+c)}. \tag{9.15}$$

Note that the first term in the numerator inside the square bracket tends to zero beyond the surface layer as $\Delta\sigma_{ym}/E \to \Delta\varepsilon_b$. Likewise the second term tends to zero as one approaches the free surface, $\Delta K \to 0$ as $a \to 0$.

The crack growth rate given by (9.15) depends on the soft surface layer properties, the crack tip stress/plastic strain range, and the bulk mechanical and fatigue properties. The prediction of (9.15) for a linearly varying yield strength from a minimum of $0.4\,\sigma_{yb}$ to $\sigma_{yb} = 300$ over a distance of $H = 9\delta^*$ with $\delta^* = 3.33 \times 10^{-5}$ m, $E = 2 \times 10^5$ MPa, $\sigma_f' = 850$ MPa, $\varepsilon_f' = 1.0$, $b = -0.12$ and $c = -0.6$, is shown in Fig. 9.15. The above material properties are representative of low and intermediate strength steels. In Figure 9.15, the abscissa is non-dimensionalized with respect to the process zone

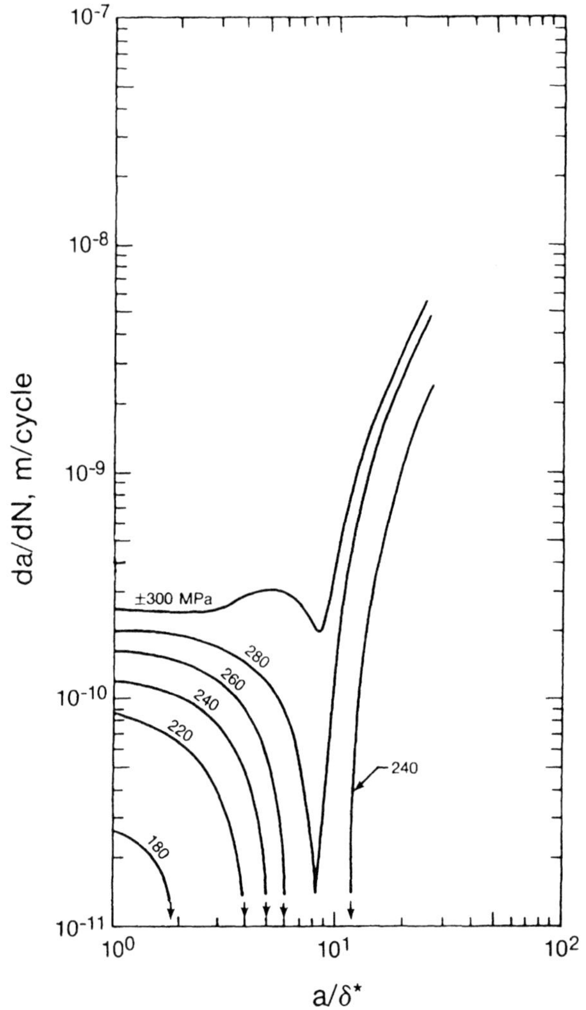

Fig. 9.15 Predicted crack growth rate versus non-dimensional crack length, a/δ^*, for various applied stresses up to the yield stress. Both short and long cracks are included in the figure.

length, a/δ^*, and the ordinate is the growth rate, da/dN. It can be seen that the crack growth depends on the applied bulk stress and crack length. For example, a short crack $a/\delta^* < 8$ will self-arrest at an applied bulk stress of less than 280 MPa, while a crack of $a/\delta^* > 10$ will grow at $\sigma_b \geqslant 240$ MPa. Cracks of any size will grow at applied stress values slightly below the yield strength of the bulk material, $\sigma_b \approx 300$ MPa.

This model not only predicts the threshold condition for small cracks but it also predicts the crack growth rate for applied stresses above and below the threshold values. In this regard it is more general than the previous two models discussed above. However, this model does not take explicitly into account crack closure, since as shown by Lam, Kujawaski and Ellyin [52], the plasticity induced closure for small cracks is negligible. It was discussed earlier that the closure development observed in small crack growth is mainly due to roughness and the tortuous path of the crack, and this cannot easily be modelled in the case of a mathematically straight crack.

9.5.4 EFFECTIVE STRESS INTENSITY FACTOR RANGE MODELS

A number of investigators have attempted to correlate the crack growth rate of a small crack with an effective stress intensity factor range, $\Delta K_{\text{eff}} = \Delta K - \Delta K_{\text{op}}$, similar to the effective stress intensity used for long cracks. The basic idea is that long and short crack growth behaviour could be unified if they were expressed in terms of ΔK_{eff}. For example, Tanaka and Nakai [53] measured the crack closure of long and small cracks emanating from a sharp notch in mild steel specimens. They reported that the difference between long and small crack growth rate diminished when the data were plotted against ΔK_{eff}. It should be noted however that crack closure meaurement by the compliance method is not a straightforward method as discussed in section 9.4.3.

Newman and co-investigators [40] have developed a crack closure model which is a modification of the Dugdale model. The model essentially predicts that a steady-state crack closure is reached once the crack length exceeds 40 μm, a value calibrated from experimental data. Obviously, this model is a highly idealized treatment of the closure development in view of the measurements reported in Fig. 9.10. The model predictions for small cracks at stress levels near the threshold regime did not agree well with the test data for two high strength aluminium alloys reported in [40].

Figure 9.16 shows the growth rate data reported in Fig. 9.6 re-plotted against ΔK_{eff}. The corner crack stress intensity factor ΔK was calculated based on the Newman *et al.* [40] analysis (see [21, 39] for details). Of course, the determination of ΔK_{eff} is dependent on the values obtained for the crack opening stress. In this figure the data were taken from the strain gauge measurement described in section 9.4.3 and [44]. The growth rate of long cracks for the same material is included in the figure, taken from [54]. It can be seen that ΔK_{eff} does not unify the long and short crack growth data. Essentially, the inclusion of ΔK_{eff} results in a decrease in the crack growth rate. While this may be true for long cracks, ΔK_{eff} cannot describe the growth rate acceleration of small cracks once they cross the microstructural barriers. Neither could ΔK_{eff} describe the early growth deceleration of

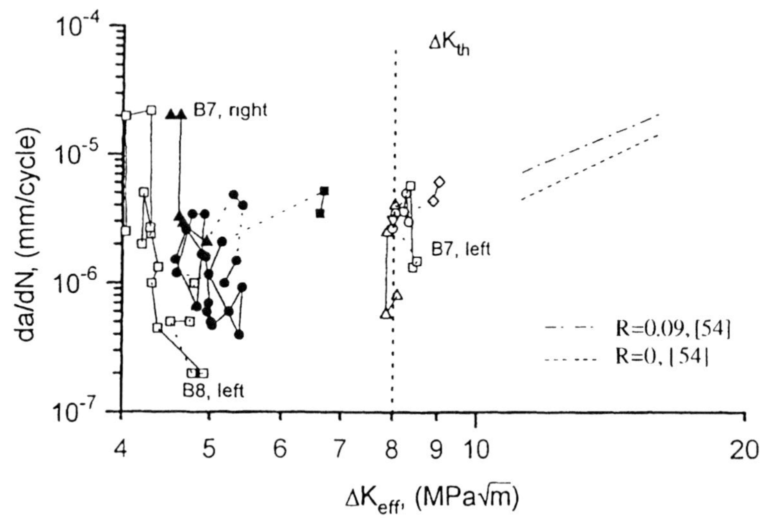

Fig. 9.16 Crack growth rate of a corner crack plotted against ΔK_{eff} for both small and long cracks. Reprinted with permission from [39].

short cracks due to surface layer properties. In effect, the observed crack closure in small cracks is due to crack tip shielding by microstructural barriers and the roughness of the crack path, rather than plasticity induced closure. Therefore, it is not surprising that models based solely on plasticity induced crack closure do not correlate well with experimental data.

REFERENCES

1. Pearson, S. (1975) Initiation of fatigue cracks in commercial aluminum alloys and subsequent propagation of very short cracks. *Engng. Fract. Mech.*, **7**, 235–47.
2. Kitagawa, H. and Takahashi, S. (1976) *Applicability of Fracture Mechanics to Very Small Cracks or Cracks in the Early Stage.* Proc. 2nd Int. Conf. on Mechanical Behaviour of Materials (ICM-2), American Soc. for Metals, Metals Park, Ohio, pp. 627–31.
3. El Haddad, M.H., Topper, T.H. and Smith, K.N. (1979) Prediction of non-propagating cracks. *Engng. Fract. Mech.*, **11**, 573–84.
4. Morris, W.L., James, M.R. and Buck, O. (1981) Growth rate models for short surface cracks in A12219-T851. *Metall. Trans.*, **12A**, 57–64.
5. Hudak, S.J., Jr. (1981) Small crack behavior and the prediction of fatigue life. *J. Engng. Mater. Technol., Trans. ASME*, **103**, 26–35.
6. Nisitani, H. and Takao, K.-I. (1981) Significance of initiation, propagation and closure of microcracks in high cycle fatigue of ductile metals. *Engng. Fract. Mech.*, **15**, 445–56.
7. Taylor, D. and Knott, J.F. (1981) Fatigue crack propagation behavior of short cracks – The effect of microstructure. *Fatigue Engng. Mater. Struct.*, **4**, 147–55.

8. Lankford, J. (1982) The growth of small fatigue cracks in 7075 – T6 aluminum. *Fatigue Engng. Mater. Struct.*, **5**, 233–48.
9. Tanaka, K., Hojo, M. and Nakai, Y. (1983) Crack initiation and early propagation in 3% silicon iron, in *Fatigue Mechanisms: Advances in Qualitative Measurement of Fatigue Damage*, ASTM STP 811, American Society for Testing and Materials, Philadelphia, PA, pp. 207–32.
10. Suresh, S. and Ritchie, R.O. (1984) Propagation of short fatigue cracks. *Int. Metals Rev.*, **24**, 445–76.
11. Miller, K.J. (1984) Initiation and growth rates of short fatigue cracks, in *Fundamentals of Deformation and Fracture* (Proc. IUTAM Symposium Eshelby Memorial Conf.), Cambridge University Press, Cambridge, pp. 477–500.
12. Leis, N.B., Hopper, A.T. and Ahmad, J. (1986) Critical review of the fatigue growth of short cracks. *Engng. Fract. Mech.*, **23**, 883–98.
13. McEvily, A.J. (1984) On the growth of small/short fatigue cracks. *JSME Int. J.*, Series I, **32**, 181–91.
14. Miller, K.J. and de los Rios, E.R. (eds) (1986) *The Behaviour of Short Fatigue Cracks*, EGF 1, Mechanical Engineering Publications, London, 560 pp.
15. Ritchie, R.O. and Lankford, J. (eds) (1986) *Small Fatigue Cracks*. Metallurgical Society, Inc., Warrendale, PA.
16. Miller, K.J. and de los Rios, E.R. (eds) (1992) *Short Fatigue Cracks*, ESIS 13, Mechanical Engineering Publications, Instn. Mech. Engrs., London.
17. Miller, K.J. (1987) The behaviour of short fatigue cracks and their initiation, Part I – A review of two recent books, and Part II – A general summary. *Fatigue Fract. Engng. Mater. Struct.*, **10**, 75–113.
18. Kitagawa, H. and Tanaka, T. (eds) (1990) *Fatigue 90*, Proc. 4th Int. Conf. on Fatigue and Fatigue Thresholds, Honolulu, Hawaii, July 1990, Materials and Components Engineering Publications, Birmingham, UK.
19. Baïlon, J.P. and Dickson, J.I. (eds) (1993) *Fatigue 93*, Proc. 5th Int. Conf. on Fatigue and Fatigue Thresholds, Montreal, Canada, May 1993, Engineering Materials Advisory Services Ltd. (EMAS), Warley, West Midlands, UK.
20. Forsyth, P.J.E. (1963) Fatigue damage and crack growth in aluminum alloys. *Acta Metall.*, **11**, 703–15.
21. Craig, D.F. (1994) Growth and behaviour of short cracks in a ferritic/pearlitic steel. M.Sc. Thesis, University of Alberta, Edmonton, Canada.
22. Koss, D.A. and Chan, K.S. (1980) Fracture along planar slip band. *Acta Metall.*, **28**, 1245–52.
23. Ma, B.-T. and Laird, C. (1989) Overview of fatigue behaviour in copper single crystals: II, Population, size distribution and growth kinetics of stage I cracks for test at constant strain amplitude. *Acta Metall.*, **37**, 337–48.
24. Li, Chingshen and Bretheau, T. (1989) The role of grain boundary compatibility in fatigue cracking of aluminum bicrystals. *Acta Metall.*, **37**, 337–48.
25. Laird, C. (1979) Mechanisms and theories of fatigue, in *Fatigue and Microstructure* (ed. M. Meshii), American Society of Metals, Metals Park, Ohio, pp. 149–204.
26. Neumann, P. (1969) Coarse slip model of fatigue. *Acta Metall.*, **17**, 1219–25.
27. Morris, W.L., James, M.R. and Zurek, A.K. (1985) The extent of crack tip plasticity for short fatigue cracks. *Scripta Metall.*, **19**, 149–53.
28. Li, Chingshen (1993) Crystallographic fatigue crack growth in incompatible aluminum bicrystals – Its dependence on secondary slip. *Metall. Trans.*, **23A**, 3293–301.

29. Ellyin, F. and Li, Chingshen (1995) The role of cyclic plasticity in crystallographic crack growth retardation. *Mater. Sci. Res. Int.*, **1**, 137–43.
30. Beevers, J.C. (ed.) (1982) *Advances in Crack Length Measurement*, Engineering Materials Advisory Service Ltd. (EMAS), West Midlands, UK.
31. Hussey, I.W., Roche, J. and Grabowski, L. (1991) A fully automated system for monitoring short crack growth. *Fatigue Fract. Engng. Mater. Struct.*, **14**, 309–15.
32. Davidson, D.L. (1992) The experimental mechanics of microcracks, in *Small Crack Test Methods*, ASTM STP 1149 (eds J.M. Larsen and J.E. Allison), American Society for Testing and Materials, Philadelphia, PA, pp. 81–91.
33. Swan, M.H. (1992) Monitoring small-crack growth by the replication methods, in *Small Crack Test Methods*, ASTM STP 1149 (eds J.M. Larson and J.E. Allison), American Society for Testing and Materials, Philadelphia, PA, pp. 34–56.
34. Silk, M.G. (1992) Ultrasonics, in *Fatigue Crack Measurement: Techniques and Applications* (eds K.J. Marsh, R.A. Smith and R.O. Ritchie), Engineering Materials Advisory Service Ltd. (EMAS), West Midlands, UK, pp. 34–56.
35. Jenkins, P.J. and Briggs, G.A.D. (1992) Acoustics microscopy, in *Fatigue Crack Measurement: Techniques and Applications* (eds K.J. Marsh, R.A. Smith and R.O. Ritchie), Engineering Materials Advisory Service Ltd. (EMAS), West Midlands, UK, pp. 147–72.
36. Newman, J.C., Jr. and Edwards, P.R. (1988) Short-Crack Growth Behaviour in an Aluminum Alloy – An AGARD Cooperative Test Programme, AGARD-R-732, Dec. 1988.
37. Kujawski, D. and Ellyin, F. (1992) Propagation of small corner cracks initiated in cyclic compression. *Engng. Fract. Mech.*, **43**, 657–62.
38. Kujawski, D. and Ellyin, F. (1993) Fatigue growth of physically small inclined cracks. *Fatigue Fract. Engng. Mater. Struct.*, **16**, 743–52.
39. Craig, D., Ellyin, F. and Kujawski, D. (1995) The behaviour of small corner cracks in a ferritic/pearlitic steel: Experiments and analysis. *Fatigue Fract. Engng. Mater. Struct.*, **18**, 861–73.
40. Newman, J.C., Jr., Wu, X.R., Venneri, S.L. and Li, C.G. (1994) Small-Crack Effects in High-Strength Aluminum Alloys, A NASA/CAE Cooperative Program, NASA Ref. Publ. 1309, National Aeronautics and Space Administration, Hampton, Virginia, May 1994, 118 pp.
41. Ogura, K., Miyoshi, Y. and Nishikawa, I. (1987) Fatigue crack growth and closure of small cracks at the notch root, in *Current Research on Fatigue Cracks* (eds T. Tanaka, M. Jono and K. Komai), Current Japanese Materials Research – Vol. 1, Elsevier, London, pp. 67–91.
42. Chen, D.L., Weiss, B. and Stickler, R. (1991) *Application of a New Evaluation Method for Near-Threshold Closure Effects of Small Surface Fatigue Cracks.* Proc. 10th Congress on Materials Testing, Gaepipari Tudomaanyos Egyeseulet, Budapest, Hungary, Vol. 2 (ed. E. Czoboly), pp. 466–71.
43. DuQuesnay, D.L. (1991) Fatigue damage accumulation in metals subjected to high mean stress and overload cycles. Ph.D. Thesis, University of Waterloo, Waterloo, Ontario.
44. Craig, D., Kujawski, D. and Ellyin, F. (1995) An experimental technique applied to study the behaviour of small corner cracks. *Int. J. Fatigue*, **17**, Elsevier Science Ltd, Oxford, 253–9.
45. Tanaka, K., Nakai, Y. and Yamashita, M. (1981) Fatigue growth threshold of small cracks. *Int. J. Fract.*, **17**, Kluwer Academic Publishers, Dordrecht, 519–33.

46. Ohuchida, H., Usami, S. and Nishioka, A. (1975) Fatigue limit of steel with cracks. *Bulletin JSME*, **18**, 1185–93.
47. Usami, S. (1982) Application of threshold cyclic-plastic-zone-size criterion to some fatigue limit problems, in *Fatigue Threshold* (eds J. Backlund, A.F. Blom and C.J. Beevers), Engineering Materials Advisory Services Ltd. (EMAS), Warley, UK, pp. 205–32.
48. Abdel-Raouf, H., DuQuesnay, D.L., Topper, T.H. and Plumtree, A. (1992) Notch-size effects in fatigue based on surface strain redistribution and crack closure. *Int. J. Fatigue*, **14**, Elsevier Science Ltd, Oxford, 157–62.
49. Kujawski, D. and Ellyin, F. (1992) A microstructurally motivated model for short crack growth rate, in *Short Fatigue Cracks*, ESIS 13 (eds K.J. Miller and E.R. de los Rios), Mechanical Engineering Publications Ltd., London, pp. 391–405.
50. Morris, W.L., Cox, B.N. and James, M.R. (1987) Microplastic surface deformation of Al 2219-T851. *Acta Metall.*, **35**, 1055–65.
51. Troshchenko, V.T., Prokopenko, A.V. and Targov, V.N. (1988) Surface cracks and the fatigue limit of metals. *Fatigue Fract. Engng. Mater. Struct.*, **11**, 123–38.
52. Lam, Y.C., Kujawski, D. and Ellyin, F. (1991) The development of crack closure with crack extension. *Scripta Metallurgica et Materialia*, **25**, 2313–18.
53. Tanaka, K. and Nakai, Y. (1983) Propagation and non-propagation of short fatigue cracks at a sharp notch. *Fatigue Engng. Mater. Struct.*, **6**, 315–27.
54. Ellyin, F. and Li, H.-P. (1984) Fatigue crack growth in large specimens with various stress ratios. *J. Pressure Vessel Technol., Trans. ASME*, **106**, 255–60.

Probabilistic fatigue crack growth

10

10.1 INTRODUCTION

In the previous chapters we have treated the crack initiation and propagation process in a deterministic manner. However, there is ample evidence which indicates a variability in strength of components/samples of the same shape manufactured in an identical fashion. This was briefly discussed in section 1.6 (see also Fig. 1.17). Thus, there is a need for statistical studies of the crack growth process to determine its central tendency and dispersion around it. The general idea for this viewpoint dates back to the well-known Griffith's theory [1]. He proposed that the difference between the calculated strengths of materials (based on a microscopic analysis) and those actually observed, resides in the fact that there are randomly sized and oriented flaws which weaken a material's bonds. This assertion then leads to the assumption that there will be a distribution of strengths in a given material in the sense that a different amount of stress will be needed to fracture it (or break the bonds) at one or another point. It should be emphasized that any statistical analysis regarding the strength of a flawed specimen has to start from deformation study near the flaw (crack) in the specimen, i.e. the functional relationship must be derived from the local (near the flaw) properties.

The existing crack propagation laws can, in principle, be integrated to give the expected life of a cyclically loaded structure if the initial crack length as well as the amplitude of the fatigue loading are known. In practice, these two input variables are usually known in probabilistic terms. In the ideal situation where the inputs are deterministic, there is often a need to quantify the probability of failure as a function of time in order to assess the risk involved, for example, in not carrying out a maintenance or repair programme [2]. In such a situation, the probability distribution of the failure time is as important as the deterministically calculated time to failure itself.

The recognition of this need has led the analysts to use some *a posteriori* probabilistic distributions on the deterministically calculated growth path.

Such a method requires extensive fracture data for the estimation of the distribution parameters. A probabilistic crack growth model would reduce the data dependency in two ways. Firstly, it would be possible to choose the family of distribution functions relevant to the situation at hand, and secondly it may be possible to obtain a functional correspondence between the probability distribution parameters and growth rate parameters. This type of analysis will be discussed later.

10.2 BACKGROUND

Failure takes place when the applied load exceeds a component's resistance. Resistance of a component diminishes with time due either to repeated loads which cause defect growth or a wear process which removes material. Thus, to determine the probability of failure at a specified service time, one has to be able to predict the residual strength of the component at that instant. The difficulty with predicting the strength decay with time has hampered progress towards reliability analysis of industrial components. In most cases, parameters entering into a deterministic analysis are assigned probability density functions. These density functions are either obtained through extensive test programs of similar components, e.g. [3], or a classical distribution is arbitrarily assumed.

The probabilistic fatigue failure studies have followed the same trend as the deterministic investigations. The 'fatigue failure' is defined in the sense described in Chapter 3, which is interpreted as the initiation and growth of a crack to a certain size, e.g. the order of a millimetre in ductile materials, or the fracture of brittle material. Earlier works were mostly concerned with the near fatigue threshold investigations where most of the life was spent in the initiation of cracks. Pioneering contributions in this area were made by Weibull [4], Freudenthal [5] and Gumbel [6].

Numerous studies have been reported whereby the scatter in fatigue data has been fitted to exponential distribution, the normal (Gaussian) distribution, the log–normal distribution, the Weibull distribution, and the Gumbel (extreme value) distributions. It should be noted that to describe the above distributions, one must identify one parameter for the exponential, two parameters for the normal, log–normal and Gumbel, and three parameters for the Weibull. A description of the above distributions can be found in most textbooks on statistics.

A list of contributions in this area may be found in Heller [7], Haugen [8], Bertrand, Lefebvre and Ellyin [9], Provan [10] and Yao *et al.* [11], among others.

It will be shown later through a study of experimental data, that once a crack grows beyond the micro-scale or a detectable macro-crack size ($a > 1$ mm), the fundamental problem then becomes the prediction of the crack growth rate and the dispersion about its central tendency. This will be illustrated in the following.

At any given service life, N, there will be a variation in crack lengths (or flaw sizes) in a population of similar components or samples as shown in Fig. 10.1(a). This is because industrial materials have defects of various shapes and sizes, and they grow at different rates under an applied cyclic load. These variations could be represented in a statistical form, i.e. as probability density functions, provided sufficient data is available. Figure 10.1(a) schematically shows probability density functions (PDF) of crack lengths at different service lives. Conversely, a specific crack length (or flaw size) may be obtained at different lives of similar components as shown in Fig. 10.1(b).

Test results of Virkler *et al.* [12] on thin-walled centre crack panels of 2024-T3 aluminium alloy, indicated that the distribution of the number of cycles, N, to reach a certain crack length (Fig. 10.1(b)) was best described by a 'three-parameter' log–normal distribution. This is in general agreement with the results of Impellizeri [3], however, the form of the log–normal distribution in the latter was more restrictive, i.e. the variance of the distribution was independent of the crack length. Referring to Fig. 1.17 of Chapter 1, the above observation leads to the corollary that for all points on the crack propagation curve, there is a constant ratio with respect to the median life, $\tilde{N}_a$ as shown in the figure, i.e.

$$N_{a,z} = Z \cdot \tilde{N}_a \tag{10.1}$$

where the probability density function of Z is log–normal with the mean $\tilde{m}_z = 1$ and its variance is independent of the crack length [2].

Attempts have been made to describe the whole crack propagation curve by incorporating the observed scatter discussed above, i.e. to generalize the discrete observation to various crack levels. Initially this was done through the extension of the Paris–Erdogan relation (7.70)

$$\frac{\mathrm{d}a}{\mathrm{d}N} = C(\Delta K)^m \tag{10.2}$$

where the coefficient C, e.g. [13]; m, e.g. [14] or both C and m, e.g. [15] were considered to be random variables.

More sophisticated crack growth models were subsequently proposed which could be represented in the form of

$$\frac{\mathrm{d}a}{\mathrm{d}N} = f(\Delta K)\zeta \tag{10.3}$$

where ζ represents a stochastic function of time, position or both. Various forms of ζ are discussed by Provan in Chapter 1 of [10]. Other crack growth models can be found in Bogdanoff and Kozin [16], Ellyin and Fakinlede

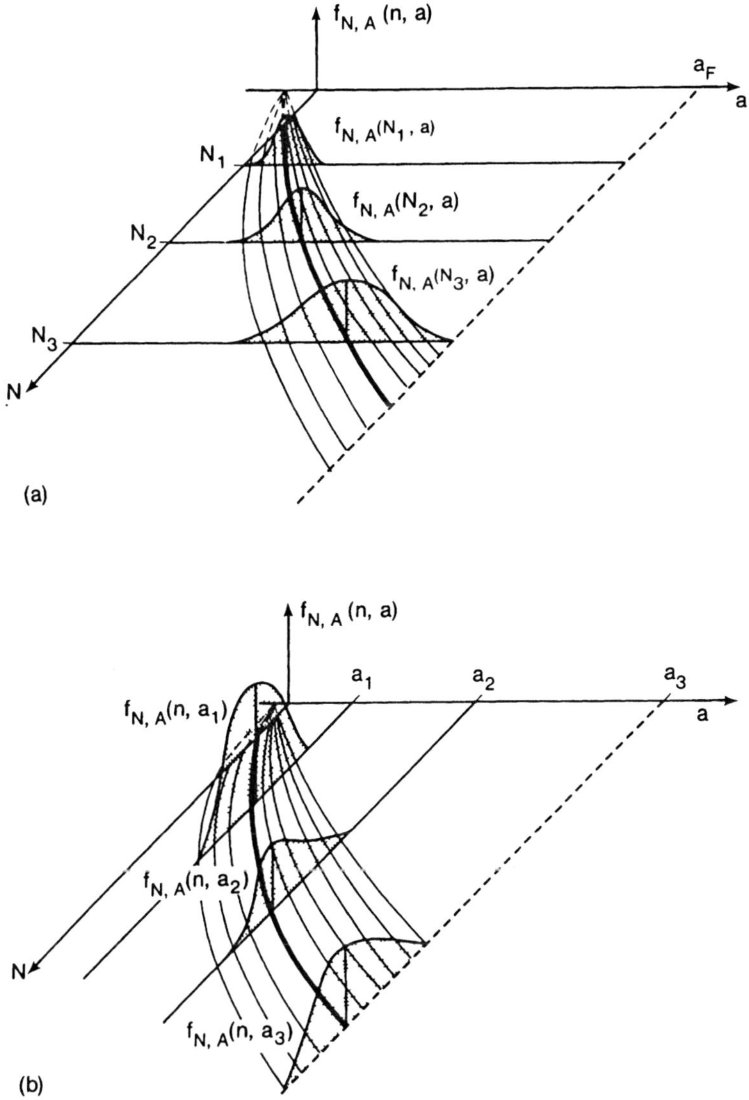

Fig. 10.1 Schematic variation of crack size with applied number of cycles. (a) statistical distribution of crack lengths at a specified number of cycles; (b) statistical distribution of number of cycles to arrive at a specific crack length. Reprinted with kind permission from [2].

[17], Winkler, Michel and Skurt [18] and Itagaki [19] where references to similar investigations are listed. In particular, [18] contains a concise survey of the probabilistic fracture mechanics and stochastic crack growth with an extensive list of survey articles and reference list.

10.3 EXPERIMENTAL OBSERVATIONS

In the following we will present the results of an experimental investigation by Bertrand *et al.* [9]. The specimens used in this study were thin-walled tubes with a profile similar to that of C in Fig. 5.2 of Chapter 5, with thickness $t = 0.76$ mm and inside diameter $d_i = 27.18$ mm. They were machined from an AISI 4340 solid bar, especially cast for this investigation to obtain, to the greatest extent, isotropic properties. The specimens were tested in a fully-reversed strain-controlled condition by subjecting them to a fluctuating axial load and a varying differential pressure across the thickness. The latter was obtained through a fluctuating internal pressure and a constant external pressure in a test facility similar to that described in Chapter 5. The imposed axial and tangential strain amplitudes were: $\Delta\varepsilon_a/2 = \pm 0.16\%$ and $\Delta\varepsilon_t/2 = \pm 0.08\%$ with a strain ratio of $\rho = \Delta\varepsilon_a/\Delta\varepsilon_t = -2$. The corresponding stresses were such that they were below the material's yield stress of 852 MPa.

Six series of biaxial tests were conducted at various life cycles up to fatigue failure. The first series were monotonic tests to determine the fracture strength, and a reference datum. The sixth series was the standard fatigue test to obtain the number of cycles to failure, while the other four series yielded the residual strength distribution at a prescribed life cycle ($T_2 = 20 \times 10^3$, $T_3 = 40 \times 10^3$, $T_4 = 60 \times 10^3$ and $T_5 = 80 \times 10^3$ cycles). The period T_2 corresponds approximately to the crack initiation time, while T_5 was approximately the time taken for the initiated crack growth to become unstable. After the application of the above load cycles, the specimens were pulled monotonically to determine the remaining residual strengths. From these results one could determine the evolution of damage in terms of the number of applied cycles. The numerical results are listed in Table 1 of [9]. The results of these series of tests show a fairly wide variation, thus necessitating a statistical study based on the variability of material strength.

The statistical theories of fatigue which attempt to describe the dispersion in the result are generally based on two hypotheses: (i) the fatigue resistance of any part of a component/specimen is a random variable, and (ii) it is always possible to separate a component/specimen into a large number of sub-elements (statistically independent) where the failure of one causes the failure of the ensemble (weakest link). (This is more appropriate for the failure of brittle materials.) Weibull proposed a statistical theory of fatigue based on the above hypotheses [4]. Weibull's cumulative probability function, $F(x)$, can be written as

$$F(x) = 1 - \exp\left[-\left(\frac{x-\gamma}{s}\right)^{\beta}\right] \tag{10.4}$$

and the corresponding probability density function (PDF) is

$$f(x) = \left[\frac{\beta}{s}\left(\frac{x-\gamma}{s}\right)^{\beta-1}\right]\exp\left[-\left(\frac{x-\gamma}{s}\right)^{\beta}\right] \tag{10.5}$$

where x represents the number of cycles to failure (N_f) or the residual strength in tension (σ_r), and γ, s, β are respectively the location, scale and form parameter of the Weibull distribution. The location parameter γ is the time or number of cycles prior to which the damage is nil. The scale parameter s represents the number of cycles for which the probability of survival is equal to 1/e or 37%. Finally, the parameter β defines the form of the Weibull distribution. Special cases of the Weibull distribution reduce to the well-known distributions when one sets $\gamma = 0$ and $s = 1$, and varies the shape parameter β. For example, the exponential distribution is obtained when $\beta = 1$, the Rayleigh distribution when $\beta = 2$, and the normal distribution when $\beta \simeq 3.25$.

For the test results mentioned above, the three parameters of the Weibull distribution were determined through a nonlinear optimization procedure described in [9]. The probability density function $f(N_f)$ and cumulative probability function $F(N_f)$ for the fatigue tests (series 6) are shown in Fig. 10.2. The abscissa indicates the number of cycles to failure, N_f, while the ordinate is $f(N_f)$ on the LHS and $F(N_f)$ on the RHS. It can be seen that $f(N_f)$ is zero up to 22 890 cycles, and has a rounded peak around a mode of

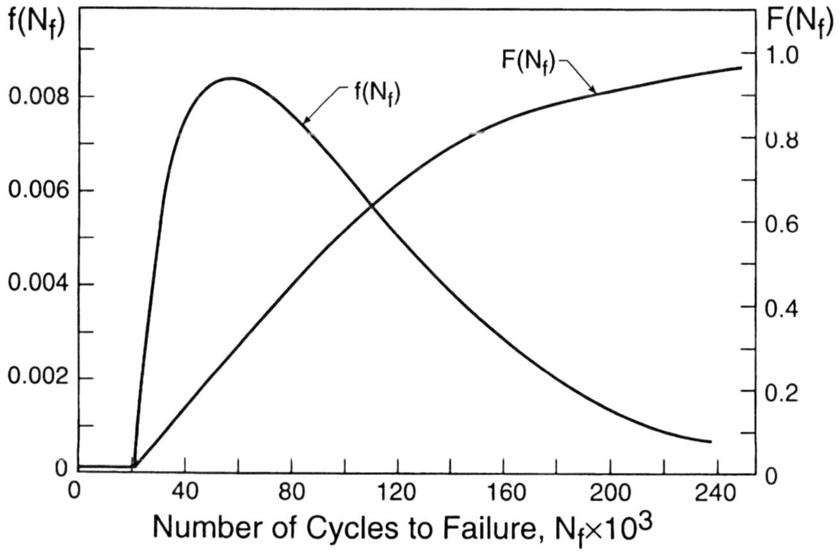

Fig. 10.2 The three-parameter Weibull probability density function $f(x)$ and cumulative distribution function, $F(x)$, fitted to the fatigue test results of series 6. Reprinted with permission from [9].

53 390 cycles. The $F(N_f)$ curve indicates a zero probability of failure up to 22 890 cycles, a probability of 0.1 at 40 000 cycles, of 0.3 at 60 000 cycles, and of 0.45 at 80 000 cycles. For this series of tests the three Weibull parameters determined from the nonlinear optimization procedure were: $\gamma = 22.8933$ MPa, $s = 84.4647$ MPa and $\beta = 1.3421$. It should be noted that the form of $f(N_f)$ in Fig. 10.2 is very close to a log–normal distribution.

The evolution of damage can be observed from Fig. 10.3, where the residual strength σ_r is plotted against the number of applied cycles, N. This figure indicates that the residual strength is unaffected by the application of up to 20 000 cycles, and it gradually decreases thereafter up to 60 000 cycles. There is an abrupt decrease in the residual strength beyond 60 000 cycles. Three stages of damage are observed: no discernible damage up to 20 000 cycles; macro-crack initiation up to 60 000 cycles; and fast crack growth thereafter. The above discussion was related to the central tendency (dotted line) in Fig. 10.3. As pointed out earlier and portrayed in Fig.

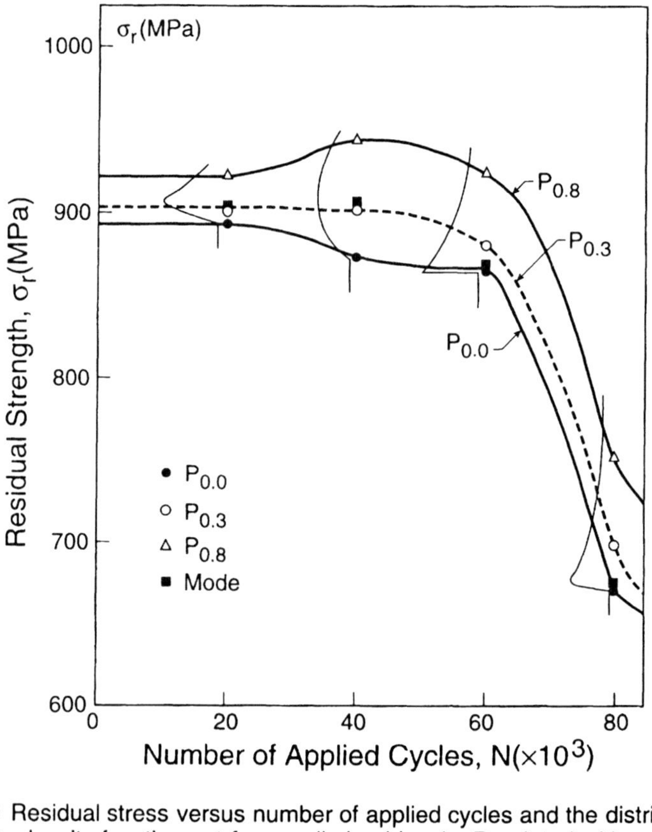

Fig. 10.3 Residual stress versus number of applied cycles and the distribution of probability density functions at four cyclic load levels. Reprinted with permission from [9].

10.1(a), for any given cycle, there will be a distribution of the residual strength. The Weibull three-parameter distribution was determined for each series of tests [9]. The probability density function for each series is shown in Fig. 10.3. The lowest bound of the residual strength for each series is identified as P_{00}; that which corresponds to a 30% probability is shown by a circle, the mode of the distribution by a filled square, and finally that which corresponds to 80% probability is depicted by a triangle in Fig. 10.3. It can be seen that for the onset of macro-crack growth, $N = 60\,000$ cycles, the probability density function is very close to the exponential function, $\beta \simeq 1.0$ [9].

10.4 A PROBABILISTIC CRACK GROWTH MODEL

In this section a probabilistic fatigue crack growth model will be presented. The model is based on the physical observations of micro- and macro-mechanical responses of industrial materials. To quantify a material's resistance to crack growth, two major inputs are required. The first is a suitable damage function which quantifies the damage accumulation rate and is directly related to the applied load. The second is a failure criterion which specifies the critical value of the damage function which causes the material failure (separation of bonds), and thus results in a crack growth.

In most investigations a linear damage rule is used which results in a reduced computational effort. However, such a damage accumulation rule has a number of disadvantages, e.g. it cannot predict the effect of loading sequence. In the case of a cracked body there is a steep stress and strain gradient along the crack plane. Thus, a point in the crack plane experiences progressively higher magnitudes of stresses and strains as the crack tip approaches it, i.e. the point is subjected to a low to high sequence of loading. In the following a probabilistic simulation by a nonlinear damage accumulation rule described in section 3.5 will be used.

10.4.1 BASIC ASSUMPTIONS

The development of the probabilistic crack growth model to be outlined here is based on the following assumptions [20]:

1. There exists a damage zone (DZ) ahead of a crack tip, beyond which the accumulation of damage is insignificant. This damage zone is taken to coincide with the reversed plastic zone (Fig. 10.4). Three zones are identified in this figure: a monotonic plastic zone which corresponds to the maximum initial load, a reversed plastic zone which results from cyclic loading, and a process zone where an intense damage occurs with its size being of the order of the grain size of the material. Generally, $\delta^* \ll R_p \simeq R_y/4$ for an elasto-plastic material with a low strain hardening exponent.

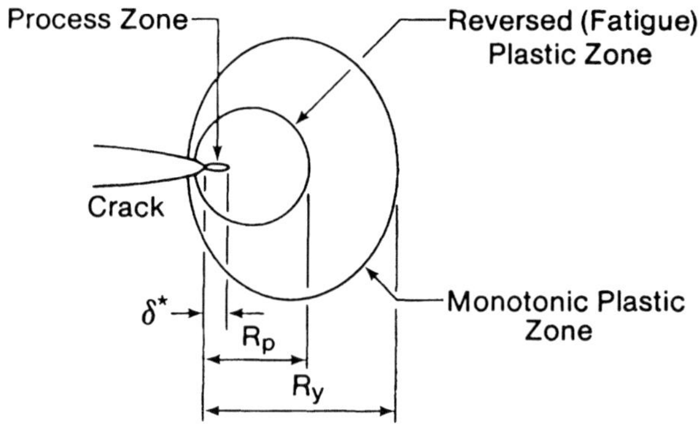

Fig. 10.4 Three zones ahead of a cyclically loaded crack: process zone, reversed plastic zone and monotonic plastic zone. The process zone defines the size where a continuum theory would apply.

2. The damage zone is divided into **finite** elements of the order of the process zone with fatigue damage resistances (FDR) that are random observations from the same probability distribution (Fig. 10.5). The fatigue damage resistance depicts major deformation barriers of microstructural elements to crack advance, e.g. grain boundaries, inclusions, banded microstructural features, etc.
3. The material elements mentioned above are so dimensioned that macroscopic (continuum) laws apply. Generally, a strain cycling tends to

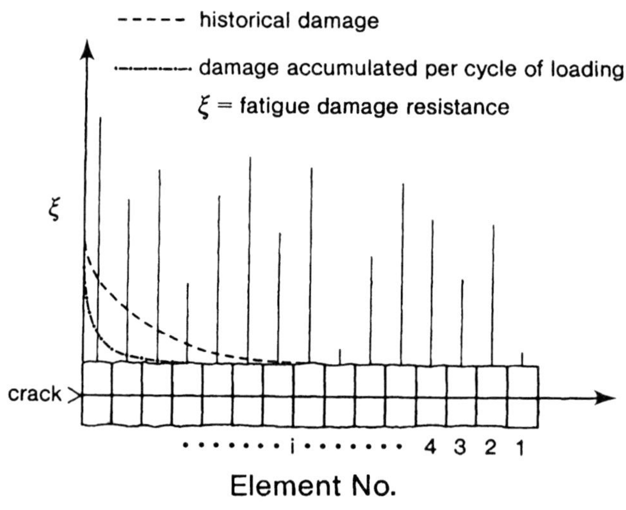

Fig. 10.5 Schematic illustration of material elements in front of a crack and their fatigue resistance and cumulative damage. Reprinted with kind permission from [17].

equalize the plastic strain distribution in a grain, thus allowing for the application of continuum theories, i.e. the material properties in the bulk can be applied to the micro-elements of Fig. 10.5 as an averaged measure.

4. At each material element damage accumulates according to the following nonlinear damage law.

The nonlinear damage model

A material element subjected to a specified level of stress/strain or energy may accumulate damage at each cycle. When the absorbed energy reaches a certain level a crack will initiate. In the plot of energy versus the number of cycles, the ensemble of the crack initiation points form a curve similar to the one shown in Fig. 3.23 of Chapter 3 which was termed the critical damage curve. The critical damage curve is the demarcation between a small initiated crack which cannot grow at an energy level corresponding to the material's fatigue limit, ΔW^{t}_{FL} in Fig. 3.23. Beyond the critical damage curve, an initiated crack grows and the small material element fails after a certain amount of crack growth.

The derivation of the cumulative damage law for multi-level loading is given in section 3.5. This nonlinear damage law for j levels of loading is given by (3.47)

$$f(\Delta W_1, \Delta W_2, \ldots, \Delta W_j) = \left\{\left[\left(\frac{n_1}{N_{f1}}\right)^{\eta_2/\eta_1} + \left(\frac{n_2}{N_{f2}}\right)\right]^{\eta_3/\eta_2} + \ldots + \frac{n_{(j-1)}}{N_{f(j-1)}}\right\}^{\eta_j/\eta_{j-1}} + \frac{n_j}{N_{f_j}} = 1.0 \qquad (10.6)$$

where

$$\eta_i = \log\left[\frac{(\Delta W_i^{t} - C)}{\Delta W^*}\right] \quad i = 1, 2, \ldots, j \qquad (10.7)$$

and

$$\Delta W_i^{t} = \Delta W_i^{p} + \Delta W_i^{e}. \qquad (10.8)$$

In the above, ΔW^{p} is the plastic strain energy per cycle, and ΔW^{e} is the elastic part of the strain energy associated with the positive stress amplitude. C is the non-damaging part of the energy corresponding to the material's fatigue (endurance) limit and is determined from a virgin specimen under constant amplitude loading, i.e. C is the elastic energy of (10.8) at the fatigue limit (crack initiation stage). For all crack propagation stages, the material has an 'apparent' (reduced) fatigue limit defined by ΔW^*, n_i is the number of cycles (time) spent at the ith load level and N_{f_i} is the corresponding number of cycles to failure in a constant amplitude test at the ith load level. For further explanation see section 3.5.3.

For each material element shown in Fig. 10.5, the energy function $f(\Delta W_1, \Delta W_2, \ldots, \Delta W_i)$ in (10.6) has to be calculated at each cycle. The manner in which these quantities can be evaluated is described in section 3.4.3, and they will not be repeated here. The failure of an element (i.e. crack advance) takes place when the value calculated for the LHS of (10.6) is equal to its fatigue resistance, ξ, i.e.

$$f(\Delta W_1, \Delta W_2, \ldots, \Delta W_i) = \xi \tag{10.9}$$

where ξ is a unit random variate replacing the deterministic unit value on the RHS of (10.6). It is the random fatigue damage resistance (FDR) of each material element shown in Fig. 10.5.

Stress and strain fields ahead of a crack

To calculate (10.9) we require the stress and strain distribution ahead of a crack. The stress and strain distribution near a crack (commonly referred to as HRR singularity fields (7.72)), are

$$\begin{aligned} \Delta\sigma_{ij} &= \Delta\sigma'_{\mathrm{y}}\left(\frac{\Delta K^2}{\alpha'\Delta\sigma'^2_{\mathrm{y}} I_{n'} r}\right)^{n'/(1+n')} \tilde{\sigma}_{ij}(\theta;n') \\ \Delta\varepsilon^{\mathrm{p}}_{ij} &= \frac{\alpha'\Delta\sigma'_{y}}{E}\left(\frac{\Delta K^2}{\alpha'\Delta\sigma'^2_{\mathrm{y}} I_{n'} r}\right)^{1/(1+n')} \tilde{\varepsilon}_{ij}(\theta;n') \end{aligned} \tag{10.10}$$

where the symbols have already been defined in Chapter 7. Once the value of ΔK is specified, the distribution of stress/strain range at each material element $\theta = 0$ and $r = x$, can be calculated. (Note that for $\theta = 0$, $\tilde{\varepsilon}_{ij} = \tilde{\sigma}_{ij} = 1$.) Subsequently, in any configuration, the energy quantity, ΔW^{t}_i in (10.6) can be evaluated as described in section 3.4.3. The exponent η_i (10.7) for each material element is then determined once ΔW^{t}_i, C and ΔW^* are known. It remains to specify the relationship between ΔW^{t}_i and the number of reversals to failure, $2N_{\mathrm{f}}$. This is given by (3.71) in Chapter 3, i.e.

$$\Delta W^{\mathrm{t}}_i = \kappa_{\mathrm{t}}(2N_{\mathrm{f}i})^{\alpha} + C. \tag{10.11}$$

From the above, $N_{\mathrm{f}i}$ for each element can be determined, thus the RHS of (10.6) is now fully determined for each material element.

In the absence of availability of material properties, κ_{t}, α and C, one may use an approximation based on the Basquin–Coffin–Masson relationships, (3.5) and (3.10)

$$\begin{aligned} \Delta\sigma_i/2 &= \sigma'_{\mathrm{f}}(2N_{\mathrm{f}i})^{b} \\ \Delta\varepsilon_i/2 &= \frac{\sigma'_{\mathrm{f}}}{E}(2N_{\mathrm{f}i})^{b} + \varepsilon'_{\mathrm{f}}(2N_{\mathrm{f}i})^{c} \end{aligned} \tag{10.12}$$

where σ_f' and ε_f' are the strength and ductility coefficients of the material and b and c are the fatigue strength and ductility exponents, discussed in sections 3.2 and 3.3. Substitution from (10.12) into (3.28) then yields

$$\Delta W_i^{\mathrm{p}} = 4\left(\frac{1-n'}{1+n'}\right)\sigma_f'\varepsilon_f'(2N_{f_i})^{b+c} \qquad (10.13)$$
$$\Delta W_i^{\mathrm{e}} = \frac{1}{2}\frac{\sigma_f'^2}{E}(2N_{f_i})^{b}$$

for fully-reversed cyclic loading.

The energy function $f(\Delta W_1, \Delta W_2, \ldots, \Delta W_j)$, (10.6), includes the damage history of each material element. For each element, ΔW_1, $\Delta W_2, \ldots$ represent strain energies at the first, second, ..., configuration after the element in question becomes part of the damage zone. The approximation involved by using (10.13) instead of (10.11) is discussed in [21].

10.4.2 THE PROBABILISTIC CRACK GROWTH RATE

The material element size in Fig. 10.5 is obtained by considering that the maximum stress or strain amplitude cannot exceed the fatigue strength coefficient σ_f', or the ductility coefficient, ε_f' of (10.12). These critical values of stress and strain can be associated with the critical stress intensity range, ΔK_C, where an unstable crack growth is observed at the instant of loading. Thus the minimum distance, x^*, from the crack tip where HRR fields (10.10) can be applied is obtained by equating the product of stress and strain amplitudes normal to the crack plane $(\Delta\sigma/2)(\Delta\varepsilon/2)$ to $\sigma_f'\varepsilon_f'$ as shown in Fig. 10.6. For small-scale yielding and neglecting the elastic term, we get

$$x^* = \frac{\Delta K_C^2}{I_n'\varepsilon_f'\sigma_f'E}. \qquad (10.14)$$

It has been shown elsewhere [17] that the value of x^* thus obtained is of the order of the grain size of polycrystalline metals.

Within the damage zone, if the accumulated damage of an element i exceeds its random fatigue damage resistance, ξ, before all the other elements, the random crack growth rate is obtained from

$$\left\langle\frac{\mathrm{d}a}{\mathrm{d}N}\right\rangle = \frac{ix^*}{n} \qquad (10.15)$$

where n is the number of cycles spent at that configuration. The ith element then becomes the new crack tip since after a few cycles the material between its position and the original crack tip will fracture. A new

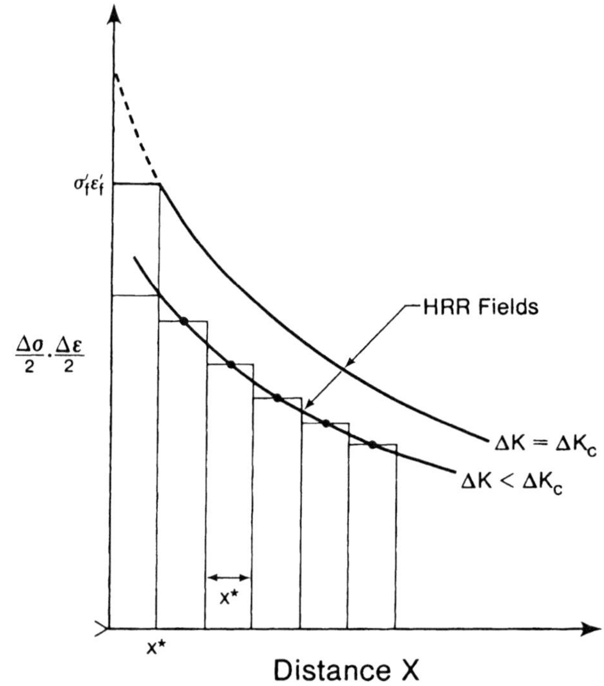

Fig. 10.6 Illustration of the method which defines the size of material elements shown in Fig. 10.5. Reprinted with kind permission from [17].

configuration is thereby attained and the whole process is repeated in the new damage zone.

The random variable $\langle da/dN \rangle$ is dependent on the random variates i and n. Even when the density function of these two are known *a priori*, the distribution $\langle da/dN \rangle$ will be complex. Here we will only attempt to compare this distribution to two popular ones used in the literature.

10.4.3 PROBABILITY DISTRIBUTIONS

Five different materials were chosen for the comparison: low alloy carbon steels A-516 Gr. 70, A302B, A533B; high alloy 10-Ni steel, and 2219-T851 aluminium alloy. The relevant material properties for these materials are reported in Table 1 of [20].

An estimate of the ensuing probability distribution of the random variable $\langle da/dN \rangle$ defined by (10.15) may be obtained by a comparison with well-known density functions. Figures 10.7 and 10.8 are probability plots of typical simulated data, using two commonly assumed

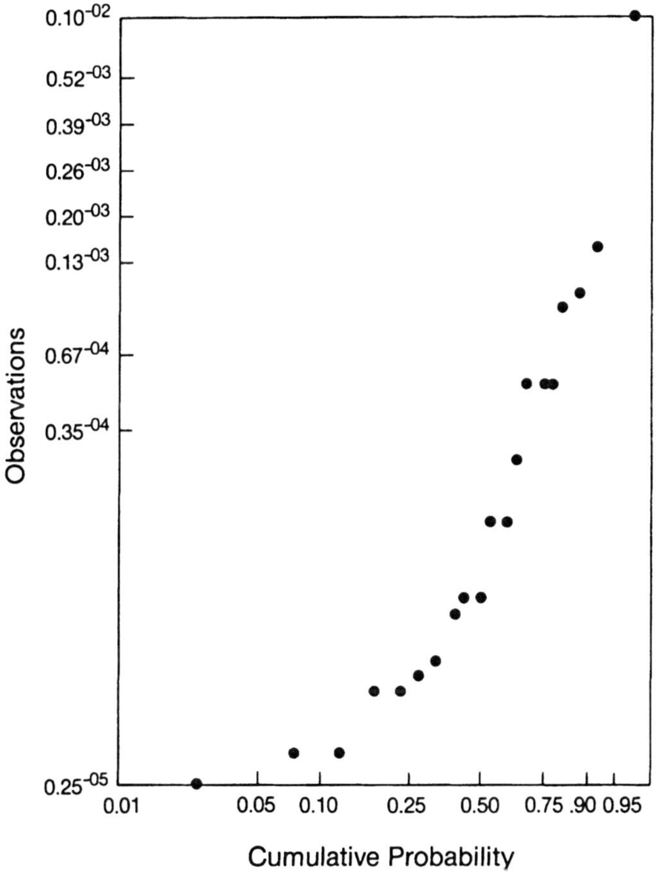

Fig. 10.7 A probability plot of the predicted crack growth results using the Weibull distribution. A straight line shows the correlation. Reprinted with permission from [20].

distributions for the crack growth data, the Weibull and log–normal, respectively. A linear graph indicates correlation with the assumed distribution. In the twenty points chosen, the log–normal distribution (Fig. 10.8) consistently gave better correlation than any other distribution because it gives the closest curve to a straight line of all the probability plots [20]. The Weibull distribution (Fig. 10.7) was the second best. However, it is not possible to conclude log–normality without further testing. For this a Shapiro–Wilk normality test was performed on the logarithm of the data. The test indicated the acceptance of the log–normal distribution at a 10% level of significance.

This conclusion is supported by the experimental results reported elsewhere, e.g. [3, 9].

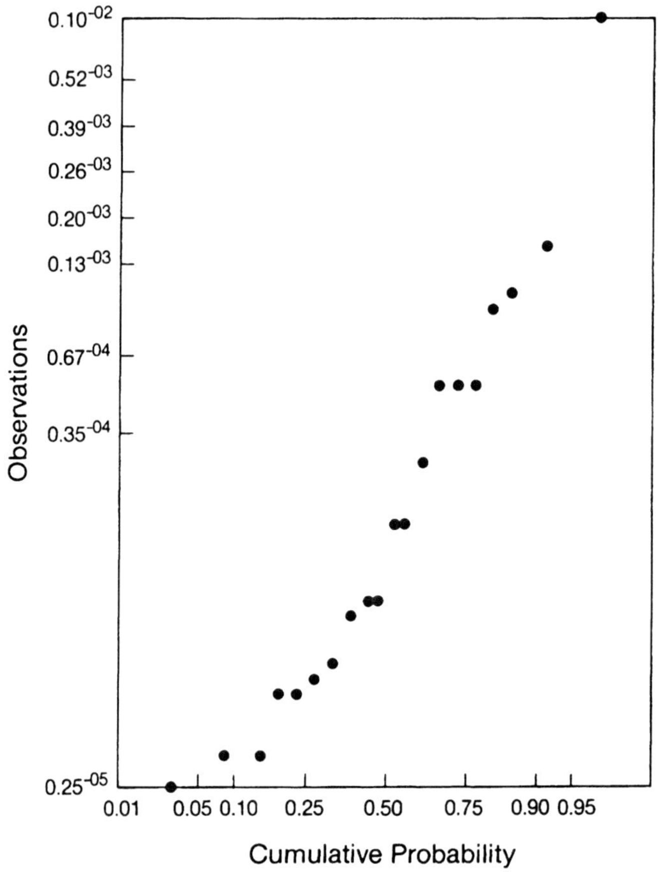

Fig. 10.8 A probability plot of the predicted crack growth results using the log-normal distribution. A straight line shows the correlation. Reprinted with permission from [20].

10.5 COMPARISON WITH CRACK GROWTH DATA

The predictions of the crack growth simulation model (10.15) were compared with the crack growth experimental results in [20]. It should be pointed out, however, that these experimental results were obtained from limited samples while the model described in section 10.4 gives a random growth rate output. The comparison therefore has to be made with some measure of central tendency of the simulated probability distribution. The median value is chosen for the comparison since it is least susceptible to the input variance. Figures 10.9 and 10.10 show the comparison between the predicted median of the probabilistic model and the experimental data of A-516 Gr. 70 steel, taken from Ellyin and Li [22], and of an aluminium alloy

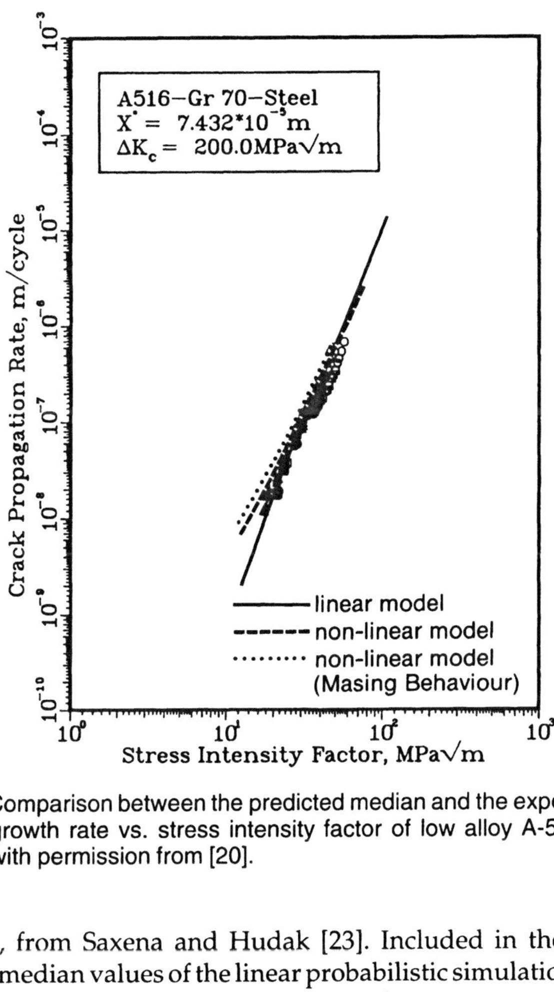

Fig. 10.9 Comparison between the predicted median and the experimental data of the crack growth rate vs. stress intensity factor of low alloy A-516 Gr. 70 steel. Reprinted with permission from [20].

2219-T851, from Saxena and Hudak [23]. Included in these figures are predicted median values of the linear probabilistic simulation of Ellyin and Fakinlede [17], using the statistical Miner's law

$$\sum_i \frac{n_i}{N_{fi}} \sim N(\mu, \sigma) \tag{10.16}$$

where N indicates a normal probability distribution, the mean value $\mu = 1$, and σ is the variance of the normal density function.

It can be seen from both figures that the predictions of the nonlinear model are closer to the experimental results than those of the linear model. In Fig. 10.9 a comparison is also made where the strain energy density ΔW_i^t (10.8) is evaluated by using either the non-Masing material description or the ideal Masing one (section 2.6). The difference between the two predictions does not appear to be very significant in this case.

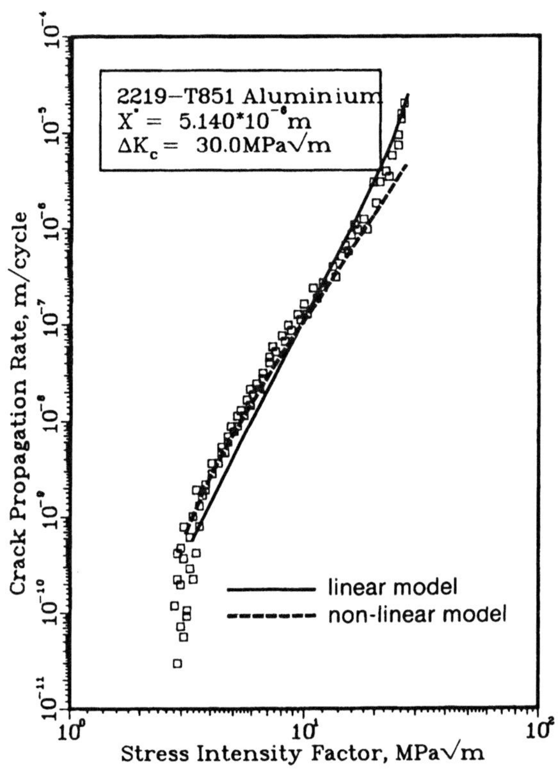

Fig. 10.10 Comparison between the predicted median and the experimental data of the crack growth rate vs. stress intensity factor of 2219-T851 aluminium alloy. Reprinted with permission from [20].

In summary, the probabilistic crack growth model described in section 10.4 is based on a physically sound nonlinear damage accumulation law and a failure criterion. The crack front is divided into material elements representative of randomly oriented grains with a randomly distributed resistance to fatigue crack growth. The model accounts for the sequence effect as the damage accumulation level varies with the element's position as it approaches the crack tip. The model predicts a rather complex probability distribution which is best approximated by a log–normal density function. It can be seen that the median values of the model prediction agree fairly well with the crack growth rate test data.

10.6 THE EFFECT OF VARIABLE AMPLITUDE LOADING

In the previous section we were concerned with the variation in the material resistance to the crack initiation and/or crack propagation. The applied load was assumed to have a constant amplitude. However, there is another source of scatter in service which is due to the load spectum, i.e. the actual load experienced by a component/structure may differ from that which is assumed assumed during the design or prototype testing stage. In section 1.2, we discussed the methods which are used to reduce service loads to well-defined load sequences. In addition to investigating the influence of different load sequences, one should also study the effect of possible variation in the load magnitudes. Studies of this type have reported scatter in the crack growth by one to two orders of magnitudes, e.g. Palmberg *et al.* [24].

The effect of occasional overloads in a constant amplitude loading on the crack growth rate was discussed in section 7.8.1. It was pointed out that the crack growth rate was strongly affected by the closure development due to the imposed overloads. In principle, using the crack growth model described in Chapter 7, one could integrate da/dN cycle by cycle for a prescribed load spectrum. Given the present-day desk top computer's speed and storage memory, this is quite feasible. However, not all cracks grow in each cycle, therefore the predicted results through the cycle by cycle integration method may not be accurate enough. Different techniques used to predict the crack growth under a spectrum loading have been compared by Engle [25]. He concluded that no single technique was superior in all aspects, however, the cycle by cycle technique was deemed to be more flexible.

In studies carried out by Topper and co-investigators, e.g. [26, 27], it was observed that the presence of intermittent compressive underloads or tensile overloads in a load history resulted in a load level interaction. In this interaction the presence of a few large cycles in a load history magnifies the effect of smaller cycles beyond that found in a constant amplitude loading. For example, long cracks grow below the threshold stress intensity in the presence of intermittent underload and overload cycles [26]. Similarly, under variable amplitude loading, the presence of large cycles increases damage done by smaller cycles even for those below the constant amplitude fatigue limit [27].

10.7 A PRACTICAL EXAMPLE

When applying the probabilistic crack growth approach to practical situations, the variability of both the material resistance and applied loads have to be considered. One may assume a central tendency derived from the experimental investigations in a laboratory setting, or a long-term observation in the field, and then assume some probability distribution around these central tendencies, e.g. Payne *et al.* [28], Ellyin [2, 29]. In [29] a pipe with a longitudinal crack was considered. This was assumed to be part of a piping system carrying pressurized hot fluid in a nuclear power plant. The probability density function of the life factor Z in (10.1), $f_Z(z)$ with $\tilde{m}_Z = 1$, was assigned three values of standard deviation $\sigma_{\ln Z}$ to study sensitivity with respect to the spread of the log–normal disribution. The residual strength of the pipe (cf. Fig. 10.3) was determined using a fracture mechanics approach [30]. The probability distribution function of the pressure was assumed to be a normal distribution with the mean and standard deviation conforming to those prescribed in the appropriate piping code. The risk of failure at any period of service life was then evaluated, and the effect of various inspection methods on the total risk were examined [29].

The reliability of the piping system and the effect of inspection methods are shown in Fig. 10.11. The abscissa in the figure is the relative life $V = N/\tilde{N}$ where $\tilde{N}$ is the median of the distribution of service life up to the formation of a detectable crack, a_0 (Fig. 1.17 in Chapter 1). The effect of the standard deviation (dispersion about the median) is shown in the figure, which indicates that the reliability decreases as the dispersion (standard deviation) increases, which intuitively is to be expected. With an inspection method in place, one would expect the reliability to improve during the service life. This is the case shown in Figure 10.11 for a relative detectable crack length of $H_d = a_0/a_F = 0.2$, where a_F is the critical crack length (onset of unstable crack growth, see Fig. 1.17), and total failure risk $h_{TOT} \leqslant 10^{-6}$. That is, a repair or replacement is carried out whenever the relative detectable crack length H_d or the total risk of failure exceeds the values specified above. The inspection interval for each value of $\sigma_{\ln z}$ is

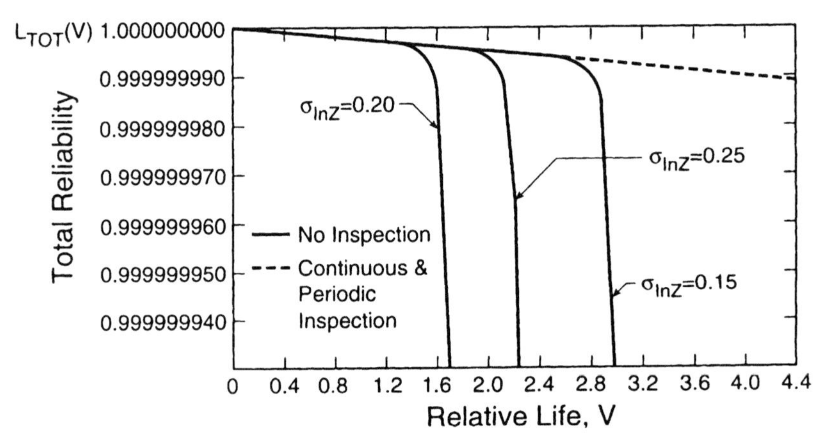

Fig. 10.11 The reliability of a pressurized pipe containing a defect with three different standard deviations of the life cycle distribution (cf. Fig. 10.1(b)). The effect of the continuous and periodic inspection for a set of defined parameters is also shown. Reprinted with kind permission from [29].

specified in [29] where further discussion on the reliability of the piping system is to be found.

Another example of a simplified aircraft wing panel following the above procedure is given in [24].

It can be seen that the fundamental studies such as those mentioned in section 10.4, enable one to obtain probabilistic crack growth curves with a central tendency and probability distribution around it. A linear or nonlinear fracture mechanics methodology can then be used to derive the residual strength of a component during its service life. The reliability of a component can subsequently be determined, assuming appropriate strength and load variations derived from the previously mentioned simulation studies.

REFERENCES

1. Griffith, A.A. (1920) The phenomenon of rupture in solids. *Phil. Trans. R. Soc. London* A, **221**, 163–98.
2. Ellyin, F. (1985) A strategy for periodic inspection based on defect growth. *Theor. Appl. Fract. Mech.*, Elsevier Science Ltd, Amsterdam, **4**, 83–96.
3. Impellizeri, L.F. (1966) Development of a scatter factor applicable to aircraft fatigue life, in *Structural Fatigue in Aircraft*, ASTM STP 404, American Society for Testing and Materials, Philadelphia, PA, pp. 136–56.
4. Weibull, W. (1949) *A Statistical Representation of Fatigue Failure in Solids*, Vol. 1, Acta Polytech., Mech. Engng. Serv., Stockholm, Sweden.
5. Freudenthal, A.M. (1946) The statistical aspect of fatigue of materials. *Proc. Roy. Soc. (London)* Series A, **187**, 416–29.
6. Gumbel, E.J. (1958) *Statistics of Extremes*, Columbia University Press, New York.
7. Heller, R.A. (ed.) (1972) *Probabilistic Aspects of Fatigue*, ASTM STP 511, American Society for Testing and Materials, Philadelphia, PA.
8. Haugen, E.B. (1980) *Probabilistic Mechanical Design*, John Wiley, New York.
9. Bertrand, M., Lefebvre, D.F. and Ellyin, F. (1983) Statistical analysis of crack initiation and fatigue fracture of thin-walled tubes using the Weibull law. *J. de Mécanique Théorique et Appliquée*, **1**, Gauthier-Villars Publishers, 493–510.
10. Provan, J.W. (ed.) (1987) *Probabilistic Fracture Mechanics and Reliability*, Martinus Nijhoff, Dordrecht, Netherlands.
11. Yao, J.T.P. *et al.* (1986) Stochastic fatigue, fracture and damage analysis. *Structural Safety*, **3**, 231–67.
12. Virkler, D.A., Hillberry, B.M. and Goel, P.K. (1979) The statistical nature of fatigue crack propagation. *J. Engng. Mater. Technol., Trans. ASME*, **101**, 148–53.
13. Lidiardi, A.B. (1979) Probabilistic fracture mechanics, in *Fracture Mechanics – Current Status, Future Prospects* (ed. R.A. Smith), Pergamon Press, Oxford, UK, pp. 149–78.
14. Engesvik, K.M. (1981) Analysis of Uncertainties in the Fatigue Capacity of Welded Joints, Tech. Report UR-82-17, Division of Marine Structures, Trondheim, Norway, Dec. 1981.
15. Varanasi, S.R. and Wittaker, I.C. (1976) Structure reliability prediction method considering crack growth and residual strength, in *Fatigue Crack Growth Under*

Spectrum Loads, ASTM STP 545, American Society for Testing and Materials, Philadelphia, PA, pp. 242–305.
16. Bogdanoff, J.C. and Kozin, F. (1985) *Probabilistic Models of Cumulative Damage*, John Wiley, New York.
17. Ellyin, F. and Fakinlede, C.O.A. (1985) Probabilistic simulation of fatigue crack growth by damage accumulation. *Engng. Fract. Mech.*, **22**, Elsevier Science Ltd, Oxford, 697–712.
18. Winkler, T., Michel, B. and Skurt, L. (1990) Fundamentals and Application of Probabilistic Fracture Mechanics – A Survey, FMC – Series No. 49, Institut für Mechanik, Akademie der Wissenschaften, Chemitz, Germany, ISSN 0232-7236, 105 pp.
19. Itagaki, H. (1991) Simulation of probabilistic fatigue crack growth, in *Mechanical Behaviour of Material – VI* (eds M. Jono and T. Inoue), Proc. 6th ICM, Kyoto, Japan, Vol. I, Pergamon Press, Oxford, UK, pp. 557–70.
20. Ellyin, F. and Fakinlede, C.O.A. (1988) Probabilistic crack growth by non-linear damage accumulation. *Int. J. Fract.*, **36**, Kluwer Academic Publishers, Dordrecht, 137–49.
21. Lefebvre, D.F. and Ellyin, F. (1984) Cyclic response and inelastic strain energy in low-cyclic fatigue. *Int. J. Fatigue*, **6**, 9–15.
22. Ellyin, F. and Li, H.-P. (1984) Fatigue crack growth in large specimens with various stress ratios. *J. Pressure Vessel Technol., Trans. ASME*, **106**, 255–60.
23. Saxena, A. and Hudak, S.J. (1979) Role of crack-tip stress relation in fatigue crack growth, in *Fracture Mechanics*, ASTM STP 677, American Society for Testing and Materials, Philadelphia, PA, pp. 215–32.
24. Palmberg, H., Blom, A.F. and Eggwertz, S. (1987) Probabilistic damage tolerance analysis of aircraft structures, in *Probabilistic Fracture Mechanics and Reliability* (ed. J.W. Provan), Martinus Nijhoff, Dordrecht, Netherlands, Chap. 2, pp. 47–130.
25. Engle, R.M. (1984) Damage accumulation techniques in damage tolerance applications, in *Damage Tolerance of Metallic Structures: Analysis Methods and Applications*, ASTM STP 842, American Society for Testing and Materials, Philadelphia, PA, pp. 25–35.
26. Topper, T.H. and Yu, M.T. (1985) The effect of overloads on threshold and crack closure. *Int. J. Fatigue*, **7**, 154–64.
27. Pompetzki, M.A., Topper, T.H. and DuQuesnay, D.L. (1990) The effect of compressive underloads and tensile overloads on fatigue damage accumulation in SAE 1045 steel. *Int. J. Fatigue*, **12**, 207–13.
28. Payne, A.O., *et al.* (1985) Reliability approach to structural safety and safety criteria, in *Probabilistic Methods in the Mechanics of Solids and Structures*, Proc. IUTAM Symposium, Stockholm 919840 Springer, Berlin, pp. 529–57.
29. Ellyin, F. (1986) Reliability of piping systems and effect of inspection methods. *Structural Safety*, **3**, Elsevier Science Ltd, Amsterdam, 85–99.
30. Ellyin, F. (1985) Residual strength of a pressurized heat transport pipe containing a defect, in *Time-Dependent Fracture* (ed. A.S. Krausz), Martinus Nijhoff, Dordrecht, Netherlands, pp. 83–92.

Index

Page numbers appearing in **bold** refer to figures and page numbers appearing in *italic* refer to tables.

Printed in the United States
104817LV00001B/67/A

9 780412 596001